Smart Mobile and Sensing Applications

Smart Mobile and Sensing Applications

Guest Editors

Chien Aun Chan
Ming Yan
Chunguo Li

MDPI

Basel • Beijing • Wuhan • Barcelona • Belgrade • Novi Sad • Cluj • Manchester

Guest Editors

Chien Aun Chan
Department of Electrical and Electronic Engineering
The University of Melbourne
Melbourne
Australia

Ming Yan
School of Information and Communications Engineering
Communication University of China
Beijing
China

Chunguo Li
School of Information Science and Engineering
Southeast University
Nanjing
China

Editorial Office
MDPI AG
Grosspeteranlage 5
4052 Basel, Switzerland

This is a reprint of the Special Issue, published open access by the journal *Sensors* (ISSN 1424-8220), freely accessible at: https://www.mdpi.com/journal/sensors/special_issues/smart_mobile.

For citation purposes, cite each article independently as indicated on the article page online and as indicated below:

Lastname, A.A.; Lastname, B.B. Article Title. *Journal Name* **Year**, *Volume Number*, Page Range.

ISBN 978-3-7258-3225-5 (Hbk)
ISBN 978-3-7258-3226-2 (PDF)
https://doi.org/10.3390/books978-3-7258-3226-2

Contents

About the Editors

Chien Aun Chan

Chien Aun Chan received a Ph.D. degree in electrical engineering from The University of Melbourne (UoM), Parkville, VIC, Australia, in 2010. He currently works in the Department of Electrical and Electronic Engineering, UoM, where he leads the commercialization of a new wireless technology for smart industry AI and robotics. He was a Research Fellow with the Centre for Energy-Efficient Telecommunications, a Lecturer with the School of Computing and Information Systems, and a Research Fellow with the Department of Electrical and Electronic Engineering, UoM. His work spans future mobile wireless systems, mobile edge computing systems, network predictive analytics, green communications and networking, the Internet of Things, and cybersecurity.

Ming Yan

Ming Yan received a Bachelor's degree in Communication Engineering from Nanjing University of Posts and Telecommunications, Nanjing, China, in 2002, and M.S. and Ph.D. degrees in Communication and Information Systems from the Communication University of China (CUC), Beijing, China, in 2006 and 2012, respectively. From 2014 to 2015, he worked as a Visiting Research Scholar with the Center for Energy-Efficient Telecommunications, The University of Melbourne, where he was involved in developing new energy models for mobile services. He is currently a professor at the School of Information and Communication Engineering, CUC. His research interests include future wireless systems, green technologies in wireless communication systems, mobile wireless networks, and mobile multimedia broadcast technologies.

Chunguo Li

Chunguo Li received a Bachelor's degree in Wireless Communications from Shandong University in 2005 and a Ph.D. degree in Wireless Communications from Southeast University in 2010. In July 2010, he joined the faculty of Southeast University, Nanjing, China, where he was an Associate Professor between 2012 and 2016 and has been a Full Professor since 2017. From June 2012 to June 2013, he was a Postdoctoral Researcher at Concordia University, Montreal, Canada. From July 2013 to August 2014, he worked with the DSL laboratory of Stanford University as a Visiting Associate Professor. From August 2017 to July 2019, he was an Adjunct Professor with Xizang Minzu University under the support of the Tibet program, organized by the China National Human Resources Ministry.

He is a Fellow of IET, a Fellow of the China Institute of Communications (CIC), a Chair of the IEEE Computational Intelligence Society Nanjing Chapter, and a Chair of the Advisory Committee for the Instruments Industry in Jiangsu Province. He has served as an editor for several international journals and as a session chair for many international conferences. His research interests include 6G cell-free distributed MIMO wireless communications, information theories, and AI-based audio signal processing.

Preface

Recent advances in mobile sensing technologies leveraged by big data analytics and machine learning have enabled a plethora of applications that have the potential to improve productivity, safety, health, and efficiency in a diverse range of use case scenarios. The following are a few examples of how this might look: the use of mobile sensing with wearable technology to assist with remote learning, especially during the pandemic; consumer mobility wireless sensing and tracking to enhance security and user experience; wearable mobile sensing and monitoring to ensure safety improve productivity in harsh working environments; mobile sensing for social behavioral research and sports analytics; and mobile sensing in AR and VR.

Therefore, this Special Issue aimed to collect the high-quality research work focusing on addressing emerging challenges in smart mobiles and sensing, as well as smart applications and use case scenarios, which could help to enhance our daily lives.

Chien Aun Chan, Ming Yan, and Chunguo Li
Guest Editors

sensors

MDPI

Article

Innovating Household Food Waste Management: A User-Centric Approach with AHP–TRIZ Integration

Shuyun Wang [1], Hyunyim Park [1,*] and Jifeng Xu [2]

1 School of Design, The Hong Kong Polytechnic University, Hong Kong, China; shuyun.wang@connect.polyu.hk
2 School of Art, Southeast University, Nanjing 211189, China; 101011429@seu.edu.cn
* Correspondence: hyunyim.park@polyu.edu.hk

Abstract: Food waste management remains a paramount issue in the field of social innovation. While government-led public recycling measures are important, the untapped role of residents in food waste management at the household level also demands attention. This study aims to propose the design of a smart system that leverages sensors, mobile terminals, and cloud data services to facilitate food waste reduction. Unlike conventional solutions that rely on mechanical and biological technologies, the proposed system adopts a user-centric approach. By integrating the analytical hierarchy process and the theory of inventive problem solving, this study delves into users' actual needs and explores intelligent solutions that are alternatives to traditional approaches to address conflicts in the problem solving phase. The study identifies five main criteria for user demands and highlights user-preferred subcriteria. It determines two physical conflicts and two technical conflicts and explores corresponding information and communications technology (ICT)-related solutions. The tangible outcomes encompass a semi-automated recycling product, a mobile application, and a data centre, which are all designed to help residents navigate the challenges regarding food waste resource utilisation. This study provides an approach that considers users' genuine demands, empowering them to actively engage in and become practitioners of household food waste reduction. The findings serve as valuable references for similar smart home management systems, providing insights to guide future developments.

Keywords: food waste management; smart system design; user-centric design; AHP–TRIZ integration; household waste reduction

Citation: Wang, S.; Park, H.; Xu, J. Innovating Household Food Waste Management: A User-Centric Approach with AHP–TRIZ Integration. *Sensors* **2024**, *24*, 820. https://doi.org/10.3390/s24030820

Academic Editor: Juan M. Corchado

Received: 22 November 2023
Revised: 11 January 2024
Accepted: 22 January 2024
Published: 26 January 2024

1. Introduction

China's urbanisation and economic development continue, especially with the expansion of the restaurant and food delivery industries in recent years. However, along with this trend, the issue of food waste generation in Chinese urban areas has become increasingly serious. Estimates showed that, in 2020, China's food waste amounted to 61.37 Mt, reflecting a significant increase of 29.67% in per capita food wastage compared to the 2016 figures, and this trend is anticipated to escalate in the future [1]. Food waste contains a large amount of organic matter. On the one hand, it has posed a potential threat of pollution to water and land resources, as well as the spread of germs that affect people's physical health. On the other hand, it has great value for resource reclamation as it can be collected and converted into value-added products [2]. Hence, investigating strategies for the collection, management and efficient conversion of food waste has emerged as a major research topic in recent years.

China is making great efforts to achieve its *zero-waste city* goal, which refers to an urban development model that aims at recycling solid waste and reducing landfills by promoting green development and green lifestyles. Garbage classification is one of the most common measures. Food waste is sorted and utilised to produce biomass fuel and natural gas [3]. Although these actions have been implemented in some pilot cities for several years, their effectiveness remains low. As a comprehensive project, waste recycling in China is currently managed through an intensive government-led approach [4], which requires overall process management, including policy measures, source-separated collection, transportation, treatment and disposal [5]. Consequently, problems such as mixed collection and low efficiency in resource utilisation are common wicked issues that affect this endeavour [6]. In addition to the government-led mode, an alternative approach worth exploring is a user-centric perspective, which is a decentralised and more direct household-focused solution to food waste management.

Many studies have proposed strategies for improving residents' awareness of effective recycling [4,7], but there is a lack of specific solutions on how to facilitate and support residents' recycling behaviour. Most Chinese households are not equipped with practical products to process food waste. Therefore, improving the household resource recycling management system to match residents' environmental awareness is important as it helps to achieve an effective combination of awareness and appropriate recycling measures [8]. Seeking suitable disposal approaches for household food waste management is a necessary and meaningful topic.

To provide a practical household disposal solution and improve residents' recycling experience, this study proposes a user-centric approach for managing household food waste in a smarter way based on the analytical hierarchy process (AHP) and the theory of inventive problem solving (TRIZ). The AHP is a structured multi-attribute decision making method (see Figure 1) in which the criteria of an issue are organised in a hierarchical structure and relatively compared to determine their order of priority [9]. The theory of inventive problem solving consists of tool sets generated from massive inventions to address the challenges within a technical system, thus improving its functionality (see Figure 2) [10]. Based on the proposed approach, modern technological components, such as wireless sensors and mobile terminals, can be integrated into the final system design solution. These technologies enable real-time data collection, enhance user engagement and streamline the management of household food waste [11]. By combining user research methods with information and communications technology (ICT), a holistic and efficient solution can be developed to effectively manage food waste while fostering residents' environmental awareness.

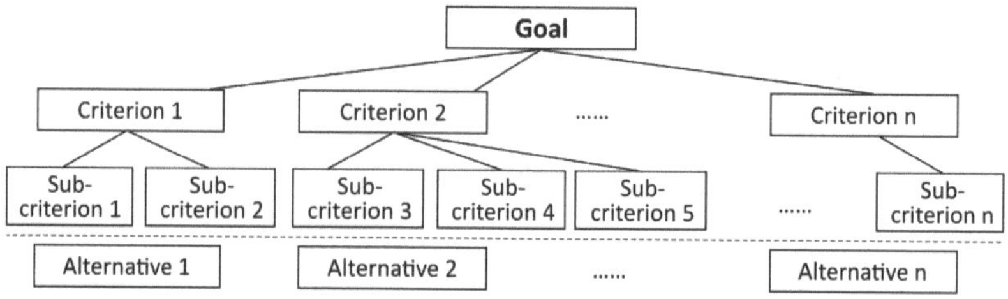

Figure 1. The AHP model.

The main contributions of this study are as follows:

- Explore factors and identify user preferences regarding the household food waste recycling issue.
- Present a theoretical approach for designing the household food waste management system, which can be applied to other household smart management issues and has the potential to evolve conventional products into smart management systems.
- Develop a practical solution incorporating smart technology to manage household food waste recycling based on the proposed approach.

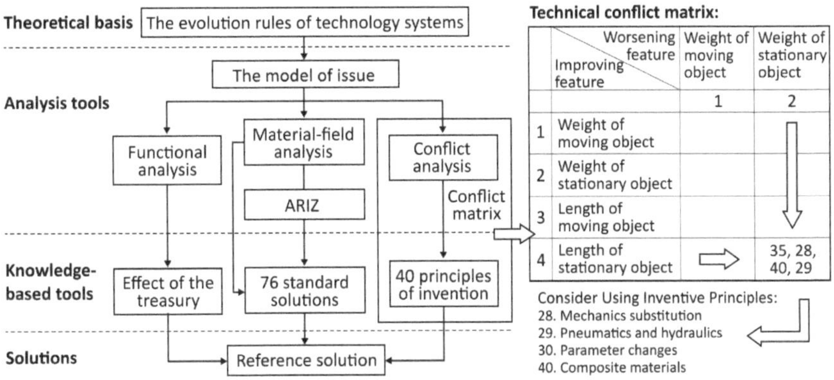

Figure 2. Flowchart of the TRIZ theory and an example of the conflict matrix.

The remaining sections are structured as follows. Section 2 discusses the background of food waste recycling in Chinese communities and similar initiatives in terms of products, biotechnology and smart systems. Section 3 contextualises the research method's architecture and discusses the approach step by step. Section 4 presents the design generated from the previous analysis. Finally, Section 5 discusses the findings from the proposed approach in the field of smart home system design.

2. Related Work

As China is experiencing a growing amount of food waste every year, the implementation of the wet waste classification policy in the country has accelerated. In 2019, 16 cities in China initiated the pilot construction of *zero waste cities*, a move aimed at promoting source reduction and resource utilisation of solid waste [12]. The most widely adopted approach involves collecting a large proportion of waste from the community or street scale, which then undergoes processing through intensive treatment by relevant authorities [13]. Consequently, research on kitchen garbage recycling equipment has primarily focused on the perspectives of large-scale public treatment facilities. Notable studies in this domain include that by Liu et al., who carried out a field investigation of in situ food waste treatment in canteens, markets and residential communities [14]. They argued for the necessity of a unified standard in equipment design to ensure treatment capacity. Hu et al. designed a compression treatment device with a modular structure to address the issues of high costs and space requirements associated with wet garbage treatment [15]. However, these centralised processing systems often overlook the management of food waste from household sources, which leads to community residents' low participation rates—an indispensable factor in food waste management initiatives. Research on household or individual recycling solutions is also relatively scarce.

Therefore, the focus of this study is on household kitchen waste recycling in Chinese urban communities, characterised by limited kitchen spaces, predominantly grain-based food waste composition and a lack of established household waste recycling habits [16]. Given that waste management practices in China are still in their early stages, with traditional centralised disposal methods prevailing and household disposal not yet mainstream, it is necessary to refer to experience from other regions around the world. Therefore, this research aims to compare works existing in a wider range of regions, specifically those from a household perspective or those related to systematic solutions. These studies will be referenced and compared on the basis of their usage scenarios and treatment methods, including mechanical processing, biological processing or service systems incorporating intelligent ICT solutions.

Regarding the existing mechanical solutions for household disposal treatments, the use of under-the-sink grinders has long been regarded as a practical way to establish source reduction, which disposes the shredded waste by routing it through connected conventional sewer systems [17]. However, the use of this product has received criticism over the years because of potential adverse effects, such as increased volumes of wastewater and sludge [18]. As a response, studies have explored how such solutions can eliminate wastewater pollutants while retaining valuable materials [19]. On the other hand, research addressing food waste disposal from a biotechnological perspective is becoming more widespread. For instance, Du et al. compared seven treatment methods for kitchen food waste, arguing that biological treatment has greater advantages over non-biological treatment [20]. Zhou et al. developed a domestic composting device that shortens the typically long composting cycle by manipulating the temperature and considering the use of microbial agents [21]. The emphasis on the study of food waste bio-degradation as an alternative to conventional and less environment-friendly landfill practices is increasing. However, these solutions predominantly focus on enhancing machine performance and biotechnology, and there is a lack of research from a user experience perspective, even though users are the primary actors in household food waste management.

With the evolution of smart sensing and mobile technology, the approach to reducing food waste has expanded beyond mechanical and biological methods. ICT solutions allow for the possibility of information recording, tracking and linkage for food waste management. For example, Marques et al. proposed a multilevel internet of things (IoT)-based management system for outdoor and indoor public bins to better manage waste separation [22]. Liegeard and Manning's research explored intelligent packaging with ink colour changing and sensors that work with intelligent appliances to remind users of the food storage conditions, thus minimising food waste [23]. Similarly, Cappelletti et al. developed an integrated smart system that guides users in their food-related daily behaviour to reduce household food waste [24]. Moreover, smart systems play a vital role in engaging stakeholders for joint management. Spyridakis et al. built a platform that connects people with surplus food to underprivileged people by using technology, volunteerism and civic engagement [25]. These research attempts imply that there are many opportunities to explore the field of integrating smart technologies into household food waste management. A comparison of the related studies is presented in Table 1.

Table 1. Comparison between related works.

Work	Scenarios	Treatment	ICT	Goal
Bernstad et al. [17]	Household: under the sink	Mechanical: grind and discharge into sewers	-	• reduce waste • create methane
Cecchi and Cavinato [19]	Public: under the sink to waste stations	Mechanical and biological: grind and then dispose in treatment plant	-	• avoid transportation • energy recovery
Zhou et al. [21]	Household: kitchen composting bins	Biological: high-temperature composting	-	• food waste reduction • make value-added products
Marques et al. [22]	Public: outdoor and indoor bins	ICT-related: sensing and recognition	RFID sensors, cloud platform, etc.	• correct separation • a simultaneous bin network
Liegeard and Manning [23]	Household: kitchen smart fridges	ICT-related: packaging for food track	Biosensors, RFID, a control unit, etc.	• manage stock control • reduce food waste
Cappelletti et al. [24]	Household: kitchen smart fridges	ICT-related: food stock track	A smart fridge an application	• food waste reduction • healthy diet
Spyridakis et al. [25]	Public: campus dining halls	ICT-related: pick-up and delivery	An open-source website	• sharing concept • reduce food waste

Through comparison, potential directions for further research can be revealed. First, many studies focus narrowly on the disposal of the product itself and lack a systematic view to address the issue with relevant stakeholders or via other technological means. Second, in terms of research involving ICT solutions, many of them concentrate on utilising intelligent technologies to manage food stock while neglecting solutions for assisting residents in dealing with the already generated food waste. Lastly, there is a lack of engagement of residents as direct actors in the system, thereby limiting their motivation, participation and long-term commitment.

Given these issues, the research gap lies in allowing residents to treat household organic waste by considering user needs and creating an ICT-enabled waste management system. As such, this study takes the AHP and the TRIZ theory as guidance. The AHP is often used to examine the degree of importance of each requirement in an issue. In Balwada's research, the AHP was used to select the best waste collection method [26]. This method is also commonly integrated with other methods, including the TRIZ theory, which systematically analyses problems to come up with innovative solutions [27,28]. Moreover, the development of wireless sensors and intelligent terminals has facilitated various smart services within urban environments [29,30], allowing for more customised requirements [31]. Leveraging the capabilities of ICT devices, this study aims to develop a smart system for urban households based on the AHP–TRIZ method. The design intends not only to focus on product efficiency in terms of weight reduction but also to provide better insights into the factors that users truly care about in the waste management process.

3. Research Methods

In the AHP–TRIZ method, the complex general goal is decomposed into multiple indexes. Each index weight is determined comparatively by following the AHP. This helps to improve the accuracy and objectivity of the identification of each weight. Then, the more vital selected indexes are transformed into specific technical requirements. The TRIZ tool is then applied to deal with the technical and physical challenges inherent in the proposition of the conceptual design. It facilitates the rapid generation of innovative concepts and helps to establish an effective thinking mode. The proposed design approach framework utilising AHP–TRIZ integration is shown in Figure 3.

3.1. Use AHP to Evaluate the Weight of Each Criterion

In the first step, a round of market research of comments on 8 prevailing household food waste disposal products on Chinese e-commerce platforms (monthly sales volume of over 100) is conducted, combined with a supplement of initial interviews with 5 users who practised household food disposal, to gain the general impression and criteria of designing household food waste management system. As a result, an overall of 22 subcriteria are obtained after using the affinity diagram [32] (shown in Figure 4). Then, the subcriteria are classified into 5 main criteria according to Donald Norman's three-level emotional design theory: the visceral, behavioural and reflective levels [33]. Similarly, Jordan defined users' needs in terms of product characteristics, including the levels of functionality, usability and pleasure [34]. These three aspects are applicable in understanding user psychology and design requirements [35,36]:

The instinct layer focuses on users' initial intuitive perceptions of the product, encompassing aesthetic factors, such as shape, colour, material and volume of the product. Therefore, the first main criterion for this layer is C1 Aesthetics. The behaviour layer focuses on functional utility and ease of operation, which include the interaction between the user and the product. Therefore, at the behavioural level, C2 Functionality aims at determining whether the provided functions of the system are effective and efficient, and C3 Operability refers to whether the operation process of the system is user-friendly. Moreover, the reflective layer shows the joint impact of instinct and behaviour. It is meant to establish emotional links through user interactions, as well as provide users with both psychological experience and conceptual value. Hence, another two main criteria, C4 Experience and C5 Value, are identified. The overall structure comprises 5 main criteria and 22 subcriteria, as depicted in Figure 5. The detailed explanations for each criterion can be found in Appendix A, which have also been informed to the evaluators in the subsequent scoring phase to ensure consistency.

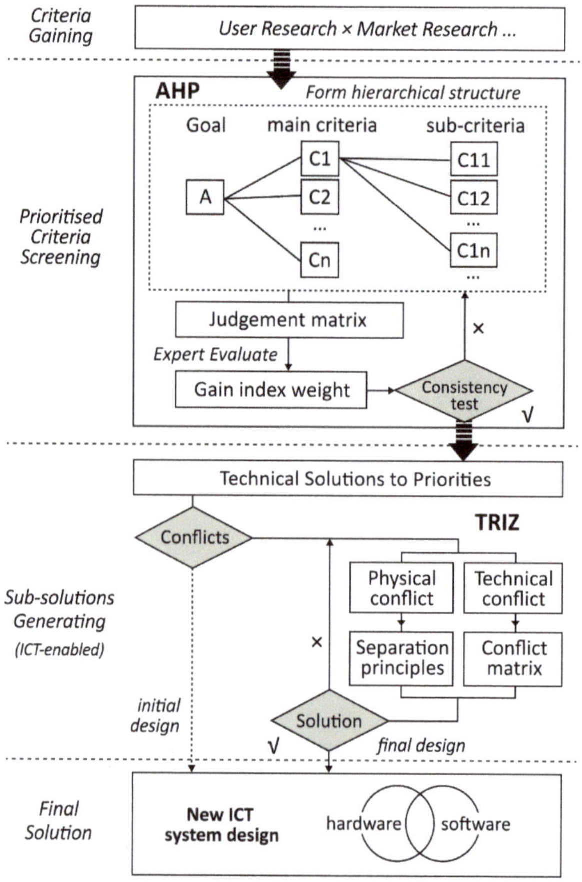

Figure 3. The AHP–TRIZ method model.

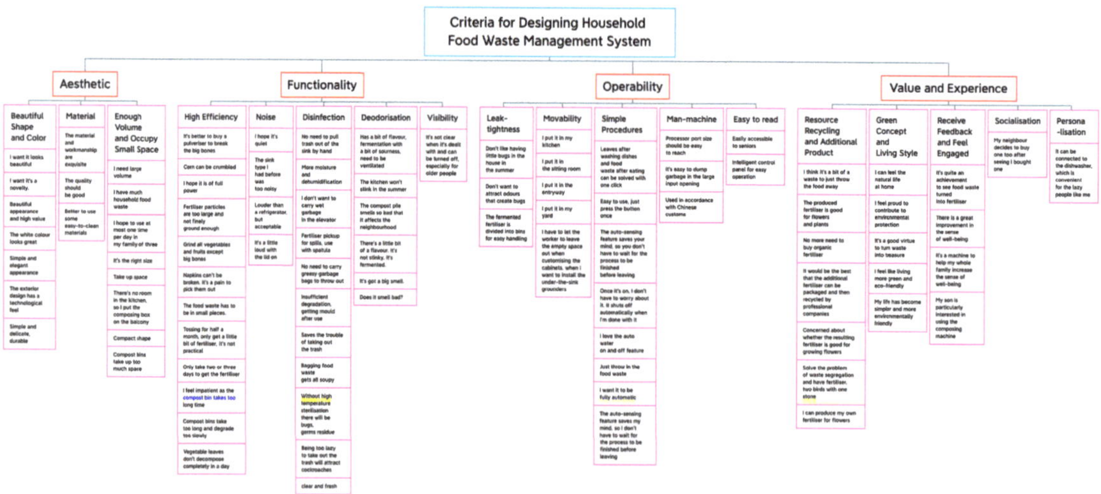

Figure 4. The categorised design criteria by using the affinity diagram.

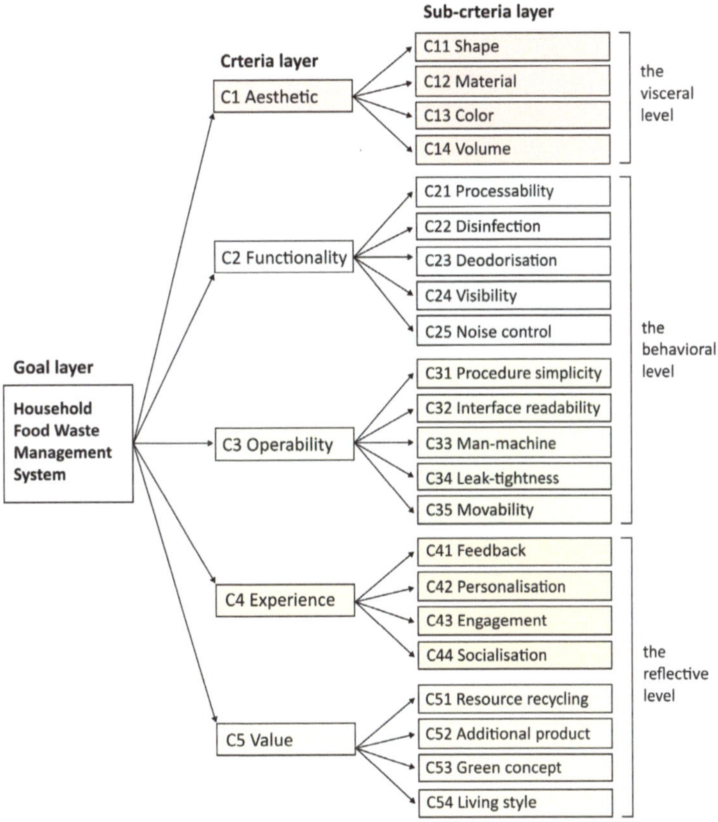

Figure 5. Hierarchical structure for household food waste management system.

After the requirements were clarified, the AHP questionnaire was developed, and the decision matrix was established. A total of 31 experts, including 6 design faculty, 8 master's degree product design students, 1 food waste disposal industry salesperson and 16 experienced users, were invited to complete the questionnaire. They evaluated the importance of each indicator of the main criteria and subcriteria layers using a nine-point scale for the decision problem. The index values of the goal layer and the five main criteria form decision matrix A. In the matrix A, the index a_{ij} refers to the relative importance value of indicator a_i compared to that of indicator a_j. n is the number of indicators:

$$A = (a_{ij})_{nn} = \begin{bmatrix} a_{11} & \cdots & a_{1n} \\ \vdots & \ddots & \vdots \\ a_{n1} & \cdots & a_{nn} \end{bmatrix} \quad (1)$$

Next, 31 importance values of the comparison of each two indicators were obtained from the 31 accomplished questionnaires. Furthermore, by using the geometric average method, these 31 indexes were aggregated into a'_{ij}. m is the number of experts, and a^m_{ij} is the importance value given by each expert.

$$a'_{ij} = \sqrt[31]{\prod_{m=1}^{31} a^m_{ij}} \quad (2)$$

A new aggregated judgment matrix A′ was then formed. The matrix A'_{C1-C5} is an example. The calculated indexes are shown in Table 2, and the calculated indexes of other subcriteria's judgement matrices are shown in Appendix B.

Table 2. Pairwise comparison of criteria in matrix A'_{C1-C5}.

	C1	C2	C3	C4	C5
C1	1.00	0.44	0.42	0.68	1.06
C2	2.29	1.00	1.89	3.28	2.74
C3	2.38	0.53	1.00	2.69	1.34
C4	1.48	0.30	0.37	1.00	1.71
C5	0.95	0.36	0.75	0.58	1.00

Then, the geometric mean of each criterion V_i and weight vector W was obtained using the geometric average calculating method:

$$V_i = \sqrt[n]{\prod_{j=1}^{n} a'_{ij}}, (i = 1, 2, 3, 4, 5) \quad (3)$$

$$w_i = \frac{V_i}{\sum_{i=1}^{n} V_i}, W = (w_1, w_2, \ldots, w_n) \quad (4)$$

To ensure the judgments made in all the matrices are reasonable, the consistency test follows the next step. Consistency ratios (CR) were identified for each of the matrices calculated using the largest eigenvalue λ_{max} and the corresponding eigenvector w_i.

$$\lambda_{max} = \frac{1}{n} \cdot \sum_{i=1}^{n} \frac{(AW)_i}{w_i} \quad (5)$$

$$CI(ConsistencyIndex) = \frac{\lambda_{max} - n}{n - 1} \quad (6)$$

$$CR(ConsistencyRatio) = \frac{CI}{RI} \quad (7)$$

The calculated CRs of each criterion were less than 0.1 (see Table 3), which means that the importance evaluation results of each index are reasonable.

Table 3. Results of the consistency ratios.

	A	C1	C2	C3	C4	C5
CI	0.040	0.004	0.012	0.036	0.034	0.023
RI	1.12	0.89	1.12	1.12	0.89	0.89
CR	0.036	0.005	0.011	0.032	0.038	0.026

Similar calculations of other matrices were conducted in the same way. The weights and rankings of each criterion are shown in Table 4:

Table 4. Results of comprehensive weights.

Criteria	Sub-Criteria	Weight	Ranking
C1: Aesthetic 0.120	C11: Shape	0.031	14
	C12: Material	0.031	15
	C13: Colour	0.015	21
	C14: Volume	0.043	10
C2: Functionality 0.374	C21: Processability	0.087	3
	C22: Disinfection	0.099	1
	C23: Deodorisation	0.091	2
	C24: Visibility	0.031	16
	C25: Noise control	0.066	5

Table 4. *Cont.*

Criteria	Sub-Criteria	Weight	Ranking
C3: Operability 0.243	C31: Procedure simplicity	0.076	4
	C32: Interface readability	0.051	8
	C33: Man–machine	0.038	12
	C34: Leak-tightness	0.060	6
	C35: Movability	0.019	19
C4: Experience 0.140	C41: Feedback	0.058	7
	C42: Personalisation	0.041	11
	C43: Engagement	0.028	17
	C44: Socialisation	0.014	22
C5: Value 0.123	C51: Resource recycling	0.045	9
	C52: Additional product	0.019	20
	C53: Green concept	0.024	18
	C54: Living style	0.035	13

The weights of C2 Functionality and C3 Operability were significantly higher than those of the other requirements in the main criterion layer, so these two factors should be given more attention. For the subcriteria, the first half were selected as key issues for the design of the system; see Figure 6.

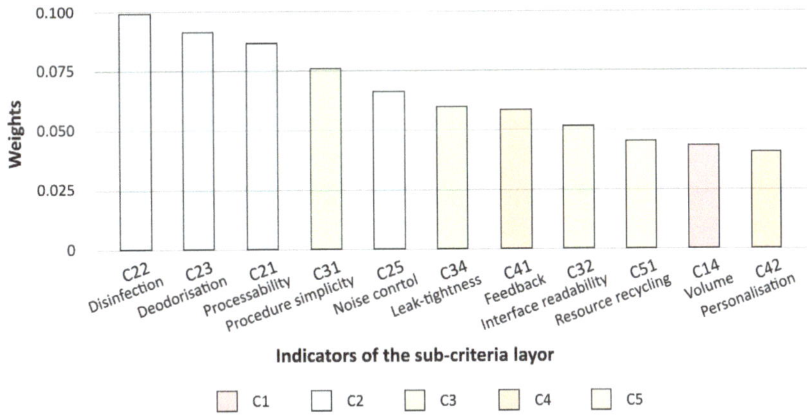

Figure 6. Weights of the first half of the indicators.

In the analysis of conventional technical solutions for the first 11 criteria, some conflicts can be identified. For instance, the product needs to incorporate modules with multiple functionalities while keeping the overall volume as compact as possible. Therefore, the TRIZ tool is introduced to address the conflicts existing among these criteria. Moreover, considering the objective of developing a smart system in this paper, there is a tendency to prioritise ICT-related solutions that align with TRIZ principles.

3.2. Use TRIZ to Solve Conflicts

There are two main types of conflicts: physical conflicts, which are addressed through four separation principles, and technical conflicts, which are solved through the invention principle and the conflict matrix composed of 39 technical parameters. Under the proposed problem of this study, two sets of physical conflicts and two sets of technical conflicts were identified and transformed into the TRIZ problem model, as shown in Table 5:

Table 5. TRIZ problem transformation of the conflicts.

No.	Type	Paradoxical Attributes	General Engineering Parameters	TRIZ Principles
1	Physical conflict	Disinfection–no germ	31 Harmful side effects	Separation upon condition
		Processability–contain germ		
2	Physical conflict	Deodorisation–let in air	32 Manufacturability	Separation in space
		Leak-tightness–air-proof		
3	Technical conflict	Requirement of multi-function	36 Complexity of device	no. 1 Segmentation
		Volume	8 Volume of non-moving object	
4	Technical conflict	Procedure simplicity	33 Convenience of use	no. 10 Preliminary
		Personalisation	24 Loss of information	

3.2.1. *Physical Conflict 1*

The addition of catalytic fungi is commonly used to accelerate food waste decomposition [37]. However, the decomposition process also leads to the rapid proliferation of harmful bacteria. A sufficient amount of catalytic fungi and as little harmful bacteria as possible should be present. This requirement of simultaneously increasing and decreasing bacterial populations creates an evident conflict.

According to the separation upon condition principle, these two types of microorganisms can be distinguished based on their differences in temperature tolerance parameters. Harmful bacteria belong to the group of thermolabile bacteria, while decomposing microorganisms belong to the group of thermophilic bacteria [38]. Therefore, a potential solution in which a temperature-sensitive heating control system is installed in the decomposition zone is proposed. This system consists of a temperature sensor and electric heating tubes uniformly arranged on the inner walls of the chamber. It continuously and intermittently heats the internal content to over 60°C during the decomposition process, thus ensuring that harmful bacteria can be effectively killed but that the catalytic fungi remain active.

3.2.2. *Physical Conflict 2*

There is a conflict between the requirements of deodorisation and leak tightness. On the one hand, deodorisation requires odours inside to be eliminated through ventilation. On the other hand, leak tightness requires the product to have good sealing properties in order to prevent internal odours and waste liquids from leaking into the external environment. This inconsistency leads to a physical conflict.

According to the separation in space principle, different modules can be set up to optimise the product's sealing structure, including a sealed chamber and an air circulation system for deodorisation and dehumidification. The air circulation system can control the flow and direction of incoming and outgoing air. Inside, an ozone disinfection device is added to oxidise and decompose the odour gas, and a high-efficiency particulate air filter is used to prevent particles in the gas from leaking out. This design prevents the diffusion of odours by filtering and treating the gas while maintaining good sealing.

3.2.3. *Technical Conflict 1*

The goal is to create a complex system that addresses multiple needs of noise reduction, dehumidification, disinfection, sterilisation and storage. This results in an increase in the product's volume and occupied space, which is not ideal for the limited space typically found in Chinese urban kitchens.

According to segmentation principle no. 1, the product can be divided into several modules based on main product functions. The modular design allows users to temporarily disassemble modules that are not currently needed. For example, they can choose whether to install a storage module based on their own requirements, effectively reducing the overall volume and space occupied by the product. This provides a more flexible and adaptable solution for Chinese households with space limitations in their kitchens.

3.2.4. *Technical Conflict 2*

The conventional manual recycling process usually includes multiple trivial steps, such as collecting, dumping, adding catalysts, regular turning and stirring, waiting and taking out. These laborious procedures need to be simplified. However, such simplification keeps users who have different needs or who enjoy the composting experience from setting parameters according to their personal habits.

Using preliminary principle no. 10, the product supports several preset options through a mobile application. Sensors and clips collect information about the recycling process, which is then transferred and displayed on the user's mobile app, merging and automating complex procedures for user convenience. Users can choose their preferred presets and select between the Quick Mode for rapid waste reduction or the Standard Mode for thorough decomposition. Based on this, the data recorded by sensors, including equipment status, waste types and disposal capacity, together with user inputs in the mobile app, are to be analysed and learned. By utilising deep learning techniques [39], the system can recognise patterns and trends in the household's dietary habits and waste disposal practices, thus automatically adjusting working modes and offering personalised recommendations for food consumption and disposal.

4. Design of the Household Food Waste Management Smart System

Based on the AHP–TRIZ model, a new food waste management system that uses intelligent techniques is illustrated in Figure 7, including a data centre, a semi-automated recycling product and a mobile application. Briefly, the hardware recycling product is responsible for fulfilling user requirements regarding hygiene and waste disposal efficiency. On the other hand, the software component, including the mobile app, the sensors that collect disposal data and the cloud centre that processes data, utilises machine learning techniques to address user demands for simplified operation and personalised settings.

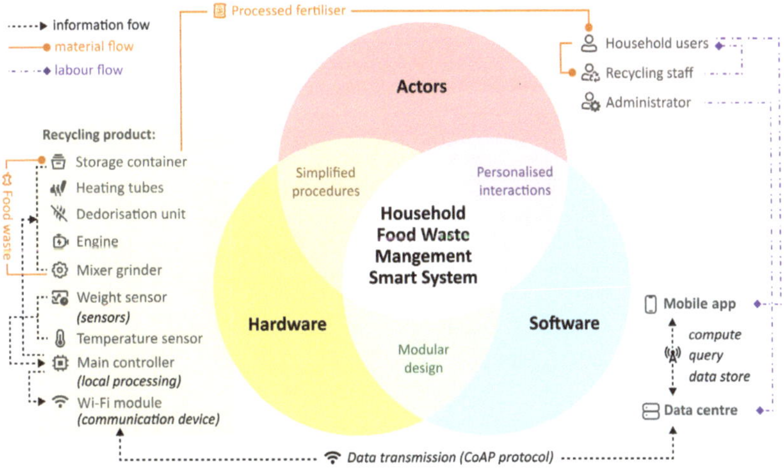

Figure 7. Proposed household food waste management system map.

The highlights of recycling management system innovation are as follows:

(1) *Modular Design:* As illustrated in Figure 8, the recycling product contains four modules arranged from top to bottom: the dropping zone, the processing zone, the fertiliser removal zone and the storage zone, enabling multiple key subfunctions. The dropping zone is designed for user input convenience with an expandable large

opening and stainless steel material to prevent stains from liquid leakage. The processing zone contains stirring and grinding blades, a small container for microbial catalysts and an embedded temperature and humidity sensor. The main controller regulates heating pipes for the sterilisation of harmful bacteria and for maintaining the activity of decomposition microorganisms. The fertiliser removal zone stores processed products and has a weight sensor that sends capacity reminders via a Wi-Fi module. The bottom storage zone stores packaged value-added products, as well as tools such as gloves, compost packages and small shovels. It is also equipped with universal wheels for movability. The modular design of the recycling product helps to reduce the space it occupies. In Figure 9, when the storage zone is removed, the product can be placed on the countertop. By attaching universal wheels at the bottom, the product can be easily moved around the kitchen, balcony or other areas. This flexibility caters to the specific environment of Chinese households.

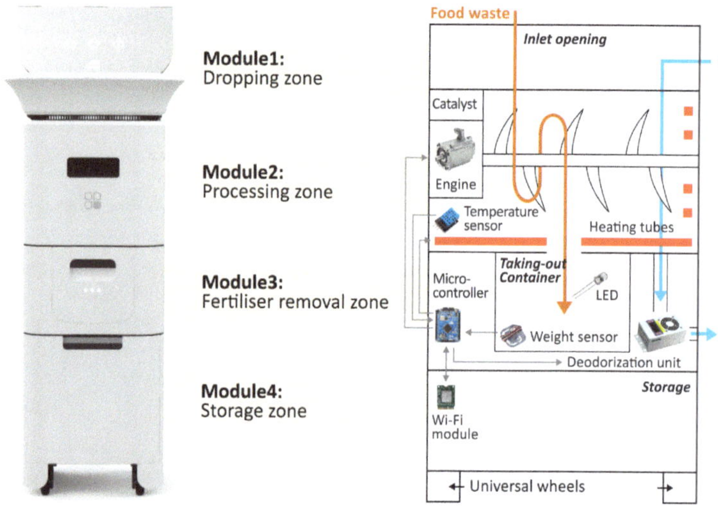

Figure 8. Structure of the recycling product.

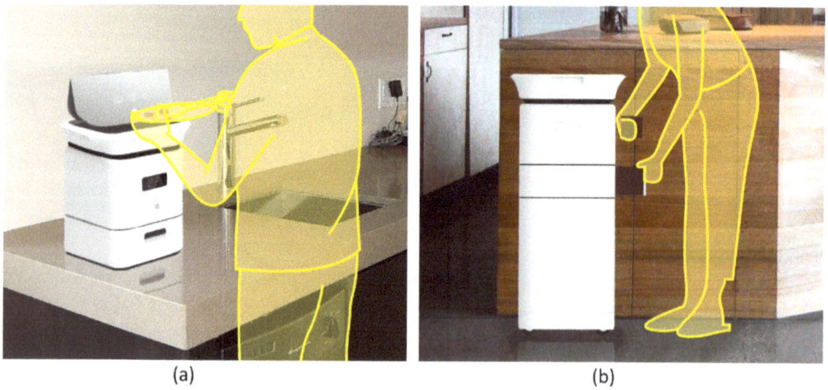

Figure 9. Usage in a household environment. (**a**) Put on the countertop. (**b**) Flexible and movable.

(2) *Simplified User Workflow:* Designed from a user-centric perspective, this system aims to facilitate long-term user engagement by minimising usage complexity and costs. Unlike traditional home composting processes in which users are required to regularly monitor, turn and control catalysts, this system incorporates sensors and automated

mechanisms to simplify the intermediate steps. As depicted in Figure 10, the process can be summarised as input–take out–store. Users begin by inputting their food waste into the recycling product. The main controller operates the stirring and grinding device, while the temperature sensor and heat pipes regulate microbial activity and decomposition efficiency. Additionally, an air circulation system connected to an odour and moisture removal device automatically maintains hygiene and cleanliness across different modules of the machine. When the value-added product fills up the container, the heavy sensor provides feedback to the main controller, triggering a lighting indicator on the machine and sending a notification to the user's mobile app. At this stage, users only need to take out the processed fertiliser and package it for storage without any check halfway through the process, awaiting scheduled collection by the recycling staff.

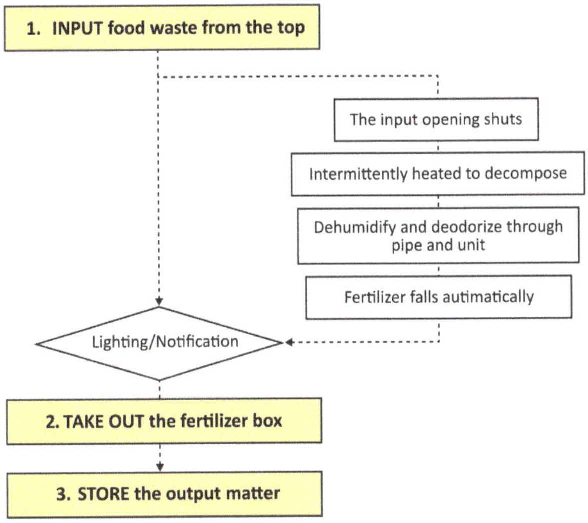

Figure 10. Steps for usage.

(3) *Personalised User Interaction:* On the other hand, simplifying the operational steps does not mean standardising user interactions during the food waste recycling process. User engagement in the recycling process can be personalised and enriched through the functions provided by the mobile application as well as adaptive product processing.
As illustrated in Figure 11, the data recognised and collected by the sensors in the hardware component can be broadly categorised into two types: basic data that can be presented to users and data analysed by the system to better understand user habits. Considering that household waste often contains sensitive personal information, data transmission is conducted using the CoAP security protocol specifically designed for IoT applications [22,40]. On the one hand, basic data, including processing count, reduction weight and recyclables weight, are made accessible to users through their mobile apps, and visually displayed. This allows users to conveniently track their disposal progress and history, imperceptibly fostering a sense of accomplishment and environmental consciousness. Additionally, users have the flexibility to choose between Quick Mode and Standard Mode, along with a do-not-disturb function, via the presetting feature, catering to their individual needs. On the other hand, data related to processing duration, time, frequency, corresponding processing modes and user input in the mobile app are recognised and processed using deep learning techniques [41,42]. This empowers the system to analyse and learn different users' waste disposal habits, enabling it to adaptively adjust different households' waste

processing modes. As a result, users can benefit from automated and customised kitchen waste disposal modes tailored to their respective food consumption habits.

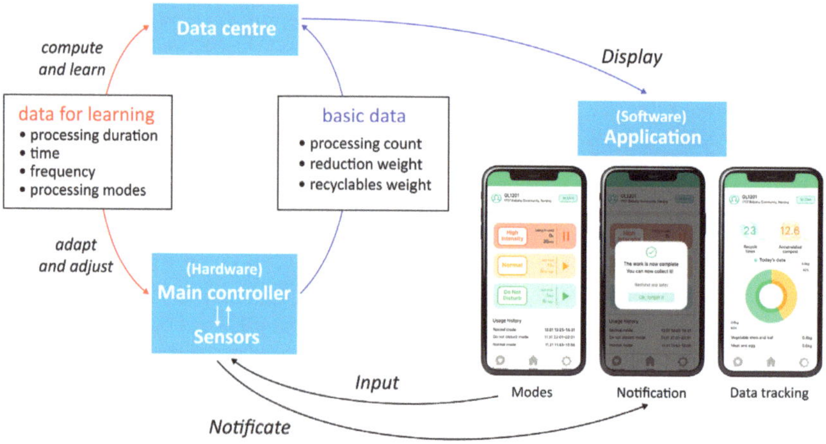

Figure 11. The software component in the system.

Compared to traditional composting methods, the new design offers significant advantages. In processing 1 kg of kitchen waste (800 g of fruit and vegetable peels, 150 g of grain and 50 g of meat residues), 175 g of fertiliser is produced within 10 h, achieving a higher reduction rate of 82% and a faster process than the traditional methods. For instance, in a month, a family of three can reduce approximately 37 kg of waste into around 6.5 kg of usable fertiliser. Furthermore, follow-up interviews with initial participants revealed a notable increase in willingness to use the new system, especially with high expectations for the ICT-related features, like tracking disposal status and adaptively streamlining processing procedures. The result demonstrates that our optimised system offers an effective and efficient solution for food waste management and has the potential for broader applications in promoting sustainable and eco-friendly living.

5. Discussion

5.1. A User-Centric Perspective to Promote Sustained Engagement

The implication of the term "user-centric" in this study is twofold. Firstly, it involves addressing users' genuine needs, which differs from conventional approaches that solely focus on mechanical or biological ways to accelerate food waste decomposition. For instance, from the AHP analysis in Section 3.1, this study recognises that users prioritise the hygiene of the recycling product as their first concern over mechanical improvements for faster decomposition. This is because, unlike centralised recycling systems, the target location of the system in this study is within households, directly impacting users' personal living environments and making hygiene their foremost consideration. By contrast, centralised recycling facilities are typically located in public spaces, leading residents to prioritise waste disposal convenience, while government institutions tend to emphasise processing efficiency. Consequently, solutions aimed at ensuring sanitation will be developed with more targeted aims. In our design, the corresponding solutions include periodic sterilisation through heating, an air circulation system equipped with a deodorisation unit and a larger input opening to facilitate user convenience. By understanding and catering to these real user needs, users become more engaged and motivated as the first practitioners to conduct household food waste management.

Secondly, the term "user-centric" means enhancing users' operational capability. This can be achieved by reducing the difficulty regarding user interaction and incorporating personalised features, such as offering preset modes and personalised adaptive tracking mechanisms. These solutions address the issue of users' unfamiliarity with household food waste recycling operations, reducing barriers that hinder their engagement in recycling, therefore fostering long-term habits and sustaining their recycling behaviour.

Adopting a user-centric perspective that addresses genuine user needs and enhances operational capability is vital in designing a food waste recycling system. By acknowledging users as active practitioners and ensuring that the design aligns with their authentic requirements, sustained engagement and effective recycling behaviour can be fostered.

5.2. ICT-Enabled Solutions to Balance Conflicts within the System

Integrating smart devices, such as sensors and mobile terminals, opens up innovative possibilities for addressing complex and conflicting requirements of household management issues. New approaches to problem solving can be explored by leveraging ICT-enabled solutions.

For example, physical conflict 1 can be resolved by utilising temperature sensors to detect changes in the status of decomposed matter. By controlling the heating tubes based on these readings, a balance between two microbial populations can be achieved, ensuring efficient decomposition. Moreover, in technical conflict 2, automation and sensor-based monitoring can simplify user interactions. By incorporating predefined control patterns and sensors, users can effortlessly select different recycling modes without complex manual adjustments. Furthermore, the system can gradually adapt to user operations by intelligently learning their usage habits, thereby directly providing disposal solutions that align with their operational preferences and food consumption.

These ICT-enabled solutions not only streamline users' management processes and improve their willingness to conduct recycling but also expand the capability of TRIZ principles in addressing conflicts within a system. These solutions offer efficient ways to balance conflicting subcriteria's solutions, resulting in optimised waste management processes and improved user experiences.

5.3. Hardware–Software Integration to Shape the New Management System

In the context of smart management, product design and manufacturing can no longer be the only sources of competitive advantage and differentiation. Integrating sensing systems and mobile service technologies offers a broader perspective for problem solving.

Traditional solutions for food waste management typically involve under-the-sink grinders or compostors, focusing on improving operational performance and efficiency from mechanical and biological perspectives. However, there has been limited exploration of utilising intelligent technologies to assist users in smooth and effortless household food waste recycling, aiming for a more comfortable and enjoyable experience. The proposed solution in this paper aims to generate a smart management system by incorporating hardware and software, guided by user-centric criteria obtained through the AHP. Through sensors embedded in the hardware, data such as the timing, frequency and disposal modes can be collected, enabling the acquisition of user behaviour patterns and preferences. User input preferences, such as specific processing methods, can be recorded through the software. By training and pattern recognition using extensive user data, the system can gradually learn and understand individualised preferences and disposal habits, thereby achieving a certain level of automation and providing personalised services to users.

By embracing a system-level perspective and leveraging the advantages of hardware-software integration, the food waste management system becomes more than just a mechanical solution. It facilitates efficiency, enhanced user experience, data-driven insights and seamless integration within the broader ecosystem of smart living.

6. Conclusions

Utilising food waste resources is a crucial step towards green development and green living. Apart from the commonly adopted centralised waste management mode, exploring how household-level food waste recycling can be facilitated is essential as the home is the most direct and decentralised source of food waste.

This study uses the AHP–TRIZ method to design an intelligent household food waste management system from a user-centric perspective. The AHP helps to identify genuine user needs and objectively prioritises requirements by calculating criterion weights. From the analysis, we found that users primarily focus on fundamental aspects, such as functionality and operability, when it comes to household recycling. Demands related to experience and value may emerge once the foundational aspects of functionality and operability are well established. Additionally, TRIZ tools are employed to address the physical and technical conflicts that arise from common solutions to these requirements. TRIZ principles have been applied and expanded in the construction of the system. Through the integration of ICT techniques, more solutions that have not widely been adopted in the field of food waste processing are generated.

The outcome is a smart food waste management system comprising a recycling product for food waste reduction, a mobile application for user engagement, and a data centre for data transmission, computing, and storage. This system effectively enhances the recycling rate of household food waste while promoting interaction between individuals and intelligent appliances.

In future research, the limitations of the current study need to be addressed and improved in two aspects. Firstly, at the method level of the system development, it is necessary to consider incorporating ICT solutions to further innovate and advance the current TRIZ part. The current TRIZ principles would break away from the predominantly mechanical development approach and facilitate a more direct reference for the development of other intelligent systems. Secondly, at the solution level, there is a need to optimise the current preliminary solutions in terms of how artificial intelligence can learn user processing disposal patterns and provide reasonable recommendations. Additionally, conducting a wider range of user testing is essential to further validate the effectiveness of the proposed solution, thus making this research a pilot for developing other household intelligent management systems.

Author Contributions: Conceptualisation, S.W. and J.X.; methodology, S.W. and J.X.; software, S.W.; validation, S.W.; writing—original draft preparation, S.W. and J.X.; writing—review and editing, S.W. and H.P.; visualisation, S.W.; supervision, H.P.; project administration, H.P.; funding acquisition, H.P. All authors have read and agreed to the published version of the manuscript.

Funding: The work described in this paper was fully supported by a grant from the Hong Kong Polytechnic University (Project No. P0036335).

Institutional Review Board Statement: Not applicable.

Informed Consent Statement: Not applicable.

Data Availability Statement: The data used to support the findings of this study are available from the corresponding author upon request.

Acknowledgments: This paper is based on our previous work [43], presented at the 2nd International Conference on Culture-oriented Science & Technology (ICCST).

Conflicts of Interest: The authors declare no conflicts of interest.

Appendix A

Table A1. Explanation for AHP criteria.

No.	Criterion	Explanation
C1	**Aesthetic**	The visual appeal of the products within the system.
C11	Shape	Physical form (round or square; curvaceous or rectilinear).
C12	Material	Choice of construction materials.
C13	Colour	Use of colours in the system's design.
C14	Volume	Capacity or size of the product.

Table A1. *Cont.*

No.	Criterion	Explanation
C2	**Functionality**	The system's ability to perform its intended tasks.
C21	Processability	Efficient handling and processing of food waste.
C22	Disinfection	Ensuring hygienic treatment of waste.
C23	Deodorisation	Elimination of unpleasant odours.
C24	Visibility	Clear visibility of the waste management process.
C25	Noise control	Minimisation of noise generated during operation.
C3	**Operability**	The ease of use and operation of the system.
C31	Procedure simplicity	The simplicity of system operation procedures.
C32	Interface readability	Clear and user-friendly interface design.
C33	Man–machine	The scale of the product lines with human operations.
C34	Leak-tightness	Prevention of any leakage or spillage.
C35	Movability	Easy movement or portability of the product.
C4	**Experience**	User's overall experience when operating the system.
C41	Feedback	Providing feedback to users (reminder, notification, etc.)
C42	Personalisation	Customisation options for individual preferences.
C43	Engagement	Encouraging user involvement and participation.
C44	Socialisation	Promoting social interactions and community engagement.
C5	**Value**	The system's social impact and conceptual value.
C51	Resource recycling	Substantial reduction in food waste.
C52	Additional product	Creation of additional valuable products from food waste.
C53	Green concept	Raising users' environmentally friendly awareness.
C54	Living Style	Changing the way users live.

Appendix B

Table A2. Pairwise comparison of criteria in matrices C1′–C5′.

matrix C1′	**C11**	**C12**	**C13**	**C14**	
C11	1.00	1.11	1.94	0.71	
C12	0.90	1.00	2.42	0.67	
C13	0.52	0.41	1.00	0.38	
C14	1.42	1.49	2.62	1.00	
matrix C2′	**C21**	**C22**	**C23**	**C24**	**C25**
C21	1.00	1.05	0.84	2.21	1.63
C22	0.95	1.00	1.38	3.31	1.41
C23	1.19	0.73	1.00	3.34	1.35
C24	0.45	0.30	0.30	1.00	0.41
C25	0.61	0.71	0.74	2.44	1.00
matrix C3′	**C31**	**C32**	**C33**	**C34**	**C35**
C31	1.00	1.50	1.99	1.29	3.81
C32	0.67	1.00	1.62	0.75	2.46
C33	0.50	0.62	1.00	0.67	2.17
C34	0.78	1.33	1.48	1.00	3.02
C35	0.26	0.41	0.46	0.33	1.00
matrix C4′	**C41**	**C42**	**C43**	**C44**	
C41	1.00	2.18	1.81	3.10	
C42	0.46	1.00	1.94	3.40	
C43	0.55	0.52	1.00	2.20	
C44	0.32	0.29	0.45	1.00	

Table A2. *Cont.*

matrix $C5'$	C51	C52	C53	C54
C51	1.00	2.61	2.74	0.92
C52	0.38	1.00	0.75	0.60
C53	0.40	1.34	1.00	0.88
C54	1.08	1.66	1.14	1.00

References

1. Ogunmoroti, A.; Liu, M.; Li, M.; Liu, W. Unraveling the environmental impact of current and future food waste and its management in Chinese provinces. *Resour. Environ. Sustain.* **2022**, *9*, 100064. [CrossRef]
2. Lo, I.M.; Woon, K.S. Food waste collection and recycling for value-added products: Potential applications and challenges in Hong Kong. *Environ. Sci. Pollut. Res.* **2016**, *23*, 7081–7091. [CrossRef]
3. Meng, M.; Wen, Z.; Luo, W.; Wang, S. Approaches and policies to promote Zero-waste City construction: China's practices and lessons. *Sustainability* **2021**, *13*, 13537. [CrossRef]
4. Xiao, L.; Zhang, G.; Zhu, Y.; Lin, T. Promoting public participation in household waste management: A survey based method and case study in Xiamen city, China. *J. Clean. Prod.* **2017**, *144*, 313–322. [CrossRef]
5. Guo, Y.; Wei, R.; Zhang, X.; Chai, F.; Zhao, Y.; Zhou, T. Positive Impacts of the Overall-Process Management Measures on Promoting Municipal Solid Waste Classification: A Case Study of Chongqing, China. *Sustainability* **2022**, *14*, 14250. [CrossRef]
6. Jiang, J.; Geng, S.; Luo, W.; Jiang, Y.; Gao, Y.; Chen, Z.; Yang, G.; Lan, T.; Meng, Y.; Ju, T.; et al. Review of hotspots of kitchen waste treatment in context of garbage classification in China in 2020. *Sci. Technol. Rev.* **2021**, *39*, 261–276.
7. Chen, F.; Chen, H.; Wu, M.; Li, S.; Long, R. Research on the driving mechanism of waste separation behavior: Based on qualitative analysis of Chinese urban residents. *Int. J. Environ. Res. Public Health* **2019**, *16*, 1859. [CrossRef] [PubMed]
8. Yu, H.; Chen, Y. Design and Application of Intelligent Waste Sorting and Recycling System. *Packag. Eng.* **2018**, *39*, 154–159.
9. Dos Santos, P.H.; Neves, S.M.; Sant'Anna, D.O.; De Oliveira, C.H.; Carvalho, H.D. The analytic hierarchy process supporting decision making for sustainable development: An overview of applications. *J. Clean. Prod.* **2019**, *212*, 119–138. [CrossRef]
10. Chou, J.R. A TRIZ-based product-service design approach for developing innovative products. *Comput. Ind. Eng.* **2021**, *161*, 107608. [CrossRef]
11. Ijemaru, G.K.; Ang, L.M.; Seng, K.P. Swarm Intelligence Internet of Vehicles Approaches for Opportunistic Data Collection and Traffic Engineering in Smart City Waste Management. *Sensors* **2023**, *23*, 2860. [CrossRef]
12. Jin, C.; Sun, S.; Yang, D.; Sheng, W.; Ma, Y.; He, W.; Li, G. Anaerobic digestion: An alternative resource treatment option for food waste in China. *Sci. Total Environ.* **2021**, *779*, 146397. [CrossRef] [PubMed]
13. Li, Y.; Jin, Y.; Li, J.; Chen, Y.; Gong, Y.; Li, Y.; Zhang, J. Current situation and development of kitchen waste treatment in China. *Procedia Environ. Sci.* **2016**, *31*, 40–49. [CrossRef]
14. Liu, D.; Ma, X.; Huang, J.; Shu, Z.; Chu, X.; Li, Y.; Jin, Y. Investigation of the aerobic biochemical treatment of food waste: A case study in Zhejiang and Jiangsu provinces in China. *Sci. Total Environ.* **2022**, *806*, 150414. [CrossRef] [PubMed]
15. Hu, Y.B.; Wang, S.F.; Xiao, W.C. Design of mechanical compressive treatment device for wet waste. *Chin. J. Eng. Des.* **2021**, *28*, 374–380.
16. Zhang, H.; Duan, H.; Andric, J.M.; Song, M.; Yang, B. Characterization of household food waste and strategies for its reduction: A Shenzhen City case study. *Waste Manag.* **2018**, *78*, 426–433. [CrossRef]
17. Bernstad, A.; Davidsson, Å.; Tsai, J.; Persson, E.; Bissmont, M.; la Cour Jansen, J. Tank-connected food waste disposer systems–Current status and potential improvements. *Waste Manag.* **2013**, *33*, 193–203. [CrossRef] [PubMed]
18. Maalouf, A.; El-Fadel, M. Effect of a food waste disposer policy on solid waste and wastewater management with economic implications of environmental externalities. *Waste Manag.* **2017**, *69*, 455–462. [CrossRef]
19. Cecchi, F.; Cavinato, C. Smart approaches to food waste final disposal. *Int. J. Environ. Res. Public Health* **2019**, *16*, 2860. [CrossRef]
20. Du, Y.; Wang, H.; Jin, Q.; Liu, W. Application status and analysis of kitchen waste treatment technology. *Energy Environ.* **2019**, *1*, 87.
21. Zhou, X.; Yang, J.; Xu, S.; Wang, J.; Zhou, Q.; Li, Y.; Tong, X. Rapid in-situ composting of household food waste. *Process Saf. Environ. Prot.* **2020**, *141*, 259–266. [CrossRef]
22. Marques, P.; Manfroi, D.; Deitos, E.; Cegoni, J.; Castilhos, R.; Rochol, J.; Pignaton, E.; Kunst, R. An IoT-based smart cities infrastructure architecture applied to a waste management scenario. *Ad. Hoc. Netw.* **2019**, *87*, 200–208. [CrossRef]
23. Liegeard, J.; Manning, L. Use of intelligent applications to reduce household food waste. *Crit. Rev. Food Sci. Nutr.* **2020**, *60*, 1048–1061. [CrossRef] [PubMed]
24. Cappelletti, F.; Papetti, A.; Rossi, M.; Germani, M. Smart strategies for household food waste management. *Procedia Comput. Sci.* **2022**, *200*, 887–895. [CrossRef]
25. Spyridakis, I.; Holbrook, M.; Gruenke, B.; Latha, S.S. Smart resource management: Civic engagement and food recovery. In Proceedings of the 2019 IEEE International Smart Cities Conference (ISC2), Casablanca, Morocco, 14–17 October 2019; pp. 378–383.
26. Balwada, J.; Samaiya, S.; Mishra, R.P. Packaging plastic waste management for a circular economy and identifying a better waste collection system using analytical hierarchy process (AHP). *Procedia CIRP* **2021**, *98*, 270–275. [CrossRef]

27. Ho, W.; Ma, X. The state-of-the-art integrations and applications of the analytic hierarchy process. *Eur. J. Oper. Res.* **2018**, *267*, 399–414. [CrossRef]
28. Ilevbare, I.M.; Probert, D.; Phaal, R. A review of TRIZ, and its benefits and challenges in practice. *Technovation* **2013**, *33*, 30–37. [CrossRef]
29. Tang, R.; Huang, C.; Zhao, X.; Tang, Y. Research on Smart Tourism Oriented Sensor Network Construction and Information Service Mode. *Sensors* **2022**, *22*, 10008. [CrossRef]
30. Basak, S.; Dey, B.; Bhattacharyya, B. Demand side management for solving environment constrained economic dispatch of a microgrid system using hybrid MGWOSCACSA algorithm. *CAAI Trans. Intell. Technol.* **2022**, *7*, 256–267. [CrossRef]
31. Huang, D.; Li, M.; Fu, J.; Ding, X.; Luo, W.; Zhu, X. P2P Cloud Manufacturing Based on a Customized Business Model: An Exploratory Study. *Sensors* **2023**, *23*, 3129. [CrossRef]
32. Plain, C. Build an affinity for KJ method. *Qual. Prog.* **2007**, *40*, 88.
33. Norman, D. *The Design of Everyday Things: Revised and Expanded Edition*; Basic Books: New York, NY, USA, 2013.
34. Jordan, P.W. Human factors for pleasure in product use. *Appl. Ergon.* **1998**, *29*, 25–33. [CrossRef] [PubMed]
35. Bhandari, U.; Chang, K.; Neben, T. Understanding the impact of perceived visual aesthetics on user evaluations: An emotional perspective. *Inf. Manag.* **2019**, *56*, 85–93. [CrossRef]
36. Yan, M.; Lou, X.; Chan, C.A.; Wang, Y.; Jiang, W. A semantic and emotion-based dual latent variable generation model for a dialogue system. *CAAI Trans. Intell. Technol.* **2023**, *8*, 319–330. [CrossRef]
37. Zhao, Y.; Cai, J.; Zhang, P.; Qin, W.; Lou, Y.; Liu, Z.; Hu, B. Core fungal species strengthen microbial cooperation in a food-waste composting process. *Environ. Sci. Ecotechnol.* **2022**, *12*, 100190. [CrossRef] [PubMed]
38. Onwosi, C.O.; Igbokwe, V.C.; Odimba, J.N.; Eke, I.E.; Nwankwoala, M.O.; Iroh, I.N.; Ezeogu, L.I. Composting technology in waste stabilization: On the methods, challenges and future prospects. *J. Environ. Manag.* **2017**, *190*, 140–157. [CrossRef] [PubMed]
39. Yan, M.; Xiong, R.; Wang, Y.; Li, C. Edge Computing Task Offloading Optimization for a UAV-assisted Internet of Vehicles via Deep Reinforcement Learning. *IEEE Trans. Veh. Technol.* **2023**, 1–12. [CrossRef]
40. Naik, N. Choice of effective messaging protocols for IoT systems: MQTT, CoAP, AMQP and HTTP. In Proceedings of the 2017 IEEE International Systems Engineering Symposium (ISSE), Vienna, Austria, 11–13 October 2017; pp. 1–7.
41. Wang, C.; Qin, J.; Qu, C.; Ran, X.; Liu, C.; Chen, B. A smart municipal waste management system based on deep-learning and Internet of Things. *Waste Manag.* **2021**, *135*, 20–29. [CrossRef] [PubMed]
42. Abdullayeva, F.J. Internet of Things-based healthcare system on patient demographic data in Health 4.0. *CAAI Trans. Intell. Technol.* **2022**, *7*, 644–657. [CrossRef]
43. Wang, S.; Xu, J. Design of Intelligent Household Food Waste Product Based on AHP-TRIZ Method. In Proceedings of the 2022 International Conference on Culture-Oriented Science and Technology (CoST), Lanzhou, China, 18–21 August 2022; pp. 95–98.

sensors

MDPI

Article

Remote Multi-Person Heart Rate Monitoring with Smart Speakers: Overcoming Separation Constraint

Thu Tran *, Dong Ma and Rajesh Balan

School of Computing and Information Systems, Singapore Management University, Singapore 178902, Singapore; dongma@smu.edu.sg (D.M.); rajesh@smu.edu.sg (R.B.)
* Correspondence: ndttran.2019@phdcs.smu.edu.sg

Abstract: Heart rate is a key vital sign that can be used to understand an individual's health condition. Recently, remote sensing techniques, especially acoustic-based sensing, have received increasing attention for their ability to non-invasively detect heart rate via commercial mobile devices such as smartphones and smart speakers. However, due to signal interference, existing methods have primarily focused on monitoring a single user and required a large separation between them when monitoring multiple people. These limitations hinder many common use cases such as couples sharing the same bed or two or more people located in close proximity. In this paper, we present an approach that can minimize interference and thereby enable simultaneous heart rate monitoring of multiple individuals in close proximity using a commonly available smart speaker prototype. Our user study, conducted under various real-life scenarios, demonstrates the system's accuracy in sensing two users' heart rates when they are seated next to each other with a median error of 0.66 beats per minute (bpm). Moreover, the system can successfully monitor up to four people in close proximity.

Keywords: heart rate monitoring; acoustic-based sensing; smart speakers; multi-person tracking; spatial localization; FMCW

Citation: Tran, T.; Ma, D.; Balan, R. Remote Multi-Person Heart Rate Monitoring with Smart Speakers: Overcoming Separation Constraint. *Sensors* **2024**, *24*, 382. https://doi.org/10.3390/s24020382

Academic Editors: Ming Yan, Chunguo Li and Chien Aun Chan

Received: 29 November 2023
Revised: 31 December 2023
Accepted: 5 January 2024
Published: 8 January 2024

1. Introduction

Heart rate is one of the key indicators used to evaluate individuals' overall health. For example, changes in heart rate can be used to assess the state of the nervous system [1] and are used as a stress indicator [2]. In addition, rapid resting heart rate has been suggested as a risk factor for cardiovascular mortality [3], and heart rate dynamics are used to infer sleep stages, as these dynamics are more conspicuous during the later Rapid Eye Movement (REM) period [4]. Although traditional cardiac monitoring approaches such as electrocardiogram (ECG) can achieve high accuracy, these are contact-based approaches that are expensive, uncomfortable to wear for prolonged periods, and cumbersome to set up and use. Thus, they are not suitable for home monitoring or for patients with skin allergies or burn injuries, where skin-contact sensors are not feasible.

Recent advancements in remote sensing have suggested various ways to leverage radio frequency (RF) signals to monitor heart rate without any sensors or probes attached to the skin. These include Frequency Modulated Continuous Wave (FMCW) radar [5–7], WiFi [8,9], and millimeter wave [10,11]. In addition, acoustic signals [12–15] have been used to extract respiration and heart rate in a contactless manner. These approaches use a speaker to emit inaudible high-frequency waves, typically above 18 kHz, use microphones to capture the reflected signal after it bounces back from targets in the nearby areas, then analyze the reflected signal to extract valuable information such as the respiration rate and HR as well as the distance and the angle of the target relative to the transceiver.

Even with extensive prior work in remote heart rate sensing, heart rate detection has focused mostly on single sensing. Only a few solutions target the sensing of multiple heart rates [6,16], and all of these only work under conditions that restrict their practical adoption. In particular, [6] requires the subjects to be 1 to 2 m separated away from each other when using RF signals; [12,16] require at least 40 to 50 cm separation as well as a 10° angular difference between each subject to achieve acceptable performance using acoustic signals. This leads to the infeasibility of tracking multiple people in real-life scenarios, such as sitting side by side or lying next to each other on the same bed, as shown in Figure 1.

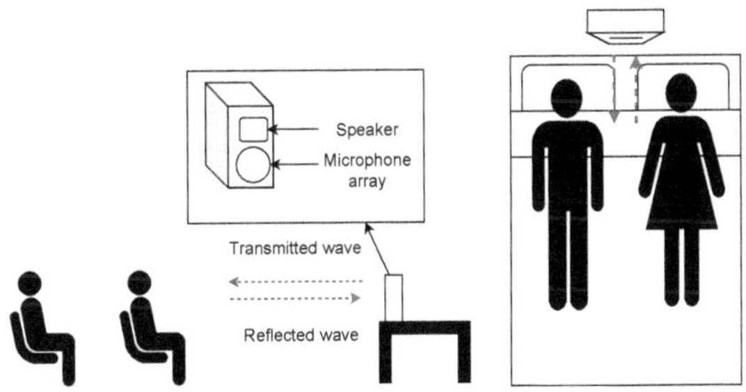

(a) (b)

Figure 1. Practical scenarios of multi-person heart rate monitoring: (**a**) two people sitting in line and (**b**) Two people sharing a bed.

In this paper, we aim to achieve multiple heart rate monitoring in such practical scenarios using a commodity smart speaker, the MiniDSP UMA-8-SP USB mic array [17], which has the same layout as the Amazon Echo Dot [18]. A smart speaker is considered as an appealing platform for contactless and acoustic-based heart rate monitoring for two key reasons. First, smart speakers have become increasingly prevalent in home environments, where they provide various voice-based services. Second, commercial smart speakers usually incorporate a microphone array design to deliver high-quality audio services. These microphone arrays offer high-resolution signals for active acoustic sensing that have been demonstrated to improve heart rate detection performance [13,19].

However, detecting and differentiating heart rates poses challenges when multiple people are in close proximity, as their acoustic reflection signals interfere with one another due to increased multipath interference. In this paper, we present an acoustic-based system that can extract multiple heart rates as well as their location with no separation requirement. To achieve this, we first separate users at different distances by processing the reflected FMCW signals. For each particular distance, we apply a Fast Fourier Transform (FFT) to extract the frequency of their corresponding heart rate. Then, we propose an algorithm to eliminate the interference and amplify the heart rate signal. Next, to further identify users' spatial information, we leverage the microphone array equipped in the smart speaker and apply beamforming to obtain their azimuth angles.

To assess the effectiveness of our approach, we conducted a study approved by the Institutional Review Board (IRB) under various realistic conditions. Using data collected from ten couples, our system shows the possibility of accurate heart rate detection when two individuals are positioned in close proximity, with a median error of 0.66 beats per minute (bpm). Moreover, we demonstrate the scalability of our technique by successfully identifying the heart rates and locations of four individuals seated next to each other.

2. Related Work

2.1. Contact-Based Heart Rate Monitoring

Contact-based heart rate monitoring approaches typically require sensors attached to the human body during the measurement process. These sensors are usually ECG or photoplethysmography (PPG). ECG is the traditional and "gold standard" contact-based technique to measure cardiac signals [20,21]. When performing ECG, electrodes are attached to the patient's skin at several spots, such as the chest or arms to record the electrical impulses of the heart when they travel across the electrodes. Both the strength and the timing of the pulses are monitored. Although ECG is highly accurate and widely used to diagnose heart-related diseases, this method is not suitable for home monitoring as it requires well-trained technicians and is done merely in clinical settings. The most common alternative monitors for home use are pulse meters or wrist bands [22,23]. They typically use PPG sensors, which contains a light source and a photodetector. The light source shines a green light onto the skin, and the changes in the light reflected back from the skin are monitored by the photodetector to extract associated heart pulses. Although this method is more convenient than ECG, it is still a contact-based approach, and is inapplicable to people with skin allergies or burn injuries.

2.2. RF-Based Heart Rate Monitoring

The past few years have witnessed growth in the number of studies that monitor vital signs, including respiration and heart rate, in a contactless manner through the use of RF waves. For example, well-studied RF-based approaches including broad-band FMCW radar [6,7], WiFi signals [8,9], RFID [24], and millimeter wave [11] have all shown accurate respiration rate detection even in the presence of more than one person. In addition to respiration sensing, [25–27] have demonstrated the possibility of heart rate detection using the these radar technologies. However, because heartbeat-induced chest displacement is subtle and always drowned out by respiration, only a few papers [6,11,28] have provided results for heart rate monitoring of more than one person; all of these papers require the subjects to be separated by a minimum of 1 to 2 m, making it infeasible to monitor two people sitting next to each other or sharing a bed. Furthermore, such RF-based systems require dedicated hardware, which generally leads to high cost.

2.3. Acoustic-Based Heart Rate Monitoring

Recently, the rise in popularity of acoustic-enabled mobile devices such as smartphones and smart speakers has introduced a potential alternative to RF-based heart rate measurement. Both RF and acoustic signals employ similar techniques to monitor respiration and heart rate. For instance, [29,30] developed an approach that utilizes the in-built speaker and microphone in smartphones and smart speakers to monitor the respiration rate in a single user. Similarly, [15] proposed a method for remotely monitoring heart rate using smartphones, while [13,14] suggested an approach using a smart speaker prototype with a single speaker–microphone pair. In order to expand the sensing range of these systems, [19] incorporated multiple microphones and a beamforming algorithm. However, due to signal interference, these studies remain unable to simultaneously monitor multiple people. Two recent studies [12,16] introduced novel beamforming algorithms that address this interference issue, enabling the monitoring of multiple heart rates; nevertheless, these algorithms require individuals to be separated by 40 to 50 cm and to be at least 10° apart. In our proposed approach, we first utilize a heatmap to identify users' distances and heart rates, then apply beamforming on multiple microphones using the known distances and heart rates. This allows us to obtain the localization of the users while eliminating the separation requirement. A summary of the differences between our proposed approach and existing studies is presented in Table 1.

Table 1. Related works on heart rate monitoring using acoustic-based approaches. Each study's median error is reported at its corresponding distance. In [13–15,19], the focus was on single users, while in [12,16] the authors monitored multiple users, for which the separation requirement is listed under *Separation*.

Study	Median Error (bpm)	Multiple Sensing	Separation	Distance (m)
[15]	0.6	No	Not applicable	0.3
[14]	0.75	No	Not applicable	0.2
[19]	0.6	No	Not applicable	0.6
[13]	1	No	Not applicable	0.4–0.6
[16]	0.8	Yes	40 cm and 10°	3
[12]	1.18	Yes	50 cm	3
Our approach	0.9	Yes	Not required	3

3. FMCW Background and Key Challenge

3.1. FMCW Background

In our proposed system, the speaker emits a sequence of FMCW chirps that are continuously modulated in frequency over a predefined time period. The reflected chirp is received by the microphone array, and the time difference between the transmitted and the reflected chirps indicates the range R between the transceiver and the object. For example, the blue line in Figure 2 shows the transmitted chirp with the frequency varying according to the sweep time T. The frequency of one single chirp is denoted as $f(t) = f_0 + \frac{B}{T}t$, where f_0 and B indicate the start frequency and the bandwidth, respectively.

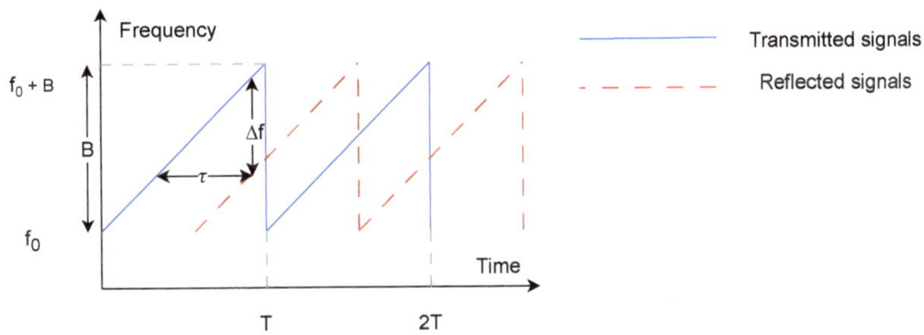

Figure 2. Transmitted and reflected FMCW signals.

Therefore, a single chirp is represented as

$$x(t) = cos\Big(2\pi \int f(t)dt\Big) = cos\Big(2\pi(f_0 t + \frac{Bt^2}{2T})\Big). \tag{1}$$

The reflected chirp from a target to the receiver is the delayed version of the transmitted chirp, as shown in Figure 2, and is expressed as

$$x'(t) = \alpha cos\Big(2\pi\Big(f_0(t-\tau) + \frac{B(t-\tau)^2}{2T}\Big)\Big), \tag{2}$$

where α and τ refer to the signal amplitude attenuation factor and time delay, respectively. With a moving target, we have $\tau = \frac{2r(t)}{c} = \frac{2(R+vt)}{c}$, where c is the speed of sound 343 m/s, v is the speed of the moving subject, and $r(t)$ is the distance of the moving subject by time. For a static object, $\tau = \frac{2R}{c}$; to compute the range R, we multiply the transmitted signal $x(t)$ by the received signal $x'(t)$. The mixed signal $x_m(t)$ is then represented as follows:

$$x_m(t) = x(t) \cdot x'(t)$$
$$= \frac{\alpha}{2}\left[cos\left(2\pi\left(f_0\tau - \frac{B(\tau^2 - 2t\tau)}{2T}\right)\right) + cos\left(2\pi\left(f_0(2t-\tau) + \frac{B(2t^2 - 2t\tau + \tau^2)}{2T}\right)\right)\right]. \quad (3)$$

The mixed signal consists of two terms. By taking the derivative of the phase by t, we have the frequency of the first term, which is a constant $\Delta f = \frac{B}{T}\tau = \frac{2BR}{cT}$. This implies that *every distance R maps to a specific frequency* Δf. The second term is a function of t with high frequency, and can be removed by a low-pass filter. In the end, after the multiplication and low pass filter, we have

$$x_{mf}(t) = \frac{\alpha}{2} \cdot exp\left[j2\pi\left(f_0\tau - \frac{B(\tau^2 - 2t\tau)}{2T}\right)\right]. \quad (4)$$

By transforming the frequency of $x_{mf}(t)$, we have

$$R = \frac{cT}{2B}\Delta f. \quad (5)$$

Given a typical audio sensing bandwidth of $B = 5k$ Hz (usually from 18 kHz to 23 kHz), according to [30], the resolution of R is $\delta R \geq \frac{cT}{2B}\delta f = \frac{cT}{2B} \cdot \frac{1}{T} = \frac{343}{2 \cdot 5000} = 3.43$ cm.

Although this resolution is sufficient to monitor centimeter-level human breathing, it is much lower than the heartbeat-induced chest displacement Δd, which is approximately 0.1–0.5 mm. Therefore, heart rate has been measured using phase-based methods [6,13–15]. It has been demonstrated that the minute chest displacement Δd can cause phase change in $x_{mf}(t)$ up to 18.9°. Specifically, $x_{mf}(t)$ can be expressed as follows.

$$\begin{aligned} x_{mf}(t) &= \frac{\alpha}{2} \cdot exp\left[j(2\pi f_0\tau - \frac{\pi B\tau^2}{T} + \frac{2\pi t\tau B}{T})\right] \\ &\approx \frac{\alpha}{2} \cdot exp\left[j(2\pi f_0\tau + \frac{2\pi t\tau B}{T})\right] \\ &= \frac{\alpha}{2} \cdot exp\left[j\left(\frac{4\pi f_0}{c}r(t) + 2\pi\Delta f t\right)\right] \\ &= \frac{\alpha}{2} \cdot exp\left[j\left(\frac{4\pi f_0}{c}\Delta d + 2\pi\Delta f t\right)\right] \end{aligned} \quad (6)$$

As such, if the chest displacement caused by a heartbeat is 0.5 mm, the phase change is calculated as

$$\frac{4\pi f_0}{c}\Delta d = \frac{4\pi \cdot 18,000 \cdot 0.0005}{343} = 0.105\pi = 18.9°. \quad (7)$$

Similar to prior works [6,29], in this paper we use distance to separate users. As each Δf indicates one R, we generate these frequency bins (which can be converted into distance bins) by applying FFT on x_{mf}, then analyzing the phase at each distance bin to determine the heart rate.

3.2. Key Challenge: Signal Interference

The reflected signals at the microphone array undergo interference when more than one person is present. Figure 3 is a heart rate–distance heatmap generated from the reflected signals, showing the heart rates and distances of two people lying down next to each other; the device is placed above their heads, with brighter points indicating higher power. There are two heart rates of 72 and 67 bpm annotated in the figure; however, they are not visibly recognizable due to interference effects.

As shown in the figure, two types of effects are observed, which we name the *distance effect* and *frequency effect*. The distance effect happens when an object or person is located at a certain distance; it makes all the frequency power at that distance higher, as the frequency power is proportional to the power of the reflection signal. This signal is shown as bright columns in Figure 3). In the heatmap, it can be observed that this effect happens mainly at 0.5 to 1 m, which is the location of the two users.

The frequency effect, depicted as bright rows in Figure 3, is caused by multipath reflections bouncing off walls or furniture and arriving at the receiver with increased arrival time. This multipath effect has been well studied in the literature [31,32]. Despite the longer travel time, these reflections carry the same frequency information and are linearly correlated to the directly reflected wave, resulting in equivalent frequencies spanning a wide range of distances. Multipath signals have lower power than true reflections due to power loss over distance. We note that while these two effects can occur with a single person, they become more severe with multiple people, as interference is more likely. Our solution to this problem is proposed in Section 4.

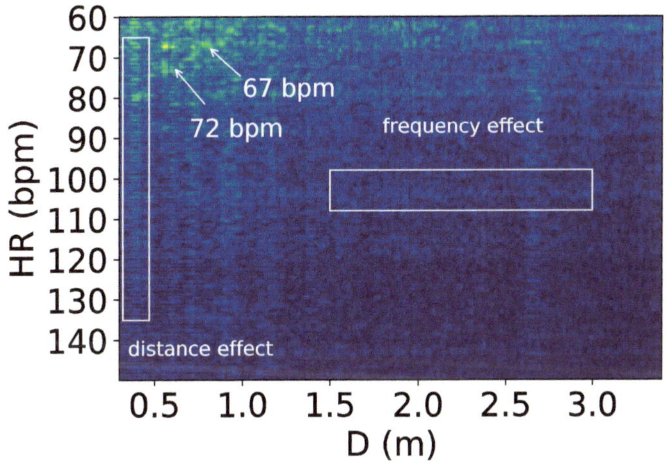

Figure 3. Heart rate–distance heatmap showing heart rates and interference from the reflected signals. The x-axis represents the distance D from the users to the device. In the figure, the two people with heart rates of 72 and 67 bpm located in front of the device cannot be distinguished visually from the heatmap.

4. System

4.1. System Overview

Our proposed system uses a commercial speaker and circular seven-microphone array with the same microphone layout and sensitivity as a commodity Amazon Echo Dot [18] (MiniDSP UMA-8-SP USB mic array [17]) as an FMCW transceiver. The speaker emits FMCW signals and the microphone array captures the signal reflected by the user, allowing all heart rates, distances, and angles of users to be identified. As shown in Figure 4, our system has four main modules:

- **Signal Processing:** This module processes raw reflected signals received by the microphones, removes noise and frequencies outside the range of f_0 and $f_0 + B$, and performs a mix operation on each chirp to generate the heart rate–distance heatmap.
- **Interference Removal:** The generated heatmap produced in the previous step is prone to distance and frequency interference effects. This interference is canceled through a two-step algorithm in order to highlight the heart rate signals.
- **Blob Detection:** Next, users' heart rates and distances are detected by applying a blob detection algorithm to the heatmap.
- **Beamforming:** Finally, beamforming is applied to detect each user's azimuth angle based on their distance and heart rate.

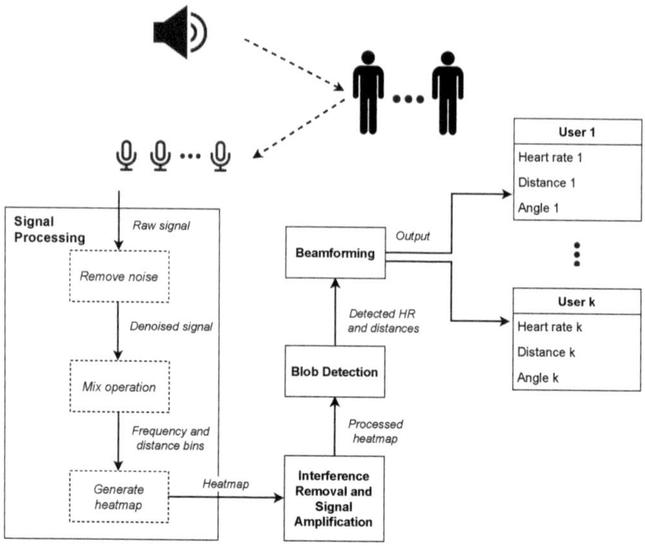

Figure 4. Overview of system for detecting the heart rates of k users.

4.2. Signal Processing

Remove Noise: The raw signal received by each microphone is filtered using a bandpass filter between f_0 to $f_0 + B$ to remove ambient noise such as human laughter and music, which are far lower than the operating FMCW frequency $f_0 \geq 18k$ Hz [33], as such noise has little impact on the system.

Mix Operation: When the raw signal has been processed, a mix operation is applied to each received chirp by multiplying them by the transmitted chirp (Equation (3)). As explained in Section 3, Δf is proportional to the distance of the target and is obtained by performing FFT on one chirp. The frequency bins after FFT are converted into distance bins using Equation (5).

In the FFT, the frequency resolution is $\frac{F_s}{N}$, where F_s is the sampling rate of the signal and N is the number of datapoints. Our system employs a sampling rate of $F_s = 48k$ Hz and a chirp length of $T = 0.04$ s. Therefore, the frequency resolution is $\frac{F_s}{T \cdot F_s} = \frac{48{,}000}{0.04 \cdot 48{,}000} =$ 25 Hz, which is converted to a *distance resolution* of 3.43 cm that is sufficient to differentiate users even when they are next to each other.

Generate Heatmap: The heart rate can be extracted using the phase changes of each distance bin over time. More particularly, we consider all distance bins within the device's working range to find the bin that contains the heartbeat signal. To illustrate this, we collected data with a subject located 1.08 m away from the speaker. Figure 5a shows the amplitude changes of the distance bins ranging from 0.58 m to 2 m. It can be observed that the signal at 1.08 m shows a periodic pattern from the user's breathing and heartbeats, while no vital signs can be seen at 0.58 m and 2 m. Figure 5b plots the frequencies for the 0.58 m, 1.08 m, and 2 m distance bins. These figures show that the bin with the highest amplitude, 0.58 m, is not guaranteed to contain heartbeat signals. This is because the distance effect (as the multipath) can cause the reflected signal to be stronger than the direct reflected signal. This phenomenon has been investigated in prior work [31,32].

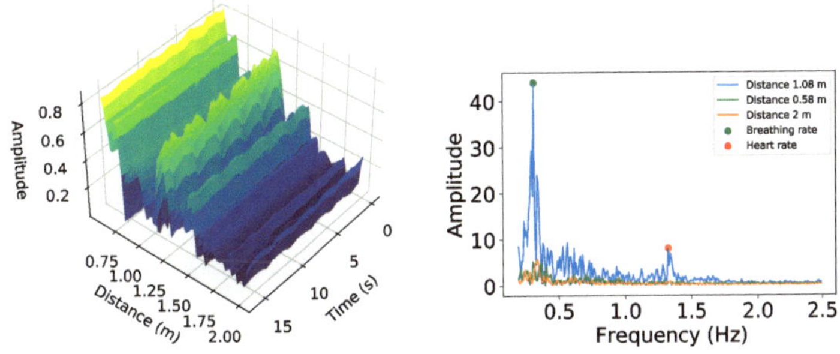

(**a**) (**b**)

Figure 5. Amplitude changes and frequency domain of distances: (**a**) FFT amplitude changes by distance and (**b**) breathing and heart rate obtained by applying FFT at 1.08 m.

Because we need to analyse the phase change of each distance bin, a chirp length of $T = 0.04$ s leads to a $F_{sh} = \frac{1}{0.04} = 25$ Hz sampling rate for the heart rate signal. Therefore, the frequency resolution of this FFT for heart rate is computed as $\frac{F_{sh}}{N} = \frac{25}{25 \cdot 60} = \frac{1}{60} \approx 0.0167$ Hz, where 60 (s) is the window length of the signal to which FFT is applied. This frequency resolution is equivalent to $0.0167 \cdot 60 = 1$ bpm. To enhance this resolution we zero-pad the signal with 4096 samples prior to the FFT. This interpolation can yield a higher display resolution, in this case, $F_{sh} = \frac{25}{25 \cdot 60 + 4096} \approx 0.004$ Hz ≈ 0.3 bpm. We note that while this approach does not truly improve the frequency resolution, it provides a smoother FFT output, which allows peak-picking algorithms to achieve better accuracy.

Finally, this module produces a heart rate–distance heatmap. If more than one microphone is used, the heatmaps from each microphone are stacked into a single map by averaging their amplitude.

4.3. Interference Removal and Heart Rate Signal Amplification

The stacked heart rate–distance heatmap generated by the last step is affected by noise from the distance and frequency effects (see Section 3.2).

To remove these unwanted effects, we apply L1-normalization, a normalization technique that modifies the dataset values to ensure that the sum of the absolute values in each row always adds up to 1, to all the rows (i.e., heart rates) and then all the columns (i.e., distances). L1-normalization balances out those cases when entire columns or rows have similar amplitude while preserving the relative ratio of rows and columns that contain heartbeat signals. Next, we apply Gaussian smoothing to the heatmap to highlight the heart rates and remove noise. The algorithm is described in Algorithm 1.

Algorithm 1: Remove interference and amplify heart rate signals

Input: Heatmap S, with n rows and m columns
Output: Interference-free and amplified heatmap S

```
Function RemoveInterferenceAndAmplify(S,n,m):
    /* Step 1                                   */
    for i ← 0 to n − 1 do
        S[i,:] = Normalize(S[i,:])
    /* Step 2                                   */
    for i ← 0 to m − 1 do
        S[:,i] = Normalize(S[:,i])
    return S
```

Figure 6b,c shows these steps to remove the effects, while Figure 6d shows the Gaussian smoothing used to highlight the heart rates.

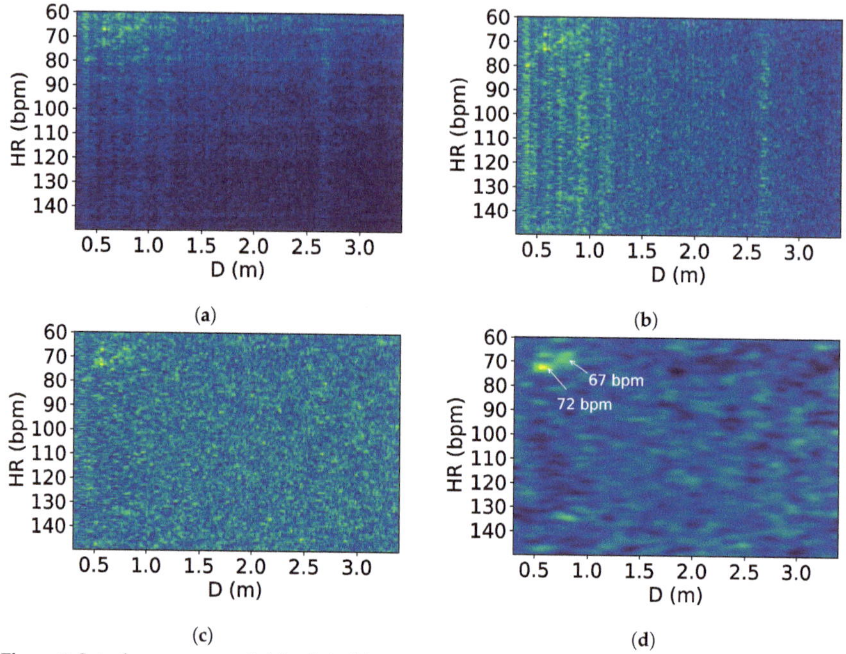

Figure 6. Interference removal: (**a**) original heatmap S; (**b**) heatmap S after step 1; (**c**) heatmap S after step 2; (**d**) smoothed heatmap S.

4.4. Blob Detection

A blob is a set of adjacent pixels that share common traits such as brightness or color. Because people can occupy many points in both distance and frequency, we detect the top brightest blobs in the heatmap instead of the top highest peaks. In this module, the input image is the heatmap from the previous step and the blobs to be detected are the elliptical bright spots on the image's dark background that indicate users' distances and heart rates. We apply the Laplacian of Gaussian (LoG) [34] as the primary blob detection method. With the blobs detected, we find the top k brightest blobs by calculating each blob's mean value in the frequency range of 0.8 Hz to 2.5 Hz, with k being the number of users. Figure 7 shows detected distances and heart rates when there are three people in front of the device.

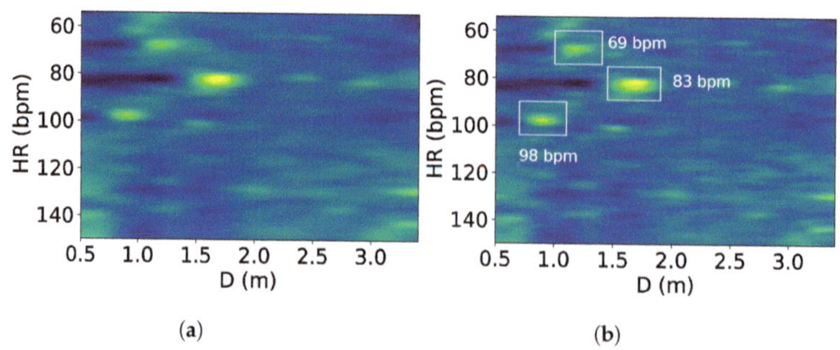

Figure 7. Three heart rates of 69 bpm, 83 bpm, and 98 bpm: (**a**) heatmap with three people and (**b**) the three brightest blobs.

4.5. Beamforming

To further identify users' locations in space, we apply digital beamforming on a circular microphone array [35] (Figures 8 and 9a) to obtain their azimuth angles from the known heart rates and distances extracted in the last module. According to Equation (6), we can rewrite x_{mf} using the nth distance bin and ith chirp:

$$x_{mf}(i,n) = \frac{\alpha}{2} \cdot exp\Big[j\Big(\frac{4\pi f_0}{c} r(f(n)+g(i)) + \frac{2\pi\tau B}{T}(f(n)+g(i))\Big)\Big]$$

where f and g are the function that linearly converts the nth distance bin and i^{th} chirp, respectively, into time t in seconds. Specifically, $g(i) = T \cdot (i-1), i \geq 1$ and $f(n) = 2\frac{ToDistance(n)}{c}, n \geq 1$, where $ToDistance(n)$ is a function to convert distance bin n to distance in meters.

Based on the azimuth angle θ provided by the circular microphone array in Figure 8, we project the source onto the x–y plane to obtain $\varphi = \frac{\pi}{2}$ with channel l; then, x_{mf} can be expressed as

$$x_{mf}(l,i,n) = \frac{\alpha}{2} \cdot exp\Big[j\Big(\frac{4\pi f_0}{c} r(f(n)+g(i)) + \frac{2\pi\tau B}{T}(f(n)+g(i)) + 2\pi\frac{r_0 \cdot cos(\theta - \Theta(l))}{c}\Big)\Big] \quad (8)$$

where r_0 is the radius of the circular array, θ is the azimuth angle of the target, and $\Theta(l) = \frac{2\pi}{L} l$ is the relative angle at microphone l. When the target is static, we have

$$x_{mf}(l,i,n) = \frac{\alpha}{2} \cdot exp\Big[j\Big(\frac{4\pi f_0}{c} R + \frac{2\pi\tau B}{T}(f(n)+g(i)) + 2\pi\frac{r_0 \cdot cos(\theta - \Theta(l))}{c}\Big)\Big]. \quad (9)$$

Because we obtained distances and heart rates in the last step, we can represent $x_{mf}(l,i,n)$ as $x_{n,l}(i)$ when fixing the distance and channel. To obtain the heartbeats across different angles for a given distance, beamforming is performed over $L = 6$ microphones:

$$y_n(i,\theta) = S^H(\theta) X_n(i) + W(i)$$

where $S(\theta) = [s_1(\theta), ..., s_L(\theta)]$ is the steering vector towards angle θ (with $s_l = exp(j2\pi\frac{r_0 \cdot cos(\theta - \Theta(l))}{c}))$, $X_n(i) = [x_{n,1}(i), ..., x_{n,L}(i)]$, and $W(i)$ is the Gaussian white noise. In our implementation, n includes a range of distances covering all users and y_n is the average of all values of n.

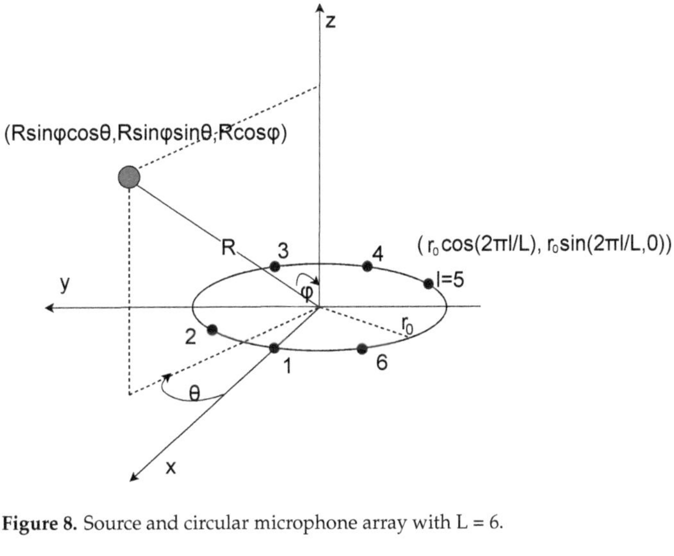

Figure 8. Source and circular microphone array with L = 6.

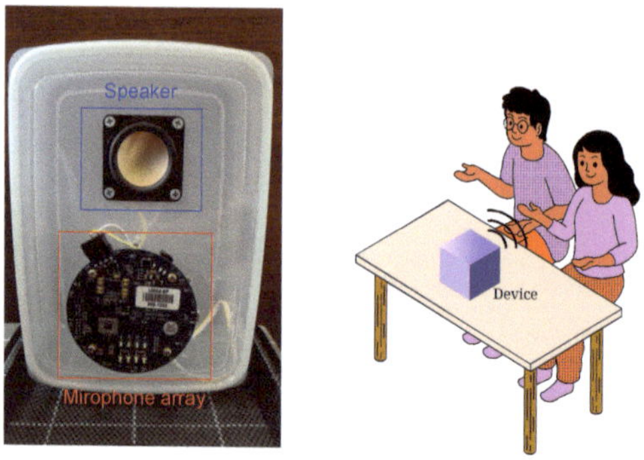

(a) (b)

Figure 9. Device and example showing the experimental setup: (**a**) the device and (**b**) one of the experimental setups.

5. Results

In this section, we report the detailed performance of the proposed system in various realistic scenarios, including sitting, lying down with different postures, in the presence of ambience noise, and with more than two users.

5.1. Experimental Setup

We prototyped our system using an off-the-shelf seven-microphone circular array [17] connected to a speaker (PUI Audio AS05308AS-R), as shown in Figure 9a. This prototype has the same microphone layout and sensitivity as the widely used Amazon Echo Dot [18]. Table 2 provides information on all of the parameters used in our experiment. In addition, we employed Polar H10 ECG sensors [36] to collect heart rates for use as ground truth. Figure 10 illustrates the heartbeats extracted from our system and corresponding heartbeats collected by the ECG sensors. The metric used to evaluate our system was the heart rate (bpm), which we compared against heart rates captured by ECG sensors worn by participants. Figure 9b describes one of our experimental setups for two people sitting next to each other with no separation requirement.

Table 2. Parameter settings for the experiments.

Parameter	Value
Chirp frequency	18 kHz to 23 kHz
Bandwidth	5 kHz
Chirp length	0.04 s
Maximum tracking distance	$\frac{cT}{4} = 3.43$ m
Sampling rate	48 kHz
Sound pressure	45 dB(A) at 0.3 m

To evaluate our system, we recruited ten couples (nine males and eleven females in the age range of 19 to 57 and with a median age of 26) to evaluate the impact of different parameters under various daily life scenarios. All experiments were approved by the IRB. We conducted further experiments with sets of three and four participants to verify that our proposed system can accurately monitor the heart rates of more than two people in close proximity.

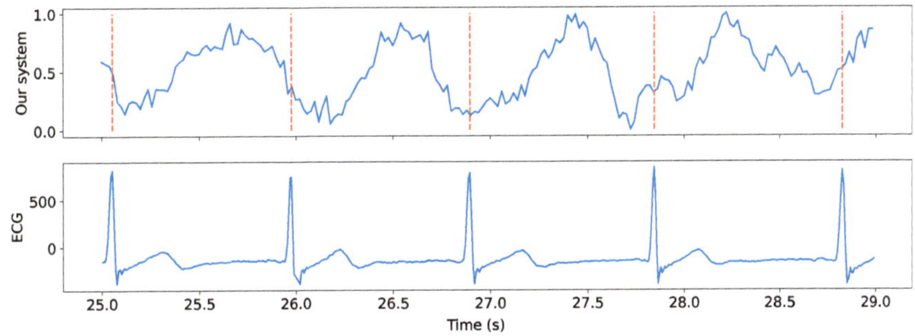

Figure 10. Extracted heartbeats of an individual located at 1 m and ground truth from ECG. The signal extracted from our system then undergoes FFT to obtain the heart rate in bpm.

5.2. Overall Performance

Initially, we assessed the system's performance in everyday scenarios involving two individuals located in front of the device at an arbitrary distance from 0.5 to 1 m. Each recording session lasted for 2 min per couple, resulting in approximately 20 min of total recording time and a total of 1124 datapoints. As depicted in Figure 11a, the observed resting heart rates ranged from 57 to 96 bpm. The grey dashed lines represent two standard deviations. Additionally, Figure 11b illustrates that the achieved median error for heart rate estimation was 0.66 bpm and 1.67 bpm at the 90th percentile.

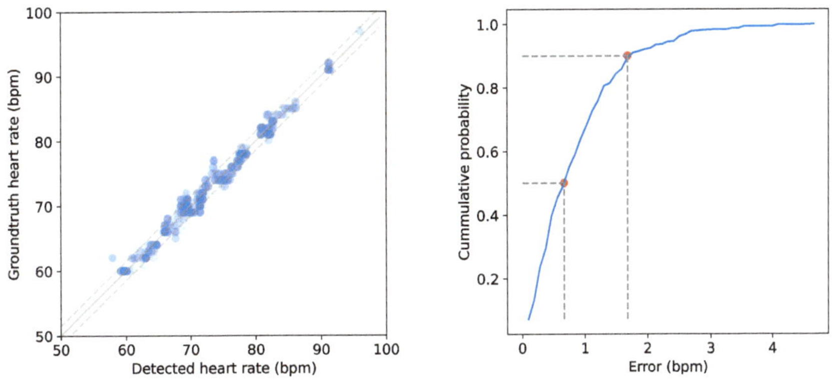

(a) (b)

Figure 11. Overall evaluation of the system: (**a**) detected and ground truth heart rates in bpm and (**b**) cumulative distribution function of the error.

5.3. Impact of Distance

We asked ten couples to evaluate the impact of distance on the system's performance when sitting next to each other in front of the device at various distances ranging from 0.5 to 3 m, with a step size of 0.5 m. The measurements were taken at each distance for 2 min while the couples are asked to remain stationary. Figure 12 shows median errors below 1 bpm when the distance was shorter than 2 m, with errors of 0.9, 0.71, 0.79, and 0.81 bpm at 0.5, 1, 1.5 and 2 m, respectively. Due to power loss, the median error increased slightly when the participants were located further from the device, with errors of 1.2 and 0.93 bpm at 2.5 and 3 m, respectively.

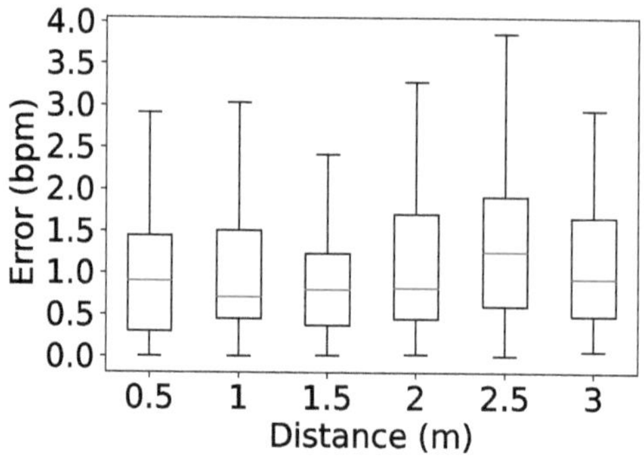

Figure 12. Impact of distance.

5.4. Impact of Angle

To assess the impact on the system of the angle between the users and the device, participants were asked to sit still at a fixed distance of 2 m with an angle φ changing from 0° to 15 ° and 30°, as shown in Figure 13. In Figure 14, it can be seen that the lowest median error when participants were seated at 0° was 0.65 bpm, while the highest median error was 0.93 bpm at 30°. We note that although the users were closer together when φ was lower, this did not decrease the accuracy very much. In fact, the error is mainly caused by the weak reflection when users are not directly facing the device, which leads to higher φ.

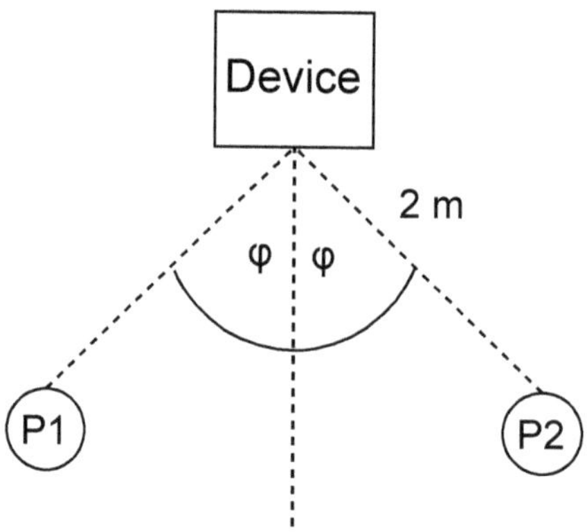

Figure 13. Results for users sitting at different angles; P1 and P2 refer to the two participants.

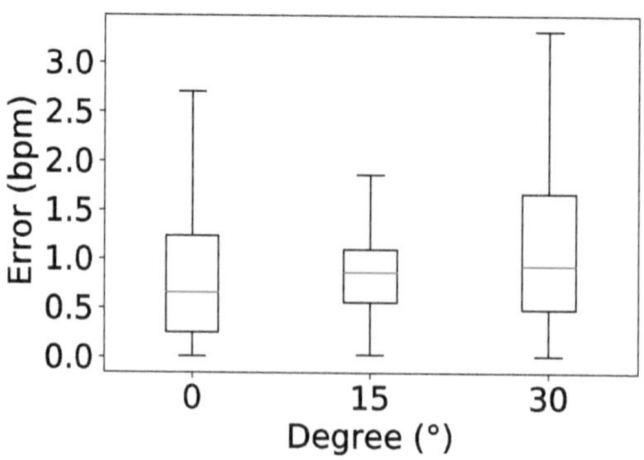

Figure 14. Impact of angle.

5.5. Impact of Ambient Noise

We investigated the impact of ambient noise by conducting the measurements in the presence of loud music. Each couple was asked to sit in front of the device at a distance of 2 m. The speaker was placed next to them while it played songs at 50 and 75 dB(A). These sound levels are comparable to normal conversation and road noise, respectively. Figure 15 shows the system performance under these two noise levels and when the average noise level when the room was quiet at 25 db(A). There was an increase in median error with higher sound pressure, especially when the sound pressure exceeded the sound pressure of the device itself, with the median error increasing to 1.5 bpm at 75 dB(A) compared to 0.8 bpm at 25 dB(A).

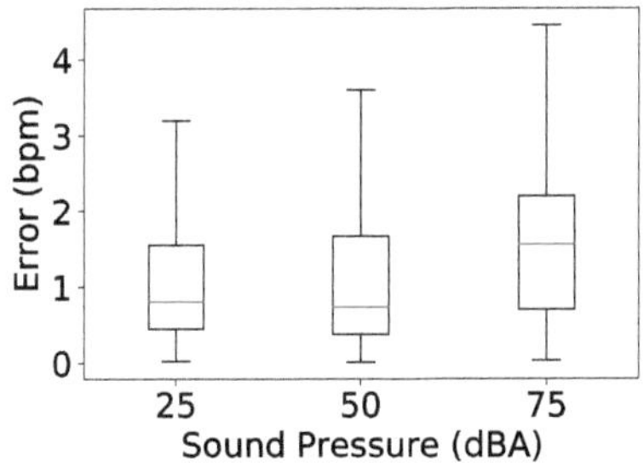

Figure 15. Impact of noise.

5.6. Lying Down with Different Postures

Different postures can lead to varying levels of accuracy. We conducted a user study involving couples lying down in the same bed in four common real-world postures: lying face-up, lying face-down, lying on the right side, and lying on the left side. To ensure that the participants could not block each others' signals when performing different postures, we positioned the device 0.5 to 1 m away from their heads, similar to Figure 1b. The results are shown in Figure 16.

In Figure 16, it can be seen that the best accuracy of 0.38 bpm was achieved with both participants lying face-up, while the lowest (1.4 bpm) was found with both lying face-down. This is because the latter posture weakens the body displacement caused by the heart. Motion signals were slightly reduced for users lying on their right and left sides as compared to facing up, with the median error increasing to 0.72 and 0.68 bpm, respectively.

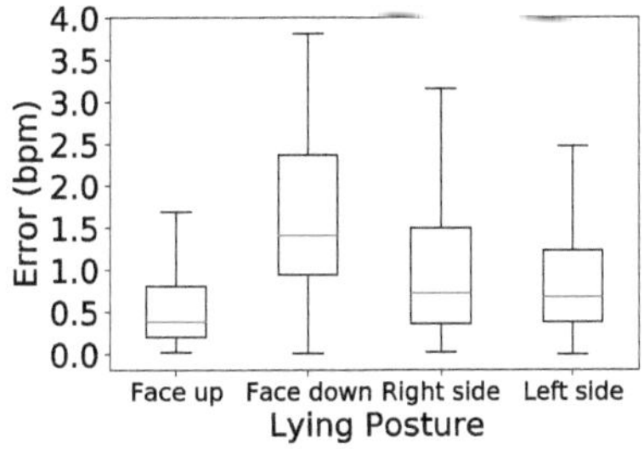

Figure 16. Impact of posture.

5.7. Lying Down with Blanket

We further evaluated our system with people lying face-up in bed covered by a blanket. As shown in Figure 17, the highest accuracy was achieved when both people were not covered by the blanket, with a median error of 0.53 bpm. Because thick cloth attenuates the signal, with the blanket there was a slight increase in the error to 0.64 bpm.

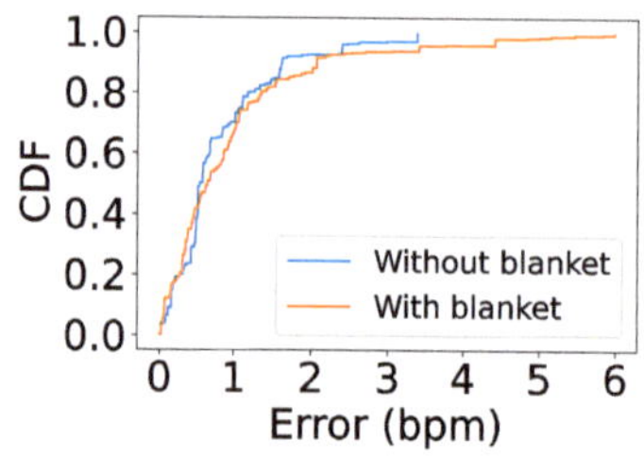

Figure 17. Impact of blanket.

5.8. Impact of Movement

Instant body movements such as posture changes, talking, or phone swiping do not lead to the same rhythm as heartbeats. Hence, such movements should have little impact on measurement. To investigate the performance of our system when users are perform such sudden movements, we asked each couple to sit next to each other while they reading a text on their phones and scrolling down or up once in a while. Participants were asked to naturally change their posture if needed; in fact, such short-time motions sometimes cause noise in the same frequency range as the heart rate, as body movements lead to much larger phase changes compared to subtle heartbeat motions. To deal with this, we applied a heuristic method that assumes the correct heart rate will change very little over time while the noise will disappear in the subsequent time intervals. Despite a slight difference in the error distribution, Figure 18 shows comparable median errors between users sitting still and users performing sudden movements, with 0.9 and 1 bpm, respectively.

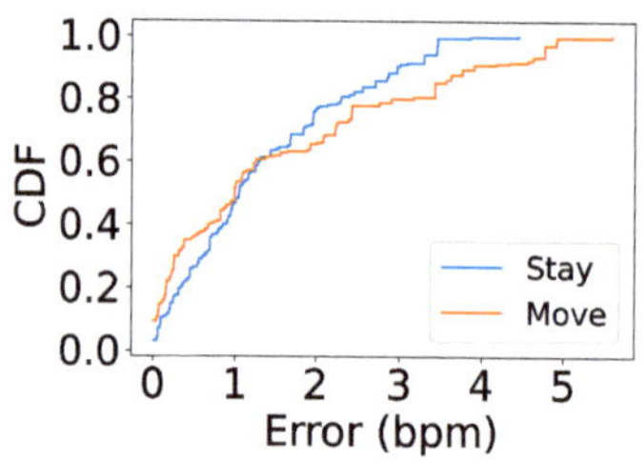

Figure 18. Impact of movement.

5.9. Impact of Number of Targets

To evaluate our solution with up to four people, we asked groups of two, three, and four people to sit side by side for a 5-min trial at a distance of 2 m. As shown in Figure 19, the errors in all four cases matched the errors in Figure 12 for two people sitting at a distance of 2 m. Thus, it can be concluded that the accuracy of the system is not impacted even when increasing the number of users to four.

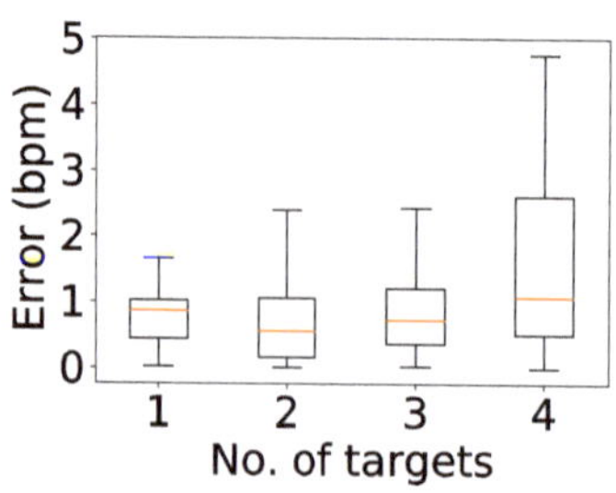

Figure 19. Impact of number of targets.

5.10. Impact of Number of Microphones

To assess the system performance under different microphone arrays, we ran the collected data with two, four, and seven microphones in the array. For the array with two microphones, we selected the ones located at 120° and 240°; for the array with four microphones, we selected the ones located at 60°, 120°, 240°, and 300° (see Figure 8); and for the array with seven microphones, all of the microphones in the array were operating. Figure 20 shows that the median heart rate error was below 1 bpm in all three cases (0.91, 0.82, and 0.85 bpm for two, four, and seven microphones, respectively). However, regarding the detectable time, which refers to the time during which all users' heart rates are visible in the heatmap, the two-microphone array had the lowest rate at 86%, compared to the seven-microphone one at 100%. This is because additional microphones improve the system's ability to capture reflected signals from more directions.

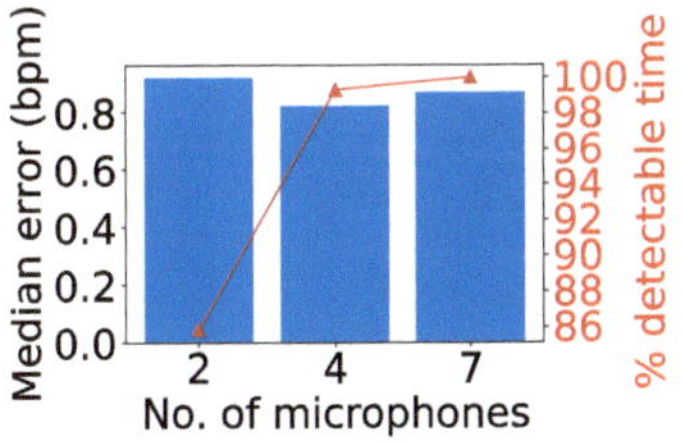

Figure 20. Impact of number of microphones.

5.11. Heart Rate Monitoring with Smartphone

To examine the approach on another platform, we implemented our system on a Samsung Galaxy S20 Plus smartphone. Due to the phone's design, it was only possible to use the single speaker–microphone pair at the bottom of the phone. Because only one microphone could be utilized, the angle information of the targets is not available. However, it remains possible to track the users by distance. We asked two volunteers to sit at distances of 2.7 m and 2.9 m in front of the smartphone. Figure 21 demonstrates that our system can achieve sensing of multiple heart rates at distances up to 3 m when deployed on a commercial smartphone with only one speaker–microphone pair; on the other hand, the sole existing smartphone-based approach [15] reports a maximum monitoring range of only 0.3 m.

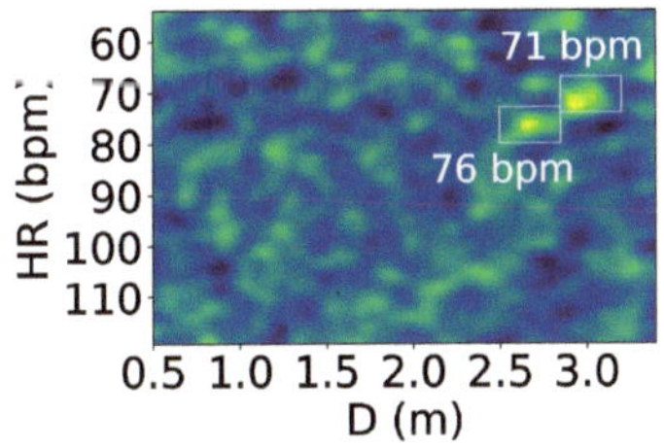

Figure 21. Heart rate detection by smartphone.

6. Discussion

The following limitations may apply to our proposed method:

- **Prone to Rhythmic Movement:** Our approach can be susceptible to the impact of body movement, which is a known challenge for handling motion noise in acoustic-based methods. Because we assume that the user position falls within a frequency of 0.8 to 2.5 Hz (i.e., the normal heartbeat frequency range), any other modulation within this frequency range that does not originate from the human heart, although very unlikely, will confuse the system. As a result, although our system works well when the motion frequency is beyond the usual range of the heart rate (e.g., if the user shakes their head during measurement), it is suggested that users remain stationary during measurement to minimize possible noise that could fall within the heart rate frequency. In addition, we assumed that the device did not vibrate and that the background within the device's working range contained no motion within the heartbeat frequency range. It is understood that voluntary or involuntary movement during signal acquisition typically reduces the fidelity of heart rate tracking [13,14].

- **Lack of Evaluation of Standing Postures**: While we conducted an extensive evaluation of our proposed system in various real-life scenarios, settings involving standing users were not included in our evaluation, as standing postures were not evaluated in any prior works in the literature. Therefore, we focused our evaluation on settings that were comparable with existing research. In fact, the standing setting is a challenge due to the difficulty of users maintaining stationary positions while in natural situations. For example, it is uncommon for an individual to remain completely still while standing for extended periods; a natural standing posture often involves walking or jogging. On the contrary, sitting or lying down naturally allows for more stationary positions. As a result, natural standing postures pose a great challenge in extracting subtle heartbeat signals, as we expect there to be significant motion noise. Consequently, we chose to leave the evaluation of different standing postures to future work.
- **Performance of Heart Rate Detection:** One assumption in our approach is that the normal heart rate falls within the range of 0.8 and 2.5 Hz, which is known as the normal heart rate range [37]. As such, any heart rate below 0.8 Hz may not be correctly detected, as the second harmonic of the respiration signal falls into this range and has significantly higher amplitude than the heartbeat.

7. Conclusions

In this paper, we present a remote approach to monitor the heart rates of multiple individuals using a commercial smart speaker with no separation requirement. Our proposed method removes interference and amplifies heart rates using a seven-microphone array on a smart speaker. This approach is able to separate heartbeat signals even when multiple users are sitting next to one another or lying down. Through our user study in various practical sitting and lying scenarios, the proposed approach is demonstrated to be highly accurate in these situations, with a median error of only 0.66 bpm. We believe that this approach can provide insightful inputs to other works, such as sleep stage classification, stress detection, and emotion classification.

Author Contributions: Conceptualization, T.T., D.M. and R.B.; methodology, T.T.; software, T.T.; validation, T.T.; formal analysis, T.T.; writing—original draft, T.T.; writing—review and editing, D.M. and R.B.; visualization, T.T.; funding acquisition, R.B.; data curation, T.T.; resources, supervision, and project administration, R.B. All authors have read and agreed to the published version of the manuscript.

Funding: This research was supported by the Singapore Ministry of Education (MOE) Academic Research Fund (AcRF) Tier 1 grant (Project ID: 22-SIS-SMU-051). Any opinions, findings and conclusions or recommendations expressed in this material are those of the author(s) and do not reflect the views of the Ministry of Education, Singapore.

Institutional Review Board Statement: IRB Approval has been approved under Category 2B: Expedited Review. The IRB approval number is IRB-21-168-A123(1021). Approval period from 12 October 2021 to 11 October 2022.

Informed Consent Statement: Informed consent was obtained from all subjects involved in the study.

Data Availability Statement: The data are not publicly available due to privacy issues.

Conflicts of Interest: The authors declare no conflicts of interest.

References

1. Acharya, U.R.; Joseph, K.P.; Kannathal, N.; Lim, C.M.; Suri, J.S. Heart rate variability: A review. *Med Biol. Eng. Comput.* **2006**, *44*, 1031–1051. [CrossRef] [PubMed]
2. Taelman, J.; Vandeput, S.; Spaepen, A.; Van Huffel, S. Influence of mental stress on heart rate and heart rate variability. In Proceedings of the 4th European Conference of the International Federation for Medical and Biological Engineering, Antwerp, Belgium, 23–27 November 2008; pp. 1366–1369.
3. Greenland, P.; Daviglus, M.L.; Dyer, A.R.; Liu, K.; Huang, C.F.; Goldberger, J.J.; Stamler, J. Resting heart rate is a risk factor for cardiovascular and noncardiovascular mortality: The Chicago Heart Association Detection Project in Industry. *Am. J. Epidemiol.* **1999**, *149*, 853–862. [CrossRef] [PubMed]
4. Snyder, F.; Hobson, J.A.; Morrison, D.F.; Goldfrank, F. Changes in respiration, heart rate, and systolic blood pressure in human sleep. *J. Appl. Physiol.* **1964**, *19*, 417–422. [CrossRef]
5. Yang, Y.; Yuan, Y.; Zhang, G.; Wang, H.; Chen, Y.C.; Liu, Y.; Tarolli, C.G.; Crepeau, D.; Bukartyk, J.; Junna, M.R.; et al. Artificial intelligence-enabled detection and assessment of Parkinson's disease using nocturnal breathing signals. *Nat. Med.* **2022**, *28*, 2207–2215. [CrossRef] [PubMed]
6. Adib, F.; Mao, H.; Kabelac, Z.; Katabi, D.; Miller, R.C. Smart homes that monitor breathing and heart rate. In Proceedings of the 33rd Annual ACM Conference on Human Factors in Computing Systems, Seoul, Republic of Korea, 18–23 April 2015; pp. 837–846.
7. Yue, S.; He, H.; Wang, H.; Rahul, H.; Katabi, D. Extracting multi-person respiration from entangled rf signals. *Proc. ACM Interact. Mob. Wearable Ubiquitous Technol.* **2018**, *2*, 86. [CrossRef]
8. Liu, J.; Wang, Y.; Chen, Y.; Yang, J.; Chen, X.; Cheng, J. Tracking vital signs during sleep leveraging off-the-shelf wifi. In Proceedings of the 16th ACM International Symposium on Mobile Ad Hoc Networking and Computing, Hangzhou, China, 22–25 June 2015; pp. 267–276.
9. Wang, H.; Zhang, D.; Ma, J.; Wang, Y.; Wang, Y.; Wu, D.; Gu, T.; Xie, B. Human respiration detection with commodity wifi devices: Do user location and body orientation matter? In Proceedings of the 2016 ACM International Joint Conference on Pervasive and Ubiquitous Computing, New York, NY, USA, 12–16 September 2016; pp. 25–36.
10. Meng, Z.; Fu, S.; Yan, J.; Liang, H.; Zhou, A.; Zhu, S.; Ma, H.; Liu, J.; Yang, N. Gait recognition for co-existing multiple people using millimeter wave sensing. In Proceedings of the AAAI Conference on Artificial Intelligence, New York, NY, USA, 7–12 February 2020; Volume 34, pp. 849–856.
11. Yang, Z.; Pathak, P.H.; Zeng, Y.; Liran, X.; Mohapatra, P. Monitoring vital signs using millimeter wave. In Proceedings of the 17th ACM International Symposium on Mobile ad hoc Networking and Computing, Paderborn, Germany, 5–8 July 2016; pp. 211–220.
12. Wang, L.; Gu, T.; Li, W.; Dai, H.; Zhang, Y.; Yu, D.; Xu, C.; Zhang, D. DF-Sense: Multi-user Acoustic Sensing for Heartbeat Monitoring with Dualforming. In Proceedings of the 21st Annual International Conference on Mobile Systems, Applications and Services, Helsinki, Finland, 18–22 June 2023; pp. 1–13.
13. Wang, A.; Nguyen, D.; Sridhar, A.R.; Gollakota, S. Using smart speakers to contactlessly monitor heart rhythms. *Commun. Biol.* **2021**, *4*, 319. [CrossRef] [PubMed]
14. Zhang, F.; Wang, Z.; Jin, B.; Xiong, J.; Zhang, D. Your Smart Speaker Can "Hear" Your Heartbeat! *Proc. ACM Interact. Mob. Wearable Ubiquitous Technol.* **2020**, *4*, 161. [CrossRef]
15. Qian, K.; Wu, C.; Xiao, F.; Zheng, Y.; Zhang, Y.; Yang, Z.; Liu, Y. Acousticcardiogram: Monitoring heartbeats using acoustic signals on smart devices. In Proceedings of the IEEE INFOCOM 2018—IEEE Conference on Computer Communications, Honolulu, HI, USA, 15–19 April 2018; pp. 1574–1582.
16. Wang, L.; Li, W.; Sun, K.; Zhang, F.; Gu, T.; Xu, C.; Zhang, D. LoEar: Push the Range Limit of Acoustic Sensing for Vital Sign Monitoring. *Proc. ACM Interact. Mob. Wearable Ubiquitous Technol.* **2022**, *6*, 145. [CrossRef]
17. UMA-8-SP USB mic Array. Available online: https://www.minidsp.com/products/usb-audio-interface/uma-8-sp-detail (accessed on 4 January 2024).
18. Amazon Echo Dot Smartspeaker. Available online: https://www.amazon.com/All-New-Amazon-Echo-Dot-Add-Alexa-To-Any-Room/dp/B01DFKC2SO (accessed on 4 January 2024).
19. Wang, Z.; Zhang, F.; Li, S.; Jin, B. Exploiting Passive Beamforming of Smart Speakers to Monitor Human Heartbeat in Real Time. In Proceedings of the 2021 IEEE Global Communications Conference (GLOBECOM), Madrid, Spain, 7–11 December 2021; pp. 1–6.
20. Israel, S.A.; Irvine, J.M.; Cheng, A.; Wiederhold, M.D.; Wiederhold, B.K. ECG to identify individuals. *Pattern Recognit.* **2005**, *38*, 133–142. [CrossRef]
21. Li, C.; Zheng, C.; Tai, C. Detection of ECG characteristic points using wavelet transforms. *IEEE Trans. Biomed. Eng.* **1995**, *42*, 21–28. [PubMed]
22. Nardelli, M.; Vanello, N.; Galperti, G.; Greco, A.; Scilingo, E.P. Assessing the quality of heart rate variability estimated from wrist and finger ppg: A novel approach based on cross-mapping method. *Sensors* **2020**, *20*, 3156. [CrossRef]
23. Temko, A. Accurate heart rate monitoring during physical exercises using PPG. *IEEE Trans. Biomed. Eng.* **2017**, *64*, 2016–2024. [CrossRef] [PubMed]
24. Chen, L.; Xiong, J.; Chen, X.; Lee, S.I.; Zhang, D.; Yan, T.; Fang, D. LungTrack: Towards contactless and zero dead-zone respiration monitoring with commodity RFIDs. *Proc. ACM Interact. Mob. Wearable Ubiquitous Technol.* **2019**, *3*, 79. [CrossRef]

25. Bakhtiari, S.; Elmer, T.W.; Cox, N.M.; Gopalsami, N.; Raptis, A.C.; Liao, S.; Mikhelson, I.; Sahakian, A.V. Compact millimeter-wave sensor for remote monitoring of vital signs. *IEEE Trans. Instrum. Meas.* **2011**, *61*, 830–841. [CrossRef]
26. Chuang, H.R.; Kuo, H.C.; Lin, F.L.; Huang, T.H.; Kuo, C.S.; Ou, Y.W. 60-GHz millimeter-wave life detection system (MLDS) for noncontact human vital-signal monitoring. *IEEE Sens. J.* **2011**, *12*, 602–609. [CrossRef]
27. Kao, T.Y.J.; Lin, J. Vital sign detection using 60-GHz Doppler radar system. In Proceedings of the 2013 IEEE International Wireless Symposium (IWS), Beijing, China, 14–18 April 2013; pp. 1–4.
28. Wang, C.; Xie, L.; Wang, W.; Chen, Y.; Bu, Y.; Lu, S. Rf-ecg: Heart rate variability assessment based on cots rfid tag array. *Proc. ACM Interact. Mob. Wearable Ubiquitous Technol.* **2018**, *2*, 85. [CrossRef]
29. Nandakumar, R.; Gollakota, S.; Watson, N. Contactless sleep apnea detection on smartphones. In Proceedings of the 13th Annual International Conference on Mobile Systems, Applications, and Services, Florence, Italy, 18–22 May 2015; pp. 45–57.
30. Wang, T.; Zhang, D.; Zheng, Y.; Gu, T.; Zhou, X.; Dorizzi, B. C-FMCW based contactless respiration detection using acoustic signal. *Proc. ACM Interact. Mob. Wearable Ubiquitous Technol.* **2018**, *1*, 170. [CrossRef]
31. Peng, C.; Shen, G.; Zhang, Y.; Li, Y.; Tan, K. Beepbeep: A high accuracy acoustic ranging system using cots mobile devices. In Proceedings of the 5th International Conference on Embedded Networked Sensor Systems, Sydney, Australia, 6–9 November 2007; pp. 1–14.
32. Wang, A.; Gollakota, S. Millisonic: Pushing the limits of acoustic motion tracking. In Proceedings of the 2019 CHI Conference on Human Factors in Computing Systems, Glasgow, UK, 4–9 May 2019; pp. 1–11.
33. Bachorowski, J.A.; Smoski, M.J.; Owren, M.J. The acoustic features of human laughter. *J. Acoust. Soc. Am.* **2001**, *110*, 1581–1597. [CrossRef] [PubMed]
34. Chen, J.S.; Huertas, A.; Medioni, G. Fast convolution with Laplacian-of-Gaussian masks. *IEEE Trans. Pattern Anal. Mach. Intell.* **1987**, *PAMI-9*, 584–590. [CrossRef] [PubMed]
35. Chan, A.; Litva, J. MUSIC and maximum likelihood techniques on two-dimensional DOA estimation with uniform circular array. *IEEE Proc. Radar Sonar Navig.* **1995**, *142*, 105–114. [CrossRef]
36. Polar H10 Heart Rate Sensor. Available online: https://www.polar.com/sg-en/sensors/h10-heart-rate-sensor/ (accessed on 4 January 2024).
37. Pulse Rate. Available online: https://www.bhf.org.uk/informationsupport/heart-matters-magazine/medical/ask-the-experts/pulse-rate (accessed on 4 January 2024).

sensors

Article

A Short Video Classification Framework Based on Cross-Modal Fusion

Nuo Pang [1], Songlin Guo [2], Ming Yan [2,*] and Chien Aun Chan [3,4]

1 School of Design, Dalian University of Science and Technology, Dalian 116052, China; pangnuo@dlust.edu.cn
2 School of Information and Communications Engineering, Communication University of China, Beijing 100024, China
3 Insta-Wireless, Notting Hill, VIC 3168, Australia; chienac@unimelb.edu.au
4 Department of Electrical and Electronic Engineering, The University of Melbourne, Parkville, VIC 3010, Australia
* Correspondence: yanm@cuc.edu.cn

Abstract: The explosive growth of online short videos has brought great challenges to the efficient management of video content classification, retrieval, and recommendation. Video features for video management can be extracted from video image frames by various algorithms, and they have been proven to be effective in the video classification of sensor systems. However, frame-by-frame processing of video image frames not only requires huge computing power, but also classification algorithms based on a single modality of video features cannot meet the accuracy requirements in specific scenarios. In response to these concerns, we introduce a short video categorization architecture centered around cross-modal fusion in visual sensor systems which jointly utilizes video features and text features to classify short videos, avoiding processing a large number of image frames during classification. Firstly, the image space is extended to three-dimensional space–time by a self-attention mechanism, and a series of patches are extracted from a single image frame. Each patch is linearly mapped into the embedding layer of the Timesformer network and augmented with positional information to extract video features. Second, the text features of subtitles are extracted through the bidirectional encoder representation from the Transformers (BERT) pre-training model. Finally, cross-modal fusion is performed based on the extracted video and text features, resulting in improved accuracy for short video classification tasks. The outcomes of our experiments showcase a substantial superiority of our introduced classification framework compared to alternative baseline video classification methodologies. This framework can be applied in sensor systems for potential video classification.

Keywords: video classification; cross-modal fusion; video features; text features; Timesformer

Citation: Pang, N.; Guo, S.; Yan, M.; Chan, C.A. A Short Video Classification Framework Based on Cross-Modal Fusion. *Sensors* **2023**, *23*, 8425. https://doi.org/10.3390/s23208425

Academic Editors: Anastasios Doulamis and Zhe-Ming Lu

Received: 22 August 2023
Revised: 30 September 2023
Accepted: 11 October 2023
Published: 12 October 2023

1. Introduction

In the past few years, the emergence of short video applications has exploded, such as Tiktok, YouTube Shorts, Likee, Bilibili. Most of the short videos in these video applications are tagged when they are released, enabling users to browse videos by category and search within a certain category [1]. These short videos are typically characterized by their brief duration, diverse content, and a wide range of topics. However, the exponential increase in the number of short videos poses significant challenges in terms of effectively classifying and managing this vast video content.

At present, the application of deep learning methods in the classification of violent videos [2] and social media videos [3,4] has achieved good results. However, due to the unique characteristics of short videos, such as the short duration, large amount, and many spliced contents, it is still a difficult task to classify short videos. Research on short video classification predominantly employs single-modal approaches which utilize either visual or textual features for classification. Visual feature extraction involves the extraction of image features from video frames, encompassing attributes such as color histograms, texture features, and shape characteristics. Studies have demonstrated the high efficacy of visual features in short video classification, as they facilitate the capture of visual information within the videos, including objects, scenes, and actions. Traditional techniques for visual feature extraction comprise the histogram of oriented gradient (HOG) [5], histogram of flow (HOF) [6], and motion boundary histograms (MBH) [7]. To leverage the complementarity of these three feature types and enhance the representational capacity of video features, researchers have introduced the dense trajectories (DT) algorithm [8] and its improved variant, the improved dense trajectories (IDT) [9]. These algorithms are both based on decision tree classification approaches, and utilize HOG descriptors as image features derived by statistically analyzing the gradient information of images.

In addition to visual features, textual features are also widely employed in the field of short video classification. These textual features typically originate from metadata information such as video titles, descriptions, and tags. The advantage of textual features lies in their ability to provide semantic information about the video content, thereby enhancing classification accuracy. Researchers have developed various methods for extracting textual features, including those based on traditional natural language processing techniques such as the bag of visual words (BOVW) model [10], as well as deep learning-based methods such as recurrent neural networks (RNNs) and attention mechanisms. The core idea of BOVW is to represent images as a collection of visual words and use the frequency of word occurrences as the image's feature vector [11]. Firstly, local features are extracted from the image, such as scale-invariant feature transform (SIFT) [12], local binary patterns (LBP) [13], color histograms, etc. Subsequently, all local features are clustered into several clusters, with each cluster corresponding to a visual word. The frequency of each word's occurrence is computed, and finally, this feature representation is employed for tasks such as training classifiers. The convolutional neural networks (CNNs) primarily decompose videos into a sequence of frames and then extract features from each frame through multiple layers of convolutional and pooling operations. These extracted features from all frames are aggregated and used for classification with the assistance of a classifier.

However, at present, users upload short videos with great randomness and divergence, and users' understanding of video categories generally varies and there is more and more false information, resulting in inconsistent categories marked by users [14]. This inconsistency not only affects the accuracy of the search and recommendation results of video content categories, but also in the face of these challenges, users are more inclined to make subjective judgments through visual content to meet their personal needs. In addition, there is a significant difference between video content features and hashtag text features. It is difficult to match accurate hashtags to meet users' content consumption needs due to insufficient video text information in the method of searching for tags with the same text in videos [15,16]. Moreover, some videos usually do not contain classification information, and video feature analysis is mainly based on understanding visual image information, but lacks text semantic mining, resulting in an underutilization of semantic information [17,18].

Thus, short video classification is essential to determine the category of a video so that videos without user-labeled categories can also be organized in the same way as videos with category labels. A distinct video classification framework is introduced herein which leverages both textual and visual features in a new way. We bring together visual features obtained from the training dataset with text features extracted from subtitles across modalities, and integrate them into joint features for downstream classification tasks. Specifically, the text feature uses the bidirectional encoder representation from the Transformers (BERT) pre-training model, adds context using the attention mechanism, and solves the parallel calculation between sentences [19–21]. Video features are extended from image space to spatio-temporal three-dimensional volume through a self-attention mechanism, which treats video as a series of patches extracted from a single frame. Like the vision Transformer (VIT), each patch undergoes linear mapping into an embedding, to which positional data are subsequently incorporated [22]. The proposed framework uses textual and visual features to classify short videos, which improves the accuracy of short video classification. The related techniques can be applied in sensor systems for potential video classification, so the subject of this paper belongs to data fusion and analysis in sensor systems. The main contributions are as summarized below:

1. We propose an improved hierarchical clustering approach for keyframe extraction. Unlike traditional keyframe extraction algorithms, hierarchical clustering does not require a predefined number of keyframes to be extracted. Instead, it adaptively determines which frames are keyframes through clustering to offer greater flexibility. This method is capable of preserving essential information from the video while effectively reducing redundant frames, resulting in more representative extracted keyframes.
2. We investigate the extraction methods of visual features and textual features within videos. The method of combining visual information and text information for video classification is used in this paper. The visual information is first processed by the key frame extraction method to divide the video into multiple images representing the main content. The pre-trained Timesformer network is used for feature extraction to obtain the visual features of the video. At the same time, the text information is also extracted by the fine-tuning-based method in BERT. Finally, these two kinds of features are fused by the feature aggregation algorithm for video classification.
3. We propose a cross-modal fusion short video classification (CFVC) framework. This framework utilizes text features and visual features in a new way, combining the integration of visual attributes extracted from the training dataset and text features extracted from subtitles to achieve cross-modal fusion and integrate them into joint features for downstream classification tasks.

The subsequent sections of this paper are structured as follows. Section 2 summarizes the existing work related to this paper. Section 3 introduces the proposed cross-modal fusion short video classification framework. Section 4 evaluates the proposed framework through experiments. Section 5 concludes our work.

2. Related Work

Despite considerable progress having been achieved for image representation architectures over recent years, the realm of video architecture remains devoid of a distinctly defined forefront structure. The current main video classification architectures are shown in Figure 1, where k represents the count of frames within a video, and N represents a subset of adjacent frames of the video. The main differences between these frameworks are: (1) The first differentiation lies in determining whether the convolution and layer operators utilize 2D (image-based) or 3D (video-based) kernels [23–25]. (2) Another key variation involves the nature of the input provided to the network. This can be limited to just an RGB video or expanded to encompass both an RGB video and pre-computed optical flow [26–28]. (3) In the context of 2D convolutions, a significant consideration is how information propagates across frames. This can be achieved through the incorporation of temporary recurrent layers such as SlowFast or the application of feature aggregation over time [29–31].

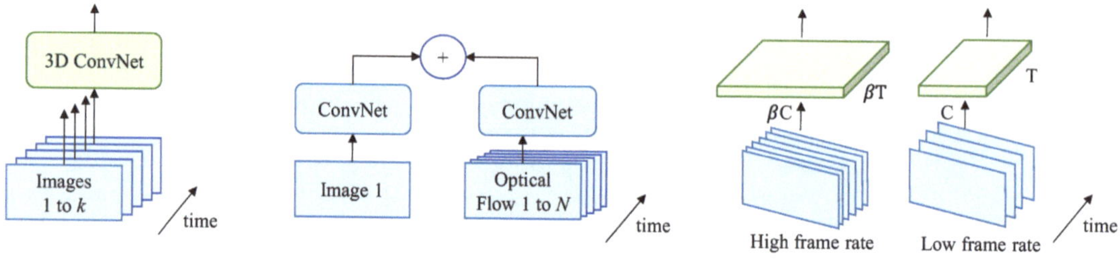

Figure 1. Different types of video classification architectures (**a**) 3D-ConvNet, (**b**) Two-Stream and (**c**) SlowFast.

2.1. I3D Networks

Traditional 2D convolutional neural networks have been a huge success in tasks such as image classification, but there are some challenges in video classification tasks. To make the most of temporal information and motion features in videos, researchers proposed a variety of three-dimensional convolutional network (3D ConvNet) models as shown in Figure 1a [23]. The inflated 3D ConvNet (I3D) model is extended on the basis of a two-dimensional convolutional network. Specifically, it constructs a three-dimensional convolutional network structure by copying and filling the weights of the pre-trained two-dimensional convolutional network in the time dimension [24]. This approach enables the I3D model to simultaneously process features in both spatial and temporal dimensions, thereby better capturing dynamic information in videos. To efficiently train the I3D model, two strategies are adopted: pre-training of the second-rate network and multi-scale cropping [25]. First, by pre-training on a large-scale video dataset, the I3D model can learn rich visual features. Then, it is fine-tuned on the dataset of the target task to improve its performance on the specific task. In addition, to take advantage of the spatio-temporal information in the video, a multi-scale cropping strategy is also introduced, which is trained by extracting multiple cropped segments of different scales from the video. Applications of I3D models have achieved remarkable results in several video understanding tasks.

2.2. Two-Stream Networks

Simulations of high-level changes can be achieved by the long short-term memory (LSTM) networks based on features extracted from the final convolutional layer, but the capturing of essential fine-grained low-level motion, pivotal in numerous scenarios, might be hindered [26]. Training also incurs significant costs, given the necessity for the network to be unrolled across multiple frames to facilitate time-based backpropagation. An enhanced methodology entails the modeling of brief temporal video snapshots, achieved by combining forecasts originating from an individual RGB frame and a compilation of 10 externally generated optical flow frames. This is subsequently followed by the traversal of two iterations of an ImageNet-pre-trained ConvNet [27]. An adapted input convolutional layer is integrated within the two-stream architecture, boasting double the number of input channels in comparison to the frames within the stream as shown in Figure 1b. During the testing phase, numerous video snapshots are sampled and subsequently aggregated to yield action predictions. Experiments validate the achievement of exceptional performance on established benchmarks, concurrently showcasing remarkable efficiency during both training and testing intervals.

Two-stream models have achieved remarkable performance in various computer vision tasks. It has been widely used in action recognition, outperforming previous methods on benchmark datasets such as UCF101 and HMDB51 [28]. Moreover, the two-stream model has also found applications in other domains such as gesture recognition, video captioning, and video segmentation, demonstrating its versatility and effectiveness. Future research directions may focus on developing more efficient architectures, exploring attention mechanisms, and using unsupervised or weakly supervised learning paradigms to further build up the performance and generalization capabilities of two-stream models.

2.3. SlowFast Networks

During the preceding years, an array of video action recognition networks has been put forth by researchers, including 2D CNN, 3D CNN, and I3D network. However, these methods have certain limitations when dealing with challenging scenarios such as long-term dependencies and fast actions. The SlowFast network as shown in Figure 1c addresses the problem of spatio-temporal scale differences in videos by introducing two branches, slow and fast [29,30]. The slow branch is used to process low-frequency information to capture long-term timing dependencies by reducing the frame rate of the input video. The fast branch is used to process high-frequency information to capture instantaneous actions by preserving the high frame rate of the input video. This design can effectively balance information on both temporal and spatial scales. It primarily comprises two main components: the slow path and the fast path [31]. The slow path is processed at a lower frame rate, typically 1/8 or 1/16 of the input video. The fast path is processed at native framerate. The two paths extract feature representations, βC and C, at different scales, respectively, and integrate information through the fusion module. Finally, after global average pooling and classification layers, βT and T, the network outputs the behavior category of the video. The SlowFast network achieves significant performance gains on video action recognition tasks [32]. Compared with the traditional 2D CNN network and 3D CNN network, the SlowFast network can better handle long-term dependencies and fast actions, and improve the accuracy and robustness of behavior recognition. In addition, the SlowFast network structure is simple and efficient, with low computing and storage overhead, and is suitable for training and reasoning on large-scale video data [33,34].

3. System Model and Problem Formulation

In the context of viewing brief video content, the assessment of the video's substance based solely on subtitles is not universally definitive. Particularly for elements devoid of auditory components, visual data assume an integral role. Consequently, a proposition emerges wherein visual attributes are incorporated within each subtitle segment to prognosticate video content. The crux of this approach pertains to the harmonious alignment of features originating from video frames and subtitle text. Subsequently, a process of multi-classification ensues, conducted upon the act of mapping subtitle spans into an equivalent vector space as their corresponding video frames. For the text mode, we input the subtitle text into the BERT pre-training model, and obtain the text features by fine-tuning the parameters. The best results across various tasks within the field of natural language processing (NLP) have been achieved by the BERT pre-training model. For the visual pattern, we extract raw frames from the video by down-sampling. Then, we use the Timesformer feature extraction method to obtain visual features, which reaches state-of-the-art results on several large datasets. We perform contextual query concatenation to jointly adjust textual and visual features for the final multi-class prediction.

In this chapter, an elaborate exposition is provided regarding the method introduced for the task of video classification. Since the approach of pre-training language has the capacity to augment the performance of semantic representation for textual subtitle queries, we designed a two-channel cross-modal fusion video classification method, and the process framework is shown in Figure 2 specifically.

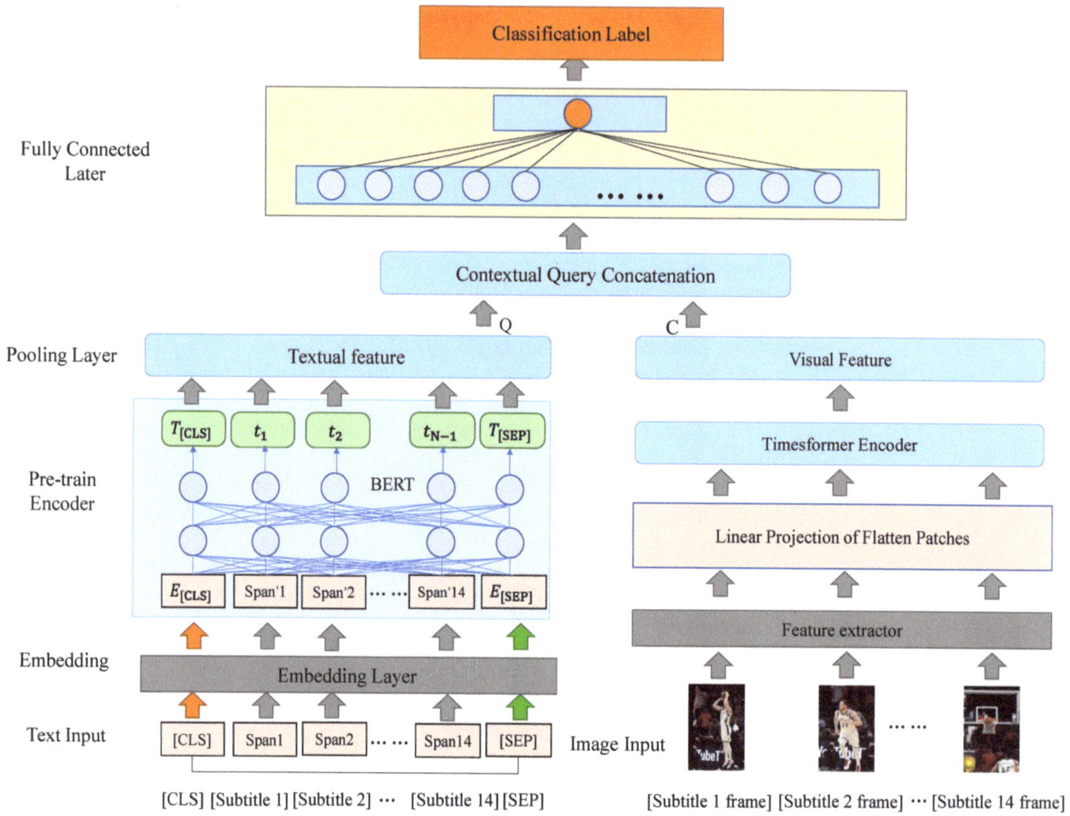

Figure 2. Two-channel classification framework based on cross-modal fusion.

During the observation of a video, the evaluation of its content based on subtitle text does not invariably constitute the sole criterion. Particularly for non-verbal components, the visual information assumes paramount importance. Therefore, for each subtitle span, we can add visual features to predict video content. As illustrated in Figure 2, an intricate cross-modal video classification model is devised. Specifically, we focus on the feature joint alignment of video frames and subtitle text. Following this alignment, the classification of videos is executed subsequent to the transformation of subtitle spans and their corresponding video frames into a unified vector space. For the textual modality, the subtitle text is introduced to a pre-trained language model to derive textual attributes. On the other hand, for the visual modality, the raw frames undergo down-sampling, with keyframes being captured at regular intervals within each video. The subsequent procedure involves the utilization of an attention mechanism to obtain visual attributes. The integration of contextual query concatenation facilitates the synergistic alignment of textual attributes (Q) and visual attributes (C), culminating in the ultimate prediction for video classification.

BERT and Timesformer have demonstrated outstanding performance in extracting text and visual features. They are capable of generating high-quality feature representations for text and images, respectively. Therefore, utilizing their output vectors can provide a powerful feature basis for video classification. At the same time, end-to-end error updates can require significant computing resources and time, while using only the output vectors of BERT and Vision Timesformer can significantly reduce computing costs. This is particularly advantageous for large-scale video classification tasks when resources are limited or efficient processing is required. In conclusion, considering only the output vectors from BERT and Vision Timesformer to find a joint feature space is feasible. This approach can provide high-quality feature representations and reduce computational costs.

3.1. Visual Feature Extraction

As deep learning continues to evolve, the architecture of the neural network exhibits as more diversified, the network structure is more complex, its feature expression ability becomes stronger and stronger, and it can learn image and video features very well.

At present, there are two main ways to extract visual features. One is to directly use the 3D convolutional network to extract the features of the entire video. The other is that overall video features are formed by feature aggregation. Due to 3D convolution, compared with 2D convolution, it adds one dimension (time dimension) to the input and then directly expands 2D convolution to 3D convolution. Although it can capture the time information of the video, at the same time it increases the parameters of the network, resulting in a larger amount of calculation, which is not conducive to real-time feature extraction. As the length of the video increases, its calculation speed will become slower and slower. Based on the above considerations, we use the method of selecting and extracting key frame features to extract video features.

Whether the selection of key frames is reasonable or not directly affects the accuracy of classification tasks. The K-means clustering algorithm is the most commonly used key frame extraction algorithm based on clustering, which has the advantages of simplicity and fast convergence speed. However, because the K-means clustering algorithm is very sensitive to the initial parameters, it is easy to fall into a local optimal solution. This paper proposes an improved hierarchical clustering algorithm based on it. This method mainly uses the characteristics of image information entropy to measure the similarity of two frames. If the similarity reaches a certain value, they will be merged into the same cluster and the extracted cluster center is used as the initial clustering result. Subsequently, the K-means algorithm is used to optimize the initial clustering result to obtain key frames.

Figure 3 is the overall frame diagram of the visual feature extraction in this experiment. The video data have the characteristics of different time lengths. This paper mainly studies the classification of short videos. At present, the videos in various application platforms are basically edited and contain information. More parts can increase the number of views of the video. Most of the video data selected in this article are about 5 s long clips, and frame images are extracted by *ffmpeg* every second. In the process of data preprocessing, for videos whose original video data length is less than 5 s, that is, the video frame is less than 300 frames, we adopt the method of filling 0 to keep it consistent. For an original video data length greater than 5 s, that is, the video frame is greater than 300 frames, we use truncation processing.

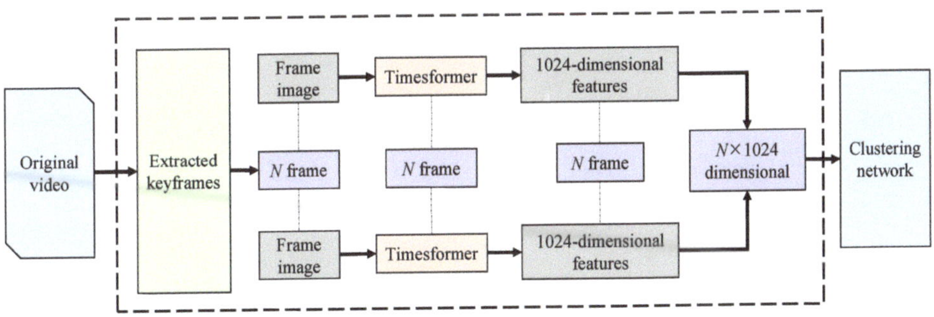

Figure 3. Video feature extraction process.

After the video is processed by the frame extraction operation, multiple pictures can be obtained, and then feature extraction needs to be performed on the images. At present, image features with generalization ability are widely used. At present, the commonly used convolutional neural networks for extracting image features mainly include the I3D network, SlowFast network, etc. 2D and 3D convolution are still the core algorithms for spatio-temporal features across different tasks. However, the convolutional structure has translation invariance and cannot link the image context information well, so we choose the Timesformer network model based on the attention mechanism. Since videos and sentences are both continuous, coupled with the intrinsic nature of word comprehension, which often necessitates contextual referencing within the sentence, an inclination arises to combine a certain frame of action in a short video with the rest. To completely disambiguate, the choice of a self-attention model is also completely effective for video modeling, and its structure is shown in Figure 4.

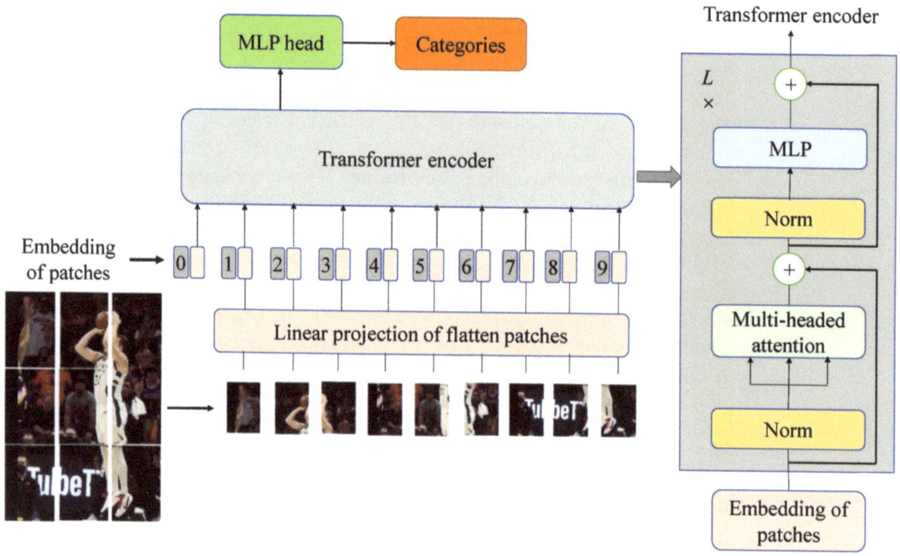

Figure 4. The Timesformer coding flow chart.

The Timesformer model stands as a video-based structure crafted exclusively upon the foundation of self-attention mechanisms. The VIT image model is adapted for video classification in the way of expanding the self-attention mechanism from its original image-based realm to spatio-temporal 3D volumes. When the VIT model has enough data for pre-training, the performance of VIT will exceed that of CNN, breaking through the limitation of the Transformer's lack of inductive bias, and a better migration effect in downstream tasks. In the context of the Timesformer model, the perception of video occurs through the lens of a sequential compilation of patches, each drawn collectively from distinct frames. Similar to VIT, the transformation of each sequence of patches undergoes linear mapping within an embedding layer, which is further enriched with positional particulars, and then each sequence is projected into a fixed-length vector and sent to the Transformer for subsequent encoder operations.

(1) Step 1: Clip input. The input to the Timesformer model consists of a clip comprising F RGB frames of size $H \times W$ sampled from the initial video input.

(2) Step 2: Break down into patches. Following the VIT method, each frame is divided into N non-overlapping patches, each with dimensions $\mathrm{P} \times \mathrm{P}$, in a manner where these N patches span the entire frame; that is, $N = HW/\mathrm{P}^2$. These patches are flattened into vector $x_{(p,t)} \in \mathbb{R}^{3\mathrm{P}^2}$, where $p = 1, \cdots, N$ represents the spatial position; $t = 1, \cdots, F$ represents an index on a frame.

(3) Step 3: Linear embedding. Each patch $x_{(p,t)}$ is linearly mapped to a vector $z_{(p,t)}^{(0)} \in \mathbb{R}^D$ by a learnable matrix $E \in \mathbb{R}^{D\times 3P^2}$:

$$z_{(p,t)}^{(0)} = E \cdot x_{(p,t)} + e_{(p,t)}^{pos} \tag{1}$$

where $e_{(p,t)}^{pos} \in \mathbb{R}^D$ denotes a location embedding that is subject to learning, and this embedding serves to encode the spatio-temporal coordinates of each individual patch. When $p = 1, \cdots, N$ and $t = 1, \cdots, F$, the obtained embedding vector $z_{(p,t)}^{(0)}$ is sent to the Transformer as the input, and its function is similar to the embedded word sequence of the input text converter in natural language processing. This paper adds a special learnable vector $z_{(0,0)}^{(0)} \in \mathbb{R}^D$ at the first position of the sequence to represent the embedded classification label.

(4) Step 4: Query (q)—key (k)—value (v) computation. The Transformer architecture employed within this paper encompasses L encoding blocks as shown in Figure 4. At each block ℓ, a vector value of (q, k, v) is computed from the representation $z_{(p,t)}^{(\ell-1)}$ encoded in the previous block.

$$q_{(p,t)}^{(\ell,\alpha)} = W_Q^{(\ell,\alpha)} \mathrm{LN}(z_{(p,t)}^{(\ell-1)}) \in \mathbb{R}^{D_h} \tag{2}$$

$$k_{(p,t)}^{(\ell,\alpha)} = W_K^{(\ell,\alpha)} \mathrm{LN}(z_{(p,t)}^{(\ell-1)}) \in \mathbb{R}^{D_h} \tag{3}$$

$$v_{(p,t)}^{(\ell,\alpha)} = W_V^{(\ell,\alpha)} \mathrm{LN}(z_{(p,t)}^{(\ell-1)}) \in \mathbb{R}^{D_h} \tag{4}$$

where W represents the weight vector, $\mathrm{LN}(\cdot)$ represents LayerNorm, $\alpha = 1, \cdots, A$ signifies an index corresponding to various attention heads, and A signifies the aggregate number of attention heads. Each attention head possesses a latent dimension set at $D_h = D/A$.

(5) Step 5: Self-attention calculation. The computation of self-attention weights is achieved through the dot product operation. The self-attention weight $\alpha_{(p,t)}^{(\ell,\alpha)} \in \mathbb{R}^{NF+1}$ of query block (p, t) is obtained by the following equation:

$$\alpha_{(p,t)}^{(\ell,\alpha)} = \mathrm{SM}\left(\frac{q_{(p,t)}^{(\ell,\alpha)^T}}{\sqrt{D_h}} \cdot \left[k_{(0,0)}^{(\ell,\alpha)} \left\{ k_{(p',t')}^{(\ell,\alpha)} \right\} \begin{matrix} p' = 1, \cdots, N \\ t' = 1, \cdots, F \end{matrix} \right] \right) \tag{5}$$

where $\mathrm{SM}(\cdot)$ signifies the activation function known as Softmax. When attention computation is confined to a dimension, such as exclusively in time or space, it culminates in substantial computational reduction. For instance, in spatial attention, the number of query key-value pair comparisons stands at only $N + 1$, wherein unique keys reference the same frame.

$$\alpha_{(p,t)}^{(\ell,\alpha)space} = \mathrm{SM}\left(\frac{q_{(p,t)}^{(\ell,\alpha)^T}}{\sqrt{D_h}} \cdot \left[k_{(0,0)}^{(\ell,\alpha)} \left\{ k_{(p',t)}^{(\ell,\alpha)} \right\}_{p'=1,\ldots,N} \right] \right) \tag{6}$$

(6) Step 6: Coding. The encoding $z_{(p,t)}^{(\ell)}$ in block ℓ is obtained by weighting the vector of values computed by the self-attention system of each attention head.

$$s_{(p,t)}^{(\ell,\alpha)} = \alpha_{(p,t),(0,0)}^{(\ell,\alpha)} v_{(0,0)}^{(\ell,\alpha)} + \sum_{p'=1}^{N} \sum_{t'=1}^{F} \alpha_{(p,t),(p',t')}^{(\ell,\alpha)} v_{(p',t')}^{(\ell,\alpha)} \tag{7}$$

Subsequently, these vectors from all attention heads are subjected to projection and directed through an MLP layer, utilizing the residual connections in each operation.

$$z'^{(\ell)}_{(p,t)} = Wo \begin{bmatrix} s^{(\ell,1)}_{(p,t)} \\ \vdots \\ s^{(\ell,A)}_{(p,t)} \end{bmatrix} + z^{(\ell-1)}_{(p,t)} \tag{8}$$

$$z^{(\ell)}_{(p,t)} = \mathrm{MLP}(\mathrm{LN}(z'^{(\ell)}_{(p,t)})) + z'^{(\ell)}_{(p,t)} \tag{9}$$

(7) Step 7: Categorical embedding. The final clip embeddings are obtained from the class-labeled final block.

$$y = \mathrm{LN}(z^{(L)}_{(0,0)}) \in \mathbb{R}^D \tag{10}$$

After being processed by multiple Transformer encoder layers, the output of the model's last position is considered as a representation of the entire image.

3.2. Text Feature Extraction

In neural machine translation, the Seq2Seq model is a widely used architecture. Typically, a Seq2Seq model consists of two recurrent neural networks (RNNS) for processing sequential data. However, such a model suffers from the obvious limitation of not being able to perform parallel computations, as it requires processing each element of the sequence in turn. The word vector model tool, Word2vec, can efficiently train on millions of dictionaries and huge datasets, and use the word vectors obtained by it to effectively determine the similarity between different words [10]. The Word2vec model is based on two algorithms, Skip-Gram and CBOW. The former predicts the surrounding context word by giving the target word, and the latter predicts the target word by given the context of the surrounding word. One disadvantage of these algorithms is that the expression of the same word in different contexts does not change after pre-trained word vectors. To solve the above problems, this paper considers using the BERT word vector model to replace the sequence model and the Word2vec word vector model, as shown in Figure 5, where x_0, x_1 and x_2 are word embeddings of different words, h_0, h_1 and h are the content streams after passing through the attention network, (q_1, k_1, v_1) and (q_2, k_2, v_2) are the query—key—value vector values, and w_q, w_k and w_v represent the weight vectors.

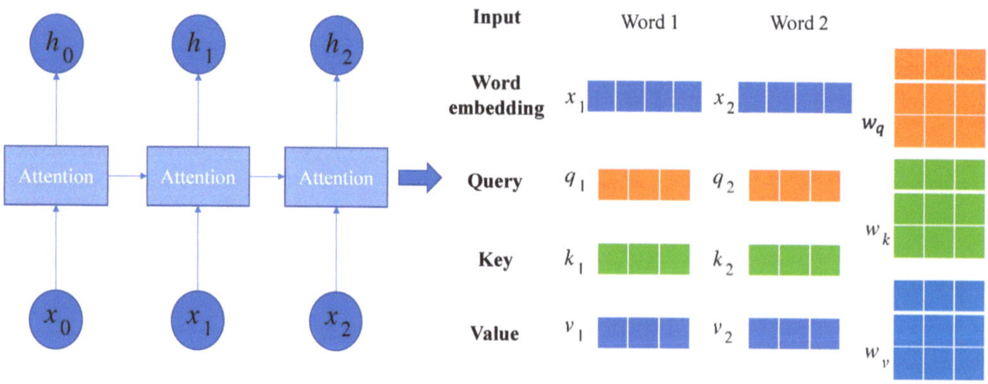

Figure 5. Word vector model.

Language model pre-training has demonstrated its effectiveness in enhancing a spectrum of natural language processing tasks encompassing a natural language inference. Presently, two strategies underpin the application of pre-trained language models to downstream tasks: feature-based and fine-tuning-based. In the feature-based approach, exemplified by ELMo [35], task-specific architectures are enhanced with supplementary pre-trained representations as additional attributes. Conversely, fine-tuning-based methods, typified by a generative pre-training Transformer (OpenAI GPT) [36], incorporate minimal task-specific parameters. These models then undergo training on downstream tasks through a straightforward fine-tuning of all pre-trained parameters.

While pursuing distinct methodologies, both strategies share the same pre-training objective function, utilizing a singular-term language model to acquire a universally applicable language representation. The text feature extraction framework is shown in Figure 5.

It is contended that the prevailing techniques impose constraints on the potential of pre-trained representations, particularly in the context of fine-tuning based methods. The primary constraint stems from the fact that conventional language models adhere to a unidirectional nature, thereby constraining the available options for pre-training architectures. These limitations are not optimal for sentence-level tasks, which ignore the incorporation of contexts from different directions of the sentence. The BERT can be used to extract text features. Its framework has two steps: pre-training and fine-tuning based methods. In this paper, we use the fine-tuning-based method in BERT as shown in Figure 6.

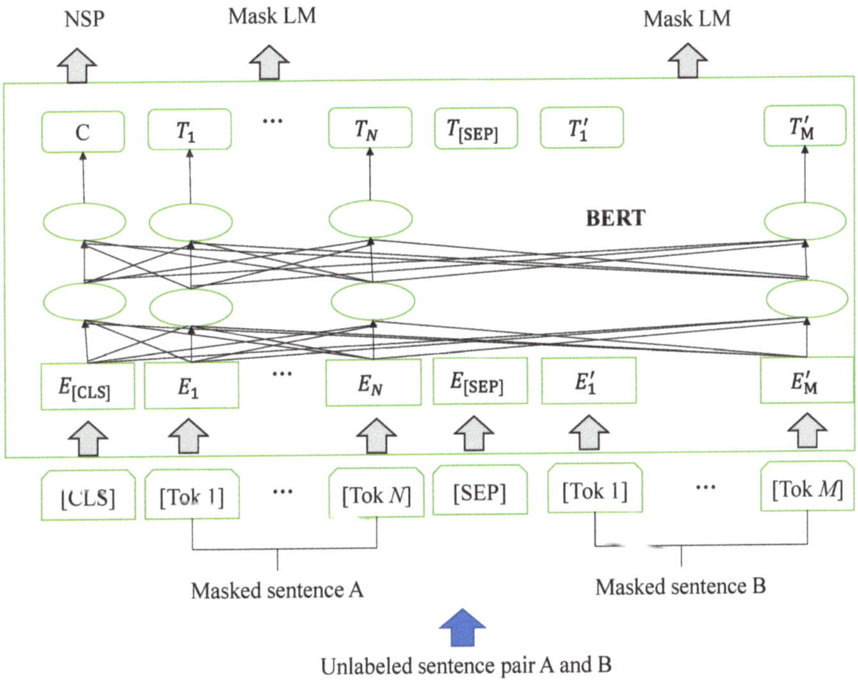

Figure 6. Text feature extraction framework.

The input text type of this article is subtitle information, which consists of a group of sentence pairs A and B. Firstly, word segmentation is performed on the sentence pair, and a piece of text is divided into N or M individual words or sub-words. The input length is fixed to 512. If the length of the input text is greater than 512, it truncates the input text, and if the length of the input text is insufficient, it takes a special symbol to fill. A special tag [CLS] is added at the beginning of the input text to indicate that the text belongs to a classification task, and [SEP] tags are used to indicate the segmentation between sentences. The BERT model optimizes its weight through multiple rounds of pre-training iterations. Finally, each input tag corresponds to a 1024-dimensional vector denoted by E or E'. These vectors constitute the hidden state representations T and T' of the last layer and can be used as feature representations, masked language modeling (LM) and next sentence prediction (NSP), for downstream tasks.

3.3. Cross-Modal Feature Fusion Framework

In this paper, cross-modal fusion techniques are introduced to fuse information from different modalities together to address the problem of multimodal information processing and analysis. In the fusion process, early fusion will inhibit the links within or between modalities, resulting in the loss of video semantics, and the interaction between different modalities cannot be achieved. Therefore, this paper adopts the late fusion method, which inputs each modality information into the clustering network and uses the dot product operation to obtain the final video feature vector. This late fusion way helps to preserve the richness of multimodal information and realizes the interaction between different modalities, which improves the model performance.

After the visual features and text features of video frame-level images are extracted by Timesformer and BERT, the frame-level features need to be aggregated to obtain video-level features before video classification. Previously, the long short-term memory network LSTM and gated recurrent unit (GRU) can obtain the timing information of the video. However, the next vector of the locally aggregated descriptors (NextVLAD) network and the AttentionCluster network, which are conducive to scene recognition, are more effective for aggregating visual features and text features.

The NextVLAD network reduces the overall parameters of the model by reducing the input dimension and splitting it into multiple groups. It first increases the dimension of y to obtain $\dot{y}$, and then divides $\dot{y}$ into groups to obtain $\widetilde{y}$, and then calculates the weights with the cluster centers, respectively. Finally, the global features are aggregated by grouping results. Assume that the video has M frames, and the feature description y of each frame is N-dimensional. For the K cluster centers included, NetVLAD first encodes the features of each frame into an $N \times K$ feature vector, as shown in Equation (11).

$$\begin{aligned} v_{ijk}{}^{g} &= \alpha_g(\dot{y}_i)\alpha_{gk}(\dot{y}_i)(\widetilde{y}_{ij}{}^{g} - c_{kj}) \\ g &\in \{1,\ldots,G\}, i \in \{1,\ldots,M\}, k \in \{1,\ldots,K\} \end{aligned} \tag{11}$$

where c_k is the N-dimensional eigenvector coordinates of the cluster center k, G is the number of groups, and the similarity measure calculation equation is as follows:

$$a_{gk}(\dot{y}) = \frac{e^{W_{gk}^T \dot{y}_i + b_{gk}}}{\sum_{S=1}^{K} e^{W_{gs}^T \dot{y}_i} + b_{gs}} \tag{12}$$

$$\alpha_g(\dot{y}_i) = \sigma(w_g^T \dot{y}_i + b_g) \tag{13}$$

where σ is the sigmoid function, $\alpha_g(\dot{y}_i)$ computes attention weights for all groups.

The encoding feature l of the entire video is expressed as follows:

$$l_{jk} = \sum_{i,j} v_{ijk}^{g} \tag{14}$$

NextVLAD divides video features into multiple groups for clustering operations, and introduces an attention mechanism to add weights to different groups. It uses AttentionCluster attention clustering while adding offset operations, thereby increasing the weight of frames strongly related to tags in video content. Finally, several local features are aggregated into a video global feature.

4. Experimental Results and Discussion

To realize the short video classification task, we extract its text information features and visual information features from the video. In our experiments, the textual attributes obtained from videos encompass elements such as video titles, subtitle information, and descriptions of videos. We stem these textual features and remove stop words using the standard BERT contextual attention mechanism. We use the filtering mechanism to perform a zero-fill operation for those that are not long enough, and directly truncate those that are too long to remove interference [37,38]. For the visual features of the video, we use the Timesformer method to extract video features using the spatio-temporal self-attention mechanism. First, the model obtained by the Timesformer network pre-trained on the ImageNet dataset is used to extract the features of each image. The 1024-dimensional vector obtained by the last fully connected layer of Timesformer is used as the feature of each image.

In the experiments, we use Pycharm as the development tool. Based on the Pytorch 1.13.0 deep learning framework, we use Python 3.7 as the development language. The main configuration of the computing server is as follows: (1) Operating system: Ubuntu21.04, (2) CPU: Intel(R) Core(TM) i7-11700K CPU @ 2.50 GHz, (3) Memory: 32 GB, (4) GPU: RTX2080Ti.

4.1. Experimental Dataset

To explore the scalability of the model, the dataset of this experiment is the BOVText dataset, which is a large-scale bilingual open video text dataset [39]. First of all, it has more than 2000 videos and more than 1,750,000 + frame fragments, which is 25 times larger than the existing largest dataset with text in videos, and the model can have a good generalization effect on it. Second, the dataset covers 31 open categories and one unknown category, with wide application options. Additionally, it contains the public dataset Kinetics-400 [40]. Kinetic stands as an extensively utilized dataset for the recognition of video actions, encompassing 400 distinct categories of human actions, with each category featuring approximately 400 video clips. These video clips are around 10 s in length and originate from real-world Internet videos. Each clip contains a single human action, such as running, jumping, cycling, etc.

4.2. Performance Evaluation Index

With the objective of assessing the efficacy of multimodal classification results, four prevalent metrics are introduced: Accuracy (*AC*), Precision (*PE*), Recall (*RE*), and F1 Score. The larger the value of these performance indicators, the better the classification effect, and their definitions are as follow equations.

$$AC = \frac{TP + TN}{n} \tag{15}$$

$$PE = \frac{TP}{TP + FP} \tag{16}$$

$$RE = \frac{TP}{TP + FN} \tag{17}$$

$$F1 = \frac{2PE \times RE}{PE + RE} \tag{18}$$

where TP means that the judgment is positive and it is actually positive, TN means that it is judged as negative and it is actually negative, FP means that it is judged as positive and it is actually negative, and FN means that it is judged as negative and it is actually positive, $n = TP + TN + FP + FN$.

4.3. Experimental Results and Analysis

The experiments mainly include single-feature experiments, multi-feature experiments, and public dataset comparison experiments. The accuracy rate commonly used in video classification datasets is the Top@*k* accuracy rate. In this experiment, the classification model performance evaluation indicators use Top@1 and Top@5. The dataset in this article is divided into 32 categories, so the model will output a one-dimensional vector containing 32 values. Each value indicates the probability that the video belongs to each category, where Top@1 refers to the correct classification of the results predicted by the model, the accuracy rate when the sample proportion is the highest. Top@5 refers to the accuracy rate in the first five categories of the predicted results when the proportion of correctly classified samples is the highest in the predicted results of the model. The accuracy requirement of the latter is wider than that of the former, so the value of the latter is generally greater than that of the former. At the same time, we use F1 Score which takes Precision and Recall into account to evaluate the model performance.

(1) Single-mode feature

The single-feature experiment is an experiment in machine learning and statistical modeling that uses only one feature to train and test a model. In single-feature experiments, other features are usually considered irrelevant or ignored, because the purpose is to understand the impact of a single feature on model performance, and it is mainly used to compare the accuracy of a single network and a combined network. Based on the outcomes in Table 1, it is evident that the accuracy by the amalgamated network model surpasses that of the individual network, and the accuracy of the combined network is 1% higher than that of the single network. It can be seen that the effect of the same feature on a single network may be good or bad, but the performance of the combined network model is better than that of the single network model regardless of the feature of that modality. At the same time, comparing the impact of different features of video data on its classification task, it is found that the most critical data is visual information, and the impact of text information on classification accuracy is slightly lower than that of visual information.

Table 1. Experimental results of single mode.

Mode	Feature	Top@1 (%)	Top@5 (%)	*F*1 (%)
NextVLAD	Video frame	60.1	70.9	65.3
NextVLAD	Subtitle	55.9	63.2	58.7
AttentionCluster	Video frame	58.2	69.4	63.3
AttentionCluster	Subtitle	52.0	62.3	56.2
NextVLAD-AttentionCluster	Video frame	61.1	79.0	67.9
NextVLAD-AttentionCluster	Subtitle	57.3	63.1	59.4

(2) Cross-modal fusion

The cross-modal fusion experiment refers to the fusion of the video-level features of the data of multiple modalities in the video through the clustering network each time, and then input them into a single network or combined network for classification. This experiment is mainly used for comparing the classification performance between a single modality and a fusion of two modalities. It can be seen from Table 2 that the effect of combining the features of video visual information and text information into joint features as video features for video classification is better than that of any single-modal feature, and the accuracy of the combined network model is the best. Compared with the accuracy rate of a single network, the accuracy rate is increased by 2%, and the effect of the NextVlAD model is better than that of the AttentionCluster model. Compared with the single-feature experiment, the accuracy rate increased by 4% to 11%.

Table 2. Experimental results of cross-modal fusion.

Mode	Feature	Top@1 (%)	Top@5 (%)	*F*1 (%)
NextVLAD	Video frame and Subtitle	64.3	72.8	68.2
AttentionCluster	Video frame and Subtitle	63.2	71.2	65.9
NextVLAD-AttentionCluster	Video frame and Subtitle	65.8	82.2	73.2

(3) Comparison of public datasets

Table 3 provides the comparison of the experimental results with public datasets. The results presented in Table 3 highlight the superiority of the CFVC model introduced in this paper over other existing models within the public dataset context. In comparison to the dual-stream network that also uses modality fusion, the accuracy rate is improved by more than 10%, because the Transformer-based Timesformer is used as the video feature extraction network. For the VIT-L model trained end-to-end [29], the accuracy rate also increased by 7.1%. To sum up, it is not difficult to see that the method proposed in this paper is superior to the current mainstream convolutional neural network-based method, because they use different structures for feature extraction, and it also proves that the Transformer-based model extraction ability is better than CNN. Although CNN has advantages in extracting low-level features and structures, how to associate with high-level semantic information is a difficult problem, and Transformer uses the attention mechanism to capture global context information to increase their relevance.

Table 3. Comparison results with other methods.

Mode	Feature	Top@1 (%)	Top@5 (%)	*F*1 (%)
I3D [23]	Video	70.1	90.1	78.1
R [2 + 1]D-Two-Stream [27]	Video + stream	73.6	90.5	81.1
Two-Stream I3D [28]	Video + stream	75.7	92.6	83.3
SlowFast [29]	Video	77.4	93.2	84.5
VIT-L (64 frames) [34]	Video	80.5	94.5	86.9
CFVC (Our method)	Video + text	87.6	96.3	91.7

5. Conclusions

In this paper, we first study the video features of different modalities and the adopted feature extraction methods. According to the information characteristics of each video modality, different network models are used to extract the corresponding features, so that it can represent the information of the modality well. Then, through the clustering algorithm, the features of the two modalities are fused to obtain the features of the video, thereby improving the representation of the overall features. The final features are made more useful for classification tasks by means of modality fusion. In experiments, classification evaluation is performed on our dataset. The comparison experiment mainly studies the difference between our model and different models. The experiment results show that visual information has the greatest impact on video classification tasks, and the accuracy of the model can be effectively improved by modality fusion, thus improving the accuracy of classification. Correlative results reveal the effectiveness of our model, which has certain advantages over other models. Due to the limited types and number of videos in the training dataset, the impact of different training datasets on the classification performance has not been further investigated. In future work, we will try to expand the type and number of videos to improve the classification performance.

Author Contributions: Conceptualization, N.P. and C.A.C.; methodology, N.P.; software, S.G.; validation, S.G. and N.P.; formal analysis, M.Y.; writing—original draft preparation, S.G.; writing—review and editing, C.A.C. and M.Y. All authors have read and agreed to the published version of the manuscript.

Funding: This research was funded by the Fundamental Research Funds for the Central Universities (Grant No. CUC220B009).

Institutional Review Board Statement: Not applicable.

Informed Consent Statement: Not applicable.

Data Availability Statement: The authors confirm that the data supporting the findings of this study are available within the article.

Conflicts of Interest: The authors declare no conflict of interest. The funders had no role in the design of the study; in the collection, analyses, or interpretation of data; in the writing of the manuscript, or in the decision to publish the results.

References

1. Jin, M.; Ning, Y.; Liu, F.; Zhao, F.; Gao, Y.; Li, D. An Evaluation Model for the Influence of KOLs in Short Video Advertising Based on Uncertainty Theory. *Symmetry* **2023**, *15*, 1594. [CrossRef]
2. Ali, A.; Senan, N. A review on violence video classification using convolutional neural networks. In *Recent Advances on Soft Computing and Data Mining, Proceedings of the Second International Conference on Soft Computing and Data Mining (SCDM-2016), Bandung, Indonesia, 18–20 August 2016*; Springer: Berlin/Heidelberg, Germany, 2016; pp. 130–140.
3. Trzcinski, T. Multimodal social media video classification with deep neural networks. In *Photonics Applications in Astronomy, Communications, Industry, and High-Energy Physics Experiments*; SPIE: Washington, DC, USA, 2018; pp. 879–886.
4. Ntalianis, K.; Doulamis, N. An automatic event-complementing human life summarization scheme based on a social computing method over social media content. *Multimed. Tools Appl.* **2016**, *75*, 15123–15149. [CrossRef]
5. Jain, A.; Singh, D. A Review on Histogram of Oriented Gradient. *IITM J. Manag. IT* **2019**, *10*, 34–36.
6. Ragupathy, P.; Vivekanandan, P. A modified fuzzy histogram of optical flow for emotion classification. *J. Ambient Intell. Hum. Comput.* **2021**, *12*, 3601–3608. [CrossRef]
7. Fan, M.; Han, Q.; Zhang, X.; Liu, Y.; Chen, H.; Hu, Y. Human Action Recognition Based on Dense Sampling of Motion Boundary and Histogram of Motion Gradient. In Proceedings of the 2018 IEEE 7th Data Driven Control and Learning Systems Conference (DDCLS), Enshi, China, 25–27 May 2018; pp. 1033–1038.
8. Wang, H.; Klaser, A.; Schmid, C.; Liu, C.-L. Action recognition by dense trajectories. In Proceedings of the CVPR 2011, Colorado Springs, CO, USA, 20–25 June 2011; pp. 3169–3176.
9. Wang, H.; Schmid, C. Action recognition with improved trajectories. In Proceedings of the IEEE International Conference on Computer Vision, Sydney, Australia, 2–8 December 2013; pp. 3551–3558.
10. Silva, F.B.; Werneck, R.d.O.; Goldenstein, S.; Tabbone, S.; Torres, R.d.S. Graph-based bag-of-words for classification. *Pattern Recognit.* **2018**, *74*, 266–285. [CrossRef]
11. Karim, A.A.; Sameer, R.A. Image Classification Using Bag of Visual Words (BoVW). *Al-Nahrain J. Sci.* **2018**, *21*, 76–82. [CrossRef]
12. Li, R.; Liu, Z.; Tan, J. Reassessing hierarchical representation for action recognition in still images. *IEEE Access* **2018**, *6*, 61386–61400. [CrossRef]
13. Singhal, S.; Tripathi, V. Action recognition framework based on normalized local binary pattern. Progress in Advanced Computing and Intelligent Engineering. *Proc. ICACIE* **2017**, *1*, 247–255.
14. Hu, Y.; Gao, J.; Xu, C. Learning dual-pooling graph neural networks for few-shot video classification. *IEEE Trans. Multimedia* **2020**, *23*, 4285–4296. [CrossRef]
15. Wang, Y.; Liu, Y.; Zhao, J.; Zhang, Q. A Low-Complexity Fast CU Partitioning Decision Method Based on Texture Features and Decision Trees. *Electronics* **2023**, *12*, 3314. [CrossRef]
16. Liu, C.; Wang, Y.; Zhang, N.; Gang, R.; Ma, S. Learning Moiré Pattern Elimination in Both Frequency and Spatial Domains for Image Demoiréing. *Sensors* **2022**, *22*, 8322. [CrossRef] [PubMed]
17. Zhang, X.; Jiang, X.; Song, Q.; Zhang, P. A Visual Enhancement Network with Feature Fusion for Image Aesthetic Assessment. *Electronics* **2023**, *12*, 2526. [CrossRef]
18. Yi, Q.; Zhang, G.; Liu, J.; Zhang, S. Movie Scene Event Extraction with Graph Attention Network Based on Argument Correlation Information. *Sensors* **2023**, *23*, 2285. [CrossRef]
19. Gudaparthi, H.; Niu, N.; Yang, Y.; Van Doren, M.; Johnson, R. Deep Learning's fitness for purpose: A transformation problem Frame's perspective. *CAAI Trans. Intell. Technol.* **2023**, *8*, 343–354. [CrossRef]
20. Luo, X.; Wen, X.; Li, Y.; Li, Q. Pruning method for dendritic neuron model based on dendrite layer significance constraints. *CAAI Trans. Intell. Technol.* **2023**, *8*, 308–318. [CrossRef]
21. Yan, M.; Lou, X.; Chan, C.A.; Wang, Y.; Jiang, W. A semantic and emotion-based dual latent variable generation model for a dialogue system. *CAAI Trans. Intell. Technol.* **2023**, *8*, 319–330. [CrossRef]
22. Han, K.; Wang, Y.; Chen, H.; Chen, X.; Guo, J.; Liu, Z.; Tang, Y.; Xiao, A.; Xu, C.; Xu, Y.; et al. A survey on vision transformer. *IEEE Trans. Pattern Anal. Mach. Intell.* **2022**, *45*, 87–110. [CrossRef]
23. Ji, S.; Xu, W.; Yang, M.; Yu, K. 3D convolutional neural networks for human action recognition. *IEEE Trans. Pattern Anal. Mach. Intell.* **2012**, *35*, 221–231. [CrossRef] [PubMed]
24. Wu, Q.; Zhu, A.; Cui, R.; Wang, T.; Hu, F.; Bao, Y.; Snoussi, H. Pose-Guided Inflated 3D ConvNet for action recognition in videos. *Signal Process. Image Commun.* **2021**, *91*, 116098. [CrossRef]
25. Chen, H.; Li, Y.; Fang, H.; Xin, W.; Lu, Z.; Miao, Q. Multi-Scale Attention 3D Convolutional Network for Multimodal Gesture Recognition. *Sensors* **2022**, *22*, 2405. [CrossRef]
26. Donahue, J.; Anne Hendricks, L.; Guadarrama, S.; Rohrbach, M.; Venugopalan, S.; Saenko, K.; Darrell, T. Long-term recurrent convolutional networks for visual recognition and description. In Proceedings of the IEEE Conference on Computer Vision and Pattern Recognition, Boston, MA, USA, 7–12 June 2015; pp. 2625–2634.

27. Simonyan, K.; Zisserman, A. Two-stream convolutional networks for action recognition in videos. *Adv. Neural Inf. Process. Syst.* **2014**, *27*. [CrossRef]
28. Feichtenhofer, C.; Pinz, A.; Zisserman, A. Convolutional two-stream network fusion for video action recognition. In Proceedings of the IEEE Conference on Computer Vision and Pattern Recognition, Las Vegas, NV, USA, 26 June–1 July 2016; pp. 1933–1941.
29. Feichtenhofer, C.; Fan, H.; Malik, J.; He, K. Slowfast networks for video recognition. In Proceedings of the IEEE/CVF International Conference on Computer Vision, Seoul, Republic of Korea, 27 October–2 November 2019; pp. 6202–6211.
30. Jin, C.; Luo, C.; Yan, M.; Zhao, G.; Zhang, G.; Zhang, S. Weakening the Dominant Role of Text: CMOSI Dataset and Multimodal Semantic Enhancement Network. *IEEE Trans. Neural Netw. Learn. Syst.* **2023**, 1–15. [CrossRef] [PubMed]
31. Patrick, S.C.; Réale, D.; Potts, J.R.; Wilson, A.J.; Doutrelant, C.; Teplitsky, C.; Charmantier, A. Differences in the temporal scale of reproductive investment across the slow-fast continuum in a passerine. *Ecol. Lett.* **2022**, *25*, 1139–1151. [CrossRef]
32. Wei, D.; Tian, Y.; Wei, L.; Zhong, H.; Chen, S.; Pu, S.; Lu, H. Efficient dual attention slowfast networks for video action recognition. *Comput. Vis. Image Underst.* **2022**, *222*, 103484. [CrossRef]
33. Jiang, Y.; Cui, K.; Chen, L.; Wang, C.; Xu, C. Soccerdb: A large-scale database for comprehensive video understanding. In Proceedings of the 3rd International Workshop on Multimedia Content Analysis in Sports, Seattle, WA, USA, 16 October 2020; pp. 1–8.
34. Dosovitskiy, A.; Beyer, L.; Kolesnikov, A.; Weissenborn, D.; Zhai, X.; Unterthiner, T.; Dehghani, M.; Minderer, M.; Heigold, G.; Gelly, S.; et al. An image is worth 16 × 16 words: Transformers for image recognition at scale. *arXiv* **2020**, arXiv:2010.11929.
35. Sarzynska-Wawer, J.; Wawer, A.; Pawlak, A.; Szymanowska, J.; Stefaniak, I.; Jarkiewicz, M.; Okruszek, L. Detecting formal thought disorder by deep contextualized word representations. *Psychiatry Res.* **2021**, *304*, 114135. [CrossRef]
36. Radford, A.; Narasimhan, K.; Salimans, T.; Sutskever, I. Improving Language Understanding by Generative Pre-Training. 2018. Available online: https://www.mikecaptain.com/resources/pdf/GPT-1.pdf (accessed on 20 August 2023).
37. Bloehdorn, S.; Basili, R.; Cammisa, M.; Moschitti, A. Semantic kernels for text classification based on topological measures of feature similarity. In Proceedings of the Sixth International Conference on Data Mining (ICDM'06), Hong Kong, China, 18–22 December 2006; pp. 808–812.
38. Hao, W.; Zhang, K.; Zhang, L.; Han, M.; Hao, W.; Li, F.; Yang, G. TSML: A New Pig Behavior Recognition Method Based on Two-Stream Mutual Learning Network. *Sensors* **2023**, *23*, 5092. [CrossRef] [PubMed]
39. Wu, W.; Zhang, D.; Cai, Y.; Wang, S.; Li, J.; Li, Z.; Tang, Y.; Zhou, H. A Bilingual, OpenWorld Video Text Dataset and End-to-End Video Text Spotter with Transformer. In Thirty-fifth Conference on Neural Information Processing Systems Datasets and Benchmarks Track (Round 2). 2021. Available online: https://openreview.net/forum?id=vzb0f0TIVlI (accessed on 20 August 2023).
40. Kay, W.; Carreira, J.; Simonyan, K.; Zhang, B.; Hillier, C.; Vijayanarasimhan, S.; Viola, F.; Green, T.; Back, T.; Natsev, P.; et al. The kinetics human action video dataset. *arXiv* **2017**, arXiv:1705.06950.

sensors

MDPI

Article

DPACFuse: Dual-Branch Progressive Learning for Infrared and Visible Image Fusion with Complementary Self-Attention and Convolution

Huayi Zhu [1], Heshan Wu [1], Xiaolong Wang [1], Dongmei He [1], Zhenbing Liu [2,*] and Xipeng Pan [1,*]

[1] School of Computer and Information Security, Guilin University of Electronic Science and Technology, Guilin 541004, China; 2001001034@mails.guet.edu.cn (H.Z.); 2000500927@mails.guet.edu.cn (H.W.); wwwxiaolong@mails.guet.edu.cn (X.W.); elh_hdm@163.com (D.H.)

[2] School of Artificial Intelligence, Guilin University of Electronic Science and Technology, Guilin 541004, China

* Correspondence: zbliu@guet.edu.cn (Z.L.); pxp201@guet.edu.cn (X.P.)

Abstract: Infrared and visible image fusion aims to generate a single fused image that not only contains rich texture details and salient objects, but also facilitates downstream tasks. However, existing works mainly focus on learning different modality-specific or shared features, and ignore the importance of modeling cross-modality features. To address these challenges, we propose Dual-branch Progressive learning for infrared and visible image fusion with a complementary self-Attention and Convolution (DPACFuse) network. On the one hand, we propose Cross-Modality Feature Extraction (CMEF) to enhance information interaction and the extraction of common features across modalities. In addition, we introduce a high-frequency gradient convolution operation to extract fine-grained information and suppress high-frequency information loss. On the other hand, to alleviate the CNN issues of insufficient global information extraction and computation overheads of self-attention, we introduce the ACmix, which can fully extract local and global information in the source image with a smaller computational overhead than pure convolution or pure self-attention. Extensive experiments demonstrated that the fused images generated by DPACFuse not only contain rich texture information, but can also effectively highlight salient objects. Additionally, our method achieved approximately 3% improvement over the state-of-the-art methods in MI, Qabf, SF, and AG evaluation indicators. More importantly, our fused images enhanced object detection and semantic segmentation by approximately 10%, compared to using infrared and visible images separately.

Keywords: multi-head self-attention; convolutional neural network; image fusion; gradient convolution; cross-modality interaction

Citation: Zhu, H.; Wu, H.; Wang, X.; He, D.; Liu, Z.; Pan, X. DPACFuse: Dual-Branch Progressive Learning for Infrared and Visible Image Fusion with Complementary Self-Attention and Convolution. *Sensors* **2023**, *23*, 7205. https://doi.org/10.3390/s23167205

Academic Editors: Ming Yan, Chunguo Li and Chien Aun Chan

Received: 6 July 2023
Revised: 3 August 2023
Accepted: 4 August 2023
Published: 16 August 2023

1. Introduction

Infrared and visible image fusion (IVF) has gained significant attention and is widely used in various applications [1]. Specifically, the effective fusion of shared and specific features from different modalities enables the generation of high-quality fused images, which, in turn, benefits downstream tasks, such as object detection [2–4], medical image processing [5,6], semantic segmentation [7–9], and pedestrian detection [10,11]. Although IVF has received much attention in various applications, IVF remains challenging due to the significant differences in appearance between these two image types. In infrared images, thermal target structures can be effectively highlighted, but these images often exhibit low contrast and blur properties. On the other hand, visible images have rich color and texture information, but they are easily affected by factors such as illumination and weather. Therefore, the effective fusion of these two different modalities into high-quality images still faces many technical difficulties.

Recently, there has been a surge in the development of deep learning-based IVF algorithms [12–16]. These methods typically involve feature extraction from the source images, the fusion of the extracted features, and the reconstruction of the fused features to obtain the final fused image [17]. Networks with strong feature extraction capabilities can usually synthesize fused images with better quality.

The main feature extraction method is currently the CNN-based auto-encoder structure, which is mainly divided into a shared encoder structure [18–20] and a dual-branch private encoder structure [21–23]. However, although the above-mentioned auto-encoder structure has good feature extraction ability, there are also some shortcomings. Firstly, the structure, based on the shared encoder, cannot distinguish the unique information in each mode. Secondly, the structure of the dual-branch private encoder often ignores the common features between modes, such as background and some large-scale features. In addition, the context-free CNN structure can only extract local features in a small service domain, and it is difficult to extract global information to generate higher-quality fusion images [15]. Moreover, many IVF networks may cause high-frequency information loss in image features when performing forward propagation [24]. Therefore, to address the above issues in IVF, this paper proposes a progressive feature extractor Cross-Modality Feature Extraction (CMFE). By introducing the CMFE module, compared with the structure of the shared encoder, our model can effectively extract the shared features between different modalities, while also better distinguishing the unique features between different modalities, to better realize IVF. Compared with the dual-branch private encoder, our structure can enhance the information interaction between different modalities to ensure thorough feature extraction, and our structure can better integrate the unique and shared features of modalities.

Vision Transformer has received extensive attention on many vision tasks. Therefore, many scholars have adopted Transformer-based methods in the feature extraction stage [13,15,25,26]. However, many transformer-based models are often limited by computational resources, the size of the input image, and weak local perception, which limits the timeliness and applicability of the IVF. To synthesize the respective advantages of Transformer and CNN architectures, we introduce ACmix [27], which integrates the flexibility of self-attention and the lightness of convolution. Therefore, we propose a complementary fusion network based on convolution and a multi-head self-attention mechanism to solve the IVF problem, integrating the advantages of CNN in extracting local information and computing portability and the ability of self-attention in context awareness and long-distance modeling. Compared with CNN-based architectures, our architecture has more powerful feature extraction capabilities and can better extract deep features in modalities. Additionally, compared with some Transformer-based architectures, our architecture has better flexibility (e.g., no fixed image size, small computational cost, etc.) and adaptability.

To this end, we propose Dual-branch Progressive learning for IVF with complementary self-Attention and Convolution (DPACFuse), such that the network can take into account the unique features and shared features of modalities, and integrate the respective advantages of CNN and a multi-head self-attention mechanism to better realize the IVF. The main contributions of this work can be summarized as follows:

- We propose a dual-branch progressive image fusion framework, based on complementary self-attention and convolution, for the IVF, which can take into account both global and local information of the source image to achieve better feature fusion.
- We propose Cross-Modality Feature Extraction (CMFE) to enhance the information interaction between modalities, while suppressing the loss of high-frequency information.
- Extensive experimental results on the public image fusion datasets, MSRS, RoadScene, and TNO, demonstrate that our method is superior to the current general fusion framework. Meanwhile, we investigate the facilitation of our fused images for object detection and semantic segmentation.

The remainder of this article is organized as follows. In Section 2, we mainly introduce the deep learning-based IVF methods. In Section 3, we introduce and describe, in detail, our fusion framework and the loss function used. In Section 4, we conduct a large number of experiments to verify the effectiveness of DPACFuse and also to explore the effect of IVF on downstream task promotion. In Section 5, we provide some conclusive proof.

2. Related Work

In this section, we briefly introduce the existing deep learning-based methods, mainly including CNN-based, AE-based, GAN-based, and Transformer-based methods.

2.1. CNN-Based and AE-Based Fusion Methods

In recent years, many CNN-based and AE-based methods have been widely used in the field of image fusion. Among them, the dual-branch structure, based on CNN and AE, greatly improves the fusion performance. For example, Tang et al. [21] proposed a dual-branch private encoder structure, namely SeAFusion, which combined a semantic segmentation network to learn more information, and to achieve better fusion results. Meanwhile, SeAFusion is a pioneer in combining IVF with downstream tasks. Additionally, Tang et al. [28] proposed an illumination-driven IVF network to solve the fusion problem in different lighting scenes. However, the designs of the network structures of the above two methods are too simple to effectively deal with complex situations. Another typical work is that of Res2Fusion, in [29], which describes a network with a dual-branch shared encoder structure, and which introduces Res2net and densely connected structures into the encoder to obtain multi-scale information. In the fusion layer, the fusion layer of double nonlocal attention models is used to realize image fusion. This method fully considers the problem of multi-scale information extraction and global modeling of the model, but it is difficult to deploy in the actual environment due to its high complexity and high computational cost. In addition, the shared encoder structure is also widely used in image fusion. For example, a typical method is DenseFuse [18]. The core concept of this method is to construct a deep neural network with dense connections, comprising an encoder (consisting of convolutional layers and dense layers) and a decoder (used for fusion). However, DenseFuse uses hand-designed fusion rules, so its results are not robust. In order to address the limitation of fusion rules that are designed manually, Li et al. proposed RFN-Nest [19] and NestFuse [20], wherein the former mainly utilized a residual fusion network to solve the problem, while the latter adopted the idea of combining spatial and channel attention mechanisms to solve the problem. Moreover, the IFCNN in [22], the PMGI in [30], and the U2Fusion in [31] proposed unified end-to-end networks to realize different fusion tasks.

Although many fusion methods based on CNN and AE have achieved good results, they usually adopt relatively simple CNN structures and hand-designed fusion rules, which limit the global modeling ability of the model and the ability to extract detailed information. At the same time, many methods lack information interaction in the process of feature extraction, which makes it impossible to fully extract more complementary information. In contrast, our method adopts the idea of cross-modality interaction to achieve image fusion, which helps to eliminate the mismatch and noise between different modalities and brings advantages in terms of improving the robustness of the fused image.

2.2. GAN-Based Fusion Methods

The Generative Adversarial Network (GAN) can learn the distribution characteristics of data and generate samples that conform to a specific distribution, which is also widely used in IVF. FusionGAN [32] was the first algorithm to apply this method to realize IVF. However, since only a single discriminator is used in FusionGAN, it cannot balance the information from the different modalities, leading to the loss of a lot of texture information in the fused images. In order to overcome the shortcomings of a single discriminator, Ma et al. [33] proposed a dual discriminator structure, namely DDcGAN, to achieve information fusion with image fusion. Moreover, Rao et al. [17] proposed AT-GAN, which introduced an intensity attention module and a semantic transition module to remove redundant information in infrared and visible images, respectively. At the same time, the quality assessment module is used to achieve the information balance between different modalities. In addition, infrared images have been greatly developed in various object detection tasks, such as pedestrian detection [11,34] and infrared small target detection [35,36]. However, due to the imaging characteristics of infrared images, the application scenarios of these methods are very limited. Therefore, there are many works [37–39] that combine IVF with object detection to overcome the limitations of using only a single modality. For example, Liu et al. [12] designed a GAN-based object perception network, TarDAL, which generated high-quality fused images and excellent detectors by including the IVF network and the object detection network in a bilevel optimization formulation.

Although GAN-based models have been widely used in the field of image fusion, the GAN-based fusion method emphasizes that discriminator learning simulates the distribution of the original image data, which may lead to poor image quality. At the same time, finding a way to balance the information from the different modalities is still a problem that needs to be studied.

2.3. Transformer-Based Fusion Methods

The transformer [40] structure is based on a multi-head self-attention mechanism, designed for sequence modeling and transduction tasks, and is known for its focus on long-term dependencies in data. Transformer has seen great success not only in NLP, but also in various visual tasks [41–43]. Many models based on Transformer have also been highly developed in the field of image fusion. For instance, Wang et al. [15] proposed a pure transformer fusion network, called SwinFuse, which used the powerful feature representation capability of the self-attention mechanism to perform image fusion. However, it uses a hand-designed fusion strategy, which does not perform well enough in handling fine-grained information. Additionally, Zhao et al. [44] introduced the Dual-branch Transformer and the structure of DenseNet (DNDT), which could consider more complete image information. In addition, inspired by the work of Swin Transformer [45] , Ma et al. [13] proposed Swin Fusion, a network architecture for multimodal fusion. In addition, Rao et al. [46] proposed TGFuse, which embedded the Transformer in a GAN-based fusion network to achieve IVF. Furthermore, Qu et al. [26] proposed TransMEF for multi-exposure image fusion, which combined CNN and Transformer to obtain powerful local modeling and global modeling capabilities. However, this method is less flexible and can only input images of fixed size.

Although many transformer-based models perform well in many fusion tasks, many methods still suffer from poor flexibility and poor ability to model trans-membrane states, such as DNDT [44], TransMEF [26], and CGTF [47]. Furthermore, many transformer-based models are computationally expensive, while our method combines the excellent computational efficiency of CNN and the excellent global modeling ability of self-attention to better realize image fusion.

3. Methodology

3.1. Network Architecture

The network architecture of DPACFuse is illustrated in Figure 1a, and is composed of three main phases: feature extraction, feature fusion, and feature reconstruction. Given a pair of aligned infrared (IR) and visible (VI) images, denoted as $I_{ir} \in R^{H \times W \times C_{in}}$ and $I_{vi} \in R^{H \times W \times C_{in}}$, respectively, the fused image $I_f \in R^{H \times W \times C_{out}}$ is obtained through these phases.

In the feature extraction phase, we extract the specific features of the respective modalities separately using a dual-branch structure. First, we obtain shallow features $\{F_{ir}^1, F_{vi}^1\}$ from the source image through a 3×3 convolutional layer. This can be expressed as:

$$\{F_{ir}^1, F_{vi}^1\} = \{H_{se}(I_{ir}), H_{se}(I_{vi})\} \tag{1}$$

where $H_{se}(\cdot)$ represents a 3×3 convolutional layer, whose activation function is Leaky Relu and the stride is 1. The convolutions usually have stable optimization performance and are very good at early visual processing. At the same time, convolution has a strong local perception ability, which can effectively mine local information and map it to high-dimensional space.

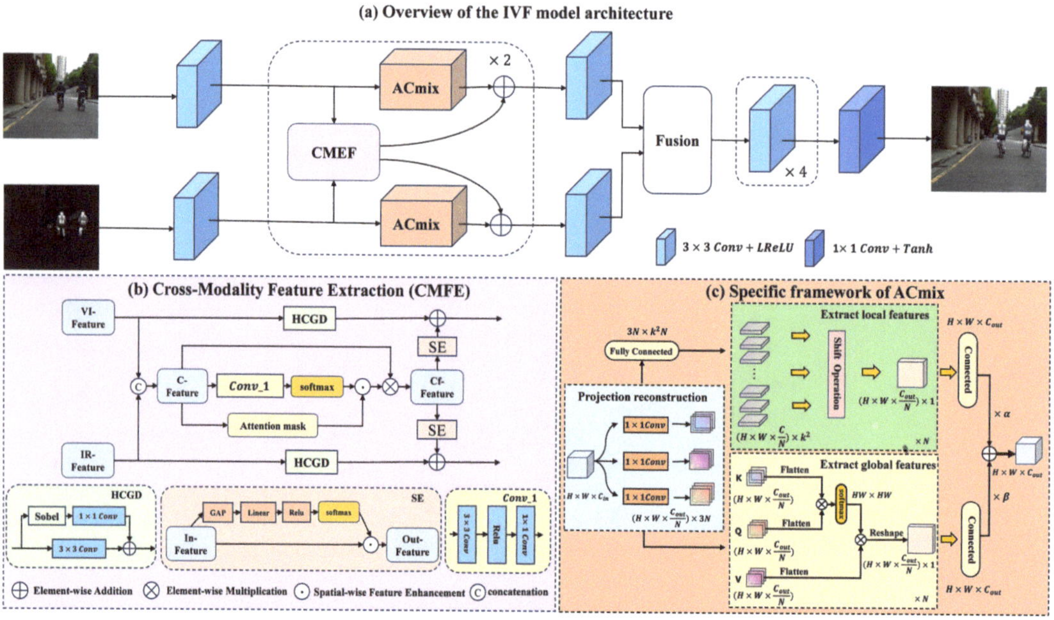

Figure 1. (**a**) Overview of the IVF model architecture. The feature extractor of the network consists of ACmix, CMEF, and 3 × 3 convolutional layers. Its feature reconstructor consists of 5 convolutional layers. (**b**) Cross-Modality Feature Extraction (CMFE). This mainly consists of a high-frequency convolution calculation based on Sobel operator (HCGD) and Squeeze-and-Excitation Network (SE). The HCGD adopts the idea of residual connection, and GAP and Linear in the SE schematic represent Global Average Pooling and Linear Function, respectively. (**c**) Specific framework of ACmix. This can be divided into three parts: Projection reconstruction, Extract local features, and Extract global features.

Then, the ACmix is embedded in the respective branches of the IR and VI images to extract their respective specific features. At the same time, CMFE is deployed between the two modalities to extract their common features, thereby guiding the network to generate better images. We represent the feature extraction of the process in two stages. The intermediate features $\{F_{ir}^2, F_{vi}^2\}$ obtained in the first stage can be expressed by the following formula:

$$
\begin{aligned}
\left\{F_{ir}^{CF_1}, F_{vi}^{CF_1}\right\} &= CMEF\left(F_{ir}^1, F_{vi}^1\right) \\
\left\{F_{ir}^{AC_1}, F_{vi}^{AC_1}\right\} &= \left\{ACmix\left(F_{ir}^1\right), ACmix\left(F_{vi}^1\right)\right\} \\
\left\{F_{ir}^2, F_{vi}^2\right\} &= \left\{\left(F_{ir}^{CF_1} \oplus F_{ir}^{AC_1}\right), \left(F_{vi}^{CF_1} \oplus F_{vi}^{AC_1}\right)\right\}
\end{aligned} \tag{2}
$$

After obtaining the intermediate features $\{F_{ir}^2, F_{vi}^2\}$ in the first stage, we use them as input for the second stage to obtain the output $\{F_{ir}^3, F_{vi}^3\}$ in a similar manner to the first stage:

$$
\begin{aligned}
\left\{F_{ir}^{CF_2}, F_{vi}^{CF_2}\right\} &= CMEF\left(F_{ir}^2, F_{vi}^2\right) \\
\left\{F_{ir}^{AC_2}, F_{vi}^{AC_2}\right\} &= \left\{ACmix\left(F_{ir}^2\right), ACmix\left(F_{vi}^2\right)\right\} \\
\left\{F_{ir}^3, F_{vi}^3\right\} &= \left\{\left(F_{ir}^{CF_2} \oplus F_{ir}^{AC_2}\right), \left(F_{vi}^{CF_2} \oplus F_{vi}^{AC_2}\right)\right\}
\end{aligned} \tag{3}
$$

Then, the features of these two modalities $\{F_{ir}, F_{vi}\}$ are obtained by a convolutional layer $H_{de}(\cdot)$. The process can be expressed by the following formula:

$$\{F_{ir}, F_{vi}\} = \{H_{de}(F_{ir}^3), H_{de}(F_{vi}^3)\} \tag{4}$$

Finally, we reconstruct the fused image through the feature fusion and image reconstruction module. Since our feature extraction network has a strong enough extraction ability, we opted for a straightforward approach by employing the cascade fusion strategy to directly fuse the F_{ir} and F_{vi}. The fusion process is represented as follows:

$$F_f = H_c(F_{ir}, F_{vi}) \tag{5}$$

where F_f represents the fused feature, and $H_c(\cdot)$ represents the cascade on the channel dimension. Finally, we can obtain the output I_f through the feature reconstructor $H_R(\cdot)$:

$$I_f = H_R(F_f) \tag{6}$$

3.2. Specific Framework of ACmix

Due to the excellent context-aware ability of the multi-head self-attention mechanism and the lightness of convolution, we introduce ACmix. As shown in Figure 1c, it can be divided into: projection reconstruction, extract local features, and extract global features.

First, image features $I_i \in R^{H\times W\times C_{in}}$ are obtained and, after the projection reconstruction, local and global features are extracted, to obtain the output $F_{out} \in R^{H\times W\times C_{out}}$. In the projection reconstruction stage, the feature map I_i is passed through three separate 1×1 convolutional layers, resulting in the generation of three feature maps $I_i^1, I_i^2, I_i^3 \in R^{H\times W\times C_{out}}$.

In the extract local feature stage, the steps of this stage are different from the traditional standard convolution, that is, we first perform a linear projection of kernel weights, then translate according to the kernel position, and finally aggregate. Firstly, the three feature maps I_i^1, I_i^2, I_i^3 in the projection reconstruction stage, which are divided into N groups in the depth direction, and then reshaped to obtain a feature map with the dimensions of $R^{N\times\frac{C_{out}}{N}\times HW}$. After partitioning the feature maps into groups and reshaping them, the resulting feature maps are concatenated to create a new feature map $X \in R^{3N\times\frac{C_{out}}{N}\times HW}$. This concatenated feature map X is then fed through a lightweight fully connected layer to generate $Z \in R^{k^2N\times\frac{C_{out}}{N}\times HW}$. Then, Z is subjected to a reshaping operation and then a shift aggregation operation, which is realized by depthwise convolution. Specifically, Z is divided into N groups, and each group $Z_l \in R^{H\times W\times\frac{C_{out}}{N}}$ is used as a basic unit of convolution. Finally, the results of N groups are spliced to obtain the output $F_{conv} \in R^{H\times W\times C_{out}}$ of the convolution path. The entire extract local feature stage can be expressed as:

$$\begin{aligned} X &= Connect(R(I_i^1), R(I_i^2), R(I_i^3)) \\ Z &= FC_k(X) \\ F_{conv} &= \mathop{\Big\|}_{l=1}^{N} C_{dev}(Z_l) \end{aligned} \tag{7}$$

where $Connect(\cdot)$, $R(\cdot)$, and $FC_k(\cdot)$ denote the concatenation, reshape, and the light fully connected layer with a kernel size of k, respectively. The value $C_{dev}(\cdot)$ represents the convolution operation in depthwise convolution and Z_l represents the input of group lth. The symbol $||$ denotes the concatenation of the results obtained from all N groups, and the entire process corresponds to depthwise convolution with kernels of size 3. The processing of the extract local feature stage is the same as the traditional convolution operation.

In the extract global feature stage, the multi-head self-attention mechanism is adopted. Specifically, we divided the three feature maps I_i^1, I_i^2, I_i^3 obtained in the projection reconstruction stage into N groups (i.e., N attention mechanism heads) in the depth direction, and obtained the $Q \in R^{H\times W\times \frac{C_{out}}{N}}$, $K \in R^{H\times W\times \frac{C_{out}}{N}}$, $V \in R^{H\times W\times \frac{C_{out}}{N}}$ of each head. Then we flattened Q, K and V to obtain $Q' \in R^{HW\times \frac{C_{out}}{N}}$, $K' \in R^{HW\times \frac{C_{out}}{N}}$ and $V' \in R^{HW\times \frac{C_{out}}{N}}$, and then used these as the inputs of the attention function:

$$\begin{aligned} Atention(Q',K',V') &= Softmax\left(\frac{Q'K'^{T}}{\sqrt{d_k}}\right)V' \\ head_j &= Atention(F(I_iW_j^q),F(I_iW_j^k),F(I_iW_j^v)) \\ F_{att} &= \overset{N}{\underset{l=1}{\Vert}} R(head_l) \end{aligned} \tag{8}$$

where W_j^q, W_j^k, and $W_j^v \in R^{C_{in}\times \frac{C_{out}}{N}}$ are corresponding input projection weights (for ease of presentation, here, we include the process of the first stage). $F(\cdot)$ and $R(\cdot)$ denote the flatten and reshape operations, respectively. The symbols d_k, $head_j \in R^{H\times W\times \frac{C_{out}}{N}}$ and $||$ denote the dimension of K', the output of the jth head, and the splicing of N heads, respectively. The symbol $F_{att} \in R^{H\times W\times C_{out}}$ represents the final output of the extract global feature stage, which is obtained by concatenating the outputs of the N self-attention heads.

Finally, the features extracted by the ACmix module can be expressed as the sum of the extract local feature path and the extract global feature path output, where the weights are determined by two learnable parameters α and β:

$$F_{out} = \alpha F_{conv} + \beta F_{att} \tag{9}$$

3.3. Specific Framework of CMEF

To enhance the information interaction between modalities, as well as to suppress the loss of high-frequency information, we propose a CMEF module, the model of which is shown in Figure 1b. We introduce the module in two stages: feature combination and feature recombination.

In the feature combination stage, the input features $F_1 \in R^{H\times W\times C_{in}}$, $F_2 \in R^{H\times W\times C_{in}}$ are given. We first concatenate F_1 and F_2 to get the fusion feature $F_{cat} \in R^{H\times W\times 2C_{in}}$, and then obtain the common feature $F_{cf} \in R^{H\times W\times C_{out}}$ through the foreground-aware spatial attention and feature-level attention mask. The specific process is as follows:

$$\begin{aligned} F_{cat} &= Concat(F_1,F_2) \\ F_{cf} &= \left[(\phi_s(F_{cat})\odot\phi_p(Conv\_1(F_{cat})))\right]F_{cat} \end{aligned} \tag{10}$$

where $Concat(\cdot)$ represents the operation of concatenating the features. The symbol $\phi_s(\cdot)$ stands for the foreground-aware spatial attention operation which is achieved by calculating the channel-wise maximum value of fusion features. The symbol $\phi_p(\cdot)$ represents the feature-level attention mask achieved by a Multiple Layer Perceptron (MLP), and $Conv\_1(\cdot)$ followed by a 2-cls softmax operation. This feature-level attention mask means that $\phi_p(\cdot)$ can predict a re-scaling score to combine features from different modalities in such a way that the combined features are independent of the specific features.

In the feature recombination stage, it is well known that useless information has a huge impact on image fusion, which misleads the fusion direction of the model, resulting in distortion of the fused image. At this stage, we hope to obtain more common features and filter the interference of useless information as much as possible. Specifically, we integrate these shared features with the fine-grained information of their respective features through channel rescaling operations:

$$\begin{aligned} F_{out}^{ir} &= SE(F_{cf})\oplus\phi_{GD}(\nabla F_1)\oplus Conv(F_1) \\ F_{out}^{vi} &= SE(F_{cf})\oplus\phi_{GD}(\nabla F_2)\oplus Conv(F_2) \end{aligned} \tag{11}$$

where $SE(\cdot)$ represents the Squeeze-and-Excitation Network [48] (its framework is shown in Figure 1b), which can assign weights to each channel to effectively filter the impact of useless information on the fusion process. The symbol $\phi_{GD}(\cdot)$ refers to the gradient convolution operation, and ∇ stands for the gradient operator. Moreover, $\oplus$ and $Conv(\cdot)$ denote the operation of element-wise addition and 3×3 convolution, respectively. Finally, $F_{out}^{ir}, F_{out}^{vi} \in R^{H \times W \times C_{out}}$, respectively, add to their respective characteristics of the backbone network.

3.4. Loss Function

To minimize information loss and improve fusion performance, this paper employs three distinct loss functions for training the network: texture loss, intensity loss, and SSIM (Structural Similarity Index) loss. These loss functions constrain the network from different perspectives. The loss function used in our network can be represented as follows:

$$L_{total} = \gamma_0 L_{int} + \gamma_1 L_{texture} + \gamma_2 L_{ssim} \tag{12}$$

where L_{int}, $L_{texture}$, and L_{ssim} represent the intensity loss, texture loss, and SSIM loss, respectively. The parameters γ_0, γ_1, and γ_2 are hyper-parameters to represent the contributions of the three losses to the entire loss, respectively.

The intensity loss emphasizes the preservation of pixel intensity information, and it helps the model better learn the overall brightness information and contrast characteristics. Intensity loss is defined as:

$$L_{int} = \frac{1}{HW} \parallel I_f - max(I_{ir}, I_{vi}) \parallel_1 \tag{13}$$

where $\parallel \cdot \parallel_1$ denotes l_1 norm, and $\max(\cdot)$ represents the maximum value in an element. By emphasizing the overall brightness and contrast characteristics, it enables the model to better understand and learn these important visual attributes.

The texture loss is a key component in image fusion, as it aims to preserve the intricate and fine-grained texture details during the fusion process. We define texture loss as:

$$L_{texture} = \frac{1}{HW} \parallel | \nabla I_f | - max(| \nabla I_{ir} |, | \nabla I_{vi} |) \parallel_1 \tag{14}$$

where the symbol ∇ represents the Sobel gradient operator. The absolute value calculation, denoted by $| \cdot |$, is applied to the gradient values to ensure that only positive magnitudes are considered. The value $\parallel \cdot \parallel_1$ represents l_1 norm, and $max(\cdot)$ selects the maximum value from the corresponding elements in the calculation.

The SSIM loss is employed to facilitate the learning of structural information by the model from the input images, and it also takes into account not only structure and contrast, but also illumination, which can be expressed as follows:

$$L_{ssim} = (1 - SSIM(I_f, I_{ir}))/2 + (1 - SSIM(I_f, I_{vi}))/2 \tag{15}$$

4. Experiments

In this section, we provide specific details of the experimental implementation. We then compare DPACFuse with seven other methods. Finally, we demonstrate the outstanding performance of DPACFuse on downstream tasks.

4.1. Experimental Configurations

Datasets. The IVF experiments used three public datasets to verify our fusion method, which were MSRS [28], RoadScene [31] and TNO [49]. We trained our IVF network on the MSRS dataset, which contained 1083 pairs of registered images with semantic labels of nine typical scenes. In addition, we employed the MSRS test set (361 pairs), RoadScene (30 pairs), and TNO (30 pairs) as test datasets to comprehensively verify the performance of DPACFuse. Among them, the RoadScene dataset contained 221 image pairs that mainly focused on capturing typical traffic scenes, including roads, pedestrians, and vehicles. The TNO dataset consisted of multispectral night and day images depicting various military-related scenes.

Evaluation metrics and comparison methods. We used EN, SD, SF, MI, VIF, AG, Qabf, and FMI_pixel as evaluation metrics. In addition, higher metrics implied that the quality of the fusion image was better. Details on these evaluation metrics can be found in [50]. At the same time, we compared DPACFuse with the state-of-the-art methods, including DenseFuse [18], IFCNN [22], U2Fusion [31], SDNet [16], GANMcC [51], SwinFusion [13], and TarDAL [12].

Experimental setup. Our experiments were conducted on a computer equipped with one NVIDIA GeForce RTX 3090 GPU. The proposed method was implemented using the PyTorch platform. Moreover, all input images were normalized to [0, 1] before training. The following values were used for the hyperparameters of the experiment: the initial parameters α and β for the balanced convolution and self-attention paths were set to 1, respectively, the total number of self-attention heads was $N = 4$, and the kernel size was $k = 3$ for the fully connected layer. In addition, the hyperparameters were $\gamma_0 = 20$, $\gamma_1 = 20$ and $\gamma_2 = 1$ for balancing each loss function. The network parameters were updated using the Adam optimizer with a momentum term of (0.9, 0.999). The training was performed with a batch size of 2, an initial learning rate of 0.001, and a weight decay of 0.0002.

4.2. *Comparative Experiment*

4.2.1. Qualitative Results

We selected two groups of images in the MSRS test set for subjective evaluation, wherein each group contained two typical scenes that were day and night.

In the daytime scene with sufficient illumination, the VI image contained abundant texture detail and fully showed the environmental information. Although the ability of IR images to display the environment was limited, they could provide semantic information about the structure of thermal targets. By integrating this complementary information, the fusion image could provide comprehensive scene information, and could effectively enrich the semantic information. As presented in Figures 2 and 3, due to the interference of useless information, the salient targets in DenseFuse, IFCNN, and U2Fusion methods weakened to varying degrees and could not maintain their original intensities. We highlighted salient regions with green boxes to illustrate this problem. Although SDNet and GANMcC could maintain the highlight intensity of infrared targets, their performances in retaining texture information was poor, and we illustrated the problem by zooming in on the areas with red boxes. In addition, compared with SwinFusion and TarDAL, DPACFuse not only retained more detailed information, but also better preserved the edge information, as can be seen from the enlarged floors, as well as the steps.

In the dark scene with insufficient illumination, due to the influence of illumination, VI images could only provide limited environmental information, and objects in them were not easy to identify, while IR images were not sensitive to illumination. Therefore, adaptive realization of the IVF in the case of the dark scene was very important, whilst being very challenging. As presented in Figures 2 and 3, all methods could effectively construct the scene information, but there were great differences between different algorithms. Excepting our DPACFuse, SwinFusion, and TarDAL, the other methods failed to maintain the highlighting of thermal targets in infrared images, which we illustrated with the green boxes. In addition, DPACFuse was better than the other methods, such as SwinFusion, in maintaining details, which we illustrated by magnifying the red area.

Overall, the experimental results highlighted the superior performance of DPACFuse in both daytime and dark scenes. It effectively preserved texture details, edge information, and the saliency of thermal targets, demonstrating its superior ability to perform IVF in a variety of environmental conditions.

Figure 2. Qualitative analysis of DPACFuse with seven methods on 00634N (**top**) and 01356N (**bottom**) images from the MSRS dataset.

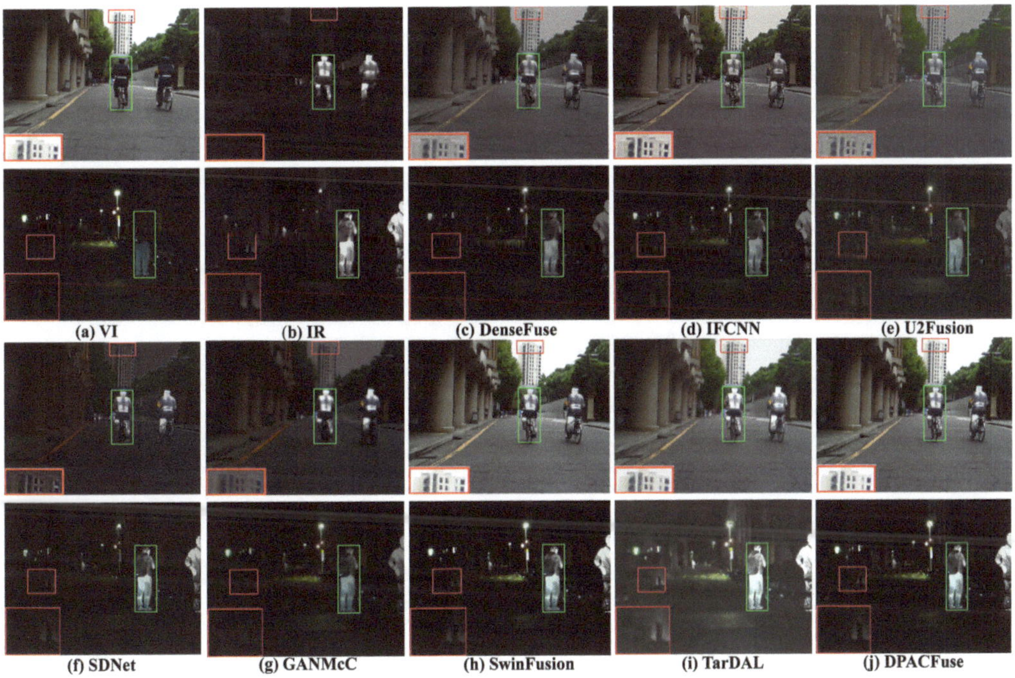

Figure 3. Qualitative analysis of DPACFuse with seven methods on 00537D (**top**) and 01012N (**bottom**) images from the MSRS dataset.

4.2.2. Quantitative Result

Figure 4 displays the quantitative results of the eight evaluation indicators on the MSRS test set. DPACFuse demonstrated superior performance in nearly all metrics, showcasing its ability to effectively extract information from the source images and its versatility across various complex scenes. The best EN, MI, and FMI_pixel indicated that our fused image contained the most information, and the highest Qabf and AG indicated that our fused image retained the most edge information. In addition, the highest SF and VIF illustrated the best visual effect was presented by our fused image. Although DPACFuse slightly lagged behind SwinFusion in terms of the SD metric, the difference was not significant, which meant that our fused images had good contrast.

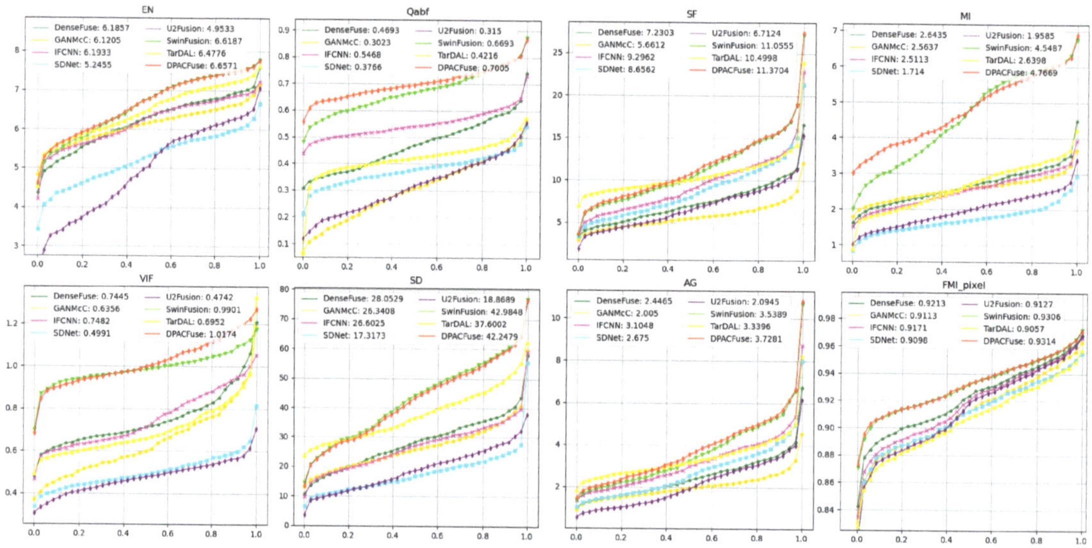

Figure 4. Quantitative comparison of eight methods on the MSRS test set. The x-axis represents cumulative distribution and the y-axis represents the values of the metric. The point (x, y) on the curve represents the measurement value of the $x \times 100$ percent of image pairs not exceeding the value of y. The average value is shown in the legend.

4.3. Generalization Experiment

Generalization ability is also an important metric to evaluate a model. We trained the model on the MSRS dataset and verified the generalization ability of DPACFuse on the RoadScene and TNO datasets.

4.3.1. Results of RoadScene

Qualitative analysis. We selected two scenes, day and night, to assess the fusion results, and the visualized results are shown in Figure 5. Observing the results of the daytime scene, we can see that almost all the algorithms suffered from the interference of useless information, among which DenseFuse, U2Fusion, SDNet, GANMcC, and TarDAL were most affected, losing a lot of texture information. We illustrated this problem by zooming in on the red area. In addition, the intensity of the infrared targets of SDNet and GANMcC also weakened to varying degrees, while SwinFusion experienced a decrease in its overall contrast, due to the influence of illumination, which we illustrate by the green box. Except for our DPACFuse and SwinFusion, the other methods weakened in the overall pixel intensity and could not maintain the original pixel intensity.

Figure 5. Qualitative analysis of DPACFuse with seven methods on FLIR_06307 (**top**) and FLIR_03952 (**bottom**) images from the RoadScene dataset.

In the dark scene, it can be seen that DenseFuse, U2Fusion, SDNet, GANMcC, and TarDAL lost a lot of texture details, such as the outline of background leaves and the zebra crossing on the ground. In addition, the salient targets of DenseFuse, SDNet, and GANMcC were severely disturbed by useless information and could not maintain the original pixel intensity. DPACFuse and SwinFusion were only disturbed by a small amount of useless information.

Quantitative analysis. As shown in Figure 6, DPACFuse achieved the highest scores in all indicators, which meant that the fused image generated by DPACFuse not only maintained a lot of information and texture details, but also had the highest contrast and the best visual quality.

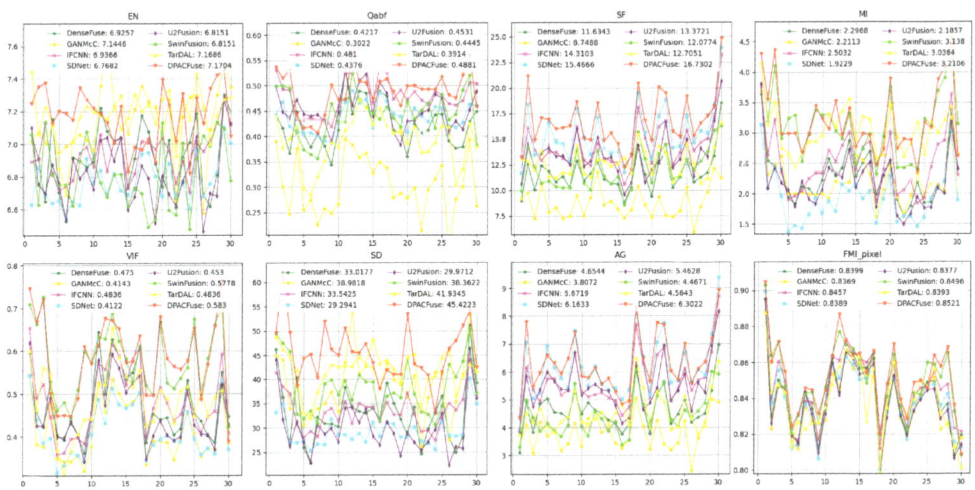

Figure 6. Quantitative comparison of eight methods on the RoadScene test set. The x-axis represents image pairs and the y-axis represents the values of the metric. The point (x, y) in the image represents the measurement y for the xth pair of images. The average value is shown in the legend.

The excellent performance of DPACFuse on the RoadScene dataset fully demonstrated the adaptability of our method to various complex traffic scenes, and also proved that DPACFuse has good generalization ability.

4.3.2. Results of TNO

Qualitative analysis. As depicted in the green boxes in Figure 7, DenseFuse, U2Fusion, and IFCNN weakened the strength of salient targets to different extents, with the first two being the most obvious. In addition, GANMcC blurred the contour of the salient targets. Excepting our method and SwinFusion, the fused images of the other methods were affected by other useless spectral information and could not effectively present the texture information, such as the bushes and fences in the red region. It is worth noting that, although SwinFusion had good fusion performance, the fused images suffered from whitening. On the whole, DPACFuse not only excelled in highlighting salient objects, but also effectively preserved the original texture information from the input images.

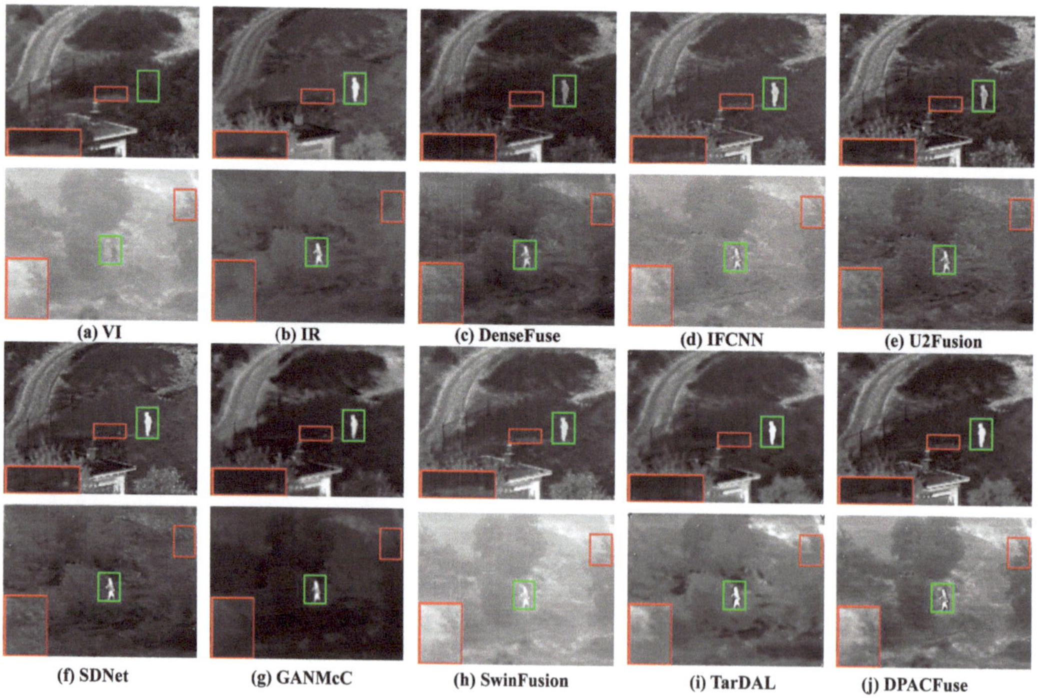

Figure 7. Qualitative analysis of DPACFuse with seven methods on two representative images from the TNO dataset.

Quantitative analysis. The results depicted in Figure 8 illustrate that DPACFuse achieved the highest scores in Qabf, MI, VIF, and FMI_pixel metrics. In addition, DPACFuse was also ahead of all the methods, excepting TarDAL, in two metrics: EN and SD. Taking the above analyses together, DPACFuse exhibited excellent performance on the TNO datasets, which further demonstrated its excellent generalization ability.

In conclusion, a large number of experiments on various datasets showed that our method can preserve a large amount of information from the source image and maintain the highlight degree of the infrared target in various complex situations. We attribute these advantages to the following aspects. On the one hand, the CMEF that we designed effectively extracts fine-grained information from source images and enhances the information interaction between different modalities. On the other hand, our network possesses a powerful feature extraction capability, and the ACmix module can effectively extract local and global information.

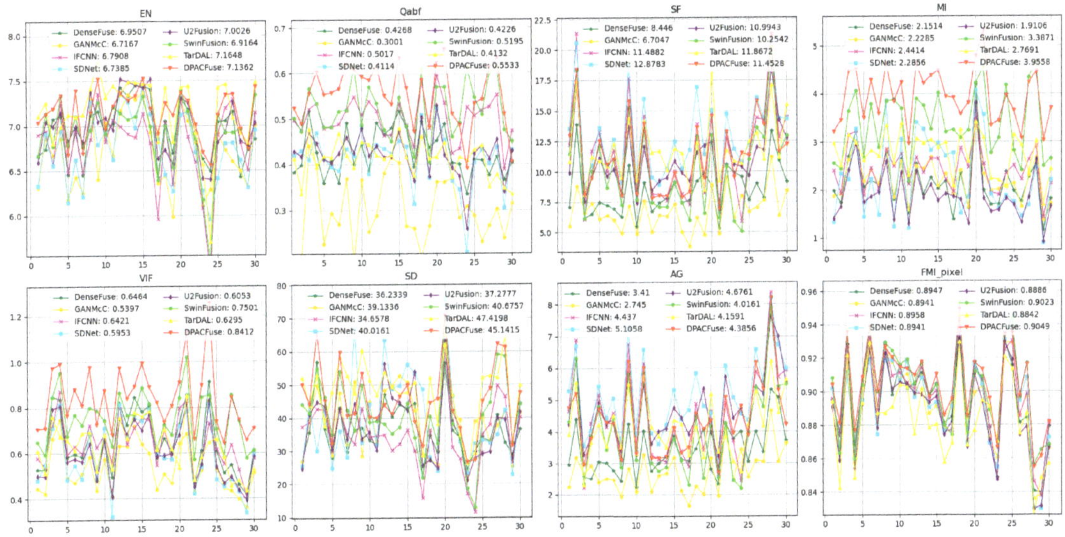

Figure 8. Quantitative comparison of eight methods on the TNO test set. The x-axis represents image pairs and the y-axis represents the values of the metric. The point (x, y) in the image represents the measurement y for the xth pair of images. The average value is shown in the legend.

4.4. Ablation Study

We performed ablation experiments to validate the efficacy of various modules and utilized EN, Qabf, SF, MI, AG, and FMI_pixel for quantitative assessment. In addition, We selected two images for qualitative analysis, one of which was 00537D from MSRS and the other was 00390 from M3FD.

Quantitative analysis. We conducted quantitative experiments on the MSRS test set and summarized the results in Table 1. The values M1 and M2 represent changing ACmix to pure self-attention and convolution, respectively. The data in the table clearly show that the removal of the ACmix led to a decrease in all the indicators. The metrics, MI, SF, and FMI_pixel, showed the most significant declines, indicating a deterioration in the network's ability to integrate complementary information between modalities. In addition, M3 and M4 denote the removal of HCGD in CMFE and the complete removal of CMFE, respectively. It can be seen that Qabf and AG experienced a large decrease when only HCGD was removed, which illustrated the effectiveness of HCGD in extracting high-frequency information. However, when the CMFE was removed, almost all the indicators significantly decreased, indicating that the performance of the network degraded a lot when there was no interaction between cross-modalities.

Table 1. Quantitative results of six indices under ablation experiments. In the evaluation results, the best-performing method is highlighted in red. The second-best result is represented in blue.

	M1	M2	M3	M4	Our
EN	6.6125	6.6146	6.6437	6.6017	6.6571
Qabf	0.6912	0.6820	0.6635	0.6507	0.7005
SF	11.2660	11.1202	10.9875	11.0457	11.3704
MI	4.5540	4.6525	4.4937	4.2439	4.7669
AG	3.6181	3.5809	3.3374	3.5419	3.7201
FMI_pixel	0.9289	0.9291	0.9277	0.9262	0.9314

Qualitative analysis. As observed in Figure 9, it is evident that the Attention was highly sensitive to whitening, leading to an overall increase in brightness and a loss of fine texture details. Since HCGD is one of the components of CMFE, removing both of them resulted in a significant loss of edges in the fused image. Moreover, removing the entire CMFE module had an even more significant impact, affecting not only the edges but also compromising the background and other critical information.

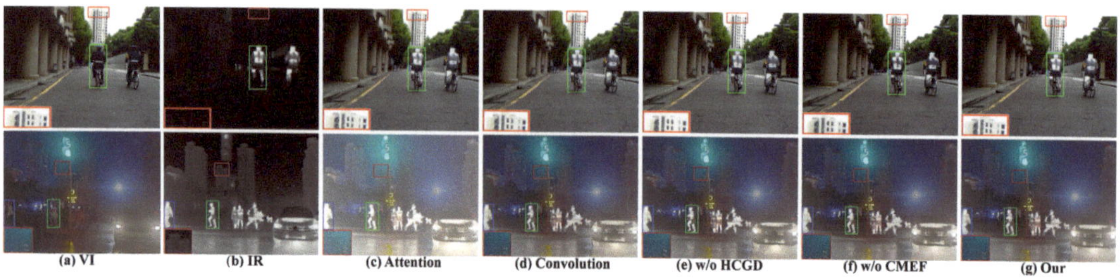

Figure 9. The results of ablation experiments.

In summary, the results in Figure 9 and Table 1 indicate the effectiveness and rationality of our designed modules, as well as of the overall network design.

4.5. Downstream IVF Applications

In this section, we applied fused images to object detection and semantic segmentation, and explored the benefits of IVF for downstream tasks.

4.5.1. Object Detection Performance

We employed the pre-trained YOLOv8 [52] detector to detect different images. We randomly selected a test set consisting of 160 images, with 80 images from the MSRS dataset and the remaining 80 images randomly selected from the M3FD dataset [12]. These 160 images contained a variety of scenes in the city, and we marked the most common objects among them, namely people and cars, as the objects to detect.

We assessed the detection performance of various methods using the mean average precision (mAP) metric. The results, indicating the mAP values at different IoU thresholds, are presented in Table 2. In addition, we calculated different mAPs, as in [53]. The prominent thermal target structure of the IR image helps the detector detect the human body, and the VI image can provide rich vehicle semantic information, so the detector can better realize the detection of vehicles. By fusing the two modalities of IR and VI images, performance in detecting both people and vehicles is enhanced. However, from the results, many algorithms tended to weaken the strength of salient objects, such as SwinFusion and SDNet, so their performance in detecting people was much lower than the recognition of source images. Taken together, our method was the best for person and car detection under almost all IoU thresholds.

At the same time, we provide the detection results for visual display. As shown in Figure 10, in the scene in the 00479D image, due to insufficient illumination in the VI image, the DenseFuse, IFCNN, and SwinFusion methods could not detect the person in the image, and a similar situation also occurred in the scene in 01348N. In the scene in 01348N, although most of the methods successfully detected people and cars in the scene, the confidence levels were very different. In Figure 11, the situation is similar to that in Figure 10. Most methods could not identify people or cars due to the influence of illumination or distance factors. Only our method completely recognized people and cars in the scene and maintained a high level of confidence. This fully shows that the images generated by DPACFuse can provide rich semantic information for the object detector.

Table 2. Object detection performance (mAP) of source images and the fused images of different fusion methods. In the evaluation results, the best performance method is highlighted in red. The second-best result is represented in blue.

	AP@0.5			AP@0.7			AP@0.9		
	Person	Car	All	Person	Car	All	Person	Car	All
IR	0.7891	0.5491	0.6691	0.7394	0.4787	0.6091	0.1961	0.1946	0.1953
VI	0.5478	0.7660	0.6569	0.3873	0.7108	0.5491	0.0359	0.3462	0.1911
DenseFuse [18]	0.7731	0.8079	0.7905	0.7296	0.7713	0.7505	0.1578	0.4583	0.3081
IFCNN [22]	0.7862	0.7768	0.7815	0.7316	0.7246	0.7281	0.1620	0.4118	0.2869
U2Fusion [31]	0.7823	0.7950	0.7937	0.7347	0.7724	0.7536	0.1599	0.4053	0.2826
SDNet [16]	0.7523	0.7649	0.7586	0.6519	0.7315	0.6917	0.1043	0.3826	0.2434
GANMcC [51]	0.7657	0.8132	0.7895	0.7250	0.7658	0.7454	0.1834	0.4415	0.3129
SwinFusion [13]	0.7699	0.7980	0.7840	0.6969	0.7499	0.7234	0.1183	0.3867	0.2525
TarDAL [12]	0.7897	0.7746	0.7822	0.7083	0.7246	0.7165	0.1646	0.4070	0.2858
DPACFuse	0.8034	0.8210	0.8122	0.7462	0.7753	0.7607	0.1862	0.4312	0.3087

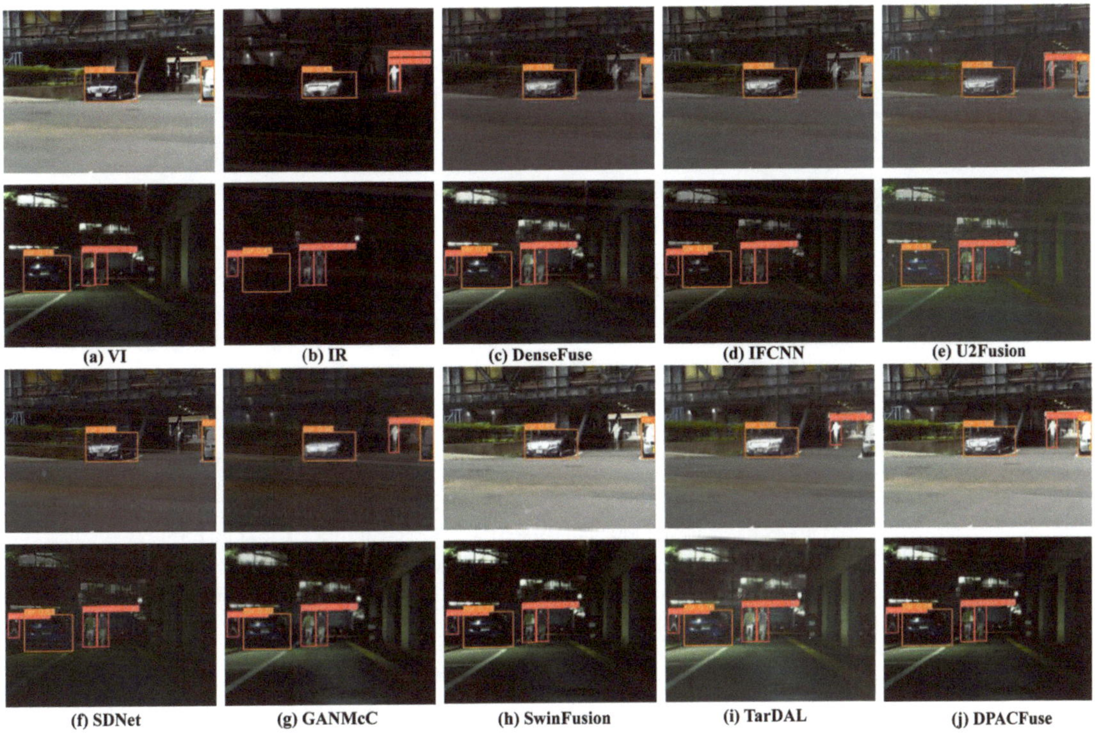

Figure 10. Object detection results on the MSRS dataset. The results are provided for two scenes from 00479D (**top**) and 01348N (**bottom**), respectively.

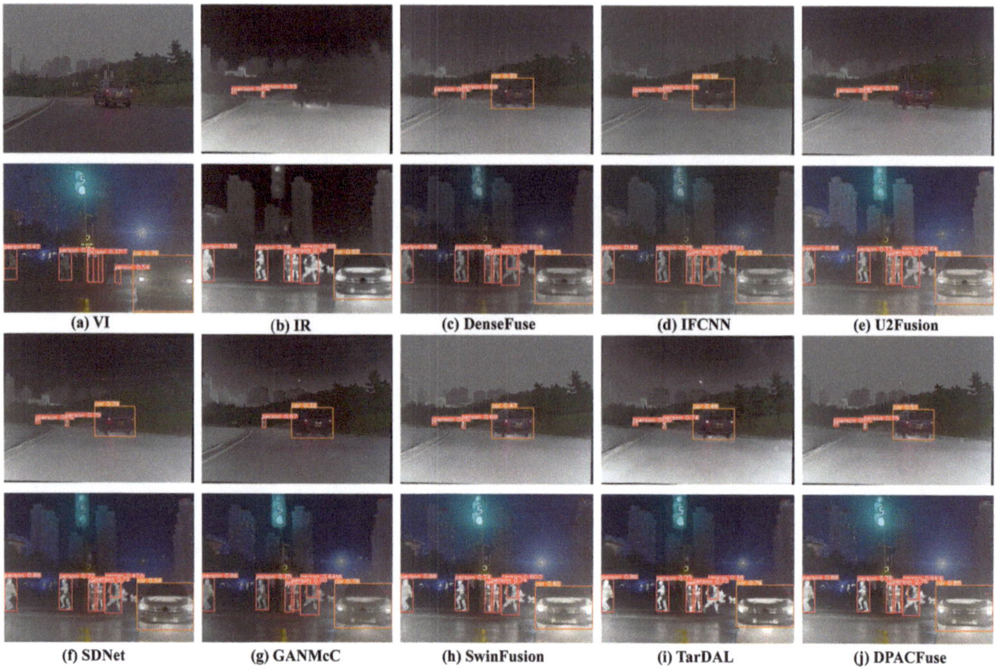

Figure 11. Object detection results on the M3FD dataset. The results are provided for two scenes from 01136 (**top**) and 00390 (**bottom**), respectively.

4.5.2. Semantic Segmentation Performance

We performed semantic segmentation on the MSRS dataset. Specifically, we utilized source images and different fused images to train semantic segmentation networks [54], respectively. For more details on the semantic segmentation network, please refer to [21]. At the same time, we evaluated model effectiveness using Intersection-over-Union (IoU). The segmentation of each object is shown in Table 3. The results clearly demonstrate that DPACFuse achieved the highest performance in all categories of IoU. This outcome strongly indicates that DPACFuse effectively integrates information from IR and VI images, thereby improving the model's ability in boundary perception, which leads to more accurate segmentation results.

Table 3. mIoU(%) values for segmentation semantics for different images on the MSRS dataset. In the evaluation results, the best performance method is highlighted in red. The second-best result is represented in blue.

Method	Background	Car	Person	Bike	Curve	Car Stop	Cuardrail	Color Tone	Bump	mIoU
IR	97.96	85.69	71.27	65.46	52.52	54.48	27.59	54.92	60.11	63.33
VI	97.69	83.07	54.67	66.27	51.97	54.94	59.69	50.35	66.18	64.98
DenseFuse [18]	98.36	89.48	73.47	69.84	57.76	63.56	65.07	62.34	66.00	71.76
IFCNN [22]	98.38	89.54	72.25	70.15	56.85	64.08	54.43	63.35	71.92	71.22
U2Fusion [31]	98.27	88.08	73.42	69.39	57.85	62.76	53.03	59.7	69.75	69.75
SDNet [16]	98.35	89.39	74.31	69.31	57.56	61.63	49.53	60.76	71.46	70.26
GANMcC [51]	98.34	88.85	73.68	69.67	56.75	65.17	57.06	61.50	71.72	71.41
SwinFusion [13]	98.25	88.08	70.74	68.66	74.31	61.70	67.34	64.06	67.86	71.67
TarDAL [12]	98.3	89.11	72.67	68.96	57.00	62.34	52.43	60.79	61.41	69.22
DPACFuse	98.6	90.38	74.58	71.94	65.39	74.44	84.66	66.03	77.18	78.13

Furthermore, we also provide the segmentation results for visual presentation. As observed in Figure 12, the IR image exhibited good segmentation performance for persons, but it performed poorly in segmenting other objects, such as color cones and curves. In addition, insufficient illumination negatively affected the segmentation performance of the VI images. From the scenes in the two images in the figure, it is evident that our DPACFuse had an excellent effect on both the segmentation of people and the segmentation of other objects, which shows that the images generated by DPACFuse can better promote semantic segmentation.

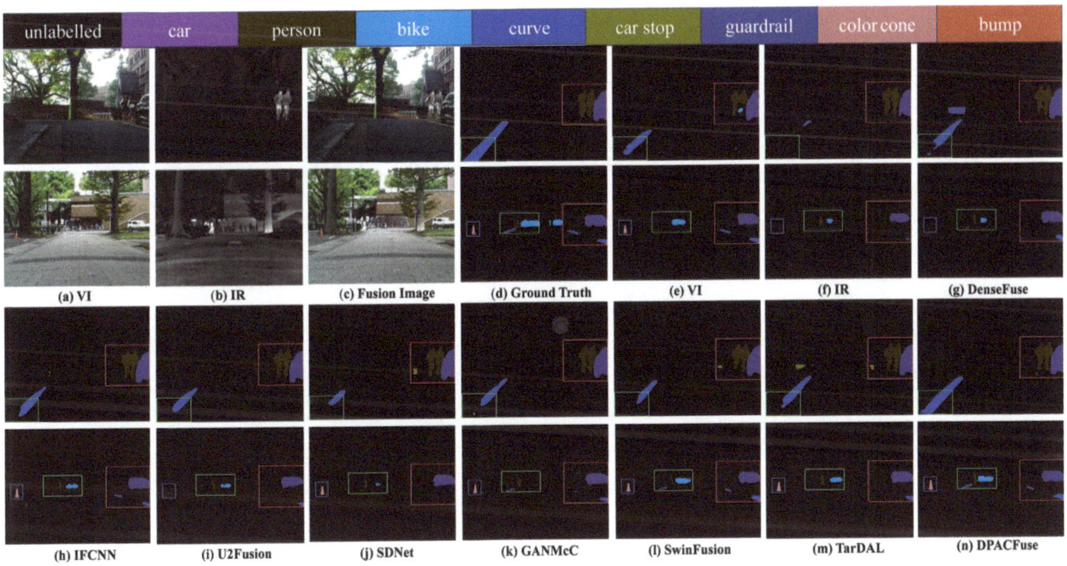

Figure 12. Visualization results of semantic segmentation on the MSRS dataset. The two scenes from top to bottom are from: 00055D and 00504N.

5. Conclusions

In this paper, we propose a dual-branch progressive fusion framework, named DPACFuse, to be used for infrared and visible image fusion. Firstly, the Cross-Modality Feature Extraction we designed extracts inter-modality shared features as well as suppresses the loss of high-frequency information. Second, with the help of the ACmix module, our architecture more fully extracts the information in the source images for fusion. Finally, extensive experiments on three publicly available datasets showed that our DPACFuse outperforms all current state-of-the-art methods. In addition, in order to evaluate our approach more comprehensively, we also conducted experiments in two downstream tasks, object detection and semantic segmentation, and the results of the experiments further demonstrated the effectiveness and superiority of our approach.

Author Contributions: Investigation, X.W. and D.H.; Methodology, H.Z.; Visualization, H.Z. and H.W.; Writing—review & editing, H.Z., Z.L. and X.P. All authors have read and agreed to the published version of the manuscript.

Funding: This research was supported in part by the National Natural Science Foundation of China (Grant No. 62002082), Guangxi Natural Science Foundation (Grant No. 2020GXNSFBA238014), and the university student innovation training program project (No. 202210595023).

Institutional Review Board Statement: Not applicable.

Informed Consent Statement: Not applicable.

Data Availability Statement: The authors confirm that the data supporting the findings of this study are available within the article.

Conflicts of Interest: The authors declare no conflict of interest.

References

1. Tang, L.; Zhang, H.; Xu, H.; Ma, J. Deep learning-based image fusion: A survey. *J. Image Graph.* **2023**, *28*, 3–36.
2. Wang, J.; Liu, A.; Yin, Z.; Liu, S.; Tang, S.; Liu, X. Dual Attention Suppression Attack: Generate Adversarial Camouflage in Physical World. *arXiv* **2021**, arXiv:2103.01050.
3. Liu, A.; Liu, X.; Yu, H.; Zhang, C.; Liu, Q.; Tao, D. Training Robust Deep Neural Networks via Adversarial Noise Propagation. *IEEE Trans. Image Process.* **2021**, *30*, 5769–5781. [CrossRef]
4. Zeng, Y.; Zhang, D.; Wang, C.; Miao, Z.; Liu, T.; Zhan, X.; Hao, D.; Ma, C. LIFT: Learning 4D LiDAR Image Fusion Transformer for 3D Object Detection. In Proceedings of the IEEE/CVF Conference on Computer Vision and Pattern Recognition (CVPR), New Orleans, LA, USA, 18–24 June 2022; pp. 17172–17181.
5. Pan, X.; Cheng, J.; Hou, F.; Lan, R.; Lu, C.; Li, L.; Feng, Z.; Wang, H.; Liang, C.; Liu, Z.; et al. SMILE: Cost-sensitive multi-task learning for nuclear segmentation and classification with imbalanced annotations. *Med. Image Anal.* **2023**, *88*, 102867. [CrossRef]
6. Jin, C.; Luo, C.; Yan, M.; Zhao, G.; Zhang, G.; Zhang, S. Weakening the Dominant Role of Text: CMOSI Dataset and Multimodal Semantic Enhancement Network. *IEEE Trans. Neural Netw. Learn. Syst.* **2023**, 1–15. [CrossRef]
7. Qin, H.; Ding, Y.; Zhang, M.; Yan, Q.; Liu, A.; Dang, Q.; Liu, Z.; Liu, X. BiBERT: Accurate Fully Binarized BERT. *arXiv* **2022**, arXiv:2203.06390.
8. Qin, H.; Zhang, X.; Gong, R.; Ding, Y.; Xu, Y.; Liu, X. Distribution-sensitive Information Retention for Accurate Binary Neural Network. *arXiv* **2022**, arXiv:2109.12338.
9. Yan, M.; Lou, X.; Chan, C.A.; Wang, Y.; Jiang, W. A semantic and emotion-based dual latent variable generation model for a dialogue system. *Caai Trans. Intell. Technol.* **2023**, *8*, 319–330. [CrossRef]
10. Wang, Z.; Feng, J.; Zhang, Y. Pedestrian detection in infrared image based on depth transfer learning. *Multimed. Tools Appl.* **2022**, *81*, 39655–39674. [CrossRef]
11. Zhang, J.; Liu, C.; Wang, B.; Chen, C.; He, J.; Zhou, Y.; Li, J. An infrared pedestrian detection method based on segmentation and domain adaptation learning. *Comput. Electr. Eng.* **2022**, *99*, 107781. [CrossRef]
12. Liu, J.; Fan, X.; Huang, Z.; Wu, G.; Liu, R.; Zhong, W.; Luo, Z. Target-aware Dual Adversarial Learning and a Multi-scenario Multi-Modality Benchmark to Fuse Infrared and Visible for Object. *arXiv* **2022**, arXiv:2203.16220.
13. Ma, J.; Tang, L.; Fan, F.; Huang, J.; Mei, X.; Ma, Y. SwinFusion: Cross-domain Long-range Learning for General Image Fusion via Swin Transformer. *IEEE/CAA J. Autom. Sin.* **2022**, *9*, 1200–1217. [CrossRef]
14. Tang, L.; Deng, Y.; Ma, Y.; Huang, J.; Ma, J. SuperFusion: A Versatile Image Registration and Fusion Network with Semantic Awareness. *IEEE/CAA J. Autom. Sin.* **2022**, *9*, 2121–2137. [CrossRef]
15. Wang, Z.; Chen, Y.; Shao, W.; Li, H.; Zhang, L. SwinFuse: A Residual Swin Transformer Fusion Network for Infrared and Visible Images. *arXiv* **2022**, arXiv:2204.11436.
16. Zhang, H.; Ma, J. SDNet: A Versatile Squeeze-and-Decomposition Network for Real-Time Image Fusion. *Int. J. Comput. Vis.* **2021**, *129*, 2761–2785. [CrossRef]
17. Rao, Y.; Wu, D.; Han, M.; Wang, T.; Yang, Y.; Lei, T.; Zhou, C.; Bai, H.; Xing, L. AT-GAN: A generative adversarial network with attention and transition for infrared and visible image fusion. *Inf. Fusion* **2023**, *92*, 336–349. [CrossRef]
18. Li, H.; Wu, X.J. DenseFuse: A Fusion Approach to Infrared and Visible Images. *IEEE Trans. Image Process.* **2019**, *28*, 2614–2623. [CrossRef] [PubMed]
19. Li, H.; Wu, X.J.; Kittler, J. RFN-Nest: An end-to-end residual fusion network for infrared and visible images. *Inf. Fusion* **2021**, *73*, 72–86. [CrossRef]
20. Li, H.; Wu, X.J.; Durrani, T. NestFuse: An Infrared and Visible Image Fusion Architecture Based on Nest Connection and Spatial/Channel Attention Models. *IEEE Trans. Instrum. Meas.* **2020**, *69*, 9645–9656. [CrossRef]
21. Tang, L.; Yuan, J.; Ma, J. Image fusion in the loop of high-level vision tasks: A semantic-aware real-time infrared and visible image fusion network. *Inf. Fusion* **2022**, *82*, 28–42. [CrossRef]
22. Zhang, Y.; Liu, Y.; Sun, P.; Yan, H.; Zhao, X.; Zhang, L. IFCNN: A general image fusion framework based on convolutional neural network. *Inf. Fusion* **2020**, *54*, 99–118. [CrossRef]
23. Ma, J.; Tang, L.; Xu, M.; Zhang, H.; Xiao, G. STDFusionNet: An Infrared and Visible Image Fusion Network Based on Salient Target Detection. *IEEE Trans. Instrum. Meas.* **2021**, *70*, 5009513. [CrossRef]
24. Zhao, Z.; Xu, S.; Zhang, J.; Liang, C.; Zhang, C.; Liu, J. Efficient and Model-Based Infrared and Visible Image Fusion via Algorithm Unrolling. *IEEE Trans. Circuits Syst. Video Technol.* **2022**, *32*, 1186–1196. [CrossRef]
25. Liang, J.; Cao, J.; Sun, G.; Zhang, K.; Gool, L.V.; Timofte, R. SwinIR: Image Restoration Using Swin Transformer. *arXiv* **2021**, arXiv:2108.10257.
26. Qu, L.; Liu, S.; Wang, M.; Song, Z. TransMEF: A Transformer-Based Multi-Exposure Image Fusion Framework using Self-Supervised Multi-Task Learning. *arXiv* **2021**, arXiv:2112.01030.
27. Pan, X.; Ge, C.; Lu, R.; Song, S.; Chen, G.; Huang, Z.; Huang, G. On the Integration of Self-Attention and Convolution. *arXiv* **2022**, arXiv:2111.14556.
28. Tang, L.; Yuan, J.; Zhang, H.; Jiang, X.; Ma, J. PIAFusion: A progressive infrared and visible image fusion network based on illumination aware. *Inf. Fusion* **2022**, *83–84*, 79–92. [CrossRef]
29. Wang, Z.; Wu, Y.; Wang, J.; Xu, J.; Shao, W. Res2Fusion: Infrared and Visible Image Fusion Based on Dense Res2net and Double Nonlocal Attention Models. *IEEE Trans. Instrum. Meas.* **2022**, *71*, 5005012. [CrossRef]

30. Zhang, H.; Xu, H.; Xiao, Y.; Guo, X.; Ma, J. Rethinking the Image Fusion: A Fast Unified Image Fusion Network based on Proportional Maintenance of Gradient and Intensity. *Proc. AAAI Conf. Artif. Intell.* **2020**, *34*, 12797–12804. [CrossRef]
31. Xu, H.; Ma, J.; Jiang, J.; Guo, X.; Ling, H. U2Fusion: A Unified Unsupervised Image Fusion Network. *IEEE Trans. Pattern Anal. Mach. Intell.* **2022**, *44*, 502–518. [CrossRef]
32. Ma, J.; Yu, W.; Liang, P.; Li, C.; Jiang, J. FusionGAN: A generative adversarial network for infrared and visible image fusion. *Inf. Fusion* **2019**, *48*, 11–26. [CrossRef]
33. Ma, J.; Xu, H.; Jiang, J.; Mei, X.; Zhang, X.P. DDcGAN: A Dual-Discriminator Conditional Generative Adversarial Network for Multi-Resolution Image Fusion. *IEEE Trans. Image Process.* **2020**, *29*, 4980–4995. [CrossRef] [PubMed]
34. Park, S.; Choi, D.H.; Kim, J.U.; Ro, Y.M. Robust thermal infrared pedestrian detection by associating visible pedestrian knowledge. In Proceedings of the ICASSP 2022—2022 IEEE International Conference on Acoustics, Speech and Signal Processing (ICASSP), Singapore, 22–27 May 2022; pp. 4468–4472.
35. Wu, X.; Hong, D.; Chanussot, J. UIU-Net: U-Net in U-Net for Infrared Small Object Detection. *IEEE Trans. Image Process.* **2023**, *32*, 364–376. [CrossRef]
36. Wang, A.; Li, W.; Wu, X.; Huang, Z.; Tao, R. Mpanet: Multi-Patch Attention for Infrared Small Target Object Detection. In Proceedings of the IGARSS 2022—2022 IEEE International Geoscience and Remote Sensing Symposium, Kuala Lumpur, Malaysia, 17–22 July 2022; pp. 3095–3098. [CrossRef]
37. Sun, Y.; Cao, B.; Zhu, P.; Hu, Q. DetFusion: A Detection-Driven Infrared and Visible Image Fusion Network. In Proceedings of the MM'22: 30th ACM International Conference on Multimedia, New York, NY, USA, 4–7 July 2022; pp. 4003–4011. [CrossRef]
38. Zhao, W.; Xie, S.; Zhao, F.; He, Y.; Lu, H. MetaFusion: Infrared and Visible Image Fusion via Meta-Feature Embedding From Object Detection. In Proceedings of the IEEE/CVF Conference on Computer Vision and Pattern Recognition, Vancouver, BC, Canada, 18–22 June 2023; pp. 13955–13965.
39. Wang, D.; Liu, J.; Liu, R.; Fan, X. An interactively reinforced paradigm for joint infrared-visible image fusion and saliency object detection. *Inf. Fusion* **2023**, *98*, 101828. [CrossRef]
40. Vaswani, A.; Shazeer, N.; Parmar, N.; Uszkoreit, J.; Jones, L.; Gomez, A.N.; Kaiser, L.; Polosukhin, I. Attention Is All You Need. *arXiv* **2017**, arXiv:1706.03762.
41. Zhu, X.; Su, W.; Lu, L.; Li, B.; Wang, X.; Dai, J. Deformable DETR: Deformable Transformers for End-to-End Object Detection. *arXiv* **2021**, arXiv:2010.04159.
42. Zhou, M.; Yan, K.; Huang, J.; Yang, Z.; Fu, X.; Zhao, F. Mutual Information-Driven Pan-Sharpening. In Proceedings of the IEEE/CVF Conference on Computer Vision and Pattern Recognition (CVPR), New Orleans, LA, USA, 18–24 June 2022; pp. 1798–1808.
43. Wang, W.; Xie, E.; Li, X.; Fan, D.P.; Song, K.; Liang, D.; Lu, T.; Luo, P.; Shao, L. Pyramid Vision Transformer: A Versatile Backbone for Dense Prediction Without Convolutions. In Proceedings of the IEEE/CVF International Conference on Computer Vision (ICCV), Montreal, BC, Canada, 11–17 October 2021; pp. 568–578.
44. Zhao, H.; Nie, R. DNDT: Infrared and Visible Image Fusion Via DenseNet and Dual-Transformer. In Proceedings of the 2021 International Conference on Information Technology and Biomedical Engineering (ICITBE), Nanchang, China, 24–26 December 2021; pp. 71–75. [CrossRef]
45. Liu, Z.; Lin, Y.; Cao, Y.; Hu, H.; Wei, Y.; Zhang, Z.; Lin, S.; Guo, B. Swin Transformer: Hierarchical Vision Transformer using Shifted Windows. *arXiv* **2021**, arXiv:2103.14030.
46. Rao, D.; Wu, X.; Xu, T. TGFuse: An Infrared and Visible Image Fusion Approach Based on Transformer and Generative Adversarial Network. *arXiv* **2022**, arXiv:2201.10147.
47. Li, J.; Zhu, J.; Li, C.; Chen, X.; Yang, B. CGTF: Convolution-Guided Transformer for Infrared and Visible Image Fusion. *IEEE Trans. Instrum. Meas.* **2022**, *71*, 5012314. [CrossRef]
48. Hu, J.; Shen, L.; Sun, G. Squeeze-and-Excitation Networks. In Proceedings of the IEEE Conference on Computer Vision and Pattern Recognition (CVPR), Salt Lake City, UT, USA, 18–23 June 2018.
49. Toet, A. TNO Image Fusion Dataset. *Figshare Data*. 2014. Available online: https://figshare.com/articles/dataset/TNO_Image_Fusion_Dataset/1008029 (accessed on 1 January 2023).
50. Ma, J.; Ma, Y.; Li, C. Infrared and visible image fusion methods and applications: A survey. *Inf. Fusion* **2019**, *45*, 153–178. [CrossRef]
51. Ma, J.; Zhang, H.; Shao, Z.; Liang, P.; Xu, H. GANMcC: A Generative Adversarial Network with Multiclassification Constraints for Infrared and Visible Image Fusion. *IEEE Trans. Instrum. Meas.* **2021**, *70*, 5005014. [CrossRef]
52. Reis, D.; Kupec, J.; Hong, J.; Daoudi, A. Real-Time Flying Object Detection with YOLOv8. *arXiv* **2023**, arXiv:2305.09972.
53. Padilla, R.; Passos, W.L.; Dias, T.L.B.; Netto, S.L.; da Silva, E.A.B. A Comparative Analysis of Object Detection Metrics with a Companion Open-Source Toolkit. *Electronics* **2021**, *10*, 279. [CrossRef]
54. Peng, C.; Tian, T.; Chen, C.; Guo, X.; Ma, J. Bilateral attention decoder: A lightweight decoder for real-time semantic segmentation. *Neural Netw.* **2021**, *137*, 188–199. [CrossRef] [PubMed]

sensors

MDPI

Article

An Asymmetric Encryption-Based Key Distribution Method for Wireless Sensor Networks

Yuan Cheng *, Yanan Liu, Zheng Zhang and Yanxiu Li

School of Network Security, Jinling Institute of Technology, Nanjing 211100, China
* Correspondence: chengyuan2018@jit.edu.cn; Tel.: +86-181-6809-2939

Abstract: Wireless sensor networks are usually applied in hostile areas where nodes can easily be monitored and captured by an adversary. Designing a key distribution scheme with high security and reliability, low hardware requirements, and moderate communication load is crucial for wireless sensor networks. To address the above objectives, we propose a new key distribution scheme based on an ECC asymmetric encryption algorithm. The two-way authentication mechanism in the proposed scheme not only prevents illegal nodes from accessing the network, but also prevents fake base stations from communicating with the nodes. The complete key distribution and key update methods ensure the security of session keys in both static and dynamic environments. The new key distribution scheme provides a significant performance improvement compared to the classical key distribution schemes for wireless sensor networks without sacrificing reliability. Simulation results show that the proposed new scheme reduces the communication load and key storage capacity, has significant advantages in terms of secure connectivity and attack resistance, and is fully applicable to wireless sensor networks.

Keywords: WSN; security; key distribution; cryptography

Citation: Cheng, Y.; Liu, Y.; Zhang, Z.; Li, Y. An Asymmetric Encryption-Based Key Distribution Method for Wireless Sensor Networks. *Sensors* **2023**, *23*, 6460. https://doi.org/10.3390/s23146460

Academic Editors: Ming Yan, Chunguo Li and Chien Aun Chan

Received: 21 June 2023
Revised: 11 July 2023
Accepted: 13 July 2023
Published: 17 July 2023

1. Introduction

Wireless sensor networks (WSNs) have been proven to be suitable for large numbers of applications, ranging from industry and security domains, such as environment monitoring, fire detection and precision agriculture, to personal use, like health supervision. WSNs are composed of a large number of sensors that work independently of each other. These sensors transmit routing information to each other and forward collected application data [1,2]. The major weakness of wireless sensor networks lies in the limitations of resources, including memory, battery capacity, data processing, and communication capabilities. Sensors and wireless channels are vulnerable to eavesdropping, physical interception, malicious attacks, message tampering, identity impersonation, and side channel attacks [3–5], and the presence of important and sensitive information in the network increases the importance of security issues. Therefore, one of the focuses of wireless sensor network research is understanding how to provide high confidentiality for the transmitted application data and control messages to prevent various illegal attacks [6–9]. At present, it is generally believed that encryption is a key technology that can provide confidentiality between the cloud and the end [10–12], which can also be used in WSNs' data exchange.

Over the years, many researchers have proposed schemes to enhance the security of wireless sensor networks. The (p, q)-Lucas polynomial-based key management scheme for WSN was proposed by Gautam et al. [13]. Their scheme outperforms other polynomials in terms of the number of keys used and efficiency. Kumar proposed a dynamic key management scheme for the clustered sensor network that supports the addition of new nodes into the network [14]. The proposed scheme has shown low energy consumption and good resiliency against node capture attacks. Moghadam et al. [15] proposed an ECDH (elliptic-curve Diffie–Hellman)-based authentication and key agreement protocol for

WSN infrastructure. The proposed protocol supports the dynamic node addition in WSN environments and uses a strong ECDH technique to generate unique symmetric and session keys for each session. The authors of [16] proposed a trust-based multipath routing protocol called TBSMR, which improved the QoS and overall performance of MANETs in cellular networks through congestion control, packet loss reduction, malicious node detection, and secure data transmission. These proposals differ from the scheme proposed in this paper as TBSMR achieves power savings from the perspective of optimized routing protocols. In MANET-based medical systems, to achieve secure communication, a logic graph-based key generation scheme hybrid and encryption scheme is proposed by Sirajuddin [17], which provides high security for MANET medical networks, as well as less computational power and shorter encryption time.

In 2018, Mishra et al. proposed an authentication scheme for multimedia communications that was designed for an IoT environment base on WSNs [18]. Wu et al. [19] designed a lightweight authentication scheme for WSNs. It addressed the common security requirements and user untraceability issues. To ensure confidentiality and security in IOT, a biometric-based authentication and key agreement protocol are proposed for wireless sensor networks [20].

In recent years, researchers have produced several more viable authentication protocols and key agreements in the field of wireless sensor network security. Naresh et al. [21] proposed a lightweight multiple shared key agreement based on the hyper-elliptic-curve Diffie–Hellman method. The protocol decreases keys exchange overhead and increases the safety of the keys. In response to the security weaknesses of the scheme in [22], Shin, S. proposed a lightweight authentication based on the three-factor technique and key agreement protocol for WSN [23]. The proposed scheme addressed several security requirements and used XOR and hash functions. A lightweight password-authenticated key exchange scheme was proposed by González et al. for heterogeneous wireless sensor networks [24]. Three 3-PAKE protocols were analyzed, and the vulnerabilities of the protocols were proposed. The new protocol provided good security features with high flexibility and efficiency.

In this paper, we present a security key management scheme for cluster-based wireless sensor networks. In our scheme, session keys can be safely distributed and updated among all sensors with the help of the base station. Both static and dynamic scenarios are studied over the hierarchical networks. In particular, in our proposed scheme, the efficient encrypting algorithm makes it possible to adopt asymmetric encryption to guarantee authentication and confidentiality during data transmission.

The rest of our paper is organized as follows: Section 2 introduces security features and design constraints in WSNs; Section 3 exhibits the details of the security key management scheme; Section 4 evaluates the performance of the proposed security protocols; and Section 5 presents the conclusion and perspectives.

2. Design Constraints and Security Issues in WSNs

2.1. Physical Characteristics and Constraints

Sensors in most of wireless sensor networks are greatly limited in terms of device size, battery capacity, computing capacity, communication capacity, and storage capacity, which make the development of applications a challenge. A feasible and efficient security protocol should minimize the number of operations needed for calculation, communication, and storage. Therefore, the following characteristics of a WSN should be taken into consideration during protocol design [25–28]:

- Limited battery capacity—Sensor networks are usually deployed in outdoor environments. Due to size limitation, each sensor is usually equipped with a small battery. As a result, a sensor is unable to calculate and communicate when the battery runs out.
- Limited memory—the cache size of a sensor is usually measured in tens of megabytes, which puts forward higher requirements for the length and number of keys stored.
- Limited bandwidth—due to power limitation, most sensors use narrowband signal transmission, and the transmission rate generally does not exceed 10 KB/s.

- Limited calculation power—In order to reduce the power consumption of CPU, most sensor nodes only use 8-bit 4-megahertz microcontrollers.
- Good scalability—Wireless sensor networks must allow new legal nodes to join the existing network at any time. At the same time, the failure of any node will not affect the normal operation of the network.
- Variability in network topology—Since sensors are often installed on mobile devices, the topology of wireless sensor networks often change. Thus, network stability and nodes connectivity should be ensured in all protocol designs.
- Environment—Some wireless sensor networks are expected to be used for remote control and reconnaissance, and they are deployed in insecure and unstable environments, which makes them subject to many attacks, such as spoofing attacks, physical damage, and any other mechanical failures associated with environmental factors.

2.2. Security Issues in WSNs

In addition to the above characteristics of wireless sensor networks, security is also an important part of the Internet of things. Since WSNs use a wireless medium for data transmission, sensors are more vulnerable to various malicious attacks based on wireless channels. The typical malicious attacks in WSNs include eavesdropping, data modification, sink hole, spoofing attacks, denial of service attacks, sybil attacks, and node capture. For example, in node capture, the attacker accesses the hardware and software of one or more sensors through the network [29]. After successful intrusion into the sensor, the attacker steals all cryptographic keys and algorithms. Thus, it is possible for the attackers to eavesdrop and tamper with messages, as well as pretend to be legal terminals to forward data to hackers.

In recent years, a lot of research work has focused on security problems in WSNs. An asymmetric key pre-distribution scheme called AP was first proposed for hierarchical sensor networks in [30]. The famous "probabilistic" schemes had low computational complexity and communication loads. However, this scheme cannot guarantee accurate sharing of pairwise keys between any two sensors. Based on the Blom matrix, a key management scheme is proposed by Boujelben in [31] to improve the resilience against node capture. However, complex matrix operation leads to that high resource consumption by ordinary sensors. Lee presented a key renewal approach for authentication based on modular exponentiation in clustered WSNs [32]. Although this scheme improved the connectivity of the network, public-key encryption brought about a large amount of computation. Tian presented a blockchain-based trusted key management approach [33], which realized key management in WSNs through a secure cluster formation algorithm and a node mobility algorithm. In the literature [34], a novel key management model for hierarchical sensor networks based on public key infrastructure (PKI) was proposed. However, the key distribution issues in case of movement were not investigated.

2.3. Aasymmetric Cryptography in WSNs

Asymmetric encryption uses key pairs to encrypt and decrypt data for both sides of communication. Any message encrypted with the public key can only be decrypted by that containing the private key. The private key is secretly held by its holder, and the public key can be obtained by the required communication entity through a public channel. Asymmetric cryptography can provide confidentiality, integrity, and authentication for different kinds of networks. Although information encryption based on asymmetric key has been proved to be applicable to sensor networks, its application is still limited by its complex computation. Furthermore, taking the actual sensor chip as an example, the time taken for asymmetric encryption is still in the order of seconds, which may not be suitable for those applications with strict real-time performance.

Fortunately, in recent years, the new cryptographic algorithms have shown great energy efficiency and reached the same security level as traditional algorithms. For example, the elliptic-curve cryptography (ECC) [35] method is the representative version of those algorithms. ECC is a cryptographic regime built on the discrete logarithm problem of elliptic curves. Using point G on an elliptic curve and integer k, it is easy to find $K = kG$. Conversely, using the points K and G on an elliptic curve, finding the integer k is a difficult task. The main advantage of ECC is that it uses smaller keys and provides a considerably higher level of security. The 164-bit key in the ECC algorithm can provide a level of security equivalent to the strength of secrecy provided by the 1024-bit key in the RSA algorithm. The ECC algorithm is less computationally intensive, is faster to process, and takes up less storage space and transmission bandwidth. Therefore, Bitcoin has also chosen ECC as its encryption algorithm.

In [36], the author proposed a new SUA-WSN scheme based on elliptic-curve cryptography (ECC) and proved that it achieves user anonymity, as well as AKE security, in the extended model. Gulen et al. implemented ECC on the MSP430 microcontroller, which is a widely used microcontroller in WSNs, using Edwards curves for point arithmetic and the number theoretic transform for the underlying finite-field multiplication and squaring operations [37]. Gulen's research shows better timing values and can be applied to ECC implementations.

From the perspective of energy consumption and computational complexity, ECC has promising uses for data encryption in WSNs. It provides comparative security with a smaller key, which also reduces the energy of computation and communication in WSNs. Based on this method, a new security key management scheme and an authentication approach are proposed in Section 3.

3. The Key Management Scheme for Cluster-Based WSNs

In this section, a security key management scheme for wireless sensor networks based on public-key cryptography is presented. To avoid long-term attacks through which attackers can analyze the encrypted traffic over the network for a long period of time, a key update approach is specifically designed.

3.1. Network Model and Assumptions

At present, wireless sensor networks commonly used in the industry mainly include two kinds of architectures, namely hierarchical structure and flat structure. A hierarchical architecture is usually used for large-scale WSNs due to its good scalability. A clustered hierarchical network is composed of base stations (BS), a large number of sensor nodes, and a small number of cluster heads (CH). BS is not limited by resources. The base station is responsible for managing all nodes of the network and receiving the service data collected via the sensor nodes. It is assumed that the cluster head has a higher configuration than the sensors, including battery capacity, memory size, communication, and computing capacity. Like the gateway, the cluster head assists in data transmission between the sensors and the base station. In the hierarchical architecture, sensors are divided into non-overlapping clusters, which collect data from the surrounding environment and send the original data to the base station. In this article, we focus on hierarchical architecture of WSNs.

In our scheme, asymmetric encryption is used to realize the authentication between the base station, the CHs, and the sensor nodes. The public key is pre-loaded into each sensor before network deployment. With the public-key system, the proposed scheme not only realizes end-to-end identity authentication, but also provides security for subsequent key distribution processes.

In our hierarchical WSN model, we make the following few assumptions:

- The base station has more energy power for calculations and communications than sensors.
- The base station owns a pair of keys (a public key and a private key).
- The network is divided into several cluster regions. In each cluster, there is only one cluster head node, and its location remains unchanged. Each cluster head can be recognized as the gateway of its cluster.

- In terms of security and ease of management, each cluster generates different session keys for dialogs between sensor nodes and cluster heads.
- Both asymmetric and symmetric cryptography are used for each sensor. The former method provides mutual authentication and key distribution, and the latter method preserves the confidentiality of traffic transmitted.
- As an optional technology in our scheme, MAC (message authentication code) provides data integrity.
- The public key is pre-loaded into each sensor and the cluster head via an off-line dealer.
- Each sensor can store at least one public key and several session keys in its memory.
- Each sensor can randomly move among different clusters at a low speed.

3.2. Network Initializtion and Definitions

In the network, there are n sensors, which are denoted as $S_{0,\ldots,n-1}$, and m cluster heads (CH), which are denoted as $CH_{0,\ldots,m-1}$. Each sensor has a unique identification code *ID_si*, which has a length of 2 bytes stored in the chip. After the initialization of the network is completed, all nodes automatically run the cluster formation algorithm (this algorithm is not discussed in this paper; for more information, please refer to [38]), which results in m clusters being formed randomly by all nodes. There is only one CH and n/m sensor in each cluster. Figure 1 shows a typical network of three clusters. Each cluster contains one CH and three sensors.

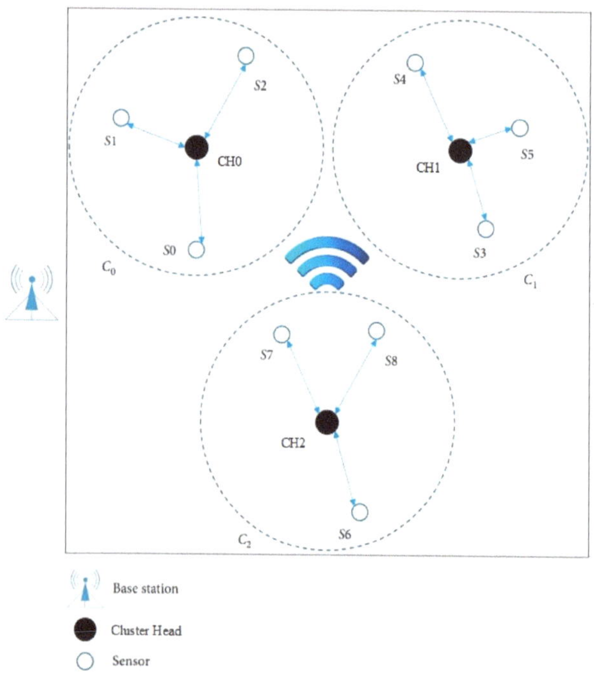

Figure 1. The network topology.

After network deployment, each CH runs a cluster forming process, and sensors are divided into clusters with no cross coverage. After a period of operation, some sensor may move into another cluster's region. In this situation, the subsequent key distribution and update process will be performed via the CH of the present cluster. In the following section, we will describe the scheme in regard to two aspects: static sensors and mobile sensors.

The following definitions will be used in our scheme and analysis:

SK_i denotes the symmetric session key with a length of 16 bytes shared by the base station and sensors located in DG_i.

PUK denotes the public key of the BS, and *PVK* denotes the corresponding private key. *PUK* can be obtained through public key infrastructure (PKI).

The function $E(x,y)$ denotes encryption (symmetric or asymmetric) operation, parameter x denotes encryption key, and parameter y denotes the plain message that needs to be encrypted. The function $D(x,y)$ denotes decryption operation.

ID_{CHi} denotes the identity code of the cluster with a length of 1 byte, and it can be acquired using the CH of that cluster. It is stored in the chip of each CH, and a tamper proof mechanism is used.

ID_si denotes the identity code of sensor S_i up to a maximum length of 2 bytes. It is stored in the chip of each sensor, and the tamper proof mechanism is used.

3.3. Static Sensors Subscheme for Hierarchical WSNs

3.3.1. Mutual Authentication and Key Distribution Process

In our clustered architecture network, the CH plays an important role in the process of key management. The key problem here is understanding how to distribute the key among the sensor nodes under many restrictions. We assume that all sensors are static and present the operations of handshake, key distribution, authentication, and key update. The handshake is destined to establish a symmetric key shared by sensors and BS. The operation of handshake includes three steps:

1. **Generation of the SK_i:** The CH_i generates a random symmetric key SK_i and a challenge R. Next, the CH_i encrypts SK_i, R, and ID_{CHi} with *PUK*, and we find

   $$\text{Cipher1} = E(PUK, SK_i||R||ID\_{CHi}||timestamp) \quad (1)$$

   The 2-byte timestamp is used to resist replay attacks. CH_i sends Cipher1 to the base station using traditional routing. Here, the *PUK* is used for authentication and preserving the confidentiality of the session key SK_i.
2. **Establishment of SK_i:** After receiving and decrypting the message, the base station finds SK_i, and R uses its *PVK* and builds a global table of all session keys of different clusters. This table is used to identify the cluster and its cluster head on the network. Meanwhile, if ID_{Chi} can be found in the database of legal CHs, the identity of the CH_i can be authenticated using BS.
3. **Completion of the handshake:** The base station encrypts R with the established session key SK_i. and finds

   $$\text{Cipher2} = E(SK_i, R) \quad (2)$$

Next, the base station sends Cipher2 to CH_i, and CH_i decrypts it. When the challenge R is correctly received, a session key is successfully established between BS and CH_i. Otherwise, CH_i will reinitiate the handshake. Considering the resource consumption caused by the computational complexity, the message authentication code (MAC) is not added to the key distribution process.

Through the above steps, the mutual authentication between the base station and CH_i is completed. After that step, each sensor in the cluster needs to achieve the session key SK_i generated using CH_i. Thus, sensor node S_i builds a message encrypted using the *PUK*, which is denoted as follows:

$$\text{Cipher3} = E(PUK, ID\_{CHi}||ID\_si||timestamp||SK\_si||R) \quad (3)$$

where *SK_si* is a symmetric key generated using sensor S_i. For sensor S_i, the Cipher3 is used to apply for the session key and identity authentication at the same time.

When the BS receives Cipher3, it picks out the corresponding session key *SK_si* according to ID_{CHi}. At the same time, if the *ID_si* can be found in the list of legal sensor nodes, the authentication of S_i is also accomplished.

To secure the session key, the base station encrypts SK_i with the session key SK_si and builds the Cipher4 as follows:

$$\text{Cipher4} = E(SK\_si, SK_i \| R). \quad (4)$$

Next, the Cipher4 is sent to S_i, and S_i will decrypt it using the symmetric key SK_si. Finally, all sensors in the same cluster have the same session key SK_i as its cluster head. Through the above key distribution subscheme, the confidentiality of traffic between the cluster head and the sensor is guaranteed. Moreover, mutual authentication between the BS and S_i is successfully performed. The detailed key distribution process is depicted in Figure 2.

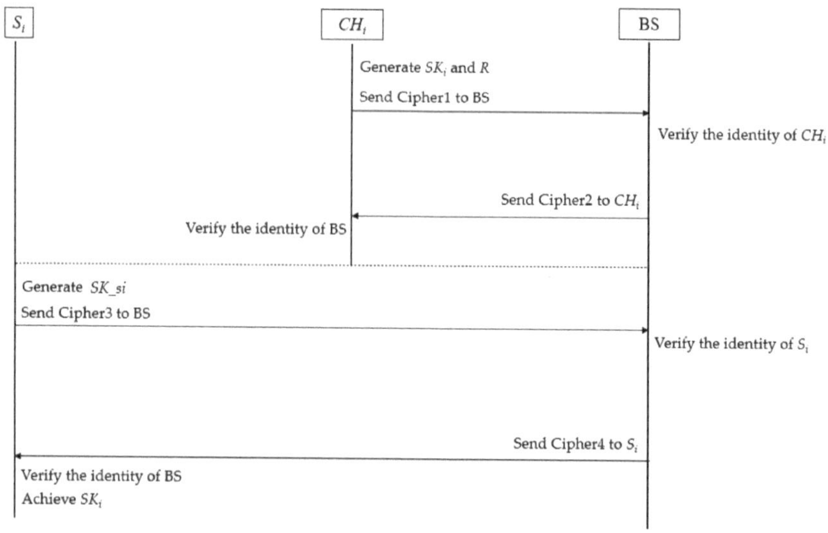

Figure 2. Flowchart of authentication and key agreement in the static scenario.

The specific implementation process of our proposed asymmetric encryption-based key distribution method in the static scenario is shown in Figures 3 and 4. In phase I, CH_1 and BS complete the two-way authentication and distribution of the session key SK_1 at the same time. In phase 2, the secure distribution of the session key between sensor S_1 and BS is realized.

3.3.2. Session Key Update Process

To protect the nodes against long-term attacks, a periodic key update mechanism is designed. The steps of the key update are given as follows.

1. The new session key SK_i' is generated via the cluster head CH_i at a certain moment.
2. CH_i notifies the base station to update the session key.
3. Using the proposed handshake operation, the new session key SK_i' is distributed between the BS and the CH_i. After that step, the CH_i notifies all sensors to update their session key in its cluster with a broadcasting message. Sensors will stop encrypting sessions until they receive the new session key SK_i'.
4. After the establishment of SK_i', the CH_i distributes SK_i' encrypted using the original session key SK_i to all sensors by broadcasting cipher5, which is denoted as follows:

$$\text{Cipher5} = E(SK_i, SK_i'). \quad (5)$$

5. Each sensor in the cluster decrypts the cipher5 using the old session key SK_i and substitutes it for the SK_i'. The subsequent dialog is decrypted using the new session key.

***Phase* I: Authentication between BS and CH_1 and allocation of session key SK_1.**

***Step1*:** CH_1 generates a random symmetric key SK_1 and a challenge R_1.

***Step2*:** CH_1 encrypts SK_1, R_1, *timestamp* and ID_{CH1} with *PUK*, and generates *Cipher1*.

$Cipher1 = E(PUK, SK_1 \| R \| ID\_{CH1} \| timestamp)$

***Step3*:** CH_1 sends *Cipher1* to the base station using traditional routing protocol.

***Step4*:** After receiving *Cipher1*, the base station decrypts it with *PVK*:

$D(PVK, Cipher1) = SK_1, R, ID\_{CH1}$. If the ID_{CH1} is a legitimate identifier, then CH_1 passes authentication. Meanwhile, BS obtains the session key SK_1.

***Step5*:** Then, BS feedbacks the challenge R_1 encrypted by session key SK_1, which is noted as $Cipher2 = E(SK_1, R_1)$.

***Step6*:** CH_1 gets $R_1 = D(SK_1, Cipher2)$. If R_1 is correctly received, the mutual authentication between CH_1 and BS is successfully completed.

Figure 3. Specific steps for phase I in an example.

Phase* II: *Sensor S_1 obtains the session key SK_1 from BS.

***Step1*:** S_1 generates a random symmetric key SK_{S1} and a challenge R_{11}.

***Step2*:** S_1 encrypts SK_{S1}, R_{11}, *timestamp*, ID_{S1} and ID_{CH1} with *PUK*, and generates *Cipher3*.

$Cipher3 = E(PUK, ID\_{CH1} \| ID\_{S1} \| timestamp \| SK\_{S1} \| R_{11})$

***Step3*:** After CH_1 forwarding, S_1 sends *Cipher3* to the base station.

***Step4*:** After receiving *Cipher3*, the base station decrypts it with *PVK*:

$D(PVK, Cipher3) = SK\_{S1}, R_{11}, ID\_{CH1}, ID\_{S1}$. If the ID_{S1} is a legitimate identifier, then S_1 passes authentication.

***Step5*:** BS picks out the required session key SK_1 *for* S_1 according to ID_{CH1}.

***Step6*:** For security reasons, the session key SK_1 is encrypted with SK_{S1} and sent to S_1, denoted as $Cipher4 = E(SK\_{S1}, SK_1 \| R_{11})$.

***Step7*:** S_1 successfully gets $SK_1 = D(SK\_{S1}, Cipher4)$. Meanwhile, if R_{11} is correctly received, the mutual authentication between S_1 and BS is completed.

***Step8*:** S_1 starts to use SK_1 to encrypt application data for confidential data transmission with BS.

Figure 4. Specific steps for phase II in an example.

3.4. Mobile Sensors Subscheme for Hierarchical WSNs

3.4.1. Mutual Authentication and Key Distribution Process

Since sensor nodes have a high probability of moving between different clusters of the network, the dynamic subscheme for hierarchical architecture is more complicated. In Figure 5, S_0 moves from the cluster C_0 into another cluster named C_2. As the location of each CH is assumed to be unchanged, the process of authentication and key distribution between CH and BS is the same as that of the static subscheme. The main difference between the static subscheme and the mobile subscheme lies in the key distribution process.

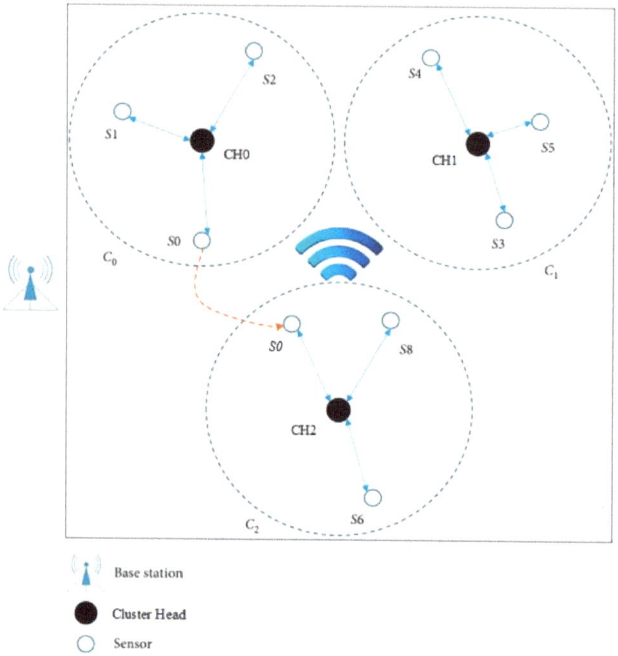

Figure 5. Sensor S_0 moves from Cluster0 to Cluster2.

The key distribution process of the mobile scene includes six steps.

1. When S_0 moves into cluster2, it will send a cluster-entry request to CH_2. The cluster forming and cluster head detection process is not described in this paper. For more information, please refer to [24].
2. CH_2 detects and receives this message. Next, CH_2 replies to S_0 with a message including its identification code ID_{CH2}.
3. S_0 updates the identification of the present cluster, replacing ID_{CH0} with ID_{CH2}.
4. S_0 applies for the latest session key SK_2 via the base station using the cipher6 denoted as follows:

$$\text{Cipher6} = E(PUK, ID\_{CH2} \| ID\_{S0} \| timestamp \| SK\_{S0} \| R) \tag{6}$$

5. The BS decrypts cipher6 with the PVK and finds ID_{CH2}, SK_{S0}, and ID_{S0} via Plain6 = $D(PVK, \text{Cipher6}) = D(PVK, E(PUK, ID\_{CH2} \| ID\_{S0} \| timestamp \| SK\_{S0} \| R))$ $= ID\_{CH2} \| ID\_{S0} \| SK\_{S0} \| R$. The latest session key SK_2 can be picked out in terms of ID_{CH2}, and the S_0 is authenticated via BS according to ID_{S0}. Next, the cipher7 will be sent to S_0. The cipher7 is built as follows:

$$\text{Cipher7} = E(SK\_{S0}, SK_2 \| R). \tag{7}$$

6. S_0 decrypts the cipher7 with the symmetric key SK_{S0} and successfully finds SK_2.

Thus, the mobile sensor can achieve the latest session key of the present cluster and send encrypted traffic to the corresponding cluster head. The detailed key agreement process in mobile subscheme is depicted in Figure 6.

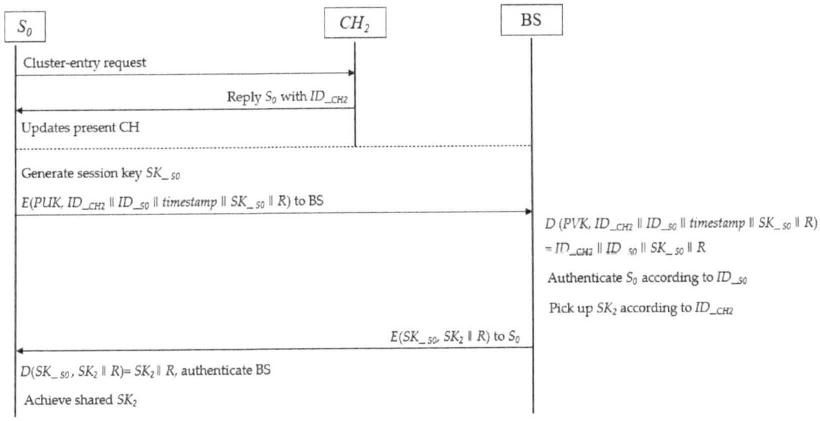

Figure 6. Flowchart of authentication and key agreement in the mobile subscheme.

3.4.2. Session Key Update Process

However, when S_0 moves to the junction of two adjacent clusters, for example C_0 and C_2 in Figure 5, it may receive key update messages from CH_0 and CH_2 at the same time. It should be noted that S_0 only knows the previous session key SK_0 of cluster0, and it is unaware of the previous session key of cluster2. Thus, S_0 can only decrypt the broadcasting message from CH_0 to update SK_0. After joining cluster2, S_0 can obtain the present session key SK_2 from the base station and wait for key updating to repeat.

4. Analysis and Comparison

Extensive simulations are provided to verify the performance of our scheme, such as memory consumption, communication overhead, connectivity, and recovery capability for node capture. Next, we compare the proposed key management scheme with other schemes from multiple dimensions.

We evaluate the performance based on NS-2 [39]. In the simulation, we randomly arranged a total of 200 sensors and 20 cluster head nodes with dimensions of 100 m by 100 m. Each sensor moves at a speed of 1–5 m/s. The signal reception range of each sensor is 10 m. The data transmission rate is 32 kbps; the traffic generation uses the CBR model, and the traffic generation interval is 30 s.

4.1. Key Storage of Sensor Nodes

In our scheme, the public key is pre-loaded into sensor's memory during the network initialization. Since the strength of encryption with the 256-bit ECC key is equal to that of the 3072-bit RSA key, a public key of 256 bits in length is used in our simulation. Moreover, two 16-byte session keys are used in the key distribution process. When a sensor receives the refreshed session key, the original key will be deleted to save the memory. Therefore, the memory overhead of each sensor is only 64 bytes, while that of the CH is 48 bytes.

The key distribution in [30] is that k keys are pre-loaded into each sensor, while m keys ($m \gg k$) are pre-loaded into each CH. If any two nodes share a pairing key, they can establish a secure link. Thus, the greater the number of keys stored, the higher probability of sharing common keys. In [40], the memory is divided into two parts. One part is used to store α pre-distributed keys, and the other part is used to store β post-deployment keys.

Table 1 presents the key storage overheads in different schemes. For large- and medium-sized wireless sensor networks, sensors in our scheme require less storage space than those of other schemes. However, our cluster heads require slightly more memory space than those of Erfani's scheme. Since the number of sensors is much larger than that of CHs, our scheme is valuable for resource-limited WSNs.

Table 1. Key storage overheads (bytes) in different schemes.

	Du [30]	Erfani [40]	Our Scheme
Sensor	32*l*	32 (α and β)	64
Cluster Head	32*M*	32	48

4.2. Communication Overhead

The communication overhead in our analysis only considers the payload related to key distribution and update, and it does not include the IP packet encapsulation of the network layer.

The length of AES-based session key is set to 16 bytes. The bytes of IP message encapsulation are not included in the calculation of the traffic generated during key distribution and update. For the static scenario, in stage 1, the effective communication load between the cluster head and the base station is 32 bytes. In stage 2, the effective communication load between the sensor node and the base station is 64 bytes. Therefore, the communication load consumed by a cluster for a complete key distribution process is 96 bytes. In the key update phase, the effective communication load between the cluster head node and the base station and the sensor nodes is 64 bytes in total, of which the load of broadcasting messages to the sensors in the cluster makes up 32 bytes. As for the dynamic scenario, the communication overhead of the CH and the sensor are the same as that of the static scenario.

As the frequency of session key update increases, the bandwidth occupied by key distribution also increases. This outcome means there is a tradeoff between security and communication load in wireless sensor networks.

4.3. Security Analysis

4.3.1. Mutual Authentication

In both subschemes, mutual authentication of BS and sensors (including CHs) is assured via the challenge–response mechanism. Terminals without legal identifiers ($ID_{\_CHi}$ or *ID_si*) cannot pass the identity authentication. Since the identifier is stored in the chip of each sensor with a tamper proof mechanism and encrypted for transmission, its confidentiality and integrity can be guaranteed. We added 10 nodes to the test network and distributed them evenly in 3 clusters. They simulated nodes that gained illegal access to the sensing network, randomly generating their identification codes *ID_si*. Since the identifiers *ID_si* used by these 10 nodes in constructing the *Ciperh3* were not included in the authorized and legitimate user list of the base station, the shared session key could not be obtained via the base station in the test. As a result, the reliability of the authentication scheme is fully demonstrated.

4.3.2. Security Connectivity

The security connectivity is defined as the probability that two nodes successfully establish a session key. Since authentication and key distribution in our proposal are cluster based, we define "inter-cluster connectivity" as the probability that a CH shares a pairwise key with the sensors in its cluster.

In our deterministic key distribution scheme, each authenticated sensor can always successfully share a session key with the present cluster head. Compared to the probabilistic key distribution approaches in [30,31,41], the inter-cluster connectivity in our scheme is 100%. Those random schemes, like AP [30], can only achieve higher security connectivity by increasing the amount of key storage. Figure 7 depicts the comparison of secure connectivity and key pool size in the AP. As the number of pre-loaded keys increases, the performance of the secure connectivity gradually improves. For fixed parameters [*l*, *M*], the security connectivity decreases significantly as the key pool increases.

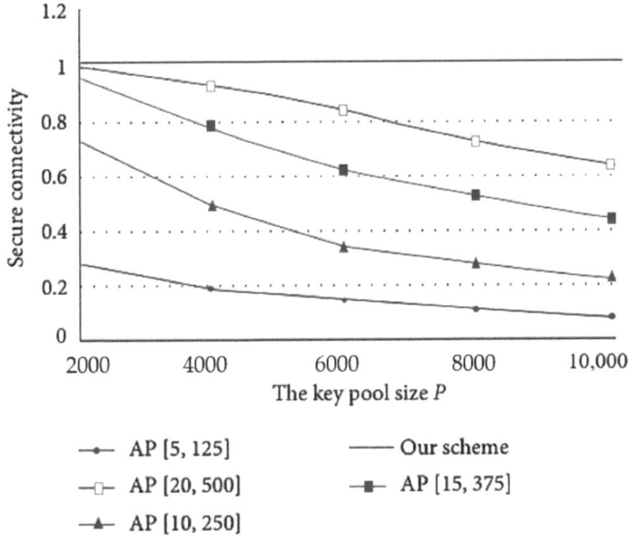

Figure 7. Secure connectivity versus key pool size P.

4.3.3. Resistance to Attacks

The new scheme provides a set of session keys to secure data exchange between the base station and sensors. Our proposal, which is based on session and public keys, can effectively resist common network attacks.

Eavesdropping can be avoided using symmetric encryption, as well as the key update mechanism proposed in this article. Spoofing attacks are avoided in our scheme through mutual authentication based on public-key encryption. Moreover, the authenticity of sensors is achieved via a challenge–response mechanism, and the identity code is preloaded before deployment.

Attacks like modification, reply, and insertion can be resisted via symmetric encryption and message authentication code added to each message. Only those authenticated nodes can send or modify data packets on the network.

Attackers obtain the secret information by capturing nodes or other physical means. We define resilience against node capture as the probability $F(x)$ that attackers obtain the key from the uncaptured node according to a certain number of captured nodes x. Thus, we find

$$F(x) = \frac{\text{number of compromised links between uncaptured nodes}}{\text{number of uncompromised links}} \tag{8}$$

Resilience against sensor capture is first evaluated. Unlike the random key pre-distribution schemes in [10,11,42], sensors only need to pre-load a public key in our approach, which saves the memory of the sensor node. Due to the periodical key update applied, it is too hard for attackers to find the constantly updated session key, despite physically capturing a sensor in our proposal. Thus, the probability of resilience against node capture is $F(x_s) = 0$, where x_s represents the number of captured sensor nodes. As shown in Figure 8, the resilience performance worsens with the increasing number of captured nodes for random key pre-distribution schemes, because of the storage of a large number of session keys. Since the sensors store matrixes instead of keys, the resilience performance of Boujelben's scheme [31] is better than that of the AP scheme [30]. Simulation results indicate that threat of sensor capture is perfectly eliminated via our scheme.

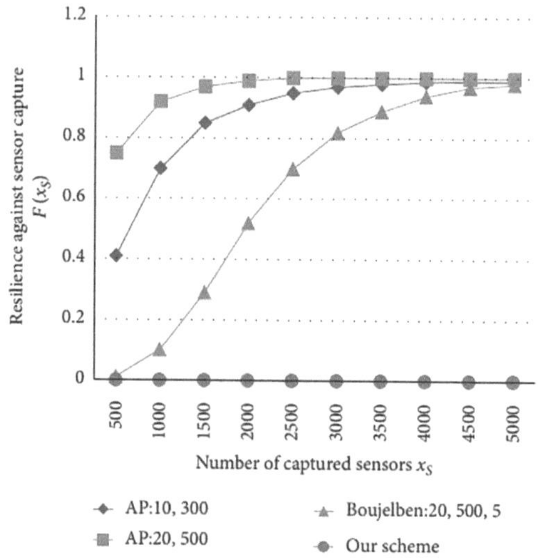

Figure 8. The probability of resilience against sensor capture in different schemes.

Finally, Table 2 presents several typical schemes of key management in WSN that emerged recent years. In our scheme, we provide a simple and feasible mutual authentication mechanism comparable to [30,34,40]. Lee, in [32], used an asymmetric encryption algorithm with more computation overhead than in [34] and our proposal. Furthermore, our scheme outperforms other schemes in terms of resilience against node capture and resistance to eavesdropping.

Table 2. Security comparisons of different key distribution solutions.

Scheme Features	Du [30]	Lee [32]	Benamar [34]	Erfani [40]	Our Scheme
Public-key encryption	—	√	√	—	√
Key pre-distribution	√	×	√	√	√
Mobility of sensors	—	×	×	√	√
Perfect resilience against node capture	×	—	—	×	√
Mutual authentication	×	√	×	×	√
Resistant to eavesdropping attacks	—	—	√	√	√

—: Not involved. √: Support. ×: Not Support.

5. Conclusions

The research work discussed in this paper focuses on key distribution schemes for static and dynamic wireless sensor networks. The novelty of this scheme is that the proposed key distribution and update strategy is particularly suitable for sensing networks in which the nodes are in motion. In addition, we evaluate the design scheme in terms of key storage capacity and the communication load generated during key exchange and security. Compared to the traditional classical key distribution scheme, our proposed new scheme is less complex to implement, reduces the cache capacity requirements of the nodes, and obtains better connection security and resistance to attacks. It can be concluded that our results are particularly suitable for wireless mobile sensing networks with high capacity, low power consumption, and high reliability requirements, such as environmental monitoring networks, energy IoT networks, and smart warehouse management systems.

Author Contributions: Conceptualization, Y.C. and Y.L. (Yanan Liu); methodology, Y.C.; software, Y.L. (Yanan Liu); validation, Y.C. and Z.Z.; formal analysis, Y.C.; investigation, Y.L. (Yanxiu Li); resources, Y.C. and Y.L. (Yanan Liu); data curation, Y.L. (Yanan Liu); writing—original draft preparation, Y.C.; writing—review and editing, Y.C.; supervision, Z.Z. All authors have read and agreed to the published version of the manuscript.

Funding: This work was funded by the Research Startup Foundation of Jinling Institute of Technology under Grant number [JIT-B-201726].

Institutional Review Board Statement: Not applicable.

Informed Consent Statement: Not applicable.

Data Availability Statement: Not applicable.

Conflicts of Interest: The authors declare no conflict of interest.

References

1. Dludla, A.G.; Abu-Mahfouz, A.M.; Kruger, C.P.; Isaac, J.S. Wireless sensor networks testbed: ASNTbed. In Proceedings of the 2013 IEEE IST-Africa Conference and Exhibition (IST-Africa), Nairobi, Kenya, 29–31 May 2013; pp. 1–10.
2. Abu-Mahfouz, A.M.; Steyn, L.P.; Isaac, S.J.; Hancke, G.P. Multi-Level Infrastructure of Interconnected Testbeds of Large-Scale Wireless Sensor Networks (MI2T-WSN). In Proceedings of the International Conference on Wireless Networks (ICWN), Athens, Greece, 1–7 January 2012; pp. 126–131.
3. Carman, D.; Kruus, P.; Matt, B. *Constraints and Approaches for Distributed sensor Network Security (Final)*; NAI Labs Technical Report; NAI Labs: Glenwood, MD, USA, 2000; pp. 1–139.
4. Ren, Y.; Leng, Y.; Qi, J.; Sharma, P.K.; Wang, J.; Almakhadmeh, Z.; Tolba, A. Multiple cloud storage mechanism based on blockchain in smart homes. *Future Gener. Comput. Syst.* **2021**, *115*, 304–313.
5. Xiong, J.; Zhao, M.; Bhuiyan, M.Z.A.; Chen, L.; Tian, Y. An AI-enabled three-party game framework for guaranteed data privacy in mobile edge crowdsensing of IoT. *IEEE Trans. Ind. Inform.* **2021**, *17*, 922–933. [CrossRef]
6. Aysal, T.C.; Barner, K.E. Sensor data cryptography in wireless sensor networks. *IEEE Trans. Inf. Forensics Secur.* **2008**, *3*, 273–289. [CrossRef]
7. Giruka, V.C.; Singhal, M.; Royalty, J.; Varanasi, S. Security in wireless sensor networks. *Wirel. Commun. Mob. Comput.* **2008**, *8*, 1–24.
8. Kundur, D.; Luh, W.; Okorafor, U.N.; Zourntos, T. Security and privacy for distributed multimedia sensor networks. *Proc. IEEE* **2008**, *96*, 112–130. [CrossRef]
9. Wang, Y.; Attebury, G.; Ramamurthy, B. A survey of security issues in wireless sensor networks. *IEEE Commun. Surv. Tutor.* **2006**, *8*, 2–23. [CrossRef]
10. Liu, G.; Yang, Q.; Wang, H. Trust assessment in online social networks. *IEEE Trans. Dependable Secur. Comput.* **2018**, *2*, 994–1007. [CrossRef]
11. Ge, C.; Susilo, W.; Baek, J.; Liu, Z.; Xia, J.; Fang, L. Revocable attribute-based encryption with data integrity in clouds. *IEEE Trans. Dependable Secur. Comput.* **2021**, *21*, 1.
12. Ge, C.; Susilo, W.; Liu, Z.; Xia, J.; Szalachowski, P.; Fang, L. Secure keyword search and data sharing mechanism for cloud computing. *IEEE Trans. Dependable Secur. Comput.* **2020**, *20*, 1. [CrossRef]
13. Gautam, A.K. A key management scheme using (p,q)-lucas polynomials in wireless sensor network. *China Commun.* **2021**, *18*, 210–228. [CrossRef]
14. Kumar, V.; Malik, N. Dynamic key management scheme for clustered sensor networks with node addition support. In Proceedings of the 2021 2nd International Conference on Intelligent Engineering and Management, London, UK, 28–30 April 2021; pp. 102–107.
15. Moghadam, M.F.; Nikooghadam, M.; Jabban, M.A.B. An efficient authentication and key agreement scheme based on ECDH for wireless sensor network. *IEEE Access* **2020**, *8*, 73182–73192.
16. Sirajuddin, M.; Rupa, C.H.; Iwendi, C.; Biamba, C. TBSMR: A trust-based secure multipath routing protocol for enhancing the QoS of the mobile Ad Hoc network. *Secur. Commun. Netw.* **2021**, *2021*, 5521713.
17. Sirajuddin, M.; Rupa, C.H.; Bhatia, S.; Thakur, R.N.; Mashat, A. Hybrid cryptographic scheme for secure communication in mobile Ad Hoc network-based E-healthcare system. *Wirel. Commun. Mob. Comput.* **2022**, *2022*, 9134036. [CrossRef]
18. Mishra, D.; Vijayakumar, P.; Sureshkumar, V.; Amin, R.; Islam, S.H.; Gope, P. Efficient authentication protocol for secure multimedia communications in IoT-enabled wireless sensor networks. *Multimed. Tools Appl.* **2018**, *77*, 18295–18325. [CrossRef]
19. Wu, F.; Li, X.; Sangaiah, A.K.; Xu, L.; Kumari, S.; Wu, L.; Shen, J. A lightweight and robust two-factor authentication scheme for personalized healthcare systems using wireless medical sensor networks. *Future Gener. Comput. Syst.* **2018**, *82*, 727–737. [CrossRef]
20. Srinivas, J.; Mishra, D.; Mukhopadhyay, S.; Kumari, S. Provably secure biometric based authentication and key agreement protocol for wireless sensor networks. *J. Ambient Intell. Hum. Comput.* **2018**, *9*, 875–895.

21. Naresh, V.S.; Reddi, S.; Murthy, N.V. Provable secure lightweight multiple shared key agreement based on hyper elliptic curve Diffie–Hellman for wireless sensor networks. *Inf. Secur. J. Glob. Perspect.* **2020**, *29*, 1–13. [CrossRef]
22. Jung, J.; Moon, J.; Lee, D.; Won, D. Efficient and security enhanced anonymous authentication with key agreement scheme in wireless sensor networks. *Sensors* **2017**, *17*, 644. [CrossRef]
23. Shin, S.; Kwon, T. A lightweight three-factor authentication and key agreement scheme in wireless sensor networks for smart homes. *Sensors* **2019**, *19*, 2012. [CrossRef]
24. Santos-González, I.; Rivero-García, A.; Burmester, M.J.; Munilla, J.; Caballero-Gil, P. Secure lightweight password authenticated key exchange for heterogeneous wireless sensor networks. *Inf. Syst.* **2020**, *88*, 101423. [CrossRef]
25. Zheng, J.; Jamalipour, A. *Wireless Sensor Networks: A Networking Perspective*; John Wiley & Sons, Inc.: Hoboken, NJ, USA, 2009.
26. Singh, S.K.; Singh, M.P.; Singh, D.K. A Survey of Energy-Efficient Hierarchical Cluster-Based Routing in Wireless Sensor Networks. *Int. J. Adv. Netw. Appl.* **2010**, *2*, 570–580.
27. Yan, M.; Chan, C.A.; Li, W.; Chih-Lin, I.; Bian, S.; Gygax, A.F.; Leckie, C.; Hinton, K.; Wong, E.; Nirmalathas, A. Network energy consumption assessment of conventional mobile services and over-the-top instant messaging applications. *IEEE J. Sel. Areas Commun.* **2016**, *34*, 3168–3180. [CrossRef]
28. Yan, M.; Li, W.; Chan, C.A.; Bian, S.; Chih-Lin, I.; Gygax, A.F. PECS: Towards personalized edge caching for future service-centric networks. *China Commun.* **2019**, *16*, 93–106. [CrossRef]
29. Gura, N.; Patel, A.; Wander, A.; Eberle, H.; Shantz, S.C. Comparing elliptic curve cryptography and RSA on 8-bit CPUs. In Proceedings of the Sixth Workshop on Cryptographic Hardware and Embedded Systems, Cambridge, MA, USA, 11–13 August 2004; pp. 119–132.
30. Du, X.; Xiao, Y.; Guizani, M.; Chen, H.H. An effective key management scheme for heterogeneous sensor networks. *Ad Hoc Netw.* **2007**, *5*, 24–34. [CrossRef]
31. Boujelben, M.; Cheikhrouhou, O.; Abid, M.; Youssef, H. Establishing pairwise keys in heterogeneous two-tiered wireless sensor networks. In Proceedings of the 3rd International Conference on Sensor Technologies and Applications Athens, Athens, Greece, 18–23 June 2009; pp. 18–23.
32. Lee, S.; Kim, K. Key renewal scheme with sensor authentication under clustered wireless sensor networks. *Electron. Lett.* **2015**, *51*, 368–369. [CrossRef]
33. Tian, Y.; Wang, Z.; Xiong, J.; Ma, J. A blockchain-based secure key management scheme with trustworthiness in DWSNs. *IEEE Trans. Ind. Inform.* **2020**, *16*, 6193–6202. [CrossRef]
34. Benamar, K.; Mohammed, F.; Abdellah, M. Architecture aware key management scheme for wireless sensor networks. *Int. J. Inf. Technol. Comput. Sci.* **2012**, *4*, 50–59.
35. Miller, V. Uses of Elliptic Curves in Cryptography. In *Advances in Cryptology—CRYPTO '85*; Williams, H.C., Ed.; Springer: Berlin, Germany, 1986; Volume LNCS 218, pp. 417–426.
36. Nam, J.; Kim, M.; Paik, J. A provably secure ECC-based authentication scheme for wireless sensor networks. *Sensors* **2014**, *14*, 21023–21044. [CrossRef]
37. Gulen, U.; Baktir, S. Elliptic Curve cryptography for wireless sensor networks using the number theoretic transform. *Sensors* **2020**, *20*, 1507. [CrossRef]
38. Mohamed, Y.; Moustafa, Y.; Khaled, A. Energy-aware management for cluster-based sensor networks. *Comput. Netw.* **2003**, *43*, 649–668.
39. University of Southern California: The Network Simulator—ns-2. September 2005. Available online: http://www.isi.edu/nsnam/ns/ (accessed on 1 June 2023).
40. Erfani, S.H.; Javadi, H.H.S.; Rahmani, A.M. A dynamic key management scheme for dynamic wireless sensor networks. *Secur. Commun. Netw.* **2015**, *8*, 1040–1049. [CrossRef]
41. Eschenauer, L.; Gligor, V.D. A key management scheme for distributed sensor networks. In Proceedings of the ACM Conference on Computer and Communication Security, Washington, DC, USA, 18–22 November 2002; pp. 41–47.
42. Chen, C.Y.; Chao, H.C. A survey of key distribution in wireless sensor networks. *Secur. Commun. Netw.* **2014**, *7*, 2495–2508. [CrossRef]

sensors

MDPI

Article

A Mobile Sensing Framework for Bridge Modal Identification through an Inverse Problem Solution Procedure and Moving-Window Time Series Models

Mohammad Talebi-Kalaleh and Qipei Mei *

Department of Civil and Environmental Engineering, University of Alberta, Edmonton, AB T6G 2H5, Canada; talebika@ualberta.ca

* Correspondence: qipei.mei@ualberta.ca

Abstract: With the rise and development of smart infrastructures, there has been a great demand for installing automatic monitoring systems on bridges, which are key members of transportation networks. In this regard, utilizing the data collected by the sensors mounted on the vehicles passing over the bridge can reduce the costs of the monitoring systems, compared with the traditional systems where fixed sensors are mounted on the bridge. This paper presents an innovative framework for determining the response and for identifying modal characteristics of the bridge, utilizing only the accelerometer sensors on the moving vehicle passing over it. In the proposed approach, the acceleration and displacement response of some virtual fixed nodes on the bridge is first determined using the acceleration response of the vehicle axles as the input. An inverse problem solution approach based on a linear and a novel cubic spline shape function provides the preliminary estimations of the bridge's displacement and acceleration responses, respectively. Since the inverse solution approach is only capable of determining the response signal of the nodes with high accuracy in the vicinity of the vehicle axles, a new moving-window signal prediction method based on auto-regressive with exogenous time series models (ARX) is proposed to complete the responses in the regions with large errors (invalid regions). The mode shapes and natural frequencies of the bridge are identified using a novel approach that integrates the results of singular value decomposition (SVD) on the predicted displacement responses and frequency domain decomposition (FDD) on the predicted acceleration responses. To evaluate the proposed framework, various numerical but realistic models for a single-span bridge under the effect of a moving mass are considered; the effects of different levels of ambient noise, the number of axles of the passing vehicle, and the effect of its speed on the accuracy of the method are investigated. The results show that the proposed method can identify the characteristics of the three main modes of the bridge with high accuracy.

Keywords: indirect bridge health monitoring; bridge mode shape identification; moving vehicles; moving-window ARX model; inverse problem; vibration-based monitoring

Citation: Talebi-Kalaleh, M.; Mei, Q. A Mobile Sensing Framework for Bridge Modal Identification through an Inverse Problem Solution Procedure and Moving-Window Time Series Models. *Sensors* **2023**, *23*, 5154. https://doi.org/10.3390/s23115154

Academic Editors: Chien Aun Chan, Chunguo Li and Ming Yan

Received: 20 April 2023
Revised: 24 May 2023
Accepted: 26 May 2023
Published: 28 May 2023

1. Introduction

The development of smart infrastructure has highlighted the need for automatic and real-time monitoring systems for critical transportation components such as bridges. The high cost associated with maintenance and reconstruction of these essential lifelines has made the implementation of such monitoring systems a pressing issue. Consequently, significant research has been conducted over the last few decades in the area of vibration-based monitoring of bridge structures using sensors installed directly on the bridges [1–3].

Previous studies have made notable contributions to bridge health monitoring. For instance, Gonzalez and Karoumi [4] proposed a damage detection method for bridges utilizing bridge weigh-in-motion (BWIM) data and machine learning techniques. Their approach employed an artificial neural network (ANN) to predict bridge health based on BWIM data, ensuring reliable evaluations over time. Similarly, Azim and Gül [5] developed a data-driven damage detection framework for truss railway bridges using operational acceleration and strain response data. Feng and Feng [6] presented a time-domain finite element (FE) model updating approach using in situ measurements of dynamic displacement responses under trainloads. Their study validated the importance of the bridge's equivalent stiffness for accurate model updating, although extracting modal information from dynamic responses proved challenging for short-span railway bridges with high natural frequencies.

Due to the high cost of real-time monitoring of bridges using fixed sensors, recent research in structural health monitoring has explored indirect methods, called indirect bridge health monitoring (iBHM), that utilize only the vertical vibration data collected by accelerometers mounted on passing vehicles. This method, first proposed by Yang et al. [7], has recently gained significant recognition. The main objective in this research direction is to find a reliable method to determine the vertical vibration responses of desired locations on a bridge without the need for expensive fixed sensors or with minimal sensor requirements [8,9]. For example, Malekjafarian and O'Brien [10] developed a method for identifying bridge mode shapes using short time frequency domain decomposition (STFDD) of responses measured in a passing vehicle. Their approach involved segmenting the bridge and employing a multi-stage procedure using frequency domain decomposition (FDD) to estimate the mode shapes. Numerical case studies validated the performance of the method, demonstrating the accurate estimation of mode shapes under low noise levels and the presence of other traffic or signal subtraction in identical axles. Furthermore, Eshkevari et al. [11] developed novel methods for modal identification of bridges using data collected by a large number of moving sensors (vehicles). Their study proposed matrix completion methods, specifically the alternating least squares algorithm, to extract modal properties from sparse and dynamic bridge response data. Three methods were evaluated: principal component analysis, structured optimization analysis, and the natural excitation technique (NExT). The results demonstrated accurate estimations of modal properties. However, the methods had limitations in terms of computational costs, modal leakage, sparse data, user-defined points, and accuracy for higher modes. The study showcased the potential of using mobile sensor networks for bridge health monitoring and system identification. Kong et al. [12] proposed a method to efficiently extract bridge modal properties using a test vehicle composed of a tractor and trailers. They verified the method on an existing bridge, considering the effects of trailer mass and stiffness. Their findings demonstrated high visibility in extracted bridge frequencies, particularly when traffic flows provided additional excitation. However, limitations were identified, such as difficulties in accurately extracting mode shapes dominated by lateral bending due to the limited modeling of trailers.

To reduce the cost of monitoring using mobile sensing, researchers have also explored the use of smartphone sensors instead of expensive and commercially graded accelerometer sensors. Smartphone data have been widely used in different fields, such as indoor positioning [13], crime prevention [14], and even agriculture [15]. In the structural health monitoring field, smartphone sensors have also demonstrated effectiveness in various applications, such as bridge seismic monitoring [16], assessment of building damage from seismic events [17], damage detection of a 3D steel frame [18], and walking vibration analysis [19]. Experimental investigations have demonstrated the reliability of smartphone technology for bridge monitoring, particularly in identifying natural frequencies, although this area of research is still in its early stages [20]. For instance, Di Matteo et al. [21] conducted a field experiment on the Corleone bridge in Palermo, Italy, to assess smartphone-based bridge monitoring through vehicle–bridge interaction. The study successfully identified the bridge's natural frequency using smartphone data with high accuracy. Shirzad-Ghaleroudkhani and Gül [22] developed a novel methodology for natural frequency identification of bridges using acceleration signals recorded by smartphones on passing vehicles. Their inverse filtering approach effectively removed the frequency content of the vehicle. Additionally, Sitton et al. [23] proposed postprocessing strategies to estimate a bridge's fundamental frequency from acceleration data recorded from a traversing vehicle without prior knowledge of bridge parameters, successfully validating their approach through finite element simulations and experimental validation on a scale-model bridge.

Despite the potential benefits of mobile sensing, challenges remain in achieving accurate bridge response prediction and mode shape identification [9,24]. Complicated mathematical techniques and principles of structural dynamics are often required. Therefore, this paper aims to explore the use of mobile sensors on crossing vehicles for bridge health monitoring, leveraging their ubiquity and potential cost savings, while addressing the need for accurate modal identification and practical solutions.

To predict the bridge response using the recorded acceleration responses from a crossing vehicle, the vehicle response needs to be spatially mapped onto some virtual fixed nodes on the bridge [25]. These estimated responses can then be used to identify the dynamic characteristics and potential damages in the structure. However, due to the interpolation of the adjacent crossing axles (moving sensors), theoretical inverse problem solutions can only determine a limited part of the response signal for each fixed node on the bridge. This interpolation results in a sparse response matrix, where each row corresponds to the response of a particular fixed node and each column corresponds to the response vector of the fixed nodes in a time stamp. This matrix contains numerous missing values (invalid regions) that require advanced statistical, mathematical, or machine learning techniques to predict or complete the response signals for the virtual fixed nodes [25,26].

A few previous studies have utilized vehicle response data to identify bridge mode shapes using the sparse response data from bridges. While some of these methods have attempted to address the issue of missing values in the bridge response matrix through soft-imputing techniques [26], short time frequency domain decomposition [10], alternating least squares technique, and principal component analysis [11], these approaches often rely on engineering judgment, involve time-consuming constrained optimization processes, and require manual parameter settings. Although these limited research works have made significant contributions to drive-by modal identification, there is a need for an innovative and automated framework to overcome these limitations.

This research work presents a novel technique for identifying the modal characteristics of bridges using only accelerometer sensors mounted on vehicles passing over them. The proposed framework consists of two stages. In the first stage, an inverse problem solution approach is employed to determine the acceleration and displacement response of virtual fixed nodes on the bridge. This is achieved by using the acceleration response of the vehicle axles as input, utilizing conventional linear interpolation functions to predict the bridge displacement responses, and introducing a novel cubic spline shape function to predict the acceleration response of the assumed fixed nodes on the bridge. However, the inverse problem solution approach yields response signals with missing parts, necessitating prediction. To address this issue, a new automated moving-window time series model based on auto-regressive exogenous (ARX) techniques is proposed in this paper. In the second stage, a novel approach combines the results of singular value decomposition (SVD) on the predicted displacement responses and frequency domain decomposition (FDD) on the predicted acceleration responses. This approach allows for the accurate identification of the first mode shape and higher mode shapes of the bridge, as well as the determination of natural frequencies. The main novelties of this research lie in the utilization of a cubic spline shape function within the inverse problem solution stage to predict the acceleration response of the fixed nodes and the application of moving-window time series models to complete the predicted incomplete signals obtained from the inverse problem solution procedure. Moreover, the method distinguishes itself by identifying mode shapes of the bridge using both acceleration and displacement responses, thereby enhancing the accuracy of identification for both lower and higher modes. Numerical simulations are conducted to evaluate the effectiveness of the proposed framework, considering different levels of ambient noise, number of axles, and vehicle speed. The future implementation of the framework in smartphone sensing-based applications is also discussed.

This paper is structured as follows: Section 1 provides the introduction and review of the literature. Section 2 focuses on the background and the inverse problem solution procedure, specifically for estimating the preliminary response signals of the bridge. Section 3 introduces the details of the proposed framework, which encompasses a two-stage approach for predicting the missing parts of the response signals of the virtual fixed nodes on the bridge, as well as identifying the mode shapes and natural frequencies of the bridge. In Section 4, numerical analyses are conducted to evaluate the performance of the proposed method. The results of the analyses are presented and discussed in Section 5, where the characteristics and limitations of the proposed method are also examined. Finally, Section 6 presents the conclusions of the study, along with recommendations for future research in this area.

2. Background and Inverse Problem Solution Procedure

Identifying moving loads is a prevalent inverse problem in the field of structural dynamics; researchers have developed several approaches to tackle this challenge [27]. These approaches can be categorized into two main types: those relying on analytical models and those formulated using finite element models, with a specific focus on solution techniques. Another type of inverse problem encountered in vehicle–bridge dynamics pertains to the identification of structural parameters using the moving load as an excitation.

In this paper, a different approach is presented, inspired by the work of Oshima et al. [25], to address this inverse problem, utilizing a FE approach with two different shape functions to estimate the bridge response by incorporating the vehicle response as the input. The proposed approach is thoroughly discussed and successfully solved within this section, with a novel shape function being employed to enhance the accuracy of the bridge's acceleration response.

2.1. Assumptions and Notations

This study assumes that the sensors are mounted on the front and rear axles of the vehicle to mitigate the effects of the suspension system [26]. For simplicity, the deformation of wheels and tires is ignored. In practice, the effect of the vehicle–bridge interaction can be eliminated by considering its empirical transfer function. Although it is assumed that the recorded data are solely accelerations, the displacements integrated twice from the accelerations will also be used as inputs for the proposed time series models. To have a linear time invariant system, the mass of the vehicle is ignored in comparison with the mass of the bridge. The speed of all traversing axles is also assumed to be identical and constant during the measurement for simplicity. The geometry of the bridge and parameters of the moving vehicle are shown in Figure 1; all other notations and variables used in this paper are introduced in Table 1.

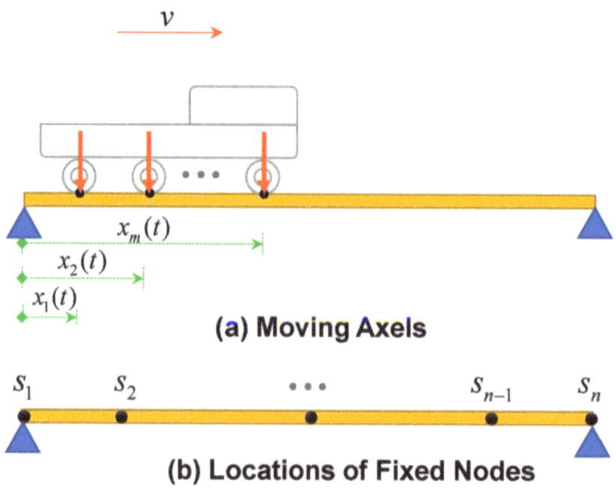

Figure 1. Illustration of the moving axles and fixed nodes.

Table 1. List of symbols and notations used in this paper.

Symbol	Description
$y(x,t)$	Continuous function of bridge vertical displacement response
$y_i^v(t)$	Vertical displacement response of the ith axle of the vehicle
$\mathbf{D}(t)$	Nodal vertical displacement vector of the bridge
$\mathbf{Y}(t)$	Vector containing the displacement responses of the moving axles
$\mathbf{N}(x)$	Interpolating shape function matrix for beam elements
$\mathbf{N}_v(t)$	Interpolating shape function matrix for estimating the response of the moving axles ($m \times n$)
S_j	jth virtual fixed node considered on the bridge
$N_j(x)$	Contribution of the jth node displacement of the displacement field
$\mathbf{Q}(t)$	Vector containing the modal coordinate responses = $[q_k(t)]$
$\mathbf{\Phi}^s$	Matrix containing the amplitudes of all mode shapes at the fixed nodes
$\phi_k(s_j)$	Amplitude of the kth mode shape at fixed node S_j
n	Total number of virtual fixed nodes considered on the bridge (mesh nodes)
m	Total number of moving axles crossing the bridge
s_j	Location of the jth fixed node from the left support of the bridge
$x_i(t)$	Location of the ith moving axle from the left support of the bridge
Δs	Mesh size of the bridge element

2.2. Inverse Problem Solution for Bridge Response Determination Utilizing Vehicle Response

In order to determine some parts of the bridge response signals at the virtual fixed nodes utilizing the vehicle response, an inverse problem solution is first employed in this section.

Based on the principles of finite element methods, the continuous vertical displacement response of a bridge can be determined by considering a proper interpolating (shape) function and using the discrete responses of the bridge at the location of the fixed nodes (shown in Figure 1b) [28].

$$y(x,t) = \begin{bmatrix} N_1(x) & \cdots & N_n(x) \end{bmatrix} \begin{Bmatrix} y_{s_1}(t) \\ \vdots \\ y_{s_n}(t) \end{Bmatrix} = \mathbf{N}(x)\mathbf{D}(t) \quad (1)$$

where $\mathbf{N}(x)$ is the interpolating shape function matrix and $\mathbf{D}(t)$ is a vector containing vertical displacements of the fixed nodes.

In Equation (1), a linear interpolating shape function is usually considered:

$$\mathbf{N}(x) = \begin{bmatrix} \overset{s_1}{0} & \cdots & \overset{s_j}{\frac{x-s_{j+1}}{s_j-s_{j+1}}} & \overset{s_{j+1}}{\frac{x-s_j}{s_{j+1}-s_j}} & \cdots & \overset{s_n}{0} \end{bmatrix}; \; s_j < x \le s_{j+1} \quad (2)$$

where S_j is the location of the jth fixed node from the left support. j can be valued from 1 to $(n-1)$.

By substituting the location history of the moving axles ($x_1(t)$ to $x_m(t)$) in Equation (1), the nodal displacement responses of the moving axles can easily be extracted [26].

$$\mathbf{Y}(t) = \begin{Bmatrix} y_1^v(t) \\ \vdots \\ y_m^v(t) \end{Bmatrix} = \begin{bmatrix} \mathbf{N}(x_1(t)) \\ \vdots \\ \mathbf{N}(x_m(t)) \end{bmatrix} \begin{Bmatrix} y_{s_1}(t) \\ \vdots \\ y_{s_n}(t) \end{Bmatrix} = \mathbf{N}_v(t)\mathbf{D}(t) \quad (3)$$

where $y_i^v(t)$ is the vertical displacement response of the ith axle of the vehicle and $\mathbf{N}_v(t)$ is an interpolating shape function matrix for estimating the response of the moving axles.

For a general condition of $m \neq n$, multiplying Equation (3) by the pseudoinverse of the matrix $\mathbf{N}_v(t)$ produces the nodal displacements of the bridge as a function of displacements of the moving axles.

$$\mathbf{D}(t) = \left(\left(\mathbf{N}_v^{tr}(t)\mathbf{N}_v(t) \right)^{-1} \mathbf{N}_v^{tr}(t) \right) \mathbf{Y}(t) \quad (4)$$

Taking the second derivation from both sides of the previous equation will produce a similar relation between the acceleration responses.

$$\ddot{\mathbf{D}}(t) = \left(\left(\mathbf{N}_v^{tr}(t)\mathbf{N}_v(t) \right)^{-1} \mathbf{N}_v^{tr}(t) \right) \ddot{\mathbf{Y}}(t) \quad (5)$$

2.3. Valid Regions of the Estimated Signals

Although in Equations (4) and (5), the pseudoinverse of the $\mathbf{N}_v(t)$ matrix is multiplied in all the responses of the moving axes, based on our numerical observations, the use of the pseudoinverse of the $\mathbf{N}_v(t)$ matrix based on the assumption of either linear or spline shape function, which is introduced in the next part, leads to an accurate estimation only for the responses of the nodes located in the vicinity of the moving axles at that time stamp; in time intervals outside of that, the prediction error will be very large, which cannot be used for structural health monitoring applications. The main reason for the error is the fact that the matrix $\mathbf{N}_v(t)$ is a sparse matrix and has lots of zeros outside of the valid region of the response, thereby its inverse can result in large errors. Although the proposed rule to determine the valid regions of the estimated response is more general and can be applied for any number of axles, Figure 2 illustrates the procedure for a case of the three moving axles crossing the bridge at an arbitrary time stamp such as t. Since there are at least two moving axles in the vicinity of the nodes S_{j-1} and S_j to contribute to their response in the given time, the estimated response by the theoretical method for these two nodes at time t is considered valid. The main reason for adopting this assumption is that, in order to determine the response of a fixed node at time t using the inverse solution of Equation (3), there must be at least two non-zero rows corresponding to that fixed node in the matrix $\mathbf{N}_v(t)$. The inverse of small or zero values in invalid regions approaches infinity and results in high error issues.

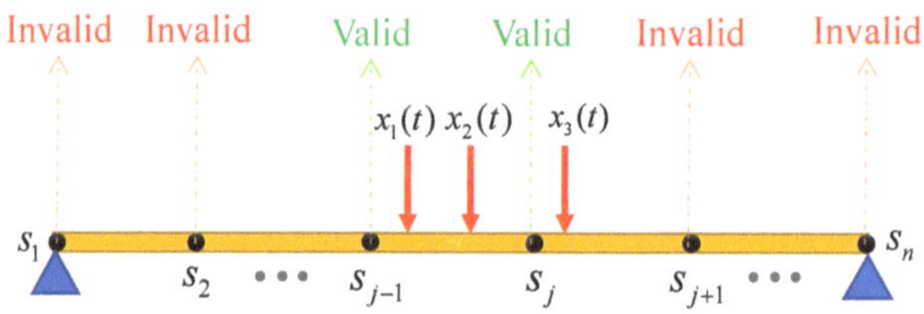

Figure 2. Definition of the valid and invalid regions of the estimated nodal displacements.

Considering this approach, the expected valid and invalid regions for the responses of each fixed node can be determined. Figure 3 shows a schematic visualization of the valid and invalid data regions in the matrix of nodal responses. It should be noted that, if the number of axles increases, the valid region of matrix **D** will increase. Therefore, the missing ratio of the matrix will be reduced. Furthermore, the more fixed nodes that are considered, the shorter valid regions and the higher missing rate we will have in matrix **D**.

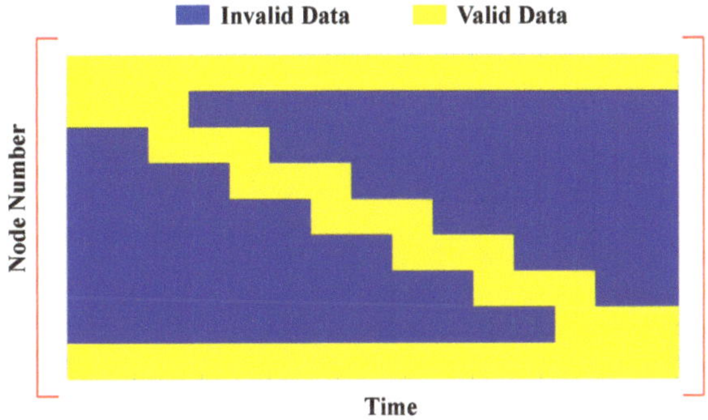

Figure 3. Visualization of valid and invalid regions in matrix **D**.

3. Proposed Response Prediction and Modal Identification Methodology

The proposed framework from collecting the acceleration response of the crossing axles to identifying the modal characteristics of the bridge is summarized in Figure 4. In this study, an inverse problem solution procedure is employed to estimate the initial displacement response signals of the bridge. Initially, a linear shape function is utilized for this purpose. However, based on numerical investigations, it has been demonstrated that incorporating a cubic spline shape function in the inverse solution procedure yields more precise results for bridge acceleration responses. The subsequent sub-section presents a detailed description of the proposed technique, which involves the use of a cubic spline interpolation function within the inverse problem solution procedure.

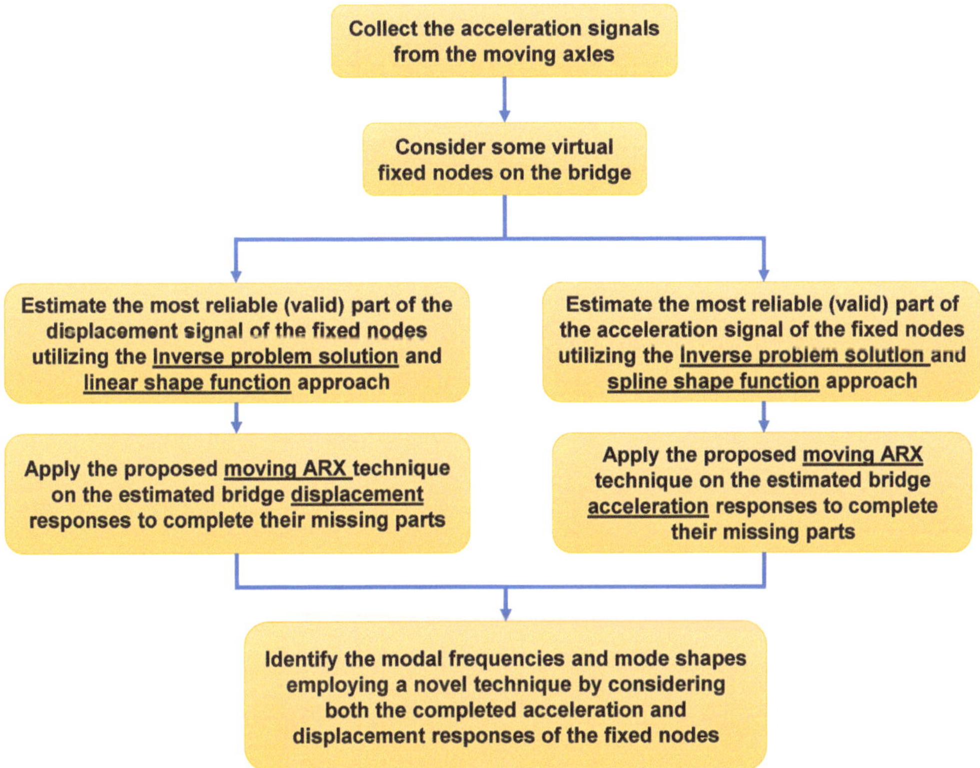

Figure 4. Flow chart of the proposed drive-by modal identification technique.

3.1. Continuous Cubic Spline as the Shape Function

A cubic spline is a piecewise polynomial in which the coefficients of each polynomial are fixed between joints [29]. In this paper, an innovative approach to estimate the bridge nodal responses at the valid regions is proposed. The novelty of the proposed approach is utilizing cubic spline polynomials as the shape function for the displacement field in lieu of the conventional discontinuous linear ones (Figure 5). Although a cubic spline interpolator is a continuous combination of some piecewise nonlinear cubic polynomials, it is shown that the linearity between the nodal responses and the response function will still be valid; the relation between the nodal responses and response field function can be written in the form of Equation (1).

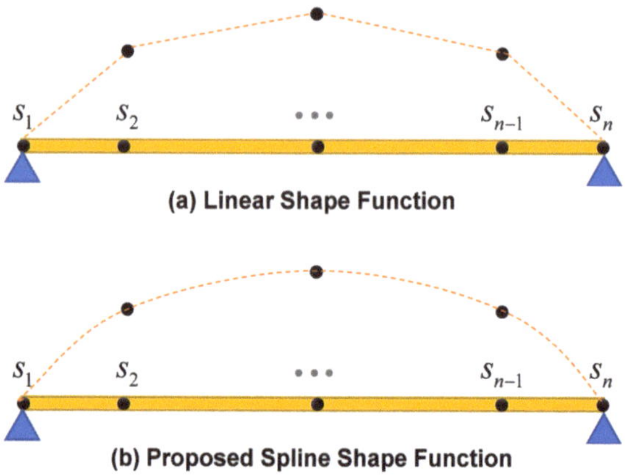

Figure 5. Linear vs. the proposed cubic spline shape function.

It should be noted that natural spline is employed in this paper due to the fact that the first and last supports of bridges are generally roller or hinge type, which releases the moment reaction at the supports and results in zero curvature at those points (i.e., $\mathbf{N}''(x = 0)$ and $\mathbf{N}''(x = L)$ are considered zero).

$$\mathbf{N}(x) = \hat{\mathbf{D}}_{j,:}(x - s_j)^3 + \hat{\mathbf{C}}_{j,:}(x - s_j)^2 + \hat{\mathbf{B}}_{j,:}(x - s_j) + \hat{\mathbf{A}}_{j,:} \ ; \ s_j < x \leq s_{j+1} \tag{6}$$

where the matrix of coefficients $\hat{\mathbf{A}}$, $\hat{\mathbf{B}}$, $\hat{\mathbf{C}}$, and $\hat{\mathbf{D}}$ are the size of n by n and to be calculated via the following equations and the row index, j, can be valued from 1 to $(n - 1)$ [29]:

$$\hat{\mathbf{A}} = \mathbf{I}_n \tag{7}$$

$$\hat{\mathbf{C}} = \mathbf{G}\mathbf{H}^{-1} \tag{8}$$

$$\hat{\mathbf{B}} = \frac{1}{\Delta s}\left[\hat{\mathbf{A}}_{2:n,:} - \hat{\mathbf{A}}_{1:n-1,:}\right] - \frac{\Delta s}{3}\left[\hat{\mathbf{C}}_{2:n,:} + 2\hat{\mathbf{C}}_{1:n-1,:}\right] \tag{9}$$

$$\hat{\mathbf{D}} = \frac{1}{3\Delta s}\left[\hat{\mathbf{C}}_{2:n,:} - \hat{\mathbf{C}}_{1:n-1,:}\right] \tag{10}$$

$$\mathbf{G} = \frac{3}{\Delta s}\begin{bmatrix} 0 & 0 & 0 & 0 & \cdots & \cdots & 0 \\ 1 & -2 & 1 & 0 & \cdots & \cdots & 0 \\ 0 & 1 & -2 & 1 & \cdots & \cdots & 0 \\ \vdots & \vdots & \vdots & \vdots & \ddots & \ddots & \vdots \\ 0 & 0 & 0 & \cdots & 1 & -2 & 1 \\ 0 & 0 & 0 & \cdots & 0 & 0 & 0 \end{bmatrix} \tag{11}$$

$$\mathbf{H} = \Delta s \begin{bmatrix} 1/\Delta s & 0 & 0 & 0 & \cdots & \cdots & 0 \\ 1 & 4 & 1 & 0 & \cdots & \cdots & 0 \\ 0 & 1 & 4 & 1 & \cdots & \cdots & 0 \\ \vdots & \vdots & \vdots & \vdots & \ddots & \ddots & \vdots \\ 0 & 0 & 0 & \cdots & 1 & 4 & 1 \\ 0 & 0 & 0 & \cdots & 0 & 0 & 1/\Delta s \end{bmatrix} \tag{12}$$

In the previous equations, Δs is the distance between two adjacent virtual fixed nodes that can be assumed constant.

After the determination of $\mathbf{N}(x)$, the continuous response function of the bridge can be calculated by Equation (1).

3.2. Proposed Moving-Window ARX Model to Complete the Missing Parts of the Estimated Responses

As discussed in Section 2, the classical method for solving the inverse problem can only estimate a small portion of the bridge displacement signal using drive-by data. In order to have the complete response for all the fixed nodes, a signal forecasting and completion approach is required.

In the present research, an innovative moving-window forecasting framework based on auto-regressive time series models with exogenous input (ARX) is introduced. The main motivation for this procedure is that the short valid part of the estimated response signal for a given fixed node can be trained by the corresponding parts of the responses of the adjacent nodes to forecast the missing response of the given node.

In other words, the proposed approach considers a unique ARX model for each of the fixed nodes. Therefore, $(n-2)$ different ARX models can be established for the whole system. The proposed procedure can be utilized to forecast the missing parts of the signal in both backward and forward directions; however, it should be noted that, for training and predicting the missing parts in the backward direction, the reverse of the signals is used (see Figure 6).

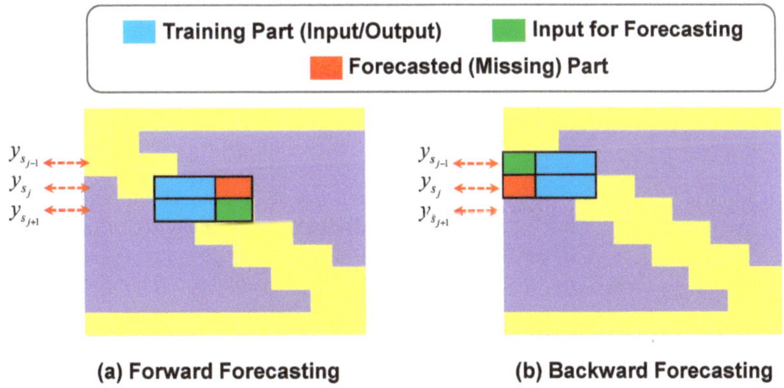

Figure 6. Explanation of the training parts and input in the ARX model for forecasting the missing parts of the response signal of node s_j.

The ARX model structure is generally given by the following equation [30]:

$$y(t) + a_1 y(t-\Delta t) + \cdots + a_{na} y(t - na\Delta t) = a_1 u(t-nk) + \cdots + a_{nb} u(t - (nb+nk-1)\Delta t) + e(t) \tag{13}$$

where $y(t)$ and $u(t)$ are the output and the input of the system, respectively; $a_1, \ldots, a_{na}$ and $b_1, \ldots, b_{nb}$ are parameters of the model, which can be identified through the least-squares optimization approach; $e(t)$ is a white-noise disturbance value.

The main reason for considering the ARX models is that the responses of two different fixed nodes located on the bridge can be decomposed to modal response components utilizing modal expansion principles for n_d modes [31]:

$$y_{s_j}(t) = \begin{bmatrix} \phi_{s_j,1} & \cdots & \phi_{s_j,n_d} \end{bmatrix} \begin{Bmatrix} q_1(t) \\ \vdots \\ q_{n_d}(t) \end{Bmatrix} \tag{14}$$

$$\mathbf{D}(t) = \begin{Bmatrix} y_{s_1}(t) \\ \vdots \\ y_{s_n}(t) \end{Bmatrix} = \begin{bmatrix} \phi_{s_1 1} & \cdots & \phi_{s_1 n_d} \\ \vdots & \ddots & \vdots \\ \phi_{s_n 1} & \cdots & \phi_{s_n n_d} \end{bmatrix} \begin{Bmatrix} q_1(t) \\ \vdots \\ q_{n_d}(t) \end{Bmatrix} = \boldsymbol{\Phi}^s \mathbf{Q}(t) \tag{15}$$

In a general case, multiplying both sides of Equation (15) by the pseudoinverse of $\boldsymbol{\phi}^s$ and substituting the resulting $\mathbf{Q}(t)$ in Equation (14) yields to Equation (17):

$$\mathbf{Q}(t) = \begin{Bmatrix} q_1(t) \\ \vdots \\ q_{n_d}(t) \end{Bmatrix} = \begin{bmatrix} \phi_{s_1 1} & \cdots & \phi_{s_1 n_d} \\ \vdots & \ddots & \vdots \\ \phi_{s_n 1} & \cdots & \phi_{s_n n_d} \end{bmatrix}^{-1} \begin{Bmatrix} y_{s_1}(t) \\ \vdots \\ y_{s_n}(t) \end{Bmatrix} \tag{16}$$

$$y_{s_j}(t) = \begin{bmatrix} \phi_{s_j,1} & \cdots & \phi_{s_j,n_d} \end{bmatrix} \begin{bmatrix} \phi_{s_1 1} & \cdots & \phi_{s_1 n_d} \\ \vdots & \ddots & \vdots \\ \phi_{s_n 1} & \cdots & \phi_{s_n n_d} \end{bmatrix}^{-1} \begin{Bmatrix} y_{s_1}(t) \\ \vdots \\ y_{s_n}(t) \end{Bmatrix} = \begin{bmatrix} b_{j,1} & \cdots & b_{j,n} \end{bmatrix} \begin{Bmatrix} y_{s_1}(t) \\ \vdots \\ y_{s_n}(t) \end{Bmatrix} \tag{17}$$

According to our numerical investigations, only the contribution of the adjacent nodes is high for the response prediction of the jth node. Therefore, we neglect the contributions of the other fixed nodes in Equation (17) and, by doing so, the pattern of the response can still be maintained.

Comparing Equation (17) with the general ARX model introduced in Equation (13), the time series model for the jth node can be simplified as follows:

$$y_{s_j}(t) + a_{j,1} y_{s_j}(t-\Delta t) + a_{j,2} y_{s_j}(t-2\Delta t) = b_{i,1} y_{s_{j+1}}(t) + b_{j,2} y_{s_{j+1}}(t-\Delta t) + b_{j,3} y_{s_{j+1}}(t-2\Delta t) + e(t) \tag{18}$$

As obtained in Equation (18), the information on external force and vehicle characteristics in the proposed model is not required. The rationale behind considering a second-order ARX model is that using acceleration response in the time series model can give better accuracy.

By performing the proposed method for all the intermediate nodes, some parts of the bridge response signals can be completed according to Figure 7. As can be seen from the figure, this approach is only able to complete the missing parts of the signal in which the input for the time series model can be provided, considering the missing gap between the two consecutive signals.

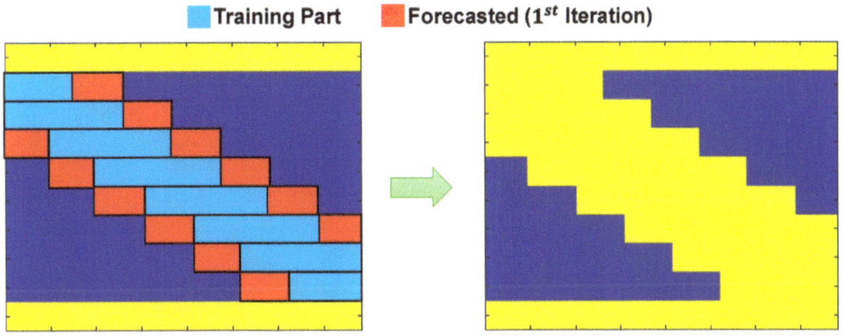

Figure 7. The first iteration of the proposed moving ARX technique to complete the missing parts of the response matrix, where the inputs are available for forecasting backward and forward values using the time series models.

In order to complete the entire responses of the fixed nodes, the moving ARX algorithm will be applied through a straight-forward iterative procedure. The proposed signal completion framework is shown in Figure 8 for the displacement response of an intermediate fixed node. It is important to highlight that the number of iterations needed depends on the quantity of moving axles and the number of virtual fixed nodes. For a model subjected to three-axle moving loads, a total of six iterations are required. As the number of axles increases, the number of iterations decreases due to a reduction in the valid regions. During each iteration of the algorithm, the predicted regions are combined with the initial valid regions, leading to an expansion of the valid regions within the response matrix. The iterative process continues until there are no remaining invalid regions in the response matrix.

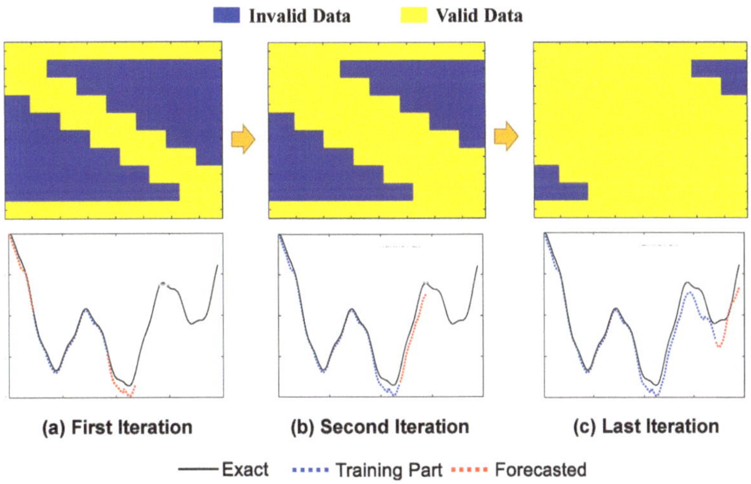

Figure 8. An illustrative example of the iterative moving ARX framework to complete the response signal of fixed nodes (applicable for either displacement or acceleration).

3.3. Integrating Displacement and Acceleration Responses for Modal Identification

In this paper, a novel mode shape identification through the response of the moving wheels only is considered. To be more precise, both displacement and acceleration of the fixed nodes are first estimated with the aid of the proposed moving ARX method, then singular value decomposition (SVD) is applied on all of the nodal displacement responses to identify the first mode shape; however, for the higher modes and identification of natural frequencies, frequency domain decomposition (FDD) is utilized, considering the acceleration responses of the fixed nodes. This is based on the fact that combining displacement and acceleration responses in modal identification improves accuracy and reliability. Displacement responses are effective for identifying lower modes, while acceleration responses capture high-frequency components and higher modes more accurately [32]. The mathematical relationship between accelerations and displacements reduces high-frequency components in dynamic displacements. This reduction is due to the fact that the integration operation acting as a low-pass filter and the accumulation of the area under the acceleration curve during integration [33].

Since the SVD can extract the orthogonal vectors of an arbitrary matrix, it can be applied to the completed response matrix (**D**) to determine $\boldsymbol{\phi}^s$:

$$\mathbf{D} = \mathbf{U}\boldsymbol{\Sigma}\mathbf{V}^{tr} \tag{19}$$

where **U** and **V** are composed of the left and right singular vectors of matrix **D**, respectively; they are orthogonal matrices. $\boldsymbol{\Sigma}$ is a diagonal matrix containing singular values of **D**.

By comparing the right side of Equations (15) and (19), identification of the mode shape can be performed with high accuracy [25]:

$$\boldsymbol{\Phi}^s = \mathbf{U}\boldsymbol{\Sigma}^{\frac{1}{2}} \tag{20}$$

We observed that SVD can identify the first mode shapes with high accuracy if applied on the displacement response matrix of the bridge.

As mentioned earlier, in order to identify the mode shapes through the acceleration response matrix of the bridge, the FDD method is employed [34]. The basis of FDD is presented in the following paragraph.

From statistics, the correlation matrix between the response of the fixed nodes can be constructed using Equation (21):

$$\mathbf{R}_{\ddot{y}\ddot{y}}(\tau) = \frac{1}{T}\int_0^T \ddot{\mathbf{D}}(t)\ddot{\mathbf{D}}^{tr}(t-\tau)dt = \begin{bmatrix} R_{\ddot{y}_{s_1}\ddot{y}_{s_1}}(\tau) & \cdots & R_{\ddot{y}_{s_1}\ddot{y}_{sn}}(\tau) \\ \vdots & \ddots & \vdots \\ R_{\ddot{y}_{sn}\ddot{y}_{s_1}}(\tau) & \cdots & R_{\ddot{y}_{sn}\ddot{y}_{sn}}(\tau) \end{bmatrix} \tag{21}$$

Considering modal expansion and substituting Equation (15) in Equation (21) gives:

$$\mathbf{R}_{\ddot{y}\ddot{y}}(\tau) = \frac{1}{T}\int_0^T \boldsymbol{\Phi}\ddot{\mathbf{Q}}(t)\ddot{\mathbf{Q}}^{tr}(t-\tau)\boldsymbol{\Phi}^{tr}dt = \boldsymbol{\Phi}\mathbf{R}_{\ddot{q}\ddot{q}}(\tau)\boldsymbol{\Phi}^{tr} \tag{22}$$

Then, taking Fourier transform from both sides of the latter equation produces the matrix containing cross/auto power spectrums of the response signals:

$$\mathbf{G}_{\ddot{q}\ddot{q}}(\omega_i) = \boldsymbol{\Phi}\mathbf{G}_{\ddot{q}\ddot{q}}(\omega_i)\boldsymbol{\Phi}^{tr} \tag{23}$$

$$\mathbf{G}_{\ddot{q}\ddot{q}}(\omega_i) = \mathbf{U}_i\boldsymbol{\Sigma}_i\mathbf{V}_i^{tr} \tag{24}$$

As can be understood from Equations (23) and (24), the SVD of matrix $\mathbf{G}_{\ddot{q}\ddot{q}}(\omega_i)$ should be calculated for each frequency, ω_i, in which the matrix of singular values ($\boldsymbol{\Sigma}_i$) is a diagonal matrix containing modal FRFs and each column of the matrix of singular vectors ($\mathbf{U}_i$ and $\mathbf{V}_i$) represents the mode shapes corresponding to the given frequency ω_i.

4. Results

4.1. Model Setup

To evaluate the effectiveness of the proposed framework, a comprehensive numerical analysis is conducted on a single-span simply supported bridge using the finite element software package, ABAQUS. The bridge under consideration has a span length of 40 m and a rectangular cross-section with dimensions of 3 m wide and 1.5 m high. The material properties of the bridge are assigned based on concrete, with a density of 2400 kg/m^3 and an elastic modulus of 27.5 GPa.

In the numerical model, the bridge is subjected to the loading of a three-axle moving vehicle. The distance between the axles is set to 2.5 m, as shown in Figure 1. The natural frequencies of the bridge model are determined to be 1.44 Hz, 5.76 Hz, and 12.95 Hz for the first three modes, respectively.

A constant speed of 60 km/h is assigned to all the moving axles. The analysis is performed using a linear implicit dynamic analysis approach, considering the contacts between the moving axles and the bridge. The simulation is terminated when the foremost moving axle reaches the right end of the bridge.

For data acquisition, the accelerometers mounted on the moving vehicle have a constant sampling frequency of 200 Hz. Although a total of nine virtual fixed nodes are defined on the bridge, for the purpose of verification and comparison, only three specific fixed nodes located at $\frac{1}{4}$, $\frac{1}{2}$, and $\frac{3}{4}$ of the span are selected to verify the displacement and acceleration responses.

Further details regarding the numerical model can be found in the previous work of the authors [26]. The established numerical setup provides a realistic representation of a bridge structure and allows for the comprehensive evaluation of the proposed framework's performance.

4.2. Interpretation of Results

As explained in Section 2, by utilizing the measured acceleration of the moving axles in Equation (5), the valid part of the acceleration response signal of the fixed nodes on the bridge can be estimated. Similarly, to determine the displacement response signal, it is enough to double integrate the measured acceleration signals of the axles and put them in Equation (4).

In order to evaluate the proposed framework, the exact response of the fixed nodes on the bridge will be used directly from the numerical model. Although the acceleration and displacement responses of all nine fixed nodes can be determined using the proposed method, only the results from the linear and spline shape functions of three validation nodes are shown in Figures 9 and 10.

The displacement and acceleration responses of the fixed nodes are determined using linear and spline shape functions, respectively, as outlined earlier. Figure 9 shows that the cubic spline shape function proposed in this study provides more accurate acceleration response estimates of the fixed nodes in the valid regions compared with the conventional linear approach. On the other hand, the linear shape function provides more precise displacement response estimates compared with the cubic spline shape function (Figure 10).

The difference observed between the linear and spline shape functions, regarding their impact on displacement response estimates, can be attributed to multiple factors. Firstly, the inherent characteristics of the displacement response itself play a significant role. Displacement responses primarily consist of low-frequency components that reflect the overall steady-state behavior of the system. The linear shape function, with its linear interpolation, aligns well with this low-frequency behavior, resulting in more precise displacement estimates.

Secondly, acceleration responses exhibit more complex dynamics and transient behaviors, often characterized by high-frequency oscillations. The cubic spline shape function, which incorporates higher-order interpolation, is better equipped to capture these intricate features, leading to more accurate estimates in the acceleration domain.

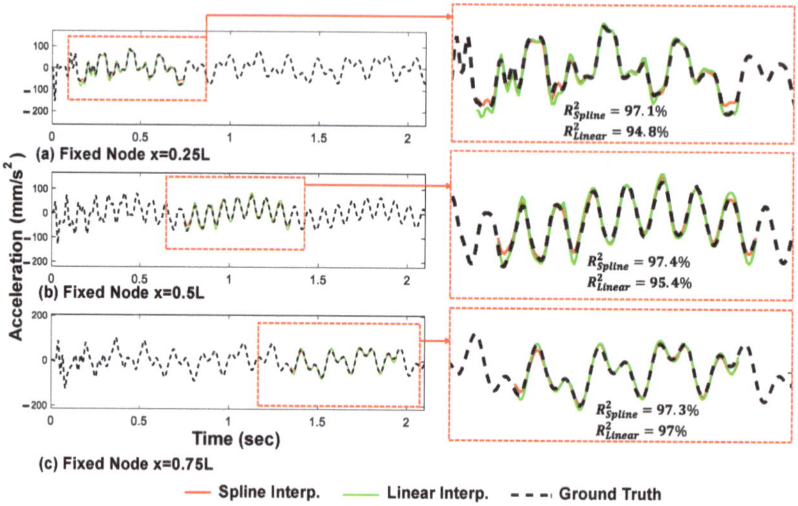

Figure 9. Estimated acceleration responses of the bridge in their valid regions through the linear and spline shape functions and inverse problem solution (three moving axles).

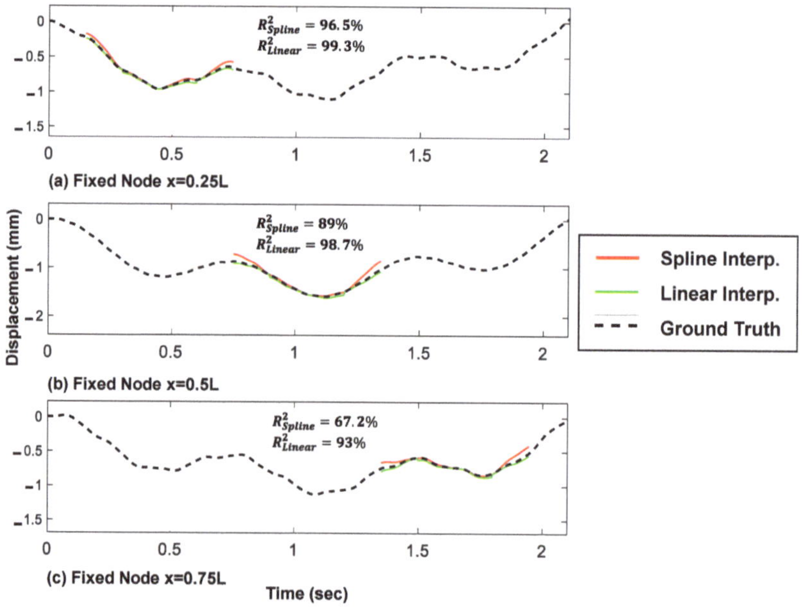

Figure 10. Estimated displacement responses of the bridge in their valid regions through the linear and spline shape functions (three moving axles).

In summary, the choice between the linear and the spline shape functions depends on the specific nature of the response being analyzed. The linear shape function excels in capturing low-frequency displacement components, while the cubic spline shape function is advantageous for accurately representing the complex dynamics and high-frequency oscillations present in acceleration responses.

Hence, both responses of the fixed nodes are utilized in the proposed modal identification method. It is worth noting that higher accuracy of the determined responses in the valid regions results in higher accuracy of the predicted whole signals from the proposed moving time series model, based on the authors' numerical observations.

The predicted displacement and acceleration responses of the bridge at the verification points, using only the measured acceleration of the moving axles and their relative amplitude errors, are presented in Figures 11 and 12, respectively. The relative error of the predicted responses outside the valid regions is high in the case of a three-axle vehicle. However, as shown in the following sections, using more moving axles reduces these errors and, for all models, the mode shape identification accuracy is very high.

It is well known that the identification of lower modes of structures is easier using displacement responses, while higher modes can be identified using acceleration responses with higher accuracy [33]. Accordingly, a hybrid mode shape identification framework is proposed, where SVD is applied to the predicted displacement responses of the fixed nodes (based on the linear shape function) to identify the first mode shape. On the other hand, the FDD technique is employed to identify the higher modes and natural frequencies by analyzing the acceleration responses (based on the cubic spline shape function) of the fixed nodes.

The results of the identified mode shapes and natural frequencies are presented in Figure 13 and Table 2, respectively. It is worth noting that the identified natural frequencies are based on the completed acceleration responses and considering the plot of the first singular value obtained from FDD.

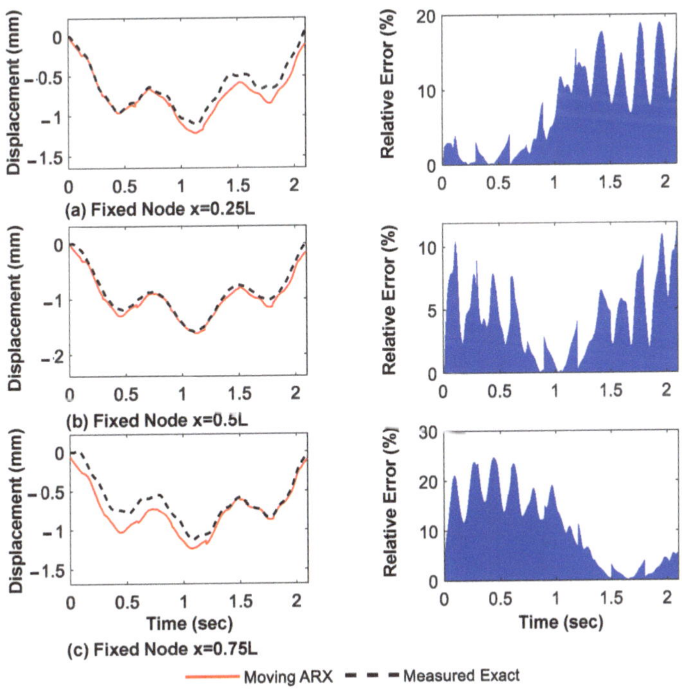

Figure 11. Predicted displacement responses extracted using the proposed framework and their relative amplitude errors (three moving axles, linear shape function).

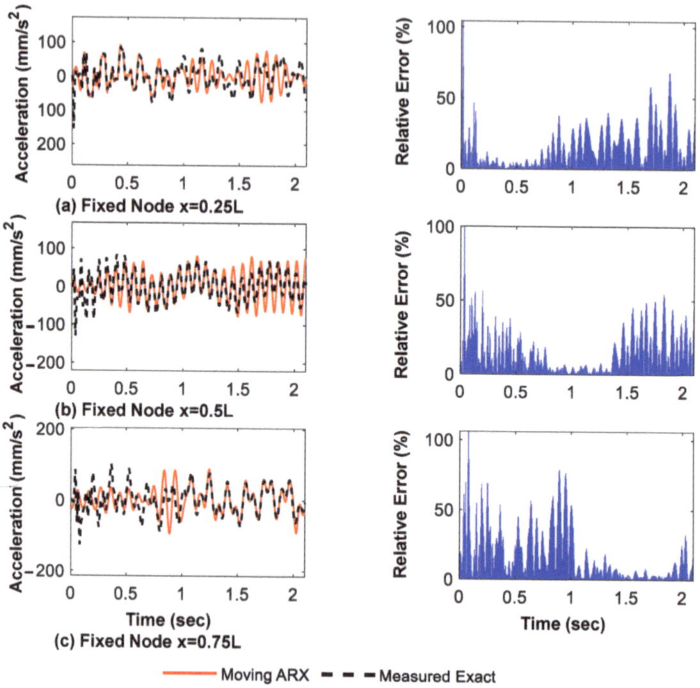

Figure 12. Predicted acceleration responses extracted using the proposed framework and their relative amplitude errors (three moving axles, cubic spline shape function).

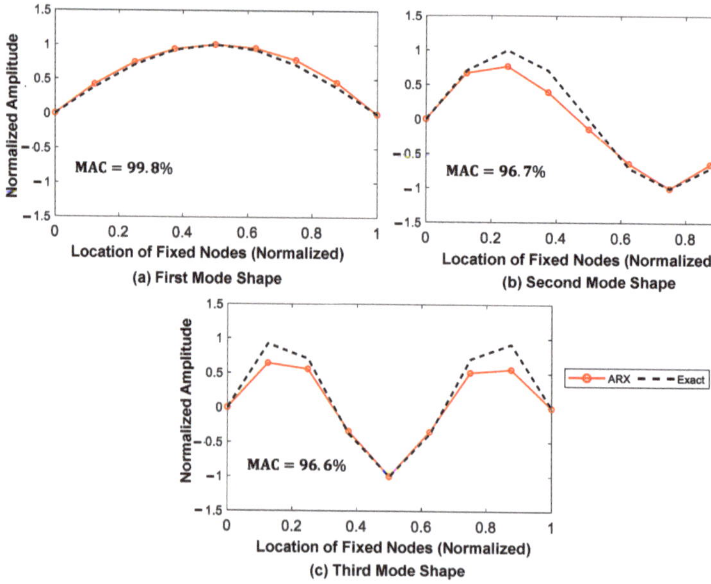

Figure 13. Identified mode shapes of the bridge using the proposed framework (three moving axles).

Table 2. Identified natural frequencies and MAC values of the mode shapes.

Mode Number	Natural Frqs (Hz) (Exact)	Identified Natural Frqs (Hz) (FDD)	Error (%)	MAC Mode Shapes (%)
Mode 1	1.44	1.56	8.33	99.8
Mode 2	5.76	5.86	1.74	96.7
Mode 3	12.95	12.5	3.47	96.6

5. Discussion

5.1. Sensitivity of the Inverse Solution to the Number of Virtual Fixed Nodes

The accuracy of the theoretical inverse solution method in Equations (4) and (5) is highly dependent on the number of virtual fixed nodes, which is equivalent to the mesh size of the bridge element. Therefore, it is crucial to investigate the sensitivity of the estimated response in the valid regions for various numbers of fixed nodes using both linear and spline shape functions. As shown in Figure 14, for the linear shape function case, increasing the number of fixed nodes improves the accuracy of the estimated acceleration and displacement responses of the mid-span point up to a certain point. Beyond this point, the accuracy of the estimated response declines steadily. The ascending branch of the curve can be attributed to the fact that the finite element method requires an increased number of intermediate nodes (number of elements) to determine the displacements of the fixed nodes with higher accuracy. However, increasing the number of fixed nodes leads to an increase in the number of columns in matrix $\mathbf{N}(t)$ and, consequently, a large computational error in calculating the pseudoinverse of $\mathbf{N}(t)$. The second part of the sensitivity graph is downward and indicates a decrease in the accuracy of the estimated response. In contrast to the linear shape function results, the cubic spline shape function can achieve high accuracy with even a few interpolating fixed nodes, such as three points, but it is more sensitive to an increase in the number of fixed nodes. To ensure a fair comparison between the two methods, this study employs a constant value of nine fixed nodes on the bridge. It is worth noting that the coefficient of determination, R^2, is used in the subsequent figures to evaluate the similarity between two signals as well as the fitting accuracy.

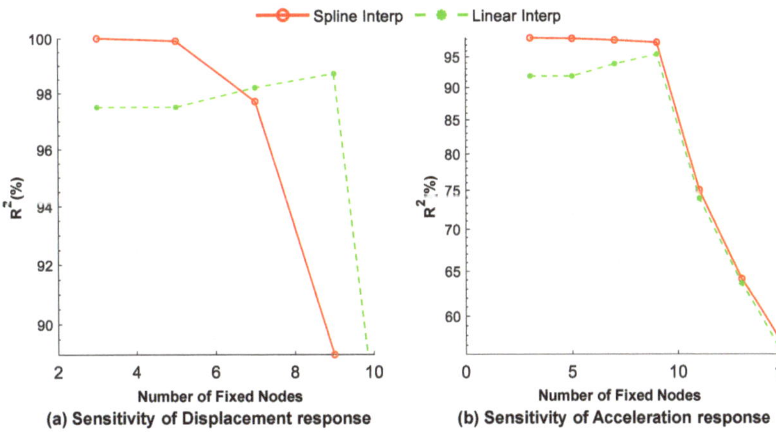

Figure 14. Sensitivity of the estimated responses using the inverse solution for the mid-span point in the valid region to the number of fixed nodes (three moving axles).

5.2. Influence of the Number of Moving Axles

As previously discussed, a larger number of axles passing over the bridge results in a longer valid region for the response signals. This provides a better measurement of the accuracy of the preliminary model fit and is expected to enhance the accuracy of the bridge response prediction. To evaluate this hypothesis, three different types of moving vehicles, each with four, six, or 8 axles, were considered; the proposed method was used to determine the displacement, acceleration response of the fixed nodes, and modal characteristics of the bridge.

Figures 15 and 16 show the predicted displacement and acceleration responses, respectively, for the mid-span of the bridge, along with their time distribution of the prediction error relative to the exact response. As anticipated, the relative prediction error of the mid-span response was significantly reduced with an increase in the number of axles. Figure 17 shows the three main identified mode shapes based on the predicted acceleration and displacement responses of all fixed nodes for all three different loading types. It can be observed that, although the accuracy of identifying higher mode shapes increased with an increase in the number of axles, the first mode shape was not significantly affected.

It is important to note that an increase in the number of axles would require a corresponding increase in the number of sensors, resulting in additional costs for monitoring the structure. However, there was no significant change in the accuracy of the identified modal characteristics. Therefore, using moving vehicles with fewer axles could reduce the costs of bridge health monitoring while maintaining an acceptable level of accuracy.

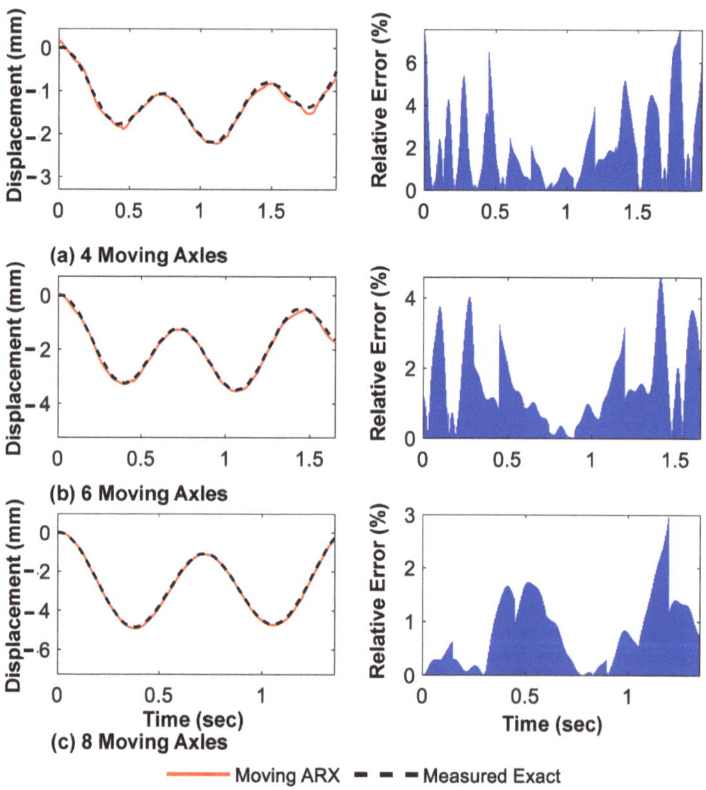

Figure 15. Predicted displacement response of mid-span using the proposed framework and their relative to the amplitude errors for different numbers of moving axles (linear shape function).

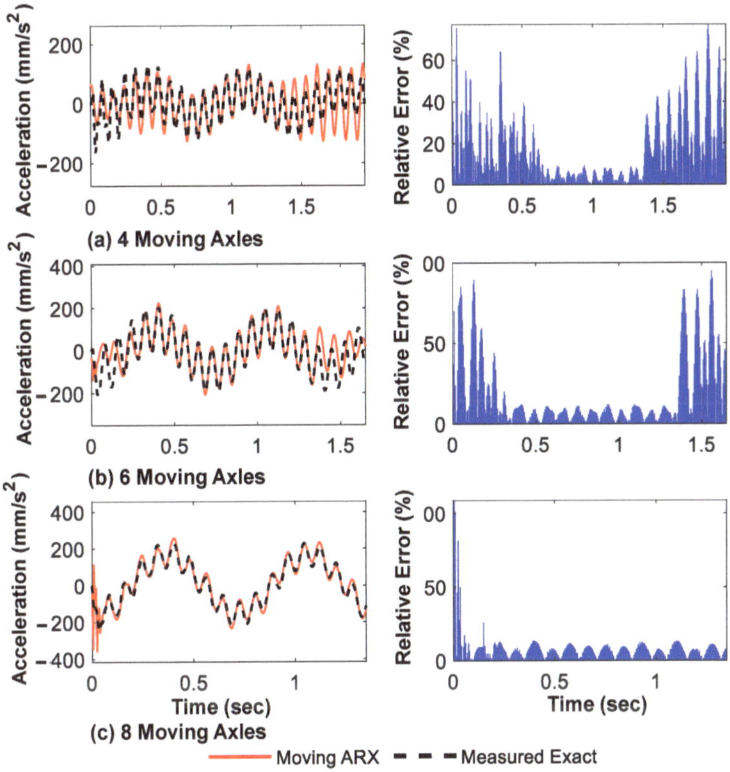

Figure 16. Predicted acceleration response of the mid-span using the proposed framework and their relative to amplitude errors for different numbers of moving axles (cubic spline shape function).

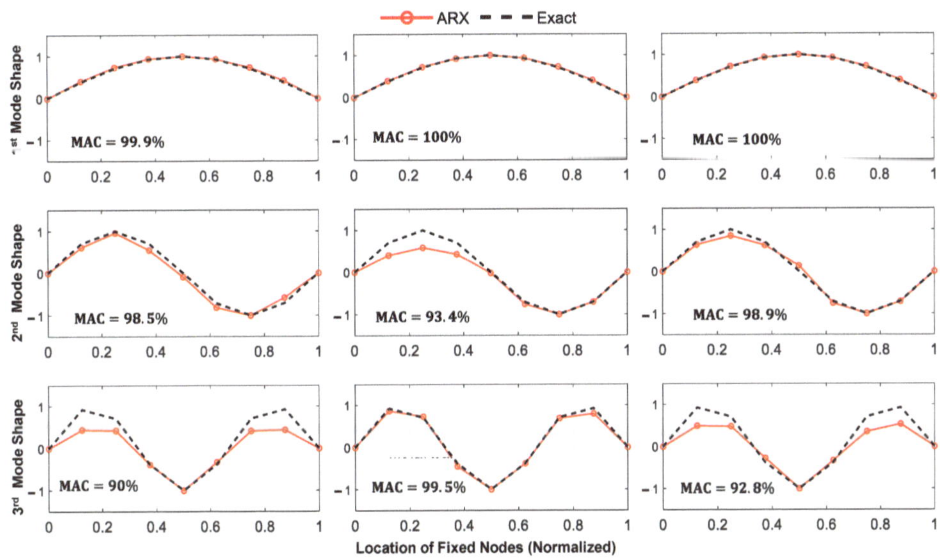

Figure 17. The first three identified mode shapes of the bridge under different number of moving axles (4, 6, and 8 moving axles) using the hybrid approach.

5.3. Influence of Vehicle Speed on the Identification Results

The speed of the vehicles passing over the bridge is one of the main parameters that may affect the accuracy of identification based on the vehicle response. In this section, the effect of the parameter on the proposed method is evaluated. The speed of the moving axles is varied between 20 and 80 km/h and its effect on the accuracy of the identified mode shapes and natural frequencies is investigated. Figure 18 compares the modal assurance criterion (MAC) values of the first three identified mode shapes at different speeds for three different vehicles with varying numbers of axles. Similarly, Figure 19 shows the relative error in identifying the natural frequencies for the first three modes at different speeds and for different numbers of axles.

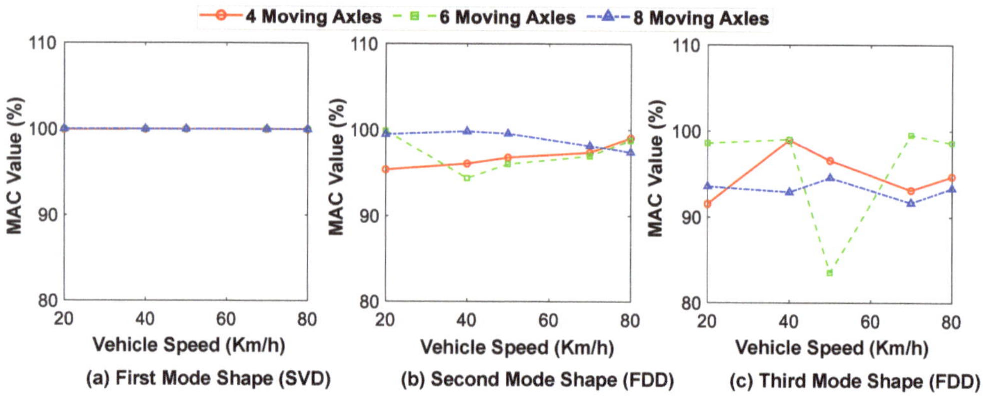

Figure 18. MAC values of the first three identified mode shapes at different speeds for different number of axles (in comparison with exact mode shapes).

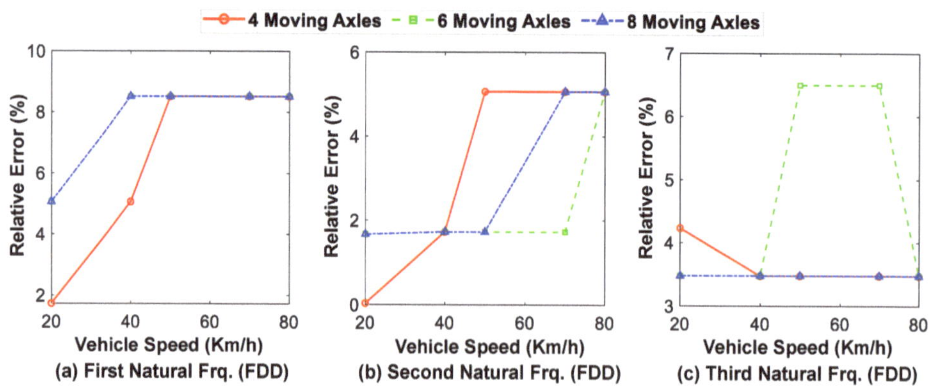

Figure 19. The relative error of the first three identified modal frequencies at different speeds for different numbers of axles (in comparison with exact natural frequencies).

Upon analyzing these figures, it can be inferred that the accuracy of mode shape identification is generally higher at lower speeds, since more time information can be obtained from the bridge response. However, the accuracy of identifying the natural frequencies reduces by almost half as the vehicle speed increases. Furthermore, increasing the number of moving axles does not significantly affect the accuracy of the identified modal characteristics through the proposed hybrid method, where the predicted displacement responses are considered for the first mode and the acceleration responses are used for the higher modes.

5.4. Investigation of Ambient Noise Effects

Measurement errors and environmental vibrations can affect the accuracy of the proposed framework. To investigate the effects of ambient noise on the identification of mode shapes and natural frequencies using the proposed technique, different levels of artificial noise were added to the measured acceleration response of the moving axles assuming a zero-mean Gaussian distribution. The applied noise amplitude was considered as a percentage of the RMS of the measured acceleration in the range of 1–5% (corresponding to the signal-to-noise ratio of 40–26 dB).

Figure 20 shows the MAC values of the identified mode shapes by the proposed hybrid method in different levels of ambient noise compared with their exact values. The presence of ambient noise reduces the accuracy of the mode shape identification, although this sensitivity to noise is more evident in some cases, such as for six moving axles. Moreover, although the sensitivity of the first mode detection in different levels of noise has decreased with an increase in the number of axles, this trend is almost inverted for higher modes.

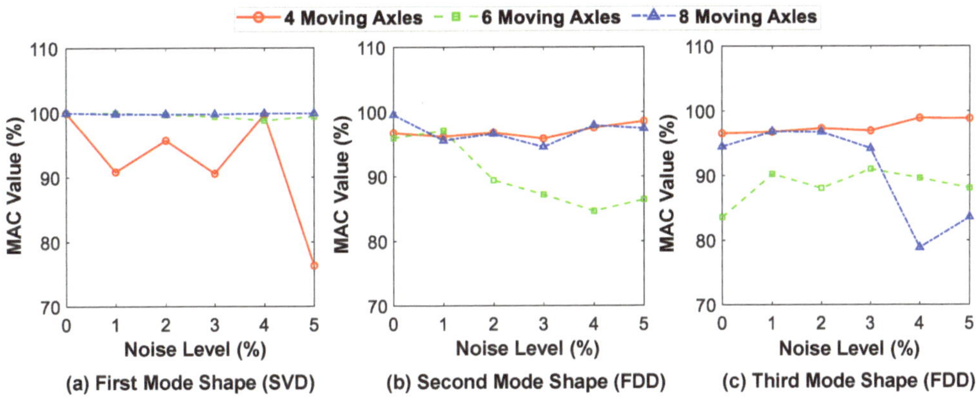

Figure 20. MAC values of the first three identified mode shapes at different noise levels for different numbers of axles (in comparison with exact mode shapes).

It should be noted that the difference in the behavior pattern between the first mode and the higher modes is due to the use of different methods. As mentioned in Section 3, the SVD method is used to identify the shape of the first mode from the predicted displacement responses of the bridge, while the FDD method is used to identify the higher modes from the acceleration responses in this study; their behavior is also different in different noise levels. In conclusion, the hybrid identification technique can detect the mode shapes reasonably accurately for all three modes while utilizing fewer axles.

Figure 21 depicts the relative error of the first three identified natural frequencies for the bridge compared with their exact values at different levels of ambient noise. The results indicate that the proposed method has high accuracy in determining the natural frequencies of the higher modes and is robust to noise. However, this behavior is not observed when investigating the effect of noise on the identified frequency of the first mode through the predicted acceleration response signal for the fixed nodes. The main reason for this is that the FDD method cannot accurately determine the position of the first peak of the first singular value diagram (first mode), utilizing the acceleration results due to the presence of ambient noise. The accuracy and robustness of the first identified natural frequency is generally higher in models with a smaller number of moving axles, highlighting the high capability of the proposed method for implementing it using the response of conventional vehicles passing over the bridge.

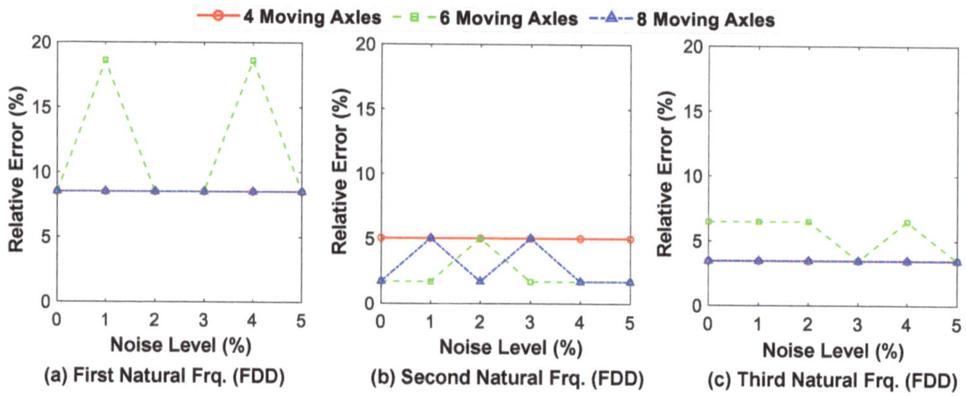

Figure 21. The relative errors of the first three identified modal frequencies at different noise levels for different numbers of axles (in comparison with exact values).

Overall, the average detection error at different levels of ambient noise for the frequency of the first mode shape is less than 10%, while for the frequency of higher modes it is less than 5%. These results indicate the promising efficiency of the proposed framework, which utilizes only the response of the axles of the moving vehicle, in identifying the mode shapes and natural frequencies of the bridge, even in the presence of ambient noise.

6. Conclusions

This paper contributes to indirect bridge health monitoring through the introduction of an automated and comprehensive framework based on an inverse problem solution approach and a novel moving-window time series model for response prediction. The major findings and contributions of this research can be summarized as follows:

- Accurate modal characteristics' identification: The proposed method demonstrates accurate identification of the modal characteristics for the first three modes of bridges within normal traffic speeds ranging from 20 to 70 km/h. This capability is crucial for assessing the structural health and integrity of bridges.
- Novel use of cubic spline shape function and moving-window time series models: The research introduces the use of a cubic spline shape function within the inverse problem solution for predicting acceleration responses. Additionally, the application of moving-window time series models to complete the predicted signals further enhances the accuracy of the framework.
- Novel approach for mode shape identification: The framework utilizes predicted displacement and acceleration responses to identify the first and higher mode shapes of the bridge, respectively. This approach enhances the accuracy and robustness of mode shape identification for lower and higher modes.
- Cost-effective solution: The method requires only one vehicle with a limited number of axles, which significantly reduces the number of sensors compared with traditional fixed sensor setups. This offers a cost-effective solution for bridge health monitoring without compromising accuracy.

However, there are certain limitations that should be acknowledged:

- Influence of the number of moving axles: The accuracy of identification improves with an increase in the number of axles passing over the bridge. However, there is no significant effect on the identification of the first mode shape. This finding highlights the need to optimize the number of axles (vehicles) used in bridge monitoring systems.
- Influence of vehicle speed on identification: Mode shape identification is more accurate at lower speeds, while the accuracy of identifying natural frequencies decreases with

higher vehicle speeds. Considering vehicle speed is important for designing effective bridge monitoring strategies.

- Sensitivity to ambient noise: As expected, the presence of ambient noise reduces the accuracy of mode shape identification, particularly for the higher modes. The accuracy and robustness of the first identified natural frequency is generally higher in models with fewer moving axles.

Further research and improvements can be pursued in the following areas:

- Enhancement of the response prediction models: exploring different models or techniques to improve the accuracy of predicted responses can lead to more precise identification of modal characteristics and better performance in the presence of noise.
- Multi-vehicle scenarios: investigating the applicability of the proposed method in scenarios with multiple vehicles of varying speeds crossing the bridge can provide a more comprehensive understanding of its capabilities and limitations.
- Experimental validation: conducting experimental investigations on real-life structures will be crucial to validate the accuracy and effectiveness of the proposed framework in practical applications.

In summary, while the proposed framework presents a promising approach for indirect bridge health monitoring, further research is needed to address the limitations and to refine the method for broader applicability and improved accuracy in real-world scenarios. These advancements will contribute to the field of bridge health monitoring, ensuring the safety and longevity of transportation infrastructure.

Author Contributions: Conceptualization, Q.M.; data curation, M.T.-K.; formal analysis, M.T.-K.; funding acquisition, Q.M.; investigation, Q.M.; methodology, M.T.-K.; project administration, Q.M.; resources, Q.M.; supervision, Q.M.; validation, M.T.-K.; visualization, M.T.-K.; writing—original draft, M.T.-K.; writing—review and editing, Q.M. All authors have read and agreed to the published version of the manuscript.

Funding: This research was funded by the Natural Sciences and Engineering Research Council of Canada (NSERC) through the Discovery Grant (RGPIN-2022-04160) and Alliance Grant (ALLRP 576826–22).

Institutional Review Board Statement: Not applicable.

Informed Consent Statement: Not applicable.

Data Availability Statement: Some or all data, models, or code that support the findings of this study are available from the corresponding author upon reasonable request.

Conflicts of Interest: The authors declare no conflict of interest.

References

1. Fan, W.; Qiao, P. Vibration-Based Damage Identification Methods: A Review and Comparative Study. *Struct. Health Monit.* **2011**, *10*, 83–111. [CrossRef]
2. Sun, L.; Shang, Z.; Xia, Y.; Bhowmick, S.; Nagarajaiah, S. Review of Bridge Structural Health Monitoring Aided by Big Data and Artificial Intelligence: From Condition Assessment to Damage Detection. *J. Struct. Eng.* **2020**, *146*, 04020073. [CrossRef]
3. Das, S.; Patro, P.S.S.K. Vibration-Based Damage Detection Techniques Used for Health Monitoring of Structures: A Review. *J. Civ. Struct. Health Monit.* **2016**, *6*, 477–507. [CrossRef]
4. Gonzalez, I.; Karoumi, R. BWIM Aided Damage Detection in Bridges Using Machine Learning. *J. Civ. Struct. Health Monit.* **2015**, *5*, 715–725. [CrossRef]
5. Azim, M.R.; Gül, M. Development of a Novel Damage Detection Framework for Truss Railway Bridges Using Operational Acceleration and Strain Response. *Vibration* **2021**, *4*, 28. [CrossRef]
6. Feng, D.; Feng, M.Q. Model Updating of Railway Bridge Using In Situ Dynamic Displacement Measurement under Trainloads. *J. Bridg. Eng.* **2015**, *20*, 04015019. [CrossRef]
7. Yang, Y.B.; Lin, C.W.; Yau, J.D. Extracting Bridge Frequencies from the Dynamic Response of a Passing Vehicle. *J. Sound Vib.* **2004**, *272*, 471–493. [CrossRef]
8. Singh, P.; Mittal, S.; Sadhu, A. Recent Advancements and Future Trends in Indirect Bridge Health Monitoring. *Pract. Period. Struct. Des. Constr.* **2023**, *28*, 03122008. [CrossRef]

9. Malekjafarian, A.; McGetrick, P.J.; Obrien, E.J. A Review of Indirect Bridge Monitoring Using Passing Vehicles. *Shock Vib.* **2015**, *2015*, 286139. [CrossRef]
10. Malekjafarian, A.; OBrien, E.J. Identification of Bridge Mode Shapes Using Short Time Frequency Domain Decomposition of the Responses Measured in a Passing Vehicle. *Eng. Struct.* **2014**, *81*, 386–397. [CrossRef]
11. Sadeghi Eshkevari, S.; Pakzad, S.N.; Takáč, M.; Matarazzo, T.J. Modal Identification of Bridges Using Mobile Sensors with Sparse Vibration Data. *J. Eng. Mech.* **2020**, *146*, 04020011. [CrossRef]
12. Kong, X.; Cai, C.S.; Deng, L.; Zhang, W. Using Dynamic Responses of Moving Vehicles to Extract Bridge Modal Properties of a Field Bridge. *J. Bridg. Eng.* **2017**, *22*, 04017018. [CrossRef]
13. Chen, Y.H.; Chang, P.Y.; Chen, Y.Y. Indoor Positioning Design for Mobile Phones via Integrating a Single Microphone Sensor and an H2 Estimator. *Sensors* **2023**, *23*, 1508. [CrossRef] [PubMed]
14. Perspectives, A. A Systematic Review of Mobile Phone Data in Crime Applications: A Coherent Taxonomy Based on Data Types and analysis perspectives, challenges, and future research directions. *Sensors* **2023**, *23*, 4350.
15. Pongnumkul, S.; Chaovalit, P.; Surasvadi, N. Applications of Smartphone-Based Sensors in Agriculture: A Systematic Review of Research. *J. Sens.* **2015**, *2015*, 195308. [CrossRef]
16. Shrestha, A.; Dang, J.; Wang, X.; Matsunaga, S. Smartphone-Based Bridge Seismic Monitoring System and Long-Term Field Application Tests. *J. Struct. Eng.* **2020**, *146*, 04019208. [CrossRef]
17. Na, Y.; El-Tawil, S.; Ibrahim, A.; Eltawil, A. Automated Assessment of Building Damage from Seismic Events Using Smartphones. *J. Struct. Eng.* **2020**, *146*, 04020076. [CrossRef]
18. Xie, B.; Li, J.; Zhao, X. Research on Damage Detection of a 3D Steel Frame Model Using Smartphones. *Sensors* **2019**, *19*, 745. [CrossRef]
19. Avnon, R. Smartphone-Based Vibration Analysis for Bridge Health Monitoring. Bachelor's Thesis, University of Twente, Enschede, The Netherlands, 2022.
20. Matarazzo, T.; Vazifeh, M.; Pakzad, S.; Santi, P.; Ratti, C. Smartphone Data Streams for Bridge Health Monitoring. *Procedia Eng.* **2017**, *199*, 966–971. [CrossRef]
21. Di Matteo, A.; Fiandaca, D.; Pirrotta, A. Smartphone-Based Bridge Monitoring through Vehicle–Bridge Interaction: Analysis and Experimental Assessment. *J. Civ. Struct. Health Monit.* **2022**, *12*, 1329–1342. [CrossRef]
22. Shirzad-Ghaleroudkhani, N.; Gül, M. Inverse Filtering for Frequency Identification of Bridges Using Smartphones in Passing Vehicles: Fundamental Developments and Laboratory Verifications. *Sensors* **2020**, *20*, 1190. [CrossRef] [PubMed]
23. Sitton, J.D.; Rajan, D.; Story, B.A. Bridge Frequency Estimation Strategies Using Smartphones. *J. Civ. Struct. Health Monit.* **2020**, *10*, 513–526. [CrossRef]
24. Shokravi, H.; Shokravi, H.; Bakhary, N.; Heidarrezaei, M.; Koloor, S.S.R.; Petrů, M. Vehicle-Assisted Techniques for Health Monitoring of Bridges. *Sensors* **2020**, *20*, 3460. [CrossRef]
25. Oshima, Y.; Yamamoto, K.; Sugiura, K. Damage Assessment of a Bridge Based on Mode Shapes Estimated by Responses of Passing Vehicles. *Smart Struct. Syst.* **2014**, *13*, 731–753. [CrossRef]
26. Mei, Q.; Shirzad-Ghaleroudkhani, N.; Gül, M.; Ghahari, S.F.; Taciroglu, E. Bridge Mode Shape Identification Using Moving Vehicles at Traffic Speeds through Non-Parametric Sparse Matrix Completion. *Struct. Control Health Monit.* **2021**, *28*, e2747. [CrossRef]
27. Zhu, X.Q.; Law, S.S. Recent Developments in Inverse Problems of Vehicle–Bridge Interaction Dynamics. *J. Civ. Struct. Health Monit.* **2016**, *6*, 107–128. [CrossRef]
28. Bathe, K.-J. *Finite Element Procedures*; Prentice-Hall, Inc.: Hoboken, NJ, USA, 1996; ISBN 9780123849847.
29. Rosenblatt, J. Cubic Splines. *FASEB J.* **1988**, *2*, 2425. [CrossRef]
30. Rune, B.; Carlos, E. *Ventura Introduction to Operational Modal Analysis*; John Wiley & Sons, Ltd.: Chichester, UK, 2015; ISBN 9781119963158.
31. Chopra, A.K. *Dynamics of Structures: Theory and Applications to Earthquake Engineering*, 4th ed.; Prentice Hall: Upper Saddle River, NJ, USA, 2012; ISBN 1292249188.
32. Kim, S.; Park, K.Y.; Kim, H.K.; Lee, H.S. Damping Estimates from Reconstructed Displacement for Low-Frequency Dominant Structures. *Mech. Syst. Signal Process.* **2020**, *136*, 106533. [CrossRef]
33. Gao, S.; Liu, F.; Jiang, C. Improvement Study of Modal Analysis for Offshore Structures Based on Reconstructed Displacements. *Appl. Ocean Res.* **2021**, *110*, 102596. [CrossRef]
34. Brincker, R.; Zhang, L.; Andersen, P. Modal Identification of Output-Only Systems Using Frequency Domain Decomposition. *Smart Mater. Struct.* **2001**, *10*, 441–445. [CrossRef]

sensors

MDPI

Article

TSML: A New Pig Behavior Recognition Method Based on Two-Stream Mutual Learning Network

Wangli Hao [1], Kai Zhang [1], Li Zhang [1], Meng Han [1], Wangbao Hao [2], Fuzhong Li [1] and Guoqiang Yang [1,*]

[1] School of Software, Shanxi Agricultural University, Jinzhong 030801, China; haowangli@sxau.edu.cn (W.H.); rheal@stu.sxau.edu.cn (K.Z.); zxcvbn@stu.sxau.edu.cn (L.Z.); qwer0932@stu.sxau.edu.cn (M.H.); lifuzhong@sxau.edu.cn (F.L.)

[2] Yuncheng National Jinnan Cattle Genetic Resources and Gene Protection Center, Yongji 044099, China; haowangbao@gmail.com

* Correspondence: sxauwangzhang@stu.sxau.edu.cn;

Abstract: Changes in pig behavior are crucial information in the livestock breeding process, and automatic pig behavior recognition is a vital method for improving pig welfare. However, most methods for pig behavior recognition rely on human observation and deep learning. Human observation is often time-consuming and labor-intensive, while deep learning models with a large number of parameters can result in slow training times and low efficiency. To address these issues, this paper proposes a novel deep mutual learning enhanced two-stream pig behavior recognition approach. The proposed model consists of two mutual learning networks, which include the red–green–blue color model (RGB) and flow streams. Additionally, each branch contains two student networks that learn collaboratively to effectively achieve robust and rich appearance or motion features, ultimately leading to improved recognition performance of pig behaviors. Finally, the results of RGB and flow branches are weighted and fused to further improve the performance of pig behavior recognition. Experimental results demonstrate the effectiveness of the proposed model, which achieves state-of-the-art recognition performance with an accuracy of 96.52%, surpassing other models by 2.71%.

Keywords: pig breeding; behavior recognition; computer vision; two stream mutual learning; animal welfare

Citation: Hao, W.; Zhang, K.; Zhang, L.; Han, M.; Hao, W.; Li, F.; Yang, G. TSML: A New Pig Behavior Recognition Method Based on Two-Stream Mutual Learning Network. *Sensors* **2023**, *23*, 5092. https://doi.org/10.3390/s23115092

Academic Editor: Francesca Antonucci

Received: 22 April 2023
Revised: 18 May 2023
Accepted: 22 May 2023
Published: 26 May 2023

1. Introduction

Behavior changes play a crucial role in the pig breeding process. Accurately monitoring and understanding pig behavior is essential for improving pig welfare [1], predicting their health status, and facilitating the development of intelligent farming. To achieve promising pig behavior recognition performance, numerous researchers have conducted extensive studies. These studies can be broadly classified into two categories: sensor-based and computer vision-based approaches.

The first group of techniques relies on sensor-based monitoring of pig behavior. Several researchers have designed automatic monitoring systems that use sensors, such as infrared-sensitive cameras for real-time monitoring of pig activities [2] and behavior measurement. Other methods employ high-frequency radiofrequency identification (HF RFID) systems for monitoring individual drinking behavior [3] or pressure pads to track lame behavior in pigs [4]. However, these techniques involve physical contact with the pigs that can lead to stress and inaccurate measurements.

The second group of methods is based on computer vision. For instance, Zhang et al. [5] proposed a two-stream convolutional neural network for pig behavior recognition, where the feature extraction network is either a residual network (ResNet) or an inception network.

Zhuang et al. [6] developed a pig feeding and drinking behavior recognition model based on three models: VGG19, Xception, and MobileNetV2. They also designed two systems to monitor pig behaviors. Their final results demonstrated that the MobileNetV2-trained model had a significant advantage in pig behavior recognition, with a recall rate above 97%.

Wang et al. [7] implemented an improved HRNet-based method for joint point detection in pigs. By employing CenterNet to determine the posture of pigs (whether they are lying or standing), and then implementing the HRST approach for joint point detection in standing pigs, they achieved an average detection accuracy of 77.4%.

Luo et al. [8] proposed a channel-based attention model for real-time detection of pig posture. They compared their model with other popular network models, such as ResNet50, DarkNet53, and MobileNetV3, and showed that their proposed model outperformed the other models in terms of accuracy. They proved that the channel-based attention model is a promising approach for real-time pig posture detection [9].

Zhang et al. [10] presented an SBDA-DL, which is a deep learning-based real-time behavior-detection algorithm for sows. They designed it to detect three typical behaviors of sows: drinking, urinating, and sitting. The algorithm utilizes a combination of convolutional neural networks (CNN) and recurrent neural networks (RNN), along with a transfer learning approach, to achieve a high level of accuracy in behavior detection.

The experimental results showed that the average detection accuracy, measured by mean average precision (mAP), reached 93.4%, indicating the effectiveness of the proposed approach. The SBDA-DL algorithm provides a non-invasive method for monitoring sow behavior, which can reduce labor costs and enhance animal welfare in pig farming.

Li et al. [11] proposed a multi-behavioral spatio-temporal network model for pigs. By comparing it with a single-stream 3D convolutional model, the proposed model achieved a top-one accuracy of 97.63% on the test set. This multi-behavioral spatio-temporal network model provides a new approach for recognizing pig behaviors [12]. It has the potential to improve the efficiency of pig farming and to ensure animal welfare [13].

In summary, sensor-based methods are vulnerable to collision damage, resulting in inaccurate recognition and causing stress to the pigs both mentally and physically. Meanwhile, although deep-learning-based methods have achieved successful recognition results, their large parameter sizes lead to lengthy training and testing times, limiting their practical deployment on low-memory and low-capacity devices.

To overcome these challenges, we propose a novel two-stream mutual-learning (TSML) model for pig behavior recognition, aiming at improving the efficiency of pig farming and ensuring animal welfare. In comparison to other methods, TSML is more accurate and efficient in recognizing pig behavior. Our method is characterized by the cooperation between the RGB and flow streams that enables it to extract both appearance and temporal information efficiently. It also allows the model to extract critical feature information while avoiding irrelevant interference. Moreover, the mutual learning strategy improves the accuracy of behavior recognition by enabling the two student networks in each stream to learn collaboratively, gaining more robust and richer features in a shorter time. Compared with other methods that use either single-stream convolutional networks or multi-stream networks, our proposed model outperforms them in terms of accuracy, while being more efficient with a smaller number of parameters. This makes it more feasible to deploy on low-memory and low-capacity devices. Additionally, our unique dataset of pig behavior videos allows for more precise and reliable behavior detection and analysis, making our method practical for use in pig farming applications. Overall, our proposed two-stream mutual-learning method offers significant improvements over existing methods in terms of accuracy and efficiency while being practical for real-world applications.

The impact of our research on pig breeding is significant. Efficient monitoring of pig behavior is essential for improving pig welfare and for increasing the economic benefits of pig farms. Accurately monitoring and understanding pig behavior also allows for the prediction of their health status and facilitates the development of intelligent farming. The proposed TSML model offers a non-invasive and efficient method for monitoring pig behavior. In addition, by utilizing our unique dataset of pig behavior videos, future pig farming can be modernized with more precise and reliable behavior detection and analysis. Overall, our proposed method and dataset could significantly impact the pig breeding industry and enhance animal welfare.

Overall, the contributions of this paper can be summarised below:

- We established a novel dataset of pig behavior recognition dataset, which contains six categories. To provide a comprehensive understanding of pig behavior recognition, we have included six categories in our dataset, with each category consisting of roughly 600 videos. Each video varies in length from 5 to 10 s, providing sufficient footage to detect and analyze behavior patterns in pigs. These videos were collected over a period of one month utilizing six Hikvision cameras capturing over 85 pigs on a farm. All of the factors mentioned above have contributed to the creation of a unique and diverse dataset, collected on this farm, that exhibits better diversity in terms of illumination, angles, and other variables. This approach ensures that the dataset accurately represents the various scenarios and environments in which pigs behave, thereby resulting in more precise and reliable behavior detection and analysis.
- We first propose a novel pig behavior recognition method based on a two-stream mutual-learning framework. This model can efficiently extract more robust and richer features via mutual learning in RGB and flow paths separately and will extract both appearance and temporal information. Simultaneously, the decisions of the RGB and flow branches can be merged to gain improved pig behavior recognition performance. Specifically, our model achieves the best performance for pig behavior recognition task, with about a 2.71% improvement in the existing model.
- Several experiments were conducted to validate the superiority of the proposed model. The experiments included evaluating the performance of the proposed models, evaluating the behavior recognition performance of different models with or without mutual learning, evaluating the performance of the proposed model based on two identical networks, and evaluating the performance of the proposed model based on two different networks.

The rest of this paper can be organized as follows: Section 2 provides a detailed description of the methods and dataset used in the study. Section 3 presents the experimental results and analysis. In Section 4, we discuss the findings of our research. Finally, we conclude the paper in Section 5.

2. Materials and Methods

2.1. Datasets

The video data were collected from a pig farm located in Xiangfen County Agricultural Green Park Agricultural Company Limited, Linfen City, Shanxi Province. The farm encompasses 20 pig barns, each of which contains drinking water and feeding equipment as shown in Figure 1. For this study, six barns were selected, housing a total of 85 three-yuan fattening pigs. To ensure effective data collection, one camera was installed on each of the six barns at a height of approximately 3 m from the ground. The cameras were angled at 45 degrees diagonally toward the aisle and recorded videos at 25 fps with a resolution of 1920 × 1080 pixels. The specific camera utilized in this research was Hikvision DS-2DE3Q120MY-T/GLSE, and the whole data collection process lasted for 45 consecutive days, from 12 August 2022 to 25 September 2022.

The final pig behavioral recognition dataset contains six categories, including fighting, drinking, eating [14], investigating, lying, and walking (as shown in Figure 2). Specifically, each category consists of approximately 600 videos, each lasting between 5 and 8 s. In total, the dataset contained 3606 videos, of which 80% (2886 samples) were utilized for training, and 20% (720 samples) were employed for testing. Further, the detailed distribution of the collected videos of different behavioral categories is shown in Table 1.

Table 1. Number of videos of different pig behaviours.

Behavior	Number
Fighting	605
Drinking	597
Eating	607
Investigating	602
Lying	601
Walking	594
Total	3606

Figure 1. Environment of the pig farm.

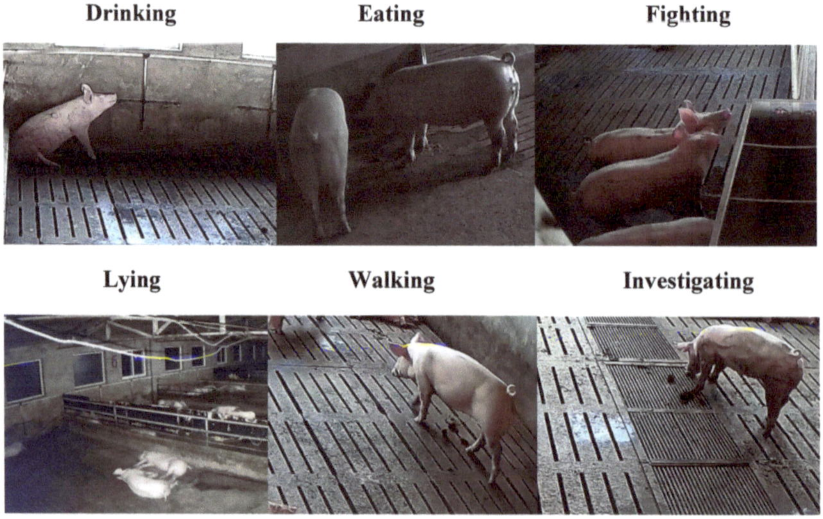

Figure 2. Sample of pig behavioural recognition dataset.

2.2. Problem Definition

This paper presents a novel TSML approach for pig behavior recognition. The model comprises two branches, spatial and temporal, each of which contains two student networks that perform mutual learning. The spatial branch extracts appearance features from still image frames while the temporal branch focuses on the optical flow motion in the video frames. The results of the two branches are subsequently weighted and fused to yield the final recognition result for pig behavior. The two-stream strategy employed in this approach effectively captures the complementary nature of the appearance and motion information underlying the video [15], while the mutual learning design further enhances the efficiency and accuracy of the model in recognizing pig behavior [16].

The framework of the proposed TSML is presented in Figure 3. The input to the framework are M videos $\mathcal{V} = \{v_i\}_{i=1}^{M}$ from C classes, with the corresponding video behavior label set denoted as $\mathcal{Y} = \{y_i\}_{i=1}^{M}$, where $y_i \in \{1,2,\ldots,C\}$.

The probability, $p_{s1}^{c}(x_i^s)$, of the RGB image x_i^s from the *ith* video v_i belonging to class c in the first student network of the spatial stream can be calculated as follows:

$$p_{s1}^{c}(x_i^s) = \frac{exp(S_{s1}^{c}(x_i^s))}{\sum_{c=1}^{C} exp(S_{s1}^{c}(x_i^s))} \tag{1}$$

Here, $S_{s1}^{c}(x_i^s)$ represents the logit output of the softmax layer from the first student network in the spatial stream for input x_i^s.

The probability, $p_{s2}^{c}(x_i^s)$, of the RGB image x_i^s from the *ith* video v_i belonging to class c in the second student network of the spatial stream can be written as follows:

$$p_{s2}^{c}(x_i^s) = \frac{exp(S_{s2}^{c}(x_i^s))}{\sum_{c=1}^{C} exp(S_{s2}^{c}(x_i^s))} \tag{2}$$

Similarly, $S_{s2}^{c}(x_i^s)$ represents the logit output of the softmax layer from the second student network in the spatial stream for input x_i^s.

The probability, $p_{t1}^{c}(x_i^t)$, of flow image x_i^t corresponding to the RGB image x_i^s from the *ith* video v_i belonging to class c in the first student network of the temporal stream can be described as follows:

$$p_{t1}^{c}(x_i^t) = \frac{exp(S_{t1}^{c}(x_i^t))}{\sum_{c=1}^{C} exp(S_{t1}^{c}(x_i^t))} \tag{3}$$

On the other hand, $S_{t1}^{c}(x_i^t)$ denotes the logit output of the softmax layer from the first student network in the flow stream for input x_i^t.

The probability, $p_{t2}^{c}(x_i^t)$, of flow image x_i^t corresponding to the RGB image x_i^s from the *ith* video v_i belonging to class c in the second student network of the temporal stream can be calculated as follows:

$$p_{t2}^{c}(x_i^t) = \frac{exp(S_{t2}^{c}(x_i^t))}{\sum_{c=1}^{C} exp(S_{t2}^{c}(x_i^t))} \tag{4}$$

Similarly, $S_{t2}^{c}(x_i^t)$ represents the logit output of the softmax layer from the second student network in the flow stream for input x_i^t.

The loss functions for the spatial and temporal two branches in the TSML can be defined as:

$$\begin{aligned} \mathcal{L}_s &= (1-\alpha) \times L_{s1} + \alpha \times L_{s2} \\ \mathcal{L}_t &= (1-\alpha) \times L_{t1} + \alpha \times L_{t2} \end{aligned} \tag{5}$$

Here, $\mathcal{L}_s$ and $\mathcal{L}_t$ represent the loss of the spatial stream and the temporal stream, respectively. The hyperparameter α controls the balance between these two loss terms. Furthermore, L_{s1} and L_{s2} denote the losses for the two student networks in the spatial stream, while L_{t1} and L_{t2} denote the losses for the two student networks in the temporal stream.

The formulations for L_{s1} and L_{t1} are as follows:

$$\begin{aligned} L_{s1} &= L_{c\_s1} + D_{KL}(p_{s1}||p_{s2}) \\ L_{t1} &= L_{c\_t1} + D_{KL}(p_{t1}||p_{t2}) \end{aligned} \tag{6}$$

Here, L_{c_s1} and L_{c_t1} represent the cross-entropy loss that measures the difference between the predicted value and the actual value. $D_{KL}(p_{s1}||p_{s2})$ represents the Kullback–Leibler (KL) divergence between the probability distributions p_{s1} and p_{s2}. L_{c_s1} and L_{c_t1} can be calculated using the following equation:

$$\begin{aligned} L_{c\_s1} &= -\sum_{i=1}^{M}\sum_{c=1}^{C} y_i^c log(p_{s1}^c(x_i^r)) \\ L_{c\_t1} &= -\sum_{i=1}^{M}\sum_{c=1}^{C} y_i^c log(p_{t1}^c(x_i^t)) \end{aligned} \tag{7}$$

Among them, y_i^c is an indicator, if $y_i = c, y_i^c = 1$; and if $y_i \neq c, y_i^c = 0$.

In the spatial stream, to enhance the generalization capacity of the first student network on testing samples, another peer network is employed to provide training experience via its posterior probability p2. The KL divergence is used to quantify the matching degree between the predictions p1 and p2. $D_{KL}(p_{s2}||p_{s1})$ indicates the KL distance from p_{s1} to p_{s2} and can be calculated using the following formula:

$$D_{KL}(p_{s2}||p_{s1}) = \sum_{i=1}^{M}\sum_{c=1}^{C} p_{s2}^c(x_i^r) log\frac{p_{s2}^c(x_i^r)}{p_{s1}^c(x_i^r)} \tag{8}$$

Here, p_{s1} and p_{s2} represent the predicted probability distributions from the first and second student networks, respectively, in the spatial stream.

In the temporal stream, $D_{KL}(p_{t2}||p_{t1})$ indicates the KL distance from p_{t1} to p_{t2} and shares a similar meaning with $D_{KL}(p_{s2}||p_{s1})$ in the spatial stream.

Moreover, the meanings of L_{s2} and L_{t2} are similar to those of L_{s1} and L_{t1}.

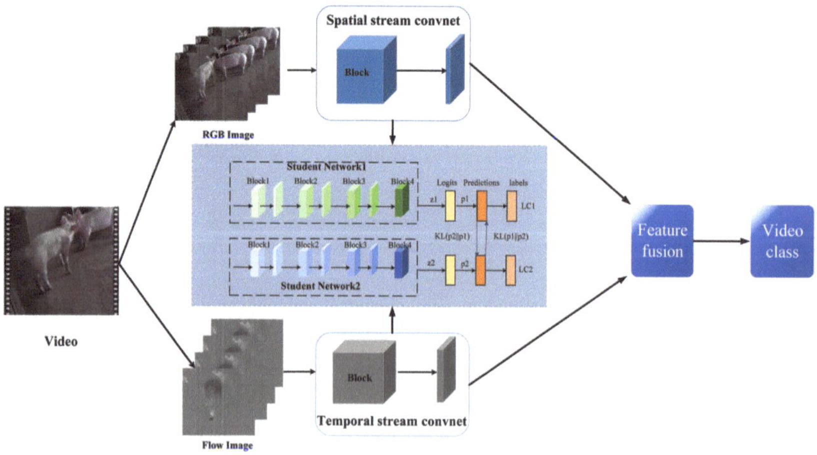

Figure 3. The structure diagram of the two-stream network model based on the idea of mutual learning.

2.3. The Implementation Details

The software and hardware system settings used in this paper are presented in Table 2. For fair comparison, we optimized all experimental models with a gradient descent algorithm using a momentum of 0.9, a batch size of 16, a learning rate of 0.001, and an Alpha value of 0.5, and we trained them for 500 epochs.

Table 2. Experimental environment configuration information.

Categories	Type or Version
Operating system	Ubuntu 18.04.5 LTS 64-bite
CPU	Intel Core i7-7800X @ 3.5 GHz*12
GPU	NVIDIA TITAN Xp
Memory	128 GB
Hard Disk	4TB SSD*3
Python	3.6.9
Pytorch	1.2.0
CUDA	11.2
CUDNN	10.0.130

2.4. Evaluation Criteria

In order to compare the performance of different models, several evaluation criteria were used, including accuracy, parameters, FLOPs (floating point operations per second), and loss. The accuracy reflects how well the model performs, while the number of parameters indicates the efficiency of the model—a smaller number of parameters is generally better. The FLOPs metric also indicates efficiency—again, a smaller number is better. It specifies the number of floating point operations required per second. All experiments were conducted on TITANX GPUs.

3. Experimental Results and Analysis

In this section, we will provide a detailed report on the experimental results and analysis. The overall experiment consists of several design parts, including evaluating the superiority of the proposed model, evaluating the efficiency of two stream mutual learning based on two identical networks, and evaluating the efficiency of two-stream mutual learning based on two different networks.

3.1. Evaluating the Superiority of the Proposed Model

To validate the superiority of the proposed TSML, several models were utilized for comparison, including ResNet18, ResNet34, ResNet50, Vgg16 [17] and MobileNetv2 [18]. The results are shown in Table 3.

Table 3. Comparison of different network models for pig behaviour recognition.

Model	Accuracy (%)
ResNet18	92.35
ResNet50	95.69
MobileNetv2	94.45
Vgg16	94.44
Ours	96.52

Table 3 shows that the proposed model outperforms other common models in pig behavior recognition. Specifically, the proposed model achieves 96.52% accuracy, which is 4.51%, 0.87%, 2.19%, 2.18% better than the accuracy rates of ResNet18 [19], ResNet50 [19], MobileNetV2, VGG16, respectively. These results demonstrate the superiority of the proposed model.

Furthermore, to provide readers with a more intuitive understanding of the superiority of TSML in pig behavior recognition, we report the accuracies and losses of different comparison models under different epochs in Figure 4. Here, Figure 4a shows the accuracy of different models under different epochs, while Figure 4b shows the loss of different models under different epochs.

The results in Figure 4 demonstrate that the accuracy and loss of TSML exceed those of other models, which further validates the effectiveness of the proposed TSML.

The outstanding performance of the TSML model can be attributed to its ability to effectively capture richer appearance and motion features [20], resulting in improved accuracy in pig behavior recognition tasks [21].

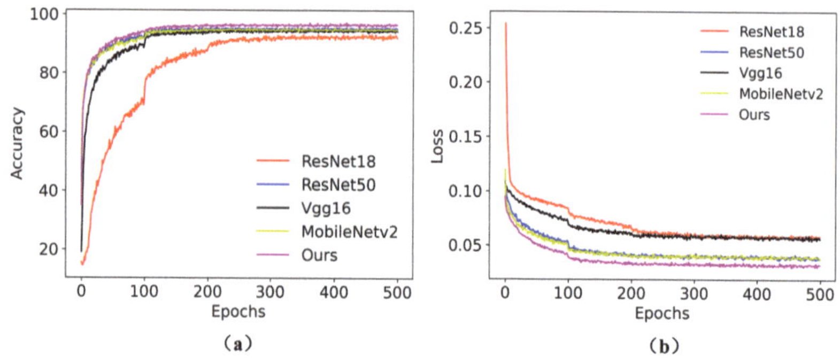

Figure 4. Comparison accuracies and losses of different models under different epochs for pig behavior recognition. (**a**) Accuracy for different models under different epochs. (**b**) indicates the Loss of different models under different epochs.

3.2. Evaluating the Efficiency of the Two-Stream Network in Pig Behavior Recognition

To validate the effectiveness of the two-stream network setting in the pig behavior [22] recognition framework, we compared the single RGB stream, single flow stream, and the fusion of two streams for several models [23], including ResNet18, ResNet34, ResNet50, Vgg16, MobileNetv2. The results of the comparison are displayed in Table 4.

Table 4. Comparison of pig behavior recognition accuracy based on two mutual learning models of the same network.

Model	Flow (%)	RGB (%)	Two-Stream Fusion (%)
ResNet18	61.47	91.24	92.35
ResNet50	86.37	93.18	95.69
MobileNetv2	86.23	93.88	94.45
Vgg16	83.31	94.02	94.44

Table 4 clearly demonstrates that the two-stream network setting consistently outperforms the single RGB and the flow networks by a significant margin.

To be more specific, the two-stream version of the ResNet18 model achieved an accuracy of 92.35%, which is 1.22% and 50.23% higher than its corresponding RGB and flow versions, respectively. The two-stream version of the ResNet50 model achieved an accuracy of 95.69%, which is 2.70% and 10.79% better than its corresponding RGB and flow versions, respectively. The two-stream version of the MobileNetv2 model achieved an accuracy of 94.45%, which is 0.60% and 9.52% better than its corresponding RGB and flow streams. The two-stream version of the Vgg16 model achieved an accuracy of 94.44%, which is 0.45% and 13.36% better than its corresponding RGB and flow versions, respectively.

Furthermore, to provide readers with a more intuitive understanding of the two-stream network, we include Figure 5. These figures illustrate the accuracies and losses of the RGB, flow, and two stream settings of different basic models at various epochs. Here, of Figure 5a denotes the comparison results based on ResNet18; Figure 5b indicates the comparison results based on ResNet50; Figure 5c represents the comparison results based on MobileNetv2; Figure 5d is the comparison results based on Vgg16.

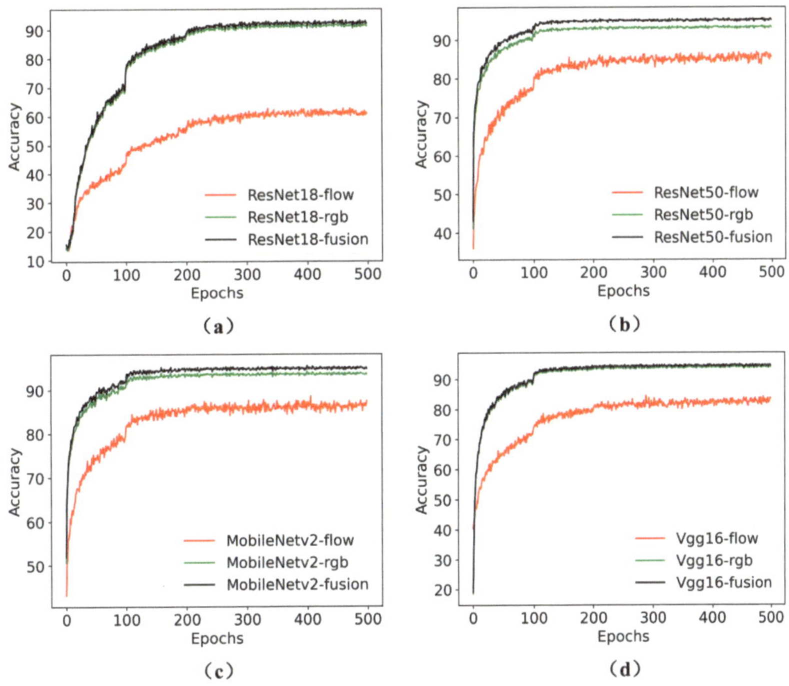

Figure 5. Comparison accuracies of RGB stream, flow stream and fusion of two streams based on different basic networks. (**a**) Comparison results based on ResNet18. (**b**) Comparison results based on ResNet50. (**c**) Comparison results based on MobileNetv2. (**d**) Comparison results based on Vgg16.

As depicted in Figure 5, the fusion of RGB and flow into two streams consistently achieved better results compared to using the RGB or flow streams alone. These results clearly demonstrate the superiority of the two-stream settings in the pig behavior recognition task. The use of both streams provides complementary information, allowing for more accurate and robust recognition of pig behaviors [24]. The fusion of multiple modalities has been a popular trend in many computer vision tasks, and our results provide evidence supporting this trend in the field of pig behavior recognition.

The results of these comparisons provide evidence of the superiority of the two-stream network in the pig behavior recognition task. The reason for this is that the two-stream network is capable of capturing both the appearance and motion information in the video, so that effective spatiotemporal features can be extracted, ultimately facilitating improved performance in pig behavior recognition. The RGB stream is capable of capturing appearance features such as color and texture, while the flow stream focuses on motion features such as the intensity and direction of movement. By combining both streams, our proposed two-stream network can effectively capture the complex spatiotemporal information for more precise and reliable recognition of pig behavior. Compared with traditional single-stream convolutional networks [25], using two streams allows for more efficient extraction of information. This approach reduces noise and irrelevant information while improving the accuracy of the recognition process. As a result, our proposed two-stream network provides a practical and viable approach for reliable pig behavior recognition in real-world applications.

In summary, the two-stream network is considered superior for pig behavior recognition tasks due to its ability to capture both appearance and motion information effectively.

By processing this information jointly, our TSML model can generate more robust and accurate feature representation, making it a promising choice for pig behavior recognition.

3.3. Evaluating the Efficiency of TSML Based on Two Identical Networks

In this section, we evaluate the performance of our proposed TSML approach based on two identical student networks. Specifically, TSML utilized different backbone architectures, including ResNet18, ResNet50, MobileNetV2, to validate the generalization of the proposed approach. To simplify the explanation, we refer to these models as Res18, Res50, and Mobilev2, respectively. The comparison results are shown in Table 5. Among Table 5, SigRes18 refers to the RGB and flow two-stream networks that comprise a single Res18 network. MulRes18(Res18) indicates that both the RGB and flow networks consist of two student networks that perform mutual learning, with each branch of the student network based on the Res18 architecture. Other single models (SigRes50 and SigMobv2) and other mutual models (MulRes18, MulRes50 and MulMobv2) share similar meanings with those of Sig18 and MulRes18 (Res18). Furthermore, MulRes18(18)-i, denotes the index of two mutual-learning [26] models.

Table 5. Comparison of pig behavior recognition accuracy based on two mutual-learning models of the same network.

Model	Flow (%)	RGB (%)	Two-Stream Fusion (%)
SigRes18	61.47	91.24	92.35
MulRes18(Res18)-1	66.20	93.60	94.44
MulRes18(Res18)-2	66.62	94.02	94.58
SigRes50	86.37	93.18	95.69
MulRes50(Res50)-1	87.67	95.97	96.52
MulRes50(Res50)-2	87.26	95.41	96.24
SigMobv2	86.23	93.88	94.45
MulMobv2(Mobilev2)-1	87.02	93.32	94.71
MulMobv2(Mobilev2)-2	87.07	92.49	94.58

Table 5 illustrates that the TSML with two identical networks achieves significantly and consistently superior performance than those of the single network. Specifically, MulRes18(Res18)-1/MulRes18(Res18)-2 obtain 2.26%/2.41% better accuracy than that of sigRes18; MulRes50(Res50)-1/MulRes50(Res50)-2 obtain 0.86%/0.57% better accuracy than that of sigRes50; and MulMobv2(Mobilev2)-1/MulMobilev2(Mobilev2)-2 obtain 0.29%/0.15% better accuracy than that of SigMobilev2. These results validate the superiority of the TSML approach, which is based on two identical student networks for both the RGB and optical flow branches.

Additionally, in order to provide readers with a more intuitive understanding and visualization of the superiority of TSML based on two identical networks, we include Figures 6 and 7 that show the accuracies and losses of the different comparison models with and without mutual learning at various epochs.

Specifically, (a1)/(a2)/(a3) of Figure 6 represent the accuracy of the RGB/flow/fusion stream on the SigRes18 and MulRes18(Res18) models under different epochs; (b1)/(b2)/(b3) of Figure 6 present the accuracy of the RGB/flow/fusion stream on the SigRes50 and MulRes50(Res50) models under different epochs.

Furthermore, (a1)/(a2) of Figure 7 represent the loss of the RGB/flow/fusion stream on the SigRes18 and MulRes18(Res18) models under different epochs; (b1)/(b2) of Figure 7 present the Loss of the RGB/flow/fusion stream on the SigRes50 and MulRes50(Res50) models under different epochs. These figures provide useful insights into the performance of each stream on different backbone networks and how they evolve over time.

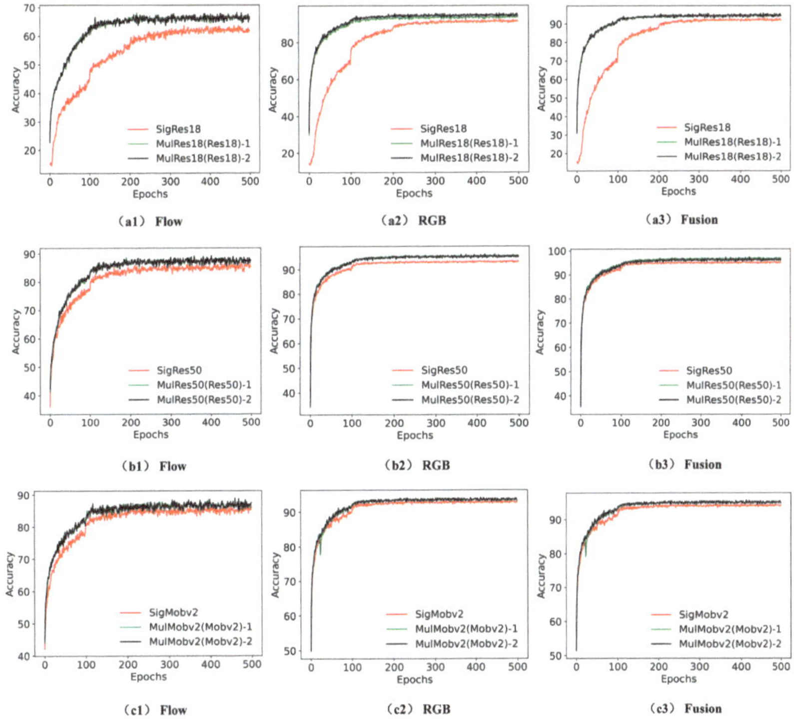

Figure 6. Comparison between accurate values of pig behaviour recognition based on mutual learning models of the two same networks. (**a1–a3**) represents the accuracy of the RGB/flow/fusion stream on the SigRes18 and MulRes18(Res18) models under different epochs. (**b1–b3**) presents the accuracy of the RGB/flow/fusion stream on the SigRes50 and MulRes50(Res50) models under different epochs. (**c1–c3**) denotes the accuracy of the RGB/flow/fusion stream on the SigMobilev2 and MulMobilev2(Mobilev2) models under different epochs.

Figures 6 and 7 demonstrate that the accuracy and the loss of MulRes18(Res18) and MulMobilev2(Mobilev2) outperform that of SigRes18 and SigMobilev2, which validates the effectiveness of the TSML based on two identical student networks.

The reason why the TSML model based on two identical student networks achieves better performance is as follows. Although the two student networks in the TSML model have the same network structure, their initial parameter values differ, resulting in the acquisition of different knowledge. Therefore, during the training process, they can obtain diverse knowledge and experience from each other, leading to the model producing better and more efficient behavior recognition performance.

3.4. Evaluating the Efficiency of TSML Based on Two Different Networks

In this section, we evaluate the performance of the proposed TSML approach using two different student networks.TSML utilized different backbone architectures, including ResNet18, ResNet34, and ResNet50. For ease of reference, we will refer to these models as Res18, Res34, Res50, and Mobilev2. The comparison results are shown in Table 6. In Table 6, the SigRes18 model refers to both the RGB and optical flow streams of TSML comprising a single Res18 network. Other single models share similar meanings as SigRes18. MulRes18(Res34) and MulRes34(Res18) indicate the two different mutual-learning student networks in the two streams of TSML that share the same structure with that of ResNet18 and ResNet34. Other mutual-learning models, such as Mul-

Res18(Res50)/MulRes50(Res18) and MulRes34(Res50)/MulRes50(Res34), share similar meanings as MulRes18(Res34)/MulRes34(Res18).

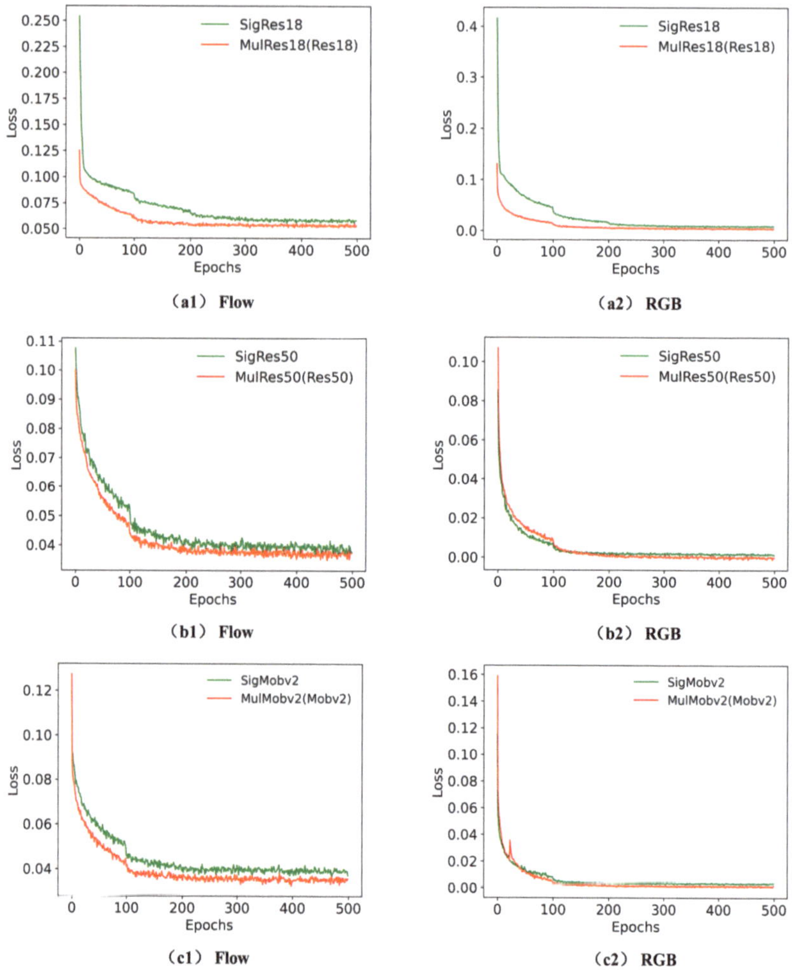

Figure 7. Comparison of loss values for pig behaviour recognition in mutual-learning models based on the two same networks. (**a1**,**a2**) represents the loss of the RGB/flow/fusion stream on the SigRes18 and MulRes18(Res18) models under different epochs. (**b1**,**b2**) presents the Loss of the RGB/flow/fusion stream on the SigRes50 and MulRes50(Res50) models under different epochs. (**c1**,**c2**) denotes the Loss of the RGB/flow/fusion stream on the SigMobilev2 and MulMobilev2(Mobilev2) models under different epochs.

Table 6 demonstrates that TSML with two different networks consistently achieves significantly superior performance compared to the corresponding single networks. Specifically, MulRes18(Res34)/MulRes34(Res18) achieve 2.41%/1.61% better accuracy than SigRes18/SigRes34; MulRes18(Res50)/MulRes50(Res18) demonstrate 2.71%/0.52% superior accuracy than SigRes18/SigRes34; and MulRes34(Res50)/MulRes50(Res34) achieve 1.77%/0.72% better accuracy than SigRes18/SigRes34. These results highlight the superiority of the TSML approach that employs two different student networks for both the RGB and optical flow branches.

In some cases, smaller student networks with mutual learning can outperform larger single neural networks.

Table 6. Comparison of the accuracy of pig behaviour recognition based on mutual-learning models of two different networks, ResNet18 and ResNet34.

Model	Flow (%)	RGB (%)	Two-Stream Fusion (%)
SigRes18	61.47	91.24	92.35
SigRes34	65.23	93.74	94.16
MulRes18(Res34)	64.67	94.44	94.58
MulRes34(Res18)	67.87	94.99	95.68
SigRes18	61.47	91.24	92.35
SigRes50	86.37	93.18	95.69
MulRes18(Res50)	71.22	94.71	94.85
MulRes50(Res18)	87.36	95.55	96.19
SigRes34	65.23	93.74	94.16
SigRes50	86.37	93.18	95.69
MulRes34(Res50)	72.34	95.55	95.83
MulRes50(Res34)	88.63	95.69	96.38

The above experimental results indicate that the TSML model based on different student networks has superiority. This is attributed to the fact that in this model, two student networks have different network structures and initial parameter values, resulting in different knowledge. Consequently, their collaborative learning allows them to obtain different knowledge and experience from their peers, thereby achieving superior performance.

4. Discussions

The proposed TSML approach leverages both the mutual-learning and two-stream network strategies to gather enhanced appearance and motion information underlying video in an interactive manner. The cooperation between the RGB and flow streams enables the TSML to achieve promising accuracy and efficiency. The mutual-learning strategy allows the two student networks in each stream to learn collaboratively, gaining more robust and richer features in a shorter time, which further enhances the accuracy of pig behavior recognition. Our approach not only improves the accuracy of pig behavior recognition, but it also enhances the efficiency of the recognition process. To validate the superiority of TSML, several experiments were designed and conducted, including evaluation of the superiority of the TSML model and evaluation of the TSML model based on two of the same or different student networks.

The experiments demonstrated that our proposed TSML model outperforms other models for pig behavior recognition, achieving an improvement of about 2.71% in accuracy. Specifically, the TSML model achieved 96.52% accuracy, which is 4.51%, 0.87%, 2.19%, 2.18% better than those of ResNet18, ResNet50, MobileNetV2, and Vgg16, respectively. To sum up, the experimental results demonstrate that our TSML model outperforms the competition in terms of accuracy when applied to the pig behavior recognition task.

The outstanding performance of the TSML model can be attributed to its ability to effectively capture richer appearance and motion features. By leveraging the two-stream mutual-learning framework, the model can efficiently extract both appearance and temporal information, leading to enhanced feature representation and improved accuracy in pig behavior recognition tasks. The RGB stream captures appearance features such as color and texture, while the flow stream captures motion features such as the intensity and direction of movement. By combining both streams and by collaboratively learning between them, our TSML model is better able to capture the complex visual cues that are critical for pig behavior recognition. In contrast to other approaches, our TSML model is specifically designed to balance the performance and efficiency trade-off in pig behavior recognition tasks. By utilizing mutual-learning and two-stream network strategies, the model can capture more robust features with fewer parameters, making it more practical for real-world applications. This approach provides a comprehensive understanding of pig behavior and further insights on the creation of a robust deep network that can be applied to various tasks.

Furthermore, our experimental results demonstrate that the TSML model with two different or same networks in both the RGB and flow streams consistently achieves significantly superior performance compared to their corresponding single network. This improvement can be attributed to several factors. Firstly, by using two student networks with unique initial parameter values or network structures, the TSML model can gain different knowledge and acquire a more comprehensive understanding of the appearance or flow of information in the videos. This approach allows the networks to learn from each other, leading to a more robust and comprehensive feature representation that enhances the accuracy of pig behavior recognition. Additionally, the collaborative learning of the student networks allows them to acquire different knowledge and experience from their peers. This approach enhances their ability to recognize pig behavior more accurately and efficiently. By combining these mechanisms, our proposed model achieves a high level of performance in pig behavior recognition. In summary, our experimental results suggest that using multiple student networks within the TSML model can significantly improve pig behavior recognition accuracy and efficiency. The benefit of mutual learning and information fusion between different networks provides a substantial gain that can be performance-driven in various domains.

However, one potential disadvantage of our TSML model is that it requires a larger amount of training data to achieve optimal performance. Nonetheless, given the significant improvement in accuracy, this method is considered suitable for practical applications in pig farming.

To further improve the accuracy and efficiency of the model, future work could explore the use of other advanced machine learning techniques such as reinforcement learning, transfer learning, and attention mechanisms. Additionally, future studies could apply our proposed approach to other domains such as wildlife conservation for animal behavior recognition.

5. Conclusions

This paper proposes a novel approach for pig behavior recognition, named TSML, which combines mutual learning with two stream neural networks that separately learn both appearance and motion information from videos. The mutual-learning strategy ensures that the basic student neural networks in the model update parameters collaboratively and gain information from each other throughout the training process. Furthermore, the two-stream network collects both appearance and motion information via its RGB and flow branches. Leveraging mutual learning and the two-stream network, the TSML model achieves excellent pig behavior recognition performance with higher efficiency and effectiveness. The experimental results show that the TSML model can greatly improve pig behavior recognition performance, delivering 2.71% higher accuracy in comparison to other models.

In terms of future work, we will explore the application of the proposed model to behavior recognition tasks for other livestock such as cattle and sheep. Additionally, we will continue to investigate more efficient and effective network structures to enhance the accuracy and efficiency of pig behavior recognition. Lastly, we will explore effective methods for identifying complex group pig behaviors.

Author Contributions: Conceptualization, W.H. (Wangli Hao) and M.H.; methodology, W.H. (Wangli Hao) and K.Z.; validation, M.H. and K.Z.; formal analysis, L.Z. and W.H. (Wangbao Hao); investigation, W.H. (Wangbao Hao) and L.Z.; resources, W.H. (Wangli Hao); data curation, F.L.; writing—original draft preparation, K.Z. and L.Z.; writing—review and editing, W.H. (Wangli Hao) and Guoqiang Yang; supervision, W.H. (Wangli Hao) and F.L.; project administration, G.Y. All authors have read and agreed to the published version of the manuscript.

Funding: This work was supported by: Shanxi Province Basic Research Program (202203021212444); Shanxi Province Education Science "14th Five-Year Plan" 2021 Annual Project General Planning Project + "Industry-University-Research"-driven Smart Agricultural Talent Training Model in Agriculture and Forestry Colleges (GH-21006); Shanxi Agricultural University Teaching Reform Project (J202098572); Shanxi Province Higher Education Teaching Reform and Innovation Project (J20220274); Shanxi Postgraduate Education and Teaching Reform Project Fund (2022YJJG094); Shanxi Agricultural University doctoral research start-up project (2021BQ88); Shanxi Agricultural University Academic Restoration Research Project (2020xshf38); Shanxi Agricultural University 2021 "Neural Network" Course Ideological and Political Project (KCSZ202133).

Institutional Review Board Statement: Not applicable.

Informed Consent Statement: Not applicable.

Data Availability Statement: The datasets generated and/or analyzed during the current study are available from the corresponding author on reasonable request.

Conflicts of Interest: We declare that this paper has no conflict of interest. Furthermore, we do not have any commercial or associative interest that represents a conflict of interest in connection with the work submitted.

References

1. Hao, W.; Han, W.; Han, M.; Li, F. A Novel Improved YOLOv3-SC Model for Individual Pig Detection. *Sensors* **2022**, *22*, 8792. [CrossRef] [PubMed]
2. Costa, A.; Ismayilova, G.; Borgonovo, F.; Leroy, T.; Berckmans, D.; Guarino, M. The use of image analysis as a new approach to assess behaviour classification in a pig barn. *Acta Vet. Brno* **2013**, *82*, 25–30. [CrossRef]
3. Maselyne, J.; Adriaens, I.; Huybrechts, T.; De Ketelaere, B.; Millet, S.; Vangeyte, J.; Van Nuffel, A.; Saeys, W. Measuring the drinking behaviour of individual pigs housed in group using radio frequency identification (RFID). *Animal* **2016**, *10*, 1557–1566. [CrossRef] [PubMed]
4. Martínez-Avilés, M.; Fernández-Carrión, E.; López García-Baones, J.M.; Sánchez-Vizcaíno, J.M. Early Detection of Infection in Pigs through an Online Monitoring System. *Transbound. Emerg. Dis.* **2015**, *64*, 364–373. [CrossRef] [PubMed]
5. Zhang, K.; Li, D.; Huang, J.; Chen, Y. Automated Video Behavior Recognition of Pigs Using two stream Convolutional Networks. *Sensors* **2020**, *20*, 1085. [CrossRef]
6. Zhuang, Y.; Zhou, K.; Zhou, Z.; Ji, H.; Teng, G. Systems to Monitor the Individual Feeding and Drinking Behaviors of Growing Pigs Based on Machine Vision. *Agriculture* **2023**, *13*, 103. [CrossRef]
7. Wang, X.; Wang, W.; Lu, J.; Wang, H. HRST: An Improved HRNet for Detecting Joint Points of Pigs. *Sensors* **2022**, *22*, 7215. [CrossRef]
8. Luo, Y.; Zeng, Z.; Lu, H.; Lv, E. Posture Detection of Individual Pigs Based on Lightweight Convolution Neural Networks and Efficient Channel-Wise Attention. *Sensors* **2021**, *21*, 8369. [CrossRef]
9. Wutke, M.; Heinrich, F.; Das, P.P.; Lange, A.; Gentz, M.; Traulsen, I.; Warns, F.K.; Schmitt, A.O.; Gültas, M. Detecting Animal Contacts—A Deep Learning-Based Pig Detection and Tracking Approach for the Quantification of Social Contacts. *Sensors* **2021**, *21*, 7512. [CrossRef]
10. Zhang, Y.; Cai, J.; Xiao, D.; Li, Z.; Xiong, B. Real-time sow behavior detection based on deep learning. *Comput. Electron. Agric.* **2019**, *163*, 104884. [CrossRef]
11. Li, D.; Zhang, K.; Li, Z.; Chen, Y. A Spatiotemporal Convolutional Network for Multi-Behavior Recognition of Pigs. *Sensors* **2020**, *20*, 2381. [CrossRef] [PubMed]
12. Tu, S.; Zeng, Q.; Liang, Y.; Liu, X.; Huang, L.; Weng, S.; Huang, Q. Automated Behavior Recognition and Tracking of Group-Housed Pigs with an Improved DeepSORT Method. *Agriculture* **2022**, *12*, 1907. [CrossRef]
13. Yang, Q.; Xiao, D.; Lin, S. Feeding behavior recognition for group-housed pigs with the Faster R-CNN. *Comput. Electron. Agric.* **2018**, *155*, 453–460. [CrossRef]
14. Chen, C.; Zhu, W.; Steibel, J.; Siegford, J.; Han, J.; Norton, T. Recognition of feeding behaviour of pigs and determination of feeding time of each pig by a video-based deep learning method. *Comput. Electron. Agric.* **2020**, *176*, 105642. [CrossRef]
15. Simonyan, K.; Zisserman, A. Two stream Convolutional Networks for Action Recognition in Videos. *arXiv* **2014**, arXiv:1406.2199.
16. Zhang, Y.; Xiang, T.; Hospedales, T.M.; Lu, H. Deep Mutual Learning. In Proceedings of the 2018 IEEE/CVF Conference on Computer Vision and Pattern Recognition, Salt Lake City, UT, USA, 18–23 June 2017; pp. 4320–4328.
17. Simonyan, K.; Zisserman, A. Very Deep Convolutional Networks for Large-Scale Image Recognition. *arXiv* **2014**, arXiv:1409.1556.

18. Sandler, M.; Howard, A.; Zhu, M.; Zhmoginov, A.; Chen, L.C. MobileNetV2: Inverted Residuals and Linear Bottlenecks. In Proceedings of the 2018 IEEE/CVF Conference on Computer Vision and Pattern Recognition, Salt Lake City, UT, USA, 18–23 June 2018; pp. 4510–4520.
19. He, K.; Zhang, X.; Ren, S.; Sun, J. Deep Residual Learning for Image Recognition. In Proceedings of the 2016 IEEE Conference on Computer Vision and Pattern Recognition (CVPR), Las Vegas, NV, USA 27–30 June 2016; pp. 770–778. [CrossRef]
20. Gan, H.; Ou, M.; Huang, E.; Xu, C.; Li, S.; Li, J.; Liu, K.; Xue, Y. Automated detection and analysis of social behaviors among preweaning piglets using key point-based spatial and temporal features. *Comput. Electron. Agric.* **2021**, *172*, 106357. [CrossRef]
21. Han, J.; Siegford, J.; Colbry, D.; Lesiyon, R.; Bosgraaf, A.; Chen, C.; Norton, T.; Steibel, J.P. Evaluation of computer vision for detecting agonistic behavior of pigs in a single-space feeding stall through blocked cross-validation strategies. *Comput. Electron. Agric.* **2023**, *204*, 107520. [CrossRef]
22. Eisermann, J.; Schomburg, H.; Knöll, J.; Schrader, L.; Patt, A. Bite-o-Mat: A device to assess the individual manipulative behaviour of group housed pigs. *Comput. Electron. Agric.* **2022**, *193*, 106708. [CrossRef]
23. He, Y.; Tiezzi, F.; Howard, J.; Maltecca, C. Predicting body weight in growing pigs from feeding behavior data using machine learning algorithms. *Comput. Electron. Agric.* **2021**, *184*, 106085. [CrossRef]
24. Yang, Q.; Xiao, D. A review of video-based pig behavior recognition. *Appl. Anim. Behav. Sci.* **2020**, 233, 105146. [CrossRef]
25. Chen, C.; Zhu, W.; Steibel, J.; Siegford, J.; Han, J.; Norton, T. Classification of drinking and drinker-playing in pigs by a video-based deep learning method. *Biosyst. Eng.* **2020**, *196*, 1–14. [CrossRef]
26. Yan, M.; Lou, X.; Chan, C.A.; Wang, Y.; Jiang, W. A semantic and emotion-based dual latent variable generation model for a dialogue system. *CAAI Trans. Intell. Technol.* **2023**, 1–12. [CrossRef]

sensors

MDPI

Article

Supporting Tremor Rehabilitation Using Optical See-Through Augmented Reality Technology

Kai Wang [1,2], Dong Tan [1], Zhe Li [3,4,*] and Zhi Sun [1,4,*]

1 School of Art and Design, Wuhan University of Technology, Wuhan 430070, China; wkaizh@gmail.com (K.W.)
2 Graduate School of Engineering Science, Osaka University, Toyonaka 5608531, Japan
3 College of Education, Fujian Normal University, Fuzhou 350117, China
4 Graduate School of Human Sciences, Osaka University, Suita 5650871, Japan
* Correspondence: lizheritetu@163.com (Z.L.); sunzhiszh@gmail.com (Z.S.);
Tel.: +86-(81)-09079697959 (Z.L.); +86-(81)-09017196167 (Z.S.)

Abstract: Tremor is a movement disorder that significantly impacts an individual's physical stability and quality of life, and conventional medication or surgery often falls short in providing a cure. Rehabilitation training is, therefore, used as an auxiliary method to mitigate the exacerbation of individual tremors. Video-based rehabilitation training is a form of therapy that allows patients to exercise at home, reducing pressure on rehabilitation institutions' resources. However, it has limitations in directly guiding and monitoring patients' rehabilitation, leading to an ineffective training effect. This study proposes a low-cost rehabilitation training system that utilizes optical see-through augmented reality (AR) technology to enable tremor patients to conduct rehabilitation training at home. The system provides one-on-one demonstration, posture guidance, and training progress monitoring to achieve an optimal training effect. To assess the system's effectiveness, we conducted experiments comparing the movement magnitudes of individuals with tremors in the proposed AR environment and video environment, while also comparing them with standard demonstrators. Participants wore a tremor simulation device during uncontrollable limb tremors, with tremor frequency and amplitude calibrated to typical tremor standards. The results showed that participants' limb movement magnitudes in the AR environment were significantly higher than those in the video environment, approaching the movement magnitudes of the standard demonstrators. Hence, it can be inferred that individuals receiving tremor rehabilitation in the AR environment experience better movement quality than those in the video environment. Furthermore, participant experience surveys revealed that the AR environment not only provided a sense of comfort, relaxation, and enjoyment but also effectively guided them throughout the rehabilitation process.

Keywords: optical see-through augmented reality; rehabilitation; tremor; yapa-PBGA; movement sensing

Citation: Wang, K.; Tan, D.; Li, Z.; Sun, Z. Supporting Tremor Rehabilitation Using Optical See-Through Augmented Reality Technology. *Sensors* **2023**, *23*, 3924. https://doi.org/10.3390/s23083924

Academic Editor: Jürgen Lorenz

Received: 16 February 2023
Revised: 14 March 2023
Accepted: 4 April 2023
Published: 12 April 2023

1. Introduction

A tremor is a type of involuntary muscle contraction that results in shaking or trembling. It is a common symptom of movement disorders that affects a large number of people [1]. There are various types of tremors, but the most prevalent ones are essential tremors (ET) and Parkinson's tremors (PT). ET affects 0.9% of the population, whereas PT is estimated to affect 0.3% of individuals under the age of 60, increasing to 1% in those over 60 years old [2]. ET is characterized by rhythmic body movements that have a frequency of 4–7 Hz [3]. On the other hand, PT typically affects non-active body parts and tends to worsen when the limbs are straightened and placed in a resting position [4].

Tremors can significantly impact an individual's quality of life by affecting their ability to perform everyday activities such as eating, writing, and using a phone. These challenges can lead to feelings of frustration and a loss of independence. Therefore, it is crucial to invest in research and technological support to develop effective treatments, Rehabilitation, and support technologies that can alleviate the effects of tremors.

Pharmacotherapy has long been the conventional approach for managing tremors in patients with ET and PT, but its effectiveness is often hindered by various factors, including drug-drug interactions, intolerable side effects, and inadequate therapeutic response. For example, dopaminergic drugs, which are commonly prescribed for PT, only alleviate tremors in about half of the patients [5]. Similarly, anticonvulsants such as gabapentin and paracetamol, as well as beta-blockers and benzodiazepines, which are often used for ET, have limited success rates, particularly for moderate to severe cases [6–8]. Surgical interventions, such as deep brain stimulation, thalamotomy with radiofrequency, radiosurgery, and focused ultrasound, have shown limited long-term efficacy, which makes them a less popular option among patients [9]. Furthermore, many individuals with mild tremors view medications as detrimental to their health, which leads them to abandon treatment, while those with mild to moderate tremors are generally hesitant to undergo surgery. For patients with severe and advanced tremors, surgery is often not a viable option due to the risks and complications associated with the procedure. Therefore, non-pharmacological interventions, such as regular exercise, must be considered as an alternative treatment approach.

Rehabilitation is an important adjunct to clinical medical treatment for tremors, particularly in patients with mild to moderate disease who are reluctant to undergo surgery and can benefit from physical therapy programs. Various approaches have been studied, including flexibility and strength training. Sajjad Farashi et al. [10] conducted a meta-analysis of a large body of literature and found that exercise significantly reduced tremor in patients with Parkinson's disease, with hand exercises showing promise for reducing distal limb tremors. Strength training involves applying resistance to the limbs of patients with tremors to stimulate neural control of muscles, and studies by G. Sequeira et al. [11], J. Kavanagh et al. [12], and M. Bilodeau et al. [13] have shown that a generalized upper limb strength program has the potential to improve stability and flexibility in patients with PT or ET. However, these studies also found that patients may experience fatigue during training and that functional capacity does not always improve after training [13]. Flexibility training is another useful method for improving execution and control in patients with tremors and gradually enhancing their self-confidence. Mona Kadkhodaie et al. used eccentric-based rehabilitation training and found a significant reduction in the amplitude of resting tremor after exercise in the intervention group, although the study had limitations in terms of assessing tremor fluctuation [14]. N.E. Vance et al. [15] used yoga to rehabilitate patients with primary tremor and demonstrated improvement in tremor assessment scales after an eight-week intervention. W. Chung et al. [16] provided behavioral relaxation training to ET and PT patients and found that regular relaxation training can reduce the effects of tremor, but noted the limitations of general rehabilitation training due to a lack of long-term follow-up. H. Rajalin [17] found that home rehabilitation training can improve not only the physical function of patients with Parkinson's disease but also their activities of daily living, but noted the lack of interventions available for use at home by patients with PT. Overall, rehabilitation has the potential to improve the quality of life of patients with tremors, but further research is needed to develop effective and accessible interventions for all types of patients. In order to promote daily rehabilitation for Parkinson's patients and reduce the pressure on rehabilitation institutions, Xiangya Hospital in China has developed a video exercise program called Yapa-PBGA (Yapa Parkinson balance and gait aerobics). This exercise program increases muscle control, improves gait and balance disorders, and upper limb flexibility in Parkinson's patients [18]. It is cost-effective and convenient for patients to perform daily rehabilitation at home. However, its limitations are that this video format cannot provide direct rehabilitation training guidance and monitoring to patients, especially considering the uncontrollable situation of Parkinson's patients. It is difficult to achieve the ideal training effect through this training method.

Virtual reality or augmented reality technology can overcome the limitations of video-based rehabilitation training methods by providing interactive visuals and personalized guidance, thus increasing patient engagement and motivation during the rehabilitation process. Hueso et al. [19] developed a virtual reality remote assistance system that allows therapists to create customized treatment plans and automatically record the patient's movement. However, the system lacks flexibility and is not suitable for developing tremor rehabilitation training. J. Cornacchioli [20] studied the use of the Oculus Rift grip as a tool to detect Parkinson's symptoms by measuring involuntary hand movements. The study tested the effectiveness of Parkinson's symptoms and concluded that the accuracy of Oculus Rift was sufficient for measurement needs but did not develop the technology for tremor rehabilitation training. G.C. Burdea et al. [21] reviewed the advantages and disadvantages of virtual reality rehabilitation applications and summarized the benefits, including improved patient motivation, adaptive data access, online data access, and reduced healthcare costs. They also revealed the limitations of virtual reality technology in rehabilitation, including a lack of supportive infrastructure, expensive equipment, and inadequate communication infrastructure for rural tele-assistance. Jiang et al. [22]. presented a multi-category gesture recognition model that uses signals from both surface electromyography and inertial measurement units. The model aims to improve the accuracy and robustness of gesture recognition in various real-world applications, such as human-computer interaction and rehabilitation.

Augmented reality refers to the combination of digital information from a virtual world with the physical environment of the real world to create a more interactive and immersive user experience. Wang et al. [23,24] conducted a study on the application of augmented reality technology to assist tremor patients in typing on a keyboard like ordinary people. Wang et al. [25] investigated the use of projection-based augmented reality technology called "Extend Hand" to help tremor patients directly interact with remote-controlled home appliances. Compared to virtual videos, augmented reality can provide more intuitive and specific rehabilitation training guidance and monitoring. By adding digital elements to real-life scenes, patients can better understand, simulate, and practice rehabilitation training skills, thus enhancing their rehabilitation outcomes. In addition, augmented reality technology can provide more personalized rehabilitation training plans and feedback, which can enhance patient engagement and motivation, leading to more active participation in rehabilitation training. Therefore, augmented reality technology has great potential to become an effective auxiliary tool for rehabilitation training. Aditya Pillai et al. [26] developed an innovative mixed reality rehabilitation tool specifically designed for upper limb injuries that utilized HoloLens 2 technology. The tool overlaid digital elements onto the real-world scene through the augmented reality feature of HoloLens 2, providing personalized rehabilitation training guidance and monitoring to help patients with upper limb injuries regain their function. However, the expensive price of HoloLens 2 limits its adoption for rehabilitation training that can be conducted at home by users. The Oculus Quest 2 by Meta is a milestone virtual reality headset that allows for free movement without the constraints of cables and is available at an affordable price. Its passthrough feature can be developed into an AR device. Additionally, the built-in controller Oculus Touch, featuring infrared emitters and inertial sensors, has been proven to be accurate and robust in detecting hand movements [27,28].

Considering the lack of a specialized augmented reality system for rehabilitation training specifically designed for patients with tremors and the need to improve the effectiveness and experience of rehabilitation training that can be conducted at home by patients with tremors, including one-on-one demonstration, posture guidance, and monitoring of training progress, we developed a low-cost augmented reality tremor rehabilitation training system using Oculus Quest 2. This study used Optitrack to record standard Yapa-PBGA movements and incorporated them into the system, creating a one-on-one virtual model for rehabilitation training. By designing natural interaction logic in the system, individuals with tremors can be guided to perform accurate limb movements during rehabilitation training. The patients' actual limb movement data can also be recorded and analyzed in real-time by the system.

2. Pilot System

2.1. System Configuration

In this study, an optical see-through AR system based on Yapa-PBGA is proposed to support individuals with tremors in rehabilitation training. The system employs Oculus Quest 2 and its self-contained controller, Oculus Touch. As illustrated in Figure 1, the virtual information to assist in tremor rehabilitation is registered in a physical space visible to the user by the development environment of the "passthrough" of Oculus. Oculus Touch was used to track the hand's position (posture) during rehabilitation training.

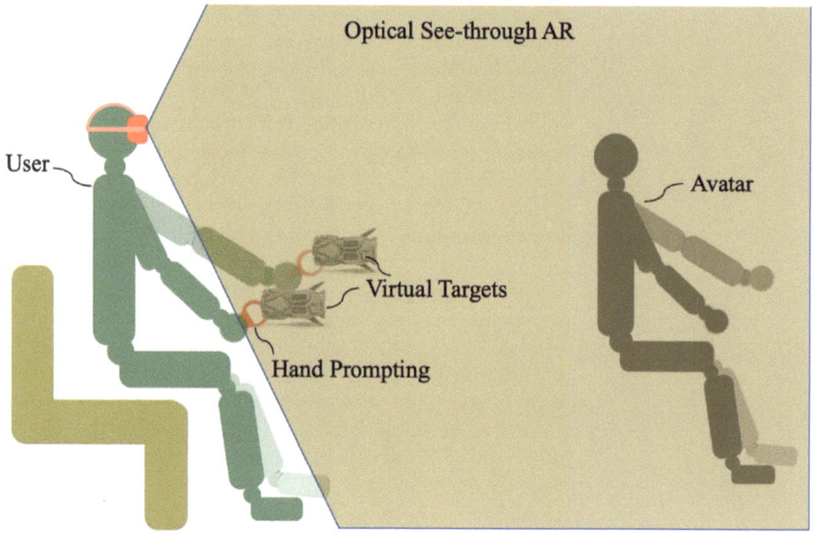

Figure 1. Tremor rehabilitation system using optical see-through AR.

2.2. System Visual Presentation

In the system environment, the user is rehabilitated as if he/she were playing Frisbee. The user can not only see his/her body and hands, but also a human-like avatar, virtual targets, and visual prompting for hand positions. This avatar is designed so that it simulates the gymnastic posture of the instructor in Yapa-PBGA and is registered in the physical environment to demonstrate the physical movements of rehabilitation training (See Figure 2a). A virtual target in the shape of a Frisbee follows the rhythm and sequence of rehabilitation gymnastics and will appear at the designated location for the user to pick up to guide their movement. (See Figure 2b). A pair of virtual bubbles synchronized with the position of the Oculus Touch informs the user in real time where the hand is located. After training, the system calculates and evaluates the user's training by comparing Oculus Touch's real-time tracking data with standard data (See Figure 2c).

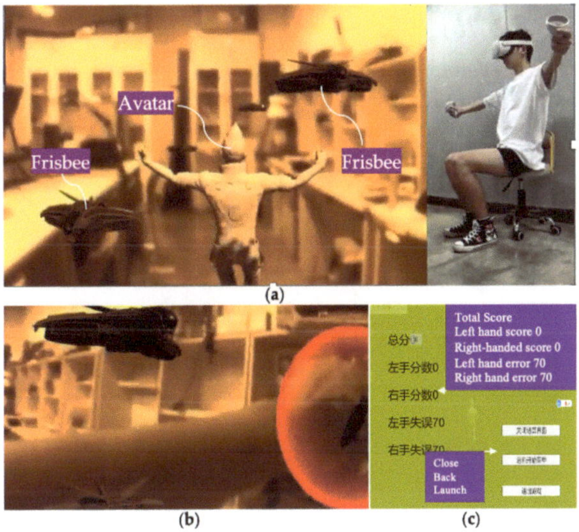

Figure 2. Under the proposed system, (**a**) An avatar, virtual targets, and physical environment can be observed, and (**b**) a user's hand can be seen and promoted. (**c**) an interface appears at the end of the training, presenting the system's evaluation of a user's rehabilitation.

3. Methods

3.1. Yapa-PBGA Rehabilitation Posture Model

Yapa Parkinson balance and gait aerobics abbreviated as Yapa-PBGA is a video-based dexterity training gymnastics for tremor patients that is proposed by the National Clinical Research Center for Geriatric Disorder, XiangYa Hospital [16], to improve gait and balance disorders as well as upper limb dexterity. This study summarized and extracted the Yapa-PBGA rehabilitation posture model, including "side bend up (G1)," "side clap (G2)," "drop up (G3)," "sun hug (G4)," "side stretch (G5)", "side flight (G6)," "spiral down finger (G7)", "greeting (G8)", and "tai chi (G9)", for use in developing the system and conducting experimental testing (Figure 3).

Figure 3. Yapa-PBGA Rehabilitation Posture Model.

3.2. Creation of Demonstrator Avatar

To create an immersive rehabilitation training experience that simulates the presence of a virtual demonstrator performing gymnastics in front of the user, we have designed a gymnastics demonstrator Avatar. According to the Yapa-PBGA posture model, a motion actor who underwent extensive Yapa-PBGA training was arranged to carefully choreograph and record the movements of Aatar, ensuring the accuracy of the postures.

In Figure 4, 10 motion capture devices, Opti-Track, were positioned around the Yapa-PBGA demonstrator in order to capture the demonstrator's poses and motion positions accurately and fully. After that, the demonstrator's postures were bound to the avatar by using Opti-Track motive software and the demonstrator's positions were transformed to the AR system by applying the spatial coordinate transformation. In addition, the demonstrator's sitting height, arm length, and distance from their heads were manually measured to adjust the system for different body types.

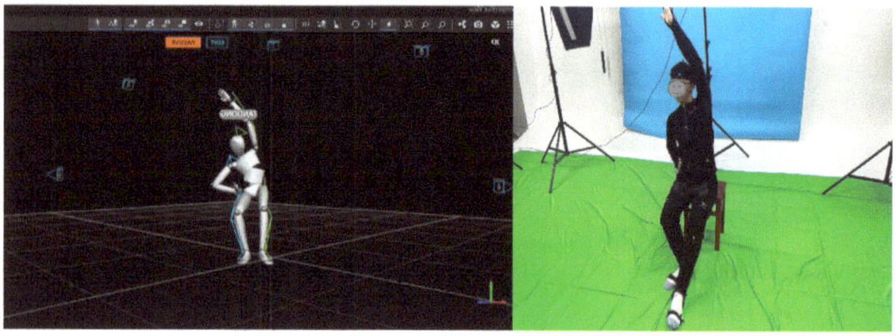

(a) (b)

Figure 4. (**a**) The demonstrator wore a marked suit and demonstrated standard Yapa-PBGA gymnastics, and (**b**) his posture and movement were recorded with Opti-Track Motive.

3.3. Guidance and Detection

To guide users to maintain standard limb postures during rehabilitation training and make the training more engaging, a natural interaction logic of catching frisbees is designed for the rehabilitation training system. Specifically, the interaction logic includes (1) The rehabilitation training system first presents the standard limb postures to guide users to maintain the correct posture. (2) Frisbees fly from a distance and reach only the hand positions corresponding to the standard limb movements. (3) Users need to catch the virtual frisbees within a specified time. (4) Users receive visual feedback from the frisbees when they successfully catch them, enhancing their perceptual experience. (5) When the user successfully catches a virtual frisbee, the system emits Vibration Feedback to enhance the user's sense of achievement. (6) The actual landing position of the frisbee is dynamically adjusted according to the user's 3D spatial coordinate position to ensure accurate landing on the user's hand position. Using this natural interaction logic can promote users to maintain standard limb postures during training, improving the quality of rehabilitation training.

The user's motion data are tracked by Oculus Touch. In light of the fact that a tremor patient's vibration frequency is generally less than 10 Hz, according to the Shannon theorem, the actual detection frequency that was set at 20 Hz was sufficient to meet the motion data collected without any distortions. User quality of movement in each gymnastic posture is measured by comparing user motion data with Frisbee positions. As shown in Equation (1), when the Euler distance between the Frisbee and the handle is less than 100 mm, i.e., when the user's limb is close to the specified position. The system offers the user both visual feedback on the frisbee's disappearance and haptic feedback through vibrations.

$$State^t = \begin{cases} Standard, \sqrt{(P_x^t - p_x^t)^2 + \left(P_y^t - p_y^t\right)^2 + (P_z^t - p_z^t)^2} \le 100\text{ m} \\ Non - standard,\ Otherwise \end{cases} \tag{1}$$

where $State^t$ is recorded whether the user's posture was standard at time t, P is the position of the virtual Frisbee relative to the helmet, and p is the position of the user's hand/Oculus Touch relative to the helmet.

4. Experiment

4.1. Experimental Setting

As part of the study, experiments were conducted to test the efficacy of the proposed system in helping people with tremor during rehabilitation. To simulate patients with Parkinson's disease (PT) and primary tremor (ET), participants were asked to wear a tremor simulator. Our study used a tremor simulator, shown in Figure 5, consisting of an Arduino Uno, a dual-channel muscle electrical stimulation module, and electrode pads. The Arduino Uno and the muscle electrical stimulation module used IIC communication to send a boosted electrical stimulation pulse current to the electrode patches. The electrode patches were placed on the lateral side of the participant's left and right hands, 3 cm from the elbow joint and 2 cm from the wrist joint. To replicate the tremor experienced by patients with PT and ET, we used a tremor simulator to randomly apply two types of electrical pulses to the participants' upper limbs. One signal caused the limb to tremble at a fixed frequency of approximately 5 Hz, while the other allowed the limb to oscillate voluntarily at a frequency of 4 to 7 Hz. We assessed their limb tremor status with Oculus Touch before each experiment to ensure that participants' involuntary limb tremors met tremor criteria.

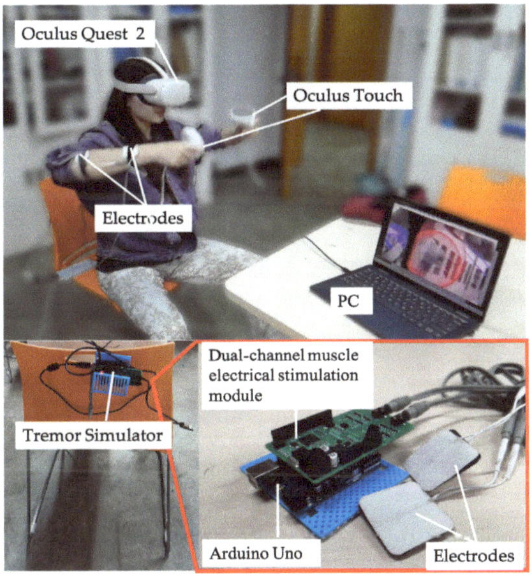

Figure 5. A participant is performing a rehabilitation task with a simulated trembling by a simulator.

The experiment will record participants' motions in the video and AR environments, respectively. To objectively compare the effects of rehabilitation training in the video and AR environments, this study further processed the sampled data and analyzed the magnitude of the body movements of the participants as they performed the rehabilitation postures in Figure 1 in both environments as shown in Equation (2).

$$M = \sum_{i=1}^{n} \sqrt{\left(\sqrt{x_i^2 + y_i^2 + z_i^2} - \sqrt{x_{i-1}^2 + y_{i-1}^2 + z_{i-1}^2} \right)^2} \quad (2)$$

where M is the magnitudes of body movements, n is the number of sampling, x, y, z are the position of the two Oculus Touch (hands) relative to the helmet.

An in-depth survey of the participants' rehab experiences was conducted through the use of the following questionnaire. The psychological feelings in relation to questions from Q1 to Q12 were rated based on a seven-point Likert scale, ranging from 0 (strongly disagree) to 7 (strongly agree).

Q1: I feel comfortable doing rehabilitation training in this environment.
Q2: I feel interested in rehabilitating in this environment.
Q3: I find it easy to do rehabilitation in this environment.
Q4: I think I can tolerate rehabilitation in this environment.
Q5: I feel unburdened by rehabilitation in that environment.
Q6: I feel that the body movements are standard in this environment.
Q7: I feel like I can follow the pace of rehabilitation training in this environment.
Q8: I feel like the environment can guide my body movements.
Q9: I feel as if rehabilitation in that environment would have good results.
Q10: I am satisfied with my rehabilitation training in this environment.
Q11 (SO): Sense of Ownership: I felt as if I was touching the virtual object directly with my hands, forgetting the existence of the handle.
Q12 (SA) SA: The human-computer interaction is easy to control without a sense of dissonance in the AR rehabilitation system.

4.2. Experimental Procedure

Firstly, we invited a physically healthy Yapa-PBGA demonstrator to perform a set of standard exercises. Throughout the entire training routine, the demonstrator wore an Oculus Touch, and the 3D positional coordinates of his upper limb movements were fully recorded. Subsequently, we extracted data from the movements in the postures shown in Figure 1 as the control group for the experiment. Next, we invited 12 healthy individuals aged between 20 and 30 to participate in the intervention study. The participants completed rehabilitation training under both the Augmented Reality (AR) condition and the video condition, in a randomized order. The tremor simulator worn by the participant was calibrated before the experiment to ensure that the participant's body trembling reached the frequency of ET or PT. Participants were required to perform a 5-min adaptation exercise to familiarize themselves with the task. In the experimental phase, each participant completed the rehabilitation task including nine rehab postures (Figure 1) using a tremor simulator and Oculus Touch. The participant answers the questionnaire after a sectional experiment and is given a rest to prepare for the next experimental condition.

5. Results

As a way of clearly describing, analyzing, and comparing the two rehabilitation modalities in the following sections, "Video" and "AR" represent the video and AR rehabilitation conditions, respectively, "Q1" to "Q10", the experience investigations, and "G1" to "G9" ("L" is the left body, "R" is the right body), the rehab postures.

A compared T-test was conducted to compare the magnitude of the mean body movement of participants making G1 to G9 under the "video" and "AR" conditions (see Figure 6). There was a significant effect on the mean body movement magnitude at the $p < 0.05$ level for G2, G3, G4, G5, G7, and G9. In groups G1 and G8, significant differences were commonly found in the L group, but not found in the R group. The results revealed that for most rehab postures, the magnitude of human motion in the AR condition was significantly greater than that in the video environment. As shown in Figure 6, the red lines indicate the actual magnitude of movement of the control group in making each rehab posture, respectively. From the results of the experiment, the AR group was remarkably closer to the control group compared to the video group. To sum up, the proposed AR method can help individuals with tremors make Yapa-PBGA postures that achieve a magnitude of body movements close to that of a standard demonstrator and can effectively promote the quality of movements in rehabilitation training for them in comparison to the video.

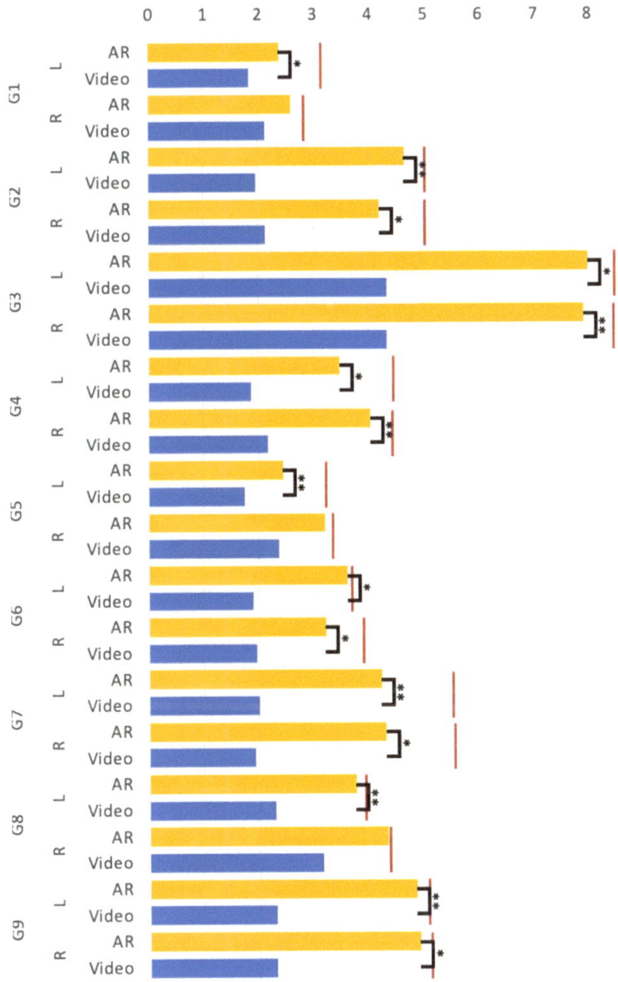

Figure 6. The mean body movement magnitude (** $p < 0.01$, * $p < 0.5$); G1 to G1 represent the rehab postures, respectively; "L" and "R" denote the left and right body part; "Video" and "AR" represent the video and AR rehabilitation conditions; the red lines are the value of the control group.

Some psychological aspects of the survey further evaluated participants' experience with "Video" and "AR" rehabilitation training. A compared T-test was used to analyze the difference in emotional experience between the conditions of "Video" and "AR" through the questions from Q1 to Q5. The results are given in Figure 7. Both conditions showed statistically significant differences with $p < 0.01$. The results showed that the participants felt comfortable and easy doing rehabilitation, did not perceive the physical burden, and were able to tolerate the intensity of the rehabilitation training. In the analysis, we compared the differences between the two conditions on the basis of Q6 to Q8 and found significant differences at $p < 0.01$. The result shows that the participants were more likely to be guided by rehabilitation training. Thus, the participants were more confident in performing standard rehabilitation training movements in the AR environment. There are no significant differences between the conditions in Q7, however, as the participants generally rated it very well (See Figure 7), it means participants could follow the pace of rehabilitation training under the AR condition. The analysis of Q9 and Q10 utilized a comparative T-test, revealing that the participants generally expressed satisfaction with the AR environment and preferred it as a rehabilitation training setting.

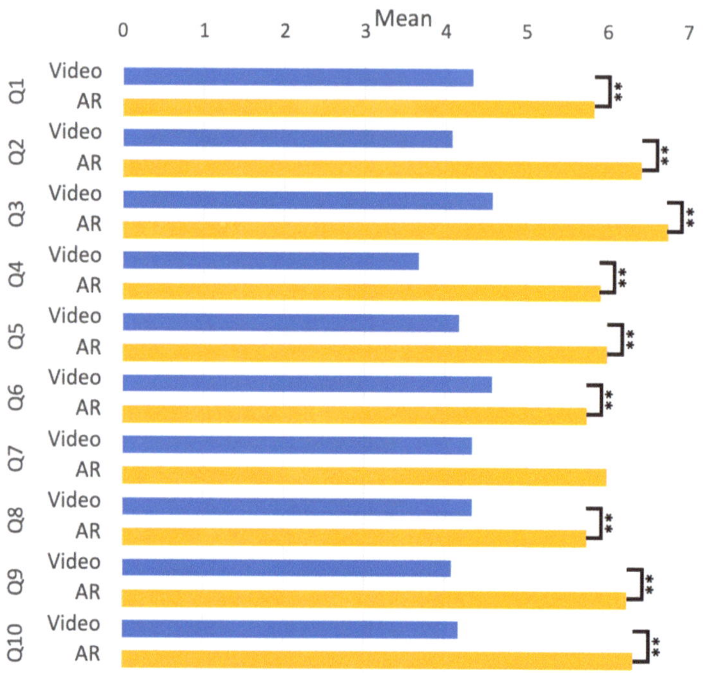

Figure 7. The mean score of the Q1 to Q10. (** $p < 0.01$, * $p < 0.5$).

In addition, the proposed methods were evaluated from the perspective of a sense of ownership and agency. The sense of ownership (SO) and the sense of agency (SA) are two central aspects of bodily self-awareness; the sense of ownership would be considered a direct perceptual experience, while the sense of agency is the sense of controlling and causing the body to act through volition, two important properties of operational logic in human-computer interaction. As we calculated the mean and standard deviation on the questionnaire of Q11 (SO) and Q12 (SA), we found a mean of 5.5 with a variation of 0.48, and a mean of 5.7, with a deviation of 0.3. This implies that participants are prone to the illusion that they are touching virtual objects, that the interaction can be easily controlled, and that users do not experience a strong sense of dissonance in the AR environment.

6. Discussion

Rehabilitation is an important way to reduce the severity and progression of tremors, and Yapa-PBGA provides a low-cost video rehabilitation program that can be used by patients at home. However, this program has limitations in providing face-to-face guidance for patients and does not track the quality of movements during rehabilitation. To address these limitations and better assist tremor patients with their rehabilitation at home, a low-cost AR system has been demonstrated to create an immersive rehab experience. This system simulates an instructor demonstrating gymnastics in front of the user, while also guiding the user's body movements during the rehabilitation process.

To evaluate the effectiveness of the proposed AR system in supporting individuals with tremors to achieve better results in Yapa-PBGA gymnastics, experiments were conducted to compare and analyze the magnitude of body movements of individuals with tremors in the video and AR environments, as well as that of the control group. The magnitude of movement is an objective measure of how well the movement achieves the correct posture, and the greater the magnitude, the better the rehabilitation will be. In most rehabilitation postures, participants demonstrated significantly greater magnitudes of limb movement in the AR environment, approaching the demonstrator's level, compared to the video environment. However, no statistically significant differences were observed in G1 (R) and G8 (R). Further analysis revealed that this was due to the fact that these two groups used poses involving hands on the waist, which are easy to assume and least influenced by experimental conditions. Therefore, G1 (R) and G8 (R) were not considered in the final experimental conclusion. It can be concluded that the proposed AR method is effective in improving the quality of movement for individuals with tremors. User experience surveys showed that participants felt more relaxed, comfortable, and engaged when using the proposed AR system compared to the video environment. The proposed AR system was also effective in providing rehabilitation guidance to individuals with tremors. Although there was no significant difference between the AR environment and the video environment in Q7, the actual evaluation revealed that participants could follow the pace of rehabilitation training under AR conditions. Additionally, an analysis of the sense of ownership and agency of participants with tremors found that the AR system was effective in providing them with a natural rehabilitation experience without causing dissatisfaction.

7. Conclusions

This study aimed to develop a low-cost augmented reality rehabilitation training system that would enable tremor patients to receive training at home with guidance. We created a set of rehabilitation posture models by extracting basic movements from Yapa-PBGA and established a one-to-one avatar demonstration training action through three-dimensional reconstruction and virtual mapping. To make rehabilitation training more engaging, we designed a natural interaction logic using a frisbee interaction mode to guide rehabilitation trainees' posture and provide necessary visual and tactile interaction feedback. We evaluated the effectiveness of our proposed tremor rehabilitation system by conducting simulated experiments and comparing it with traditional video rehabilitation methods. The results showed that our system significantly improved the movement quality of tremor patients while providing a more relaxed, comfortable, guided, and controllable rehabilitation experience than video rehabilitation methods. However, due to the limited number of patients tested, further optimization and discussion are still necessary for clinical settings. Long-term tracking and evaluation of the system's effectiveness are also essential. Therefore, we plan to continue improving the system and expanding the sample size in future research to further validate its effectiveness and feasibility.

Author Contributions: Conceptualization, K.W. and D.T.; methodology, K.W., D.T. and Z.S.; software, K.W. and D.T.; validation, K.W., D.T. and Z.S.; formal analysis, Z.S. and Z.L.; investigation, D.T.; resources, Z.L.; data curation, Z.S.; writing—original draft preparation, K.W. and D.T.; writing—review and editing, K.W., Z.S. and Z.L.; visualization, Z.S.; supervision, Z.S. and Z.L.; project administration, K.W.; funding acquisition, K.W. All authors have read and agreed to the published version of the manuscript.

Funding: This research was funded by the National Natural Science Foundation of China, grant number 61902287.

Institutional Review Board Statement: Not applicable.

Informed Consent Statement: Informed consent was obtained from all subjects involved in the study.

Data Availability Statement: Not applicable.

Conflicts of Interest: The authors declare no conflict of interest.

References

1. Louis, E.D. Tremor. *Contin. Lifelong Learn. Neurol.* **2019**, *25*, 959. [CrossRef] [PubMed]
2. Lenka, A.; Jankovic, J. Tremor Syndromes: An Updated Review. *Front. Neurol.* **2021**, *12*, 684835. [CrossRef] [PubMed]

3. Welton, T.; Cardoso, F.; Carr, J.A.; Chan, L.-L.; Deuschl, G.; Jankovic, J.; Tan, E.-K. Essential Tremor. *Nat. Rev. Dis. Prim.* **2021**, *7*, 83. [CrossRef] [PubMed]
4. Pan, M.-K.; Kuo, S.-H. Essential Tremor: Clinical Perspectives and Pathophysiology. *J. Neurol. Sci.* **2022**, *435*, 120198. [CrossRef]
5. Marjama-Lyons, J.; Koller, W. Tremor-Predominant Parkinson's Disease. *Drugs Aging* **2000**, *16*, 273–278. [CrossRef]
6. Louis, E.D.; Rios, E.; Henchcliffe, C. How Are We Doing with the Treatment of Essential Tremor (ET)? *Eur. J. Neurol.* **2010**, *17*, 882–884. [CrossRef]
7. Louis, E. Treatment of Essential Tremor: Are There Issues We Are Overlooking? *Front. Neurol.* **2012**, *2*, 91. [CrossRef]
8. Zesiewicz, T.A.; Elble, R.J.; Louis, E.D.; Gronseth, G.S.; Ondo, W.G.; Dewey, R.B.; Okun, M.S.; Sullivan, K.L.; Weiner, W.J. Evidence-Based Guideline Update: Treatment of Essential Tremor: Report of the Quality Standards Subcommittee of the American Academy of Neurology. *Neurology* **2011**, *77*, 1752–1755. [CrossRef]
9. Dallapiazza, R.F.; Lee, D.J.; Vloo, P.D.; Fomenko, A.; Hamani, C.; Hodaie, M.; Kalia, S.K.; Fasano, A.; Lozano, A.M. Outcomes from Stereotactic Surgery for Essential Tremor. *J. Neurol. Neurosurg. Psychiatry* **2019**, *90*, 474–482. [CrossRef]
10. Farashi, S.; Kiani, L.; Bashirian, S. Effect of Exercise on Parkinson's Disease Tremor: A Meta-Analysis Study. *Tremor Other Hyperkinet. Mov.* **2021**, *11*, 15. [CrossRef]
11. Sequeira, G.; Keogh, J.W.; Kavanagh, J.J. Resistance Training Can Improve Fine Manual Dexterity in Essential Tremor Patients: A Preliminary Study. *Arch. Phys. Med. Rehabil.* **2012**, *93*, 1466–1468. [CrossRef]
12. Kavanagh, J.J.; Wedderburn-Bisshop, J.; Keogh, J.W.L. Resistance Training Reduces Force Tremor and Improves Manual Dexterity in Older Individuals With Essential Tremor. *J. Mot. Behav.* **2016**, *48*, 20–30. [CrossRef] [PubMed]
13. Bilodeau, M.; Keen, D.A.; Sweeney, P.J.; Shields, R.W.; Enoka, R.M. Strength Training Can Improve Steadiness in Persons with Essential Tremor. *Muscle Nerve.* **2000**, *23*, 771–778. [CrossRef]
14. Kadkhodaie, M.; Sharifnezhad, A.; Ebadi, S.; Marzban, S.; Habibi, S.A.; Ghaffari, A.; Forogh, B. Effect of Eccentric-Based Rehabilitation on Hand Tremor Intensity in Parkinson Disease. *Neurol. Sci.* **2020**, *41*, 637–643. [CrossRef] [PubMed]
15. Vance, N.E.; Ulanowski, E.A.; Danzl, M.M. Yoga Led by a Physical Therapist for Individuals with Essential Tremor: An Explorative Pilot Study. *Complement. Ther. Clin. Pract.* **2019**, *34*, 17–22. [CrossRef]
16. Chung, W.; Poppen, R.; Lundervold, D.A. Behavioral Relaxation Training for Tremor Disorders in Older Adults. *Biofeedback Self-Regul.* **1995**, *20*, 123–135. [CrossRef] [PubMed]
17. Vaartio-Rajalin, H.; Rauhala, A.; Fagerström, L. Person-Centered Home-Based Rehabilitation for Persons with Parkinson's Disease: A Scoping Review. *Int. J. Nurs. Stud.* **2019**, *99*, 103395. [CrossRef]
18. Training and Promotion Meeting for the Project on the Construction and Application of the Assessment and Intervention System for the Elderly with Movement Disorders (Parkinson's Disease, Etc.) (Gait and Balance Disorders) (Service Category) Was Successfully Held. Available online: https://ncrcgdxy.csu.edu.cn/info/1024/2061.htm (accessed on 7 March 2023).
19. Pedraza-Hueso, M.; Martín-Calzón, S.; Díaz-Pernas, F.J.; Martínez-Zarzuela, M. Rehabilitation Using Kinect-Based Games and Virtual Reality. *Procedia Comput. Sci.* **2015**, *75*, 161–168. [CrossRef]
20. Cornacchioli, J.; Galambos, A.; Rentouli, S.; Canciello, R.; Marongiu, R.; Cabrera, D.; Njie eMalick, G. Virtual Reality Tremor Reduction in Parkinson's Disease. *Preprints.org* **2020**, 2020020452.
21. Burdea, G.C. Virtual Rehabilitation–Benefits and Challenges. *Methods Inf. Med.* **2003**, *42*, 519–523. [CrossRef]
22. Jiang, Y.; Song, L.; Zhang, J.; Song, Y.; Yan, M. Multi-Category Gesture Recognition Modeling Based on sEMG and IMU Signals. *Sensors* **2022**, *22*, 5855. [CrossRef]
23. Wang, K.; Takemura, N.; Iwai, D.; Sato, K. A Typing Assist System Considering Involuntary Hand Tremor. *J. Pap. Jpn. Virtual Real. Soc.* **2016**, *21*, 227–233. [CrossRef]
24. Wang, K.; Iwai, D.; Sato, K. Supporting Trembling Hand Typing Using Optical See-Through Mixed Reality. *IEEE Access* **2017**, *5*, 10700–10708. [CrossRef]
25. Wang, K.; Matsukura, H.; Iwai, D.; Sato, K. Stabilizing Graphically Extended Hand for Hand Tremors. *IEEE Access* **2018**, *6*, 28838–28847. [CrossRef]
26. Pillai, A.; Sunny, M.S.H.; Wang, I.; Rahman, M. Innovative Mixed Reality Based Rehabilitative Tool for Upper Limb Injuries Utilizing HoloLens 2. *Arch. Phys. Med. Rehabil.* **2022**, *103*, e71. [CrossRef]
27. Shum, L.C.; Valdés, B.A.; Loos, H.M.V. der Determining the Accuracy of Oculus Touch Controllers for Motor Rehabilitation Applications Using Quantifiable Upper Limb Kinematics: Validation Study. *JMIR Biomed. Eng.* **2019**, *4*, e12291. [CrossRef]
28. Carnevale, A.; Mannocchi, I.; Sassi, M.S.H.; Carli, M.; De Luca, G.; Longo, U.G.; Denaro, V.; Schena, E. Virtual Reality for Shoulder Rehabilitation: Accuracy Evaluation of Oculus Quest 2. *Sensors* **2022**, *22*, 5511. [CrossRef] [PubMed]

sensors

Article

P2P Cloud Manufacturing Based on a Customized Business Model: An Exploratory Study

Dian Huang, Ming Li, Jingfei Fu, Xuefei Ding, Weiping Luo and Xiaobao Zhu *

School of Information Engineering, Nanchang Hangkong University, Nanchang 330063, China
* Correspondence: zxb@nchu.edu.cn

Abstract: To overcome the problems of long production cycle and high cost in the product manufacturing process, a P2P (platform to platform) cloud manufacturing method based on a personalized custom business model has been proposed in this paper by integrating different technologies such as deep learning and additive manufacturing (AM). This paper focuses on the manufacturing process from a photo containing an entity to the production of that entity. Essentially, this is an object-to-object fabrication. Moreover, based on the YOLOv4 algorithm and DVR technology, an object detection extractor and a 3D data generator are constructed, and a case study is carried out for a 3D printing service scenario. The case study selects online sofa photos and real car photos. The recognition rates of sofa and car were 59% and 100%, respectively. Retrograde conversion from 2D data to 3D data takes approximately 60 s. We also carry out personalized transformation design on the generated sofa digital 3D model. The results show that the proposed method has been validated, and three unindividualized models and one individualized design model have been manufactured, and the original shape is basically maintained.

Keywords: personalized business model; P2P cloud manufacturing; reverse engineering; deep learning; 3D reconstruction; 3D printing

Citation: Huang, D.; Li, M.; Fu, J.; Ding, X.; Luo, W.; Zhu, X. P2P Cloud Manufacturing Based on a Customized Business Model: An Exploratory Study. *Sensors* **2023**, *23*, 3129. https://doi.org/10.3390/s23063129

Academic Editor: Antonio Fernandez-Caballero

Received: 4 February 2023
Revised: 3 March 2023
Accepted: 3 March 2023
Published: 15 March 2023

1. Introduction

The ever-increasing customization and personalization demands of customers and the ever-shortening product life cycle have brought severe challenges to the manufacturing industry. Ubiquitous connectivity, digitization and sharing provide opportunities for personalized production to meet the burgeoning demand for personalized goods [1]. In the framework for personalized production based on digital twins, blockchain and AM [1], and the consensus-oriented cloud manufacturing framework based on blockchain technology [2], professional designers may be required to design the entire product model, or traditional reverse engineering may be used to obtain the 3D data. These processes have problems such as long cycle and high cost. The development of extremely flexible cloud services [3] and novel artificial intelligence technology allows this to be realized at low cost, in high quality and quickly.

With the improvement of current manufacturing intelligence and productivity, computer-aided design and manufacturing (CAD/CAM) and rapid prototyping (RP) have become hot words in the manufacturing field. Traditionally, the two behaviors were handled separately. However, as customer demands continue to increase, there is a growing trend of combining the two, which leads to concurrent engineering [4]. Manually creating 3D models is time-consuming and expensive. For this reason, techniques for automatically reconstructing 3D objects have been developed. This technique is the process of capturing the shape of an object through surface data sampling and generating a CAD model of the part, known as reverse engineering [5]. Reverse engineering is the process of 3D scanning and data acquisition of the original physical shape, followed by data processing and 3D reconstruction to build a 3D model with the same shape and structure. Then, on the basis of the original shape, to copy or redesign the original shape to achieve innovation. These techniques can be subdivided into active and passive approaches [6]. The drawback of active methods (e.g., structured light, laser scanners, laser range maps and medical MRI) is that the reconstruction process can be a costly project [7]. Hence, the described methods are passive methods, which require less equipment and can be more widely applied. As soon as a CAD model is obtained through reverse engineering, a large amount of information can be exported and some operations can be performed, such as mechanical design, finite element (FEM) mesh generation, command code generation for CNC machines, overall property calculation, tolerance analysis, accessibility analysis, etc. This provides great support for personalization. At present, many methods [8–10] for reconstructing 3D objects can recover the 3D model of the object with only a single shot, which enables fast and low-cost acquisition of 3D data in reverse engineering.

AM is also known as layer manufacturing, rapid prototyping or 3D printing [11]. Different from subtractive manufacturing techniques, such as milling and grinding. It manufactures designed parts by removing material. Additive manufacturing describes the manufacturing process of joining materials to create parts from 3D model data, usually layer by layer [12]. The appeal of additive manufacturing to companies and industries is clear, as it has not only revolutionized the way final part shapes are obtained, but also offers a promising way to develop highly customized and personalized products [13]. AM empowers intelligent manufacturing, and on-demand personalized customization becomes a new direction of development [14]. With the gradual emergence of commercial value such as easy molding, personalized customization, and rapid manufacturing, the application scenarios of 3D printing are becoming more and more diverse. At present, 3D printing has been widely used in construction, footwear, industrial design, jewelry, engineering, aerospace, dentistry, automobiles and other fields. Some manufacturers have also begun to use 3D printing to manufacture aircraft seats, car engines, etc. [15]. After the production of products with the help of cutting-edge 3D printing technology, the innovation of the products' production process has been accelerated, and its appearance, design, and internal functions have also been further improved.

In order to cope with the ever-changing demand for personalized services, high design costs, long product manufacturing life cycle and other issues, a p2p cloud manufacturing method is proposed based on the personalized business model [1] and cloud manufacturing framework [2]. The difference between this study and these manufacturing frameworks is that it pays more attention to the entity-to-entity manufacturing process, which is used to solve the problems brought about by the time and cost of product manufacturing. This paper is a complete and complementary work to these frameworks. Based on the proposed method, long-distance transmission of physical objects can be realized. When customers see the products they want in multimedia such as video, they only need to take a screenshot to quickly generate the corresponding entity. With this method, only one photo is needed to get the entity in the photo. First, the YOLOv4 [16] is employed to detect and identify all objects in the photo. The targets are cropped to generate a new image. Then the differentiable volume rendering (DVR) [10] technology is optimized to restore the 3D model of the object based on the new image. A digital model file is produced. Finally, the obtained 3D data can be customized for customers. The entity is produced with 3D printers.

In this research, we propose and implement a novel P2P reverse manufacturing method that combines deep learning and AM technology. such that the method is compatible with fast, low-cost and personalized customization features. By using YOLOv4, object detection and recognition is realized. The conversion of 2D data to 3D data is realize by DVR technology. The production printing of 3D digital models is done by employing AM technology. A further distinction of our work from the limited existing work is the overall improvement of the scheme for 3D data acquisition during reverse engineering. The method is applied to the P2P printing service scenario, and the feasibility of the method is verified through a case study. The contributions of this paper can be summarized as follows: (1) A P2P cloud manufacturing method based on the personalized business model is proposed, which can support on-demand manufacturing and long-distance transmission. the method is an extended study of [1,2], bringing them closer to reality. This will be a fast, low-cost, and convenient P2P cloud manufacturing method in the future. (2) Add object recognition and extraction to the original 3D reconstruction method to improve the clarity of the 3D digital model. (3) Based on the proposed method, the feasibility of the proposed method is verified by using photos from the Internet and reality to produce small solid models.

The remainder of this paper is organized as follows. Section 2 briefly reviews key relevant research streams in personalized business models across various industries, deep learning-based reconstruction methods, and additive manufacturing. In Section 3, a P2P cloud manufacturing method based on the personalized business model is presented In Section 4, according to the customer-centered production model, the small models of the objects are generated from two aspects of network pictures and real photos to verify the feasibility of the proposed scheme. They are employed to verify the feasibility of the proposed scheme. Section 5 discusses the contributions of this paper as well as future research.

2. Literature Review

Personalized business model: As early as 20 years ago, various industries had a business paradigm of personalized customization. For example: personalized interactive TV advertising [17], personalized medicine [18,19], and personalized web system frameworks [20]. After the introduction of Industry 4.0, the intelligent manufacturing industry has moved towards personalized customization. Wang et al. [21] propose cloud-based manufacturing of personalized packaging. Egon [22] proposes a management tool to guide business model innovation in the direction of personalized products: the business model radar template of personalized products. Qin et al. [23] proposed the paradigm of large-scale personalized intelligent manufacturing. Zhang et al. [24] propose a flexible intelligent manufacturing system under the large-scale personalized manufacturing mode. Personalized, mass-manufactured models are gradually becoming the production paradigm of our generation. Guo et al. [1] propose a personalized production framework based on digital twins, blockchain, and additive manufacturing in the context of Industry 4.0, providing useful guidance and reference for the personalized production paradigm. Zhu et al. [2] propose a framework for cloud manufacturing by integrating blockchain technology. Inspired by [1,2], this paper proposes A P2P cloud manufacturing method that provides a quick, easy, and low-cost solution to reversely obtain 3D digital models.

3D Reconstruction: In computer vision, 3D reconstruction refers to the process of reconstructing 3D information from single-view or multi-view images or video streams. Ref. [25] is the pioneering work of using deep learning for depth map estimation. Eigen et al. divide the network into a global rough estimation and local fine estimation, estimate the depth from coarse to fine, and propose a scale-invariant loss function. For 3D reconstruction of singular or multi-view images with voxels, Choy et al. [26] combined LSTM, if the input is only one image, the input is one, and the output is also a result. If it is multi-view, consider the multi-view as a sequence, input it into LSTM [27], and output multiple results. In summary, a 2D-image-to-3D voxel model mapping is established through the network structure of the Encoder-3DLSTM-Decoder. Its disadvantage is that it needs to consider the voxel resolution, the size of the calculation time and the size of the accuracy. Fan H et al. [28] used a deep network to directly generate a point cloud from a single image, solved the problem of generating 3D geometry based on a single image object, and created a precedent for single-view 3D reconstructed point cloud representation.

Wang N et al. [29] propose an end-to-end neural network and realized the direct generation of 3D information of objects represented by mesh from a single color image, without the need for point clouds, depth or other more informative data. They used graph convolutional neural networks(GCNNs) to represent the 3D mesh information, using the features mentioned from the input image to gradually deform the ellipse to produce the correct geometry, The core idea of this paper is to use an ellipsoid as the initial shape of any object, and then gradually turn this shape into a target object.

For differentiable rendering, Chen et al. [30] propose DIB-Render, through which the gradient can be analyzed and calculated, which can be used to solve the basic rasterization steps of discrete allocation operations, with a non-differentiable rendering pipeline. The key to their approach is to treat rasterization as weighted interpolation, allowing image gradients to be back-propagated through a variety of standard vertex shaders within a single frame, resulting in single-image 3D object prediction and 3D texture object generation, both using specialized 2D supervision for training. Niemeyer M et al. [10] propose a differentiable rendering formulation that can represent continuously 3D information for implicit shape and texture representations. They can learn implicit shape and texture representations directly from single or multiple RGB images without 3D supervision and result in watertight meshes.

Additive manufacturing: Additive manufacturing is defined as the process of building 3D objects by joining materials layer by layer [20]. It is one of the most promising methods, which offers clear advantages in reducing material waste, time bottlenecks, and setup costs compared to conventional methods [31]. Due to the advancement of new technologies, the application of additive manufacturing in various industries, such as [32–34], is increasing. As a developing technology to manufacture precise and intensified complex objects by increasing production speed, it may offer an alternative to conventional manufacturing techniques in the near future [35]. Compared with traditional building material manufacturing, additive manufacturing can be manufactured according to design [36]. It provides strong support for personalized customization with higher customer participation. The integration of additive and subtractive manufacturing [37,38] has enormous potential to revolutionize how products are designed, manufactured, and delivered to customers in the form of products.

3. A Proposed P2p Cloud Manufacturing Method

Personalized production is a promising model towards the pursuit of expressing individual characteristics of human nature. AI and additive manufacturing can truly transform individual needs and preferences into personalized products and services at an affordable cost through ubiquitous connectivity, digitization, and sharing throughout the product lifecycle. In this section, a P2P cloud manufacturing method based on a customized business model is proposed.

As shown in Figure 1, customers are involved in the entire product life cycle from design to manufacturing. Customers can take pictures with digital cameras, or download screenshots on fixed and mobile terminals such as tablets and smartphones. This process involves long-distance transmission. AI-powered reverse engineering integrates image preprocessing and single-view reconstruction in the product, linking customer and model production. After the model is produced, the customer participates in the customization process of the model, which is a process of mutual feedback. The printing and production of products is also a process that requires customers and manufacturers to communicate their needs with each other, which is equivalent to the completion of the final product.

Two situations are considered: the object the customer wants is not local; the customer sees the object he wants on the Internet but has no model data. In the first case, simply take a photo of the product remotely. In the second case, just download a screenshot of the product you like. This process is entirely based on images provided by customers based on their needs and preferences. It provides customers with the greatest freedom of choice. Ubiquitous connections and sharing enable long-distance transmission of pictures.

The captured pictures may contain multiple objects, and it is difficult for the current 3D reconstruction technology based on deep learning to reply to the 3D information of each object picture. In view of this, preprocessing of the target image is necessary. YOLOv4 is used for object recognition and detection in pictures. The model output object contains the top, bottom, left, and right coordinates of all detected objects. Since cropping starts at the origin of the original image, the new coordinates are defined as follows:

$$Top_n = Max(0, top_r - 4.5), \tag{1}$$

$$Left_n = Max(0, top_r - 4.5), \tag{2}$$

$$Bottom_n = Min(W, Bottom_r + 5.5), \tag{3}$$

$$Right_n = Min(L, Right_r + 5.5), \tag{4}$$

where $\{top/Left/Bottom/Right\}_n$ denote the new coordinates of the top, left, bottom, and right, respectively. $\{top/Left/Bottom/Right\}_n$ denote the top, left, bottom, and right coordinates returned by the YOLOv4 model, respectively. $Max()$ denotes the max function. Min denotes the minimum function. W and L denote the width and length of the original image, respectively. For better calculation in the neural network, square pictures are required. The Algorithm 1 is as follows:

Algorithm 1 Square image generator.

top, bottom, left, right = 0, 0, 0, 0
fill = round(abs(L − W) /2)
if The length of the original image is greater than or fixed to the width **then**
  top, bottom = fill, fill
else
  left, right = fill, fill
end if

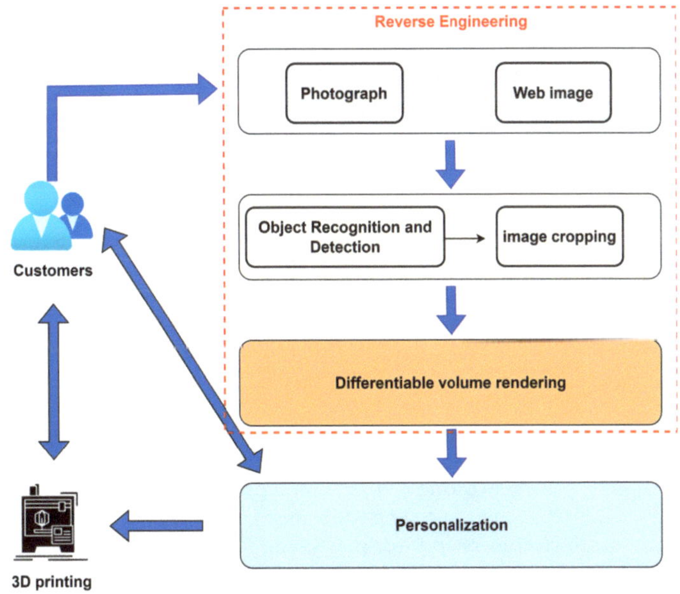

Figure 1. A P2P cloud manufacturing method based on personalized business model.

Algorithm 1 calculates the part that needs to be filled, which is filled with white. The 3D data of the object can be recovered from this image. A digital model can be obtained simply by determining the shape and texture of the object. DVR technology implicitly represents the shape f_θ and texture t_θ of the 3D model. The gradient from the surface depth is:

$$\frac{\partial \hat{d}}{\partial \theta} = \left(\frac{\partial f_\theta(\hat{p})}{\partial \hat{p}} \cdot w \right)^{-1} \frac{\partial f_\theta(\hat{p})}{\partial \theta}, \tag{5}$$

where f_θ denotes the occupancy network [39], which outputs the occupancy probability of each point in the 3D space. θ denotes the network parameter, which only involves computing the gradient at the point $\hat{p} \in R^3$. w denotes the vector of the camera pointing to a certain pixel point, and its intersection with $f_\theta(p)$ is $\hat{p}$. The input image i is encoded using the ResNet18 [40] network g_θ:

$$g_\theta(i) = Z, \tag{6}$$

where Z is a latent vector of 256 dimensions. The shape and texture of the 3D model are represented as:

$$f_\theta(p, z) = T, \tag{7}$$

$$t_\theta(p, z) = RGB, \tag{8}$$

where $p \in R^3$ denotes a point in space. $z \in Z$ denotes the encoder output vector. 3D surfaces are implicitly determined by the occupancy probability $T \in [0, 1]$. The texture of the object is given by the RGB values on the surface of the object. Five fully connected ResNet blocks and ReLu activation functions are used to implement the combined network. The output dimension of the last layer of the model is 4, one of which is occupancy probability, and the three dimensions are texture.

After reverse engineering the initial 3D model, in order to design a product model for individual needs and preferences, it is necessary to develop an effective information recommendation strategy. Designers integrate customer preferences into product design and continuously communicate with customers. Additive manufacturing also provides designers with many design-assisted design tools. Generative design, for example, is achieved through a combination of topology optimization and additive manufacturing, while optimizing topology and material distribution [41]. A digital model of the product (STL, Gcode, etc.) will be generated prior to additive manufacturing.

A designed 3D digital model is imported into the 3D printer. Many 3D printer manufacturers provide specialized model slicing software, which can adjust the actual size of the model, add suitable support structures, etc., before the model is printed. The printed product can be combined with subtractive manufacturing technology to obtain the final shape of the product. Likewise, the printed product is a personalized entity that interacts with customers.

Personalized customization is a customer-centric product manufacturing process. Introducing deep learning methods in the reverse engineering stage can reduce costs, shorten design time, and provide customers with long-distance transmission services. The generative design provides designers with more model styles, as well as topology-optimized structures. In the product production stage, additive manufacturing and material manufacturing can be combined. Manufacturers must interact and communicate with customers in real time to ensure product visibility and build connections and trust between customers and manufacturers. The customer-centric customized production model of on-demand manufacturing is shown in Figure 2.

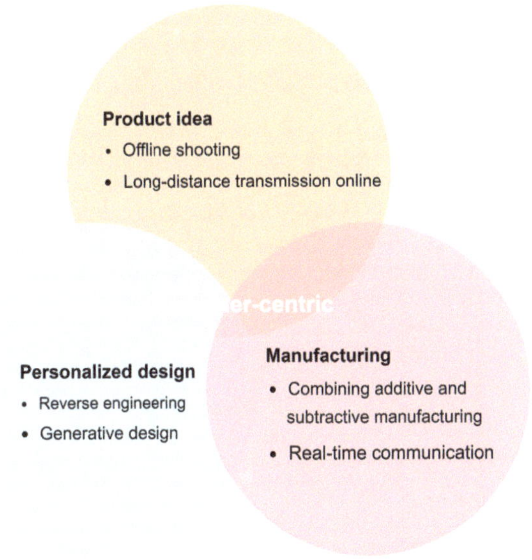

Figure 2. Customer-centric on-demand personalized production model.

4. The Case Studies of the Proposed Cloud Manufacturing Method

Two case studies are utilized to verify the feasibility of the proposed method. Assume that the customer finds the entity he wants while browsing the web or watching a video, but the customer cannot obtain the 3D scanning data of the object, only a screenshot of the website containing the object. Or if the customer sees the object he wants in the real world, he only needs to take a photo containing the object with a digital device to get the object model. The following is to produce the real small objects required by customers from online images and real photos.

4.1. Hardware and Software Environment

All procedures are coded in Python 3.8 with Pycharm IDE on a computer of Ubuntu OS with 2.2 GHz Intel i7 CPU, NVIDIA GeForce GTX 1070 GPU, and 16 GB DDR4 RAM. The real-life photos are taken with an iPhone12 with 3.0 GHz CPU, A14 Bionic chip, and 12 million front pixels. The 3D printer model used in the production of the entity is DF3, which is produced by Zhejiang Hangzhou DediBot Intelligent Technology Co., Ltd. [42] in China. Its printing method is FDM (Fused Deposition Modeling), the printing accuracy is 0.1 mm, and the printing speed is 30–100 mm/s. It supports digital model printing such as stl and obj. The specific parameters of the printer are shown in Table 1.

Table 1. DF3 printer parameter table.

Parameters	Values
Printer model	MOIRA DF3
Forming size	Φ150 × 175 mm
Printer weight	7.2 kg
Printing material	PLA
Printing method	FDM
Printing accuracy	0.1 mm
Printing speed	30–100 mm/s

4.2. Generating Small Solid Models from the Images

A picture from a webpage [43] is downloaded with a resolution of 960 × 1440 and is named Picture1. A photo with a resolution of 4032 × 3024 is taken by iphone12 in the real world and is named Picture2. Other information of the pictures are shown in Table 2. Picture1 is a four-seater sofa, as shown in Figure 3a, and Picture2 contains two cars of different shapes, as shown in Figure 3b. After detection and identification by the YOLOv4 network, a small sofa is extracted from Picture1. A new 819 pixel × 819 pixel size sofa picture is generated, as shown in Figure 4a. The picture is input into the YOLOv4 network for detection and recognition. The outputs of YOLOv4 are shown in Table 3. The probability of being identified as a sofa in the original image is 59%. The generated new picture is used as the input of the DVR network to construct the 3D model of the modified sofa, and the produced 3D model is shown in Figure 4(b1). Designers get the size and shape of the sofa, as well as personalized custom design. As shown in Figure 4(b2), a four-seater sofa can be turned into a single sofa. This one-seater sofa has the feature of being more portable. Two small sofas of different shapes have been produced. Figure 4(c1) is the sofa without any modification from the original picture Figure 4a, which is longer; Figure 4(c2) is the sofa modified by modeling customization, which is shorter. The printing parameters of the small solid model are shown in Table 4.

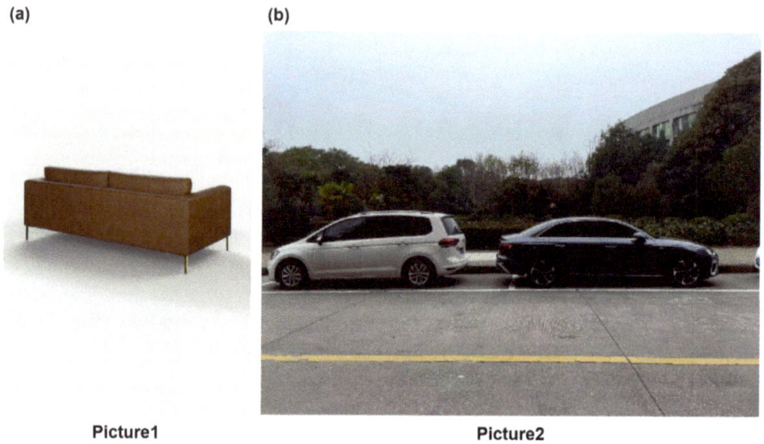

Figure 3. The images used for cloud manufacturing.

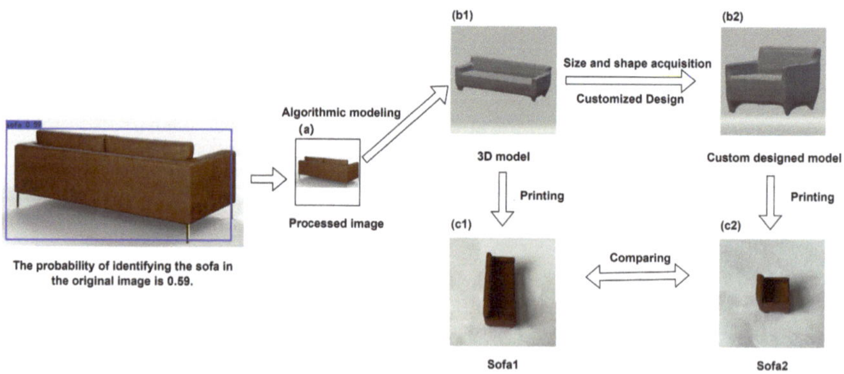

Figure 4. Two small sofa models with different shapes are manufactured using the proposed method.

Similarly, the Picture2 is input into the YOLOv4 network, and the output results are shown in Table 3. Since two cars were detected in the original picture, two new car pictures are generated. The resolutions of the new pictures are 1591 × 1591 and 1881 × 1881 respectively. The new images are fed into the DVR network, which generates 3D models of the two cars. The 3D models are imported into the DF3 printer to produce two small cars. The manufacturing process of the two cars is shown in Figure 5.

MOIRA DF3 is used for printing physical objects. The print samples of the sofa and two cars are showed in Figures 6 and 7. The Sofa2 printing process is taken as an example. The model is imported into a 3D printer. Model b2 in Figure 4 automatically adds supports, see Figure 6a. The model is sliced as shown in Figure 6b. The next step is to print (Figure 6c) and remove the supports (Figure 6d) to form the small sofa. Due to the limitations of current 3D printing technology, the size of the sofa is scaled by 233 times, and the setting is 15.00 mm × 15.89 mm × 13.65 mm. It takes 1.24 h to print the model. Parameters such as printing size and printing time of other models are listed in Table 4. The time to produce a 3D model from a 2D image is shown in Table 5, where Mesh represents the total time used to produce a 3D mesh, and other indicators represent the reconstruction time of each part. It can be seen that it only takes about a minute to recover the 3D structure from a picture. Due to the current limitations of our printer equipment and technology, the small models of sofas and cars are printed, and were not put into actual production. Nevertheless, from these two cases, it can be seen that the sofa and car models have basically been produced. The feasibility of the proposed method is verified.

Table 2. The image parameters.

Picture	Resolution	Width	High	Horizontal Resolution	Vertical Resolution	Bit Depth	Size	Inclusions
Picture1	960 × 1440	960 pixel	1440 pixel	96 dpi	96 dpi	24	238 KB	Sofa
Picture2	4032 × 3024	4032 pixel	3024 pixel	72 dpi	72 dpi	24	6.51 MB	Cars

Table 3. Probability and location of object recognition.

Object	Probability	Top	Bottom	Left	Right
Sofa1	59%	507	48	902	867
Car1	100.00%	1521	216	2153	1807
Car2	100.00%	1591	1958	2164	3840

Table 4. Object print parameter settings.

Object	Model Size	Production Time (3D Printing)
Sofa1	15.00 mm × 3.52 mm × 13.65 mm	1.36h
Sofa2	15.00 mm × 15.89 mm × 13.65 mm	1.24h
Car1	30.00 mm × 12.45 mm × 12.86 mm	0.31h
Car2	30.00 mm × 9.80 mm × 11.82 mm	0.26h

Table 5. Time for DVR to produce object 3D model (unit: s).

Object	Mesh	Time (Eval Points)	Time (Marching Cubes)	Time (Refine)	Time (Color)
Sofa1	64.897	10.463	0.993	50.421	2.829
Car1	62.468	8.483	0.989	50.648	2.186
Car2	61.729	8.851	0.991	49.573	2.314

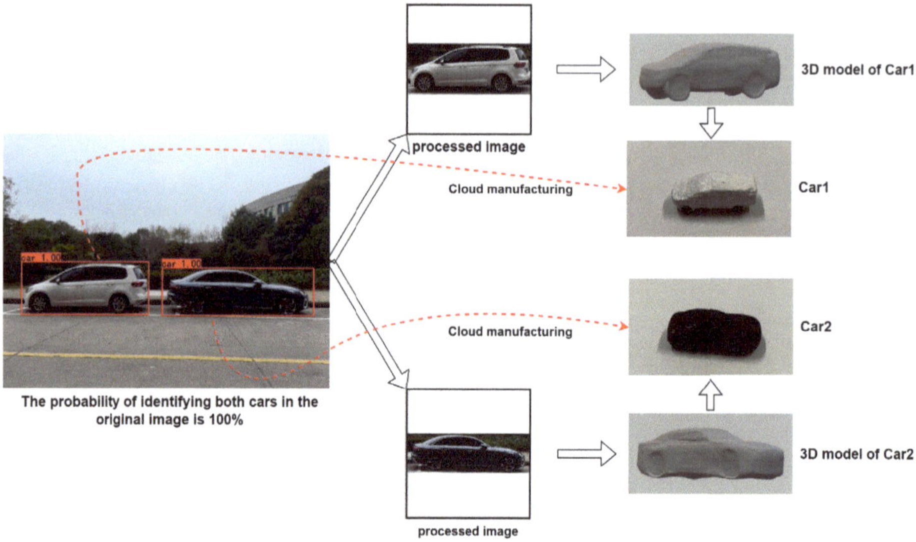

Figure 5. Two small car models with different shapes are manufactured through the proposed method.

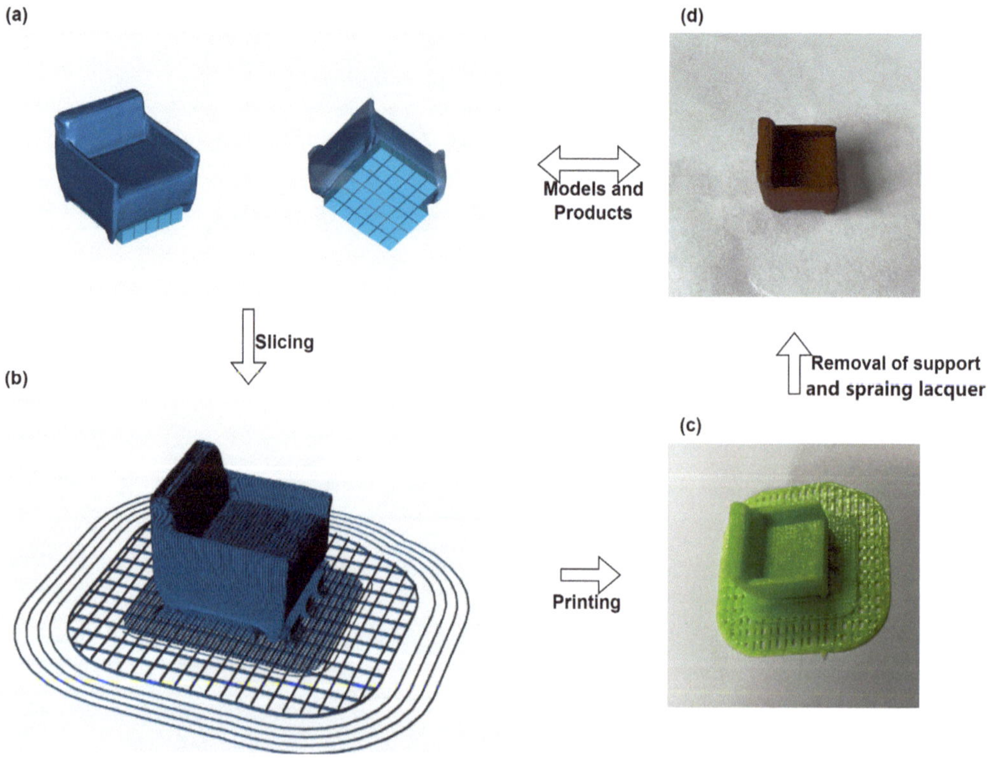

Figure 6. The printing process of Sofa2.

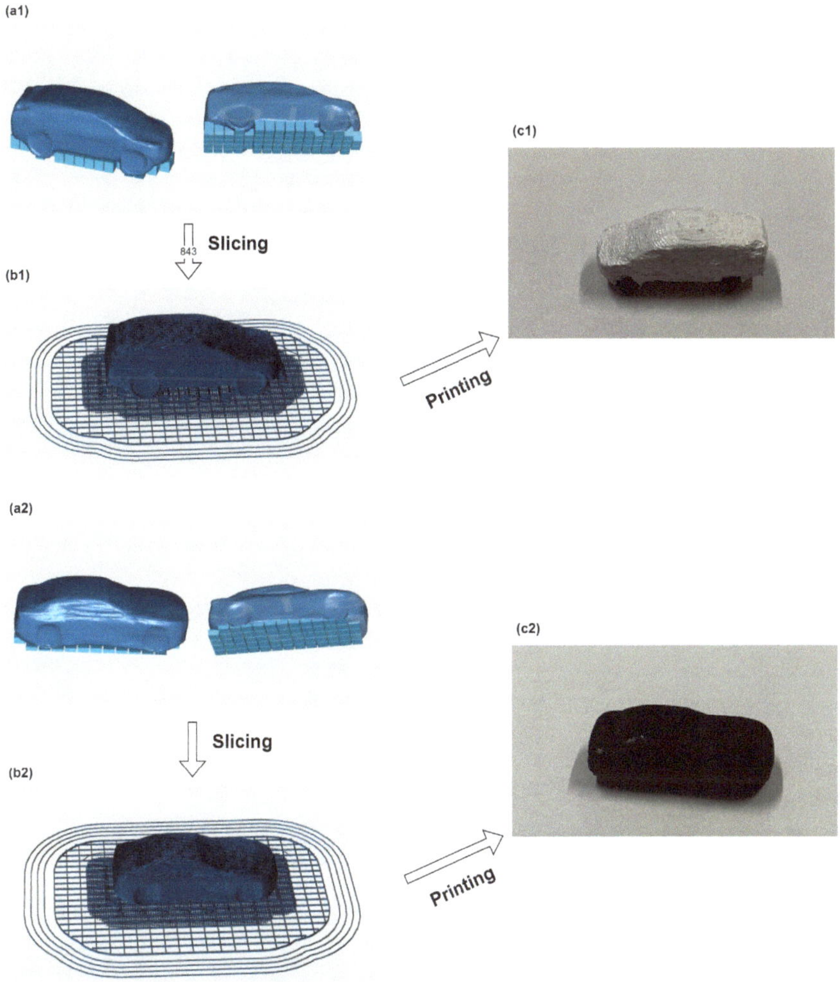

Figure 7. The printing process of two cars. Among them, (**a1**,**a2**) respectively represent the 3D digital models of Picture1 and Picture2 after adding supports, (**b1**,**b2**) represent their sliced models respectively, and (**c1**,**c2**) correspond to the printed small solid models respectively.

5. Conclusions

To cope with the ever-changing product demand in personalized services, high design costs, long product manufacturing life cycle and other issues, a P2P cloud manufacturing method based on the personalized business model is proposed. This method inherits the on-demand feature of personalized service. Based on the YOLOv4 algorithm and DVR technology, we built an object detection extractor and a 3D data generator, and conducted a case study on a 3D printing service scenario. In the case study, Internet sofa photos and real car photos are selected; the recognition rates of sofa and car are 59% and 100%, respectively. It takes about 60 s to retrogradely convert from 2D data to 3D data. We also carry out a personalized transformation design on the generated digital 3D model of the sofa. Two small sofas and two small car models are printed based on the generated 3D digital models. Judging by the printed results, the proposed method is validated and the prototypes of the sofa and the car were successfully produced. Among them, Sofa2 is transformed from the sofa in the original picture. Sofa1, Car1 and Car2 are all manufactured in their original proportions.

Although the integration of deep learning and additive manufacturing technology overcomes the time and cost problems of traditional reverse manufacturing, more detailed work is required in the future, e.g., applying more powerful printing equipment and technology to realize the value of manufactured products, enriching training data to support the generation of more 3D data to make our method easier to market, and optimizing algorithms to support the generation of objects with more complex structures.

Author Contributions: Conceptualization, X.Z. and M.L.; methodology, X.Z. and D.H.; software, D.H.; validation, X.Z. and D.H.; formal analysis, X.Z. and D.H.; investigation, X.Z., J.F. and X.D.; resources, X.D. and W.L.; data curation, J.F.; writing—original draft preparation, D.H.; writing—review and editing, X.Z.; visualization, W.L.; supervision, X.Z.; project administration, X.Z.; funding acquisition, W.L. All authors have read and agreed to the published version of the manuscript.

Funding: This research received no external funding.

Institutional Review Board Statement: Not applicable.

Informed Consent Statement: Not applicable.

Data Availability Statement: The data that support the findings of this study are available on request from the corresponding author. The data are not publicly available due to privacy or ethical restrictions.

Acknowledgments: The research carried out in this paper is supported by the Anshi Asia-Pacific Additive Research Institute of Nanchang Hangkong University, thanks to the 3D printing equipment provided by the institution.

Conflicts of Interest: The authors declare no conflicts of interest.

References

1. Guo, D.; Ling, S.; Li, H.; Ao, D.; Zhang, T.; Rong, Y.; Huang, G.Q. A framework for personalized production based on digital twin, blockchain and additive manufacturing in the context of Industry 4.0. In Proceedings of the 2020 IEEE 16th International Conference on Automation Science and Engineering (CASE), Hong Kong, China, 20–21 August 2020 ; pp. 1181–1186.
2. Zhu, X.; Shi, J.; Huang, S.; Zhang, B. Consensus-oriented cloud manufacturing based on blockchain technology: An exploratory study. *Pervasive Mob. Comput.* **2020**, *62*, 101113. [CrossRef]
3. Ozyurt, O.; Gurcan, F.; Dalveren, G.G.M.; Derawi, M. Career in Cloud Computing: Exploratory Analysis of In-Demand Competency Areas and Skill Sets. *Appl. Sci.* **2022**, *12*, 9787. [CrossRef]
4. Puntambekar, N.V.; Jablokow, A.G.; Sommer III, H.J. Unified review of 3D model generation for reverse engineering. *Comput. Integr. Manuf. Syst.* **1994**, *7*, 259–268. [CrossRef]
5. Bradley, C. The application of reverse engineering in rapid product development. *Sens. Rev.* **1998**, *18*, 115–120. [CrossRef]
6. Niem, W.; Wingbermuhle, J. Automatic reconstruction of 3D objects using a mobile monoscopic camera. In Proceedings of the International Conference on Recent Advances in 3-D Digital Imaging and Modeling (Cat. No. 97TB100134), Ottawa, ON, Canada, 12–15 May 1997 ; pp. 173–180.
7. Peng, L.W.; Shamsuddin, S.M. 3D object reconstruction and representation using neural networks. In Proceedings of the 2nd International Conference on Computer Graphics and Interactive Techniques in Australasia and South East Asia, Singapore, 15–18 June 2004; pp. 139–147.
8. Kato, H.; Ushiku, Y.; Harada, T. Neural 3d mesh renderer. In Proceedings of the IEEE conference on Computer Vision and Pattern Recognition, Salt Lake City, UT, USA, 18–26 June 2018; pp. 3907–3916.
9. Biggs, B.; Boyne, O.; Charles, J.; Fitzgibbon, A.; Cipolla, R. Who left the dogs out? 3d animal reconstruction with expectation maximization in the loop. In Proceedings of the European Conference on Computer Vision, Glasgow, UK, 23–28 August 2020; pp. 195–211.
10. Niemeyer, M.; Mescheder, L.; Oechsle, M.; Geiger, A. Differentiable volumetric rendering: Learning implicit 3d representations without 3d supervision. In Proceedings of the IEEE/CVF Conference on Computer Vision and Pattern Recognition, Seattle, WA, USA, 13–19 June 2020; pp. 3504–3515.
11. Levy, G.N.; Schindel, R.; Kruth, J.P. Rapid manufacturing and rapid tooling with layer manufacturing (LM) technologies, state of the art and future perspectives. *CIRP Ann.* **2003**, *52*, 589–609. [CrossRef]
12. Manfredi, D.; Calignano, F.; Krishnan, M.; Canali, R.; Paola, E.; Biamino, S.; Ugues, D.; Pavese, M.; Fino, P. Chapter Additive Manufacturing of Al Alloys and Aluminium Matrix Composites (AMCs). 2014. Available online: https://library.oapen.org/bitstream/handle/20.500.12657/49127/1/46882.pdf (accessed on 11 November 2022).

13. Thompson, M.K.; Moroni, G.; Vaneker, T.; Fadel, G.; Campbell, R.I.; Gibson, I.; Bernard, A.; Schulz, J.; Graf, P.; Ahuja, B.; et al. Design for Additive Manufacturing: Trends, opportunities, considerations, and constraints. *CIRP Ann.* **2016**, *65*, 737–760. [CrossRef]
14. Mehrpouya, M.; Dehghanghadikolaei, A.; Fotovvati, B.; Vosooghnia, A.; Emamian, S.S.; Gisario, A. The potential of additive manufacturing in the smart factory industrial 4.0: A review. *Appl. Sci.* **2019**, *9*, 3865. [CrossRef]
15. Joshi, S.C.; Sheikh, A.A. 3D printing in aerospace and its long-term sustainability. *Virtual Phys. Prototyp.* **2015**, *10*, 175–185. [CrossRef]
16. Bochkovskiy, A.; Wang, C.Y.; Liao, H.Y.M. Yolov4: Optimal speed and accuracy of object detection. *arXiv* **2020**, arXiv:2004.10934.
17. Pramataris, K.C.; Papakyriakopoulos, D.A.; Lekakos, G.; Mylonopoulos, N.A. Personalized interactive tv advertising: The imedia business model. *Electron. Mark.* **2001**, *11*, 17–25. [CrossRef]
18. Koelsch, C.; Przewrocka, J.; Keeling, P. Towards a balanced value business model for personalized medicine: An outlook. *Pharmacogenomics* **2013**, *14*, 89–102. [CrossRef] [PubMed]
19. Carlson, B. In Search of the Perfect Business Model: As personalized medicine moves into the mainstream, makers of diagnostics must face a new economic reality. How to develop a value proposition in a healthcare market that is becoming increasingly elastic? *Biotechnol. Healthc.* **2012**, *9*, 20.
20. Ardissono, L.; Felfernig, A.; Friedrich, G.; Goy, A.; Jannach, D.; Petrone, G.; Schafer, R.; Zanker, M. A framework for the development of personalized, distributed web-based configuration systems. *Ai Mag.* **2003**, *24*, 93.
21. Wang, S.; Wan, J.; Imran, M.; Li, D.; Zhang, C. Cloud-based smart manufacturing for personalized candy packing application. *J. Supercomput.* **2018**, *74*, 4339–4357. [CrossRef]
22. Lüftenegger, E. Achieving business model innovation with the personalized product business model radar template. In Proceedings of the IFIP International Conference on Advances in Production Management Systems, Novi Sad, Serbia, 30 August–3 September 2020; Springer: Berlin/Heidelberg, Germany, 2020; pp. 130–137.
23. Qin, Z.; Lu, Y. Self-organizing manufacturing network: A paradigm towards smart manufacturing in mass personalization. *J. Manuf. Syst.* **2021**, *60*, 35–47. [CrossRef]
24. Zhang, X.; Ming, X.; Bao, Y. A flexible smart manufacturing system in mass personalization manufacturing model based on multi-module-platform, multi-virtual-unit, and multi-production-line. *Comput. Ind. Eng.* **2022**, *171*, 108379. [CrossRef]
25. Eigen, D.; Puhrsch, C.; Fergus, R. Depth map prediction from a single image using a multi-scale deep network. *Adv. Neural Inf. Process. Syst.* **2014**, *27* .
26. Choy, C.B.; Xu, D.; Gwak, J.; Chen, K.; Savarese, S. 3d-r2n2: A unified approach for single and multi-view 3d object reconstruction. In Proceedings of the European Conference on Computer Vision, Amsterdam, The Netherlands, 16 September 2016; Springer: Berlin/Heidelberg, Germany, 2016; pp. 628–644.
27. Graves, A. Long short-term memory. Supervised Sequence Labelling with Recurrent Neural Networks. Doctoral Dissertation, Technical University of Munich, Munich, Germany, 2012 ; pp. 37–45.
28. Fan, H.; Su, H.; Guibas, L.J. A point set generation network for 3d object reconstruction from a single image. In Proceedings of the IEEE Conference on Computer Vision and Pattern Recognition, Honolulu, HI, USA, 21–26 July 2017; pp. 605–613.
29. Wang, N.; Zhang, Y.; Li, Z.; Fu, Y.; Liu, W.; Jiang, Y.G. Pixel2mesh: Generating 3d mesh models from single rgb images. In Proceedings of the European Conference on Computer Vision (ECCV), Munich, Germany, 8–14 September 2018; pp. 52–67.
30. Chen, W.; Ling, H.; Gao, J.; Smith, E.; Lehtinen, J.; Jacobson, A.; Fidler, S. Learning to predict 3d objects with an interpolation-based differentiable renderer. *Adv. Neural Inf. Process. Syst.* **2019**, *32*.
31. Wong, K.V.; Hernandez, A. A review of additive manufacturing. *Int. Sch. Res. Not.* **2012**, *2012*. [CrossRef]
32. Ulkir, O.; Ertugrul, I.; Akkus, N.; Ozer, S. Fabrication and Experimental Study of Micro-gripper with Electrothermal Actuation by Stereolithography Method. *J. Mater. Eng. Perform.* **2022**, *31*, 8148–8159. [CrossRef]
33. Khorasani, M.; Ghasemi, A.; Rolfe, B.; Gibson, I. Additive manufacturing a powerful tool for the aerospace industry. *Rapid Prototyp. J.* **2022**, *28*, 87–100. [CrossRef]
34. Michi, R.A.; Plotkowski, A.; Shyam, A.; Dehoff, R.R.; Babu, S.S. Towards high-temperature applications of aluminium alloys enabled by additive manufacturing. *Int. Mater. Rev.* **2022**, *67*, 298–345. [CrossRef]
35. Dilberoglu, U.M.; Gharehpapagh, B.; Yaman, U.; Dolen, M. The role of additive manufacturing in the era of industry 4.0. *Procedia Manuf.* **2017**, *11*, 545–554. [CrossRef]
36. Cui, W.; Yang, Y.; Di, L.; Dababneh, F. Additive manufacturing-enabled supply chain: Modeling and case studies on local, integrated production-inventory-transportation structure. *Addit. Manuf.* **2021**, *48*, 102471. [CrossRef]
37. Moussa, M.; ElMaraghy, H. Multiple platforms design and product family process planning for combined additive and subtractive manufacturing. *J. Manuf. Syst.* **2021**, *61*, 509–529. [CrossRef]
38. Chudpooti, N.; Savvides, G.; Duangrit, N.; Akkaraekthalin, P.; Robertson, I.D.; Somjit, N. Harmonized Rapid Prototyping of Millimeter-Wave Components using Additive and Subtractive Manufacturing. *IEEE Trans. Components, Packag. Manuf. Technol.* **2022**, *12*, 1241–1248. [CrossRef]
39. Mescheder, L.; Oechsle, M.; Niemeyer, M.; Nowozin, S.; Geiger, A. Occupancy networks: Learning 3d reconstruction in function space. In Proceedings of the IEEE/CVF Conference on Computer Vision and Pattern Recognition, Long Beach, CA, USA, 15–20 June 2019; pp. 4460–4470.

40. He, K.; Zhang, X.; Ren, S.; Sun, J. Deep residual learning for image recognition. In Proceedings of the IEEE conference on computer Vision and Pattern Recognition, Las Vegas, NA, USA, 27–30 June 2016; pp. 770–778.
41. Zhan, T. Progress on different topology optimization approaches and optimization for additive manufacturing: A review. In Proceedings of the Journal of Physics: Conference Series, Diwaniyah, Iraq, 21–22 April 2021; IOP Publishing: Bristol, UK, 2021; Volume 1939, p. 012101.
42. DediBot. DF3. Available online: http://www.dedibot.com/en/product/detail/10 (accessed on 1 January 2022).
43. ANTHROPOLOGIE. Edlyn Four-Seat Sofa, Leather. Available online: https://www.anthropologie.com/en-gb/shop/edlyn-four-seat-sofa-leather?color=026&type=REGULAR&size=One%20Size&quantity=1 (accessed on 11 November 2022).

sensors

MDPI

Article

Slicing Resource Allocation Based on Dueling DQN for eMBB and URLLC Hybrid Services in Heterogeneous Integrated Networks

Geng Chen [1,*], Rui Shao [1], Fei Shen [2] and Qingtian Zeng [1]

1 College of Electronic and Information Engineering, Shandong University of Science and Technology, Qingdao 266590, China
2 Shanghai Institute of Microsystem and Information Technology, Chinese Academy of Sciences, Shanghai 200050, China
* Correspondence: gengchen@sdust.edu.cn

Abstract: In 5G/B5G communication systems, network slicing is utilized to tackle the problem of the allocation of network resources for diverse services with changing demands. We proposed an algorithm that prioritizes the characteristic requirements of two different services and tackles the problem of allocation and scheduling of resources in the hybrid services system with eMBB and URLLC. Firstly, the resource allocation and scheduling are modeled, subject to the rate and delay constraints of both services. Secondly, the purpose of adopting a dueling deep Q network (Dueling DQN) is to approach the formulated non-convex optimization problem innovatively, in which a resource scheduling mechanism and the ϵ-greedy strategy were utilized to select the optimal resource allocation action. Moreover, the reward-clipping mechanism is introduced to enhance the training stability of Dueling DQN. Meanwhile, we choose a suitable bandwidth allocation resolution to increase flexibility in resource allocation. Finally, the simulations indicate that the proposed Dueling DQN algorithm has excellent performance in terms of quality of experience (QoE), spectrum efficiency (SE) and network utility, and the scheduling mechanism makes the performance much more stable. In contrast with Q-learning, DQN as well as Double DQN, the proposed algorithm based on Dueling DQN improves the network utility by 11%, 8% and 2%, respectively.

Keywords: 5G/B5G; network slicing; deep reinforcement learning; dueling deep Q network (Dueling DQN); resource allocation and scheduling

Citation: Chen, G.; Shao, R.; Shen, F.; Zeng, Q. Slicing Resource Allocation Based on Dueling DQN for eMBB and URLLC Hybrid Services in Heterogeneous Integrated Networks. *Sensors* **2023**, *23*, 2518. https://doi.org/10.3390/s23052518

Academic Editors: Chien Aun Chan, Chunguo Li and Ming Yan

Received: 26 January 2023
Revised: 15 February 2023
Accepted: 21 February 2023
Published: 24 February 2023

1. Introduction

With the explosive growth of data in mobile networks, 5G mobile communication technologies have matured to meet a wide variety of traffic needs. The two most typical types of services in 5G mobile networks are ultra-reliable and low-latency communication (URLLC) and enhanced mobile broadband (eMBB) [1]. The 5G network provides resources for the two types of users mentioned above in a sliced manner [2,3]. When slicing is performed, the allocation of resources is adjusted by the base station (BS) according to the dynamic demands of user services and adapts to different network states [4]. Slicing of network resources enables data triage management and flexible resource allocation in 5G networks [5,6], and it is also necessary to achieve a high data transmission rate, low latency and high capacity [7,8].

Due to the intense growth of network traffic and the densification of devices, there are multiple problems and great challenges in the allocation and scheduling of resources between different services [9]. For example, in the 5G scenario, when there are users of both eMBB and URLLC service types, it is necessary to allocate a lot of bandwidth resources to users of the eMBB service type within a time slot to ensure that their images and voice have high and stable quality, and it is also necessary to successfully transmit the data packets requested by URLLC service type users within the range of very short delay to meet the characteristics of ultra-high reliability and ultra-low delay [10]. If there is a sudden increase in URLLC traffic in the same area, it will quickly occupy these bandwidth resources to reach its required transmission rate, resulting in an ultra-low latency performance [11]. When bandwidth resources are insufficient, existing works typically prioritize the performance requirements of URLLC by sacrificing the quality of experience (QoE) of eMBB. The rational allocation and scheduling of slicing resources facilitate efficient resource use in hybrid services systems.

In recent years, reinforcement learning (RL) has become a potential solution to the resource allocation problem. The resource allocation algorithm based on RL has improved resource utilization efficiency [12]. As deep reinforcement learning (DRL) has evolved, many research works based on DRL approaches have also been achieved [13,14]. For example, DRL is applied to solve problems with resource allocation [15], network optimization, routing, scheduling and radio control. Chen et al. [16] modeled the problem of auctioning a finite number of channels across scheduling slots to multiple service providers as a stochastic game, then linearly decomposed the Markov decision process for each service provider and derived an online solution based on deep reinforcement learning. In [17], the stochastic decision process in vehicular networking is modeled as a discrete-time single-intelligent Markov decision process (MDP) to address the partial observability and high dimensionality curse of the local network state space faced by each vehicular user device and to make optimal band allocation and group scheduling decisions in a decentralized manner in each scheduling time slot. In [18], a deep Q-network (DQN) algorithm based on discrete normalized advantage functions (DNAF) was studied, and the advantage function was decomposed into two function terms to reduce the computational complexity of the algorithm. In addition, the simulation results verify that the deep Q-learning (DQL) based on the K-nearest neighbor algorithm can converge faster in discrete environments. Sciancalepore et al. [19] proposed the reinforcement learning-based network slice broker (RL-NSB) framework to effectively improve the utilization of the system by considering factors such as traffic flow, mobility, and optimal access control decision. The distributed idea [20] and the effect of randomness noise for spectrum efficiency (SE) and service level agreement (SLA) satisfaction ratio (SSR) are referred to [21]. Hua et al. [21] introduced the generative adversarial network and used it to allocate physical resources among multiple network slices of a single BS, which performs well in terms of demand-aware resource management. Furthermore, Li et al. [22] considered the user mobility based on [19] and utilized the actor–critic based on long short-term memory (LSTM-A2C) algorithm to follow the mobility of users, improving the practicality of the resource allocation system. Yuan et al. [23] provide a DRL-based resource-matching distributed method to maximize energy efficiency (EE) and device-to-device (D2D) capacity through a decentralized approach, match multi-user communication resources to double DQN, and optimize radio channel matching and power allocation. Sun et al. [24] distinguished resource granularity, utilized virtualized coarse resources to obtain provisioning solutions and used fine resources for dynamic slicing, proposing a dueling DQN-based algorithm customized to the diverse needs of users to improve user satisfaction and resource utilization. Chen et al. [25] used an algorithm based on dueling deep Q network (Dueling DQN) combined with bidding for bandwidth resource allocation in two layers to improve the QoE of users and verify the advantages of Dueling DQN over Double DQN in resource allocation. Boateng et al. [26] proposed a new hierarchical framework for autonomous resource slicing in 5G RANs, modeling the seller and buyer pricing and demand problems as a two-stage Stackelberg game to design fair incentives and designing a Dueling DQN scheme to achieve optimal pricing and demand strategies for autonomous resource allocation in negotiated intervals. Zhao et al. [27] performed joint optimization and obtained a great policy by proposing an algorithm that combines multiple agents with D3QN to maximize network utility and guarantee the quality of service (QoS).

Various schemes have been studied in relation to the problem of resource scheduling between different services. An innovative overlay/perforation framework [28,29] is based on the principle of overlaying a part of eMBB services when sporadic uRLLC services occur, although this approach may lead to significant degradation of the QoE of eMBB. Feng et al. [30] and Han et al. [31] designed a long and short dual time-scale algorithm for bandwidth allocation and service control, respectively, using Lyapunov optimization to centrally guarantee the latency of URLLC service while improving the quality of the eMBB continuous service. Han et al. [32] presented a dynamic framework of Q-learning based to improve the latency QoS and energy consumption ratio between URLLC and eMBB traffic in 5G/B5G. Moreover, Wang et al. [33] proposed a deep deterministic policy gradient (DDPG) algorithm to optimize the hole punch location and bandwidth allocation of URLLC services, and realize the QoS trade-off between URLLC and eMBB in 5G/B5G. Alsenwi et al. [34] used the DRL-based optimization auxiliary framework to solve the resource slicing problem in the dynamic reuse scenario of eMBB and URLLC services, achieved the ideal data rate of eMBB under the reliability constraints of URLLC and reduced the impact of URLLC traffic that was immediately scheduled on the reliability of eMBB. The time slots occupied with eMBB are split into the small slot and URLLC traffic pre-overlap at the small slot so that the proportional fairness of eMBB users can be maximized while satisfying the URLLC constraint [35]. Almekhlafi et al. [36] introduced a new technology that can reuse URLLC and eMBB services to reduce the size of perforated eMBB symbols, and improved the reliability of eMBB, symbol error rate (SER) and SE in the study to meet the delay constraints and reliability of URLLC. In [37], a new hybrid punching and coverage strategy is used to enhance the compromise between the acceptable number of URLLC users and the throughput of eMBB users.

As described in the abovementioned literature, RL is used to solve the dynamic resource allocation problem in various scenarios and has shown good performance. However, the performance requirements of URLLC are not prioritized and the resource scheduling problem among different services is not addressed. In addition, the traditional optimization algorithm and RL algorithm can be used to solve the resource scheduling problem between eMBB and URLLC services, but they still face a series of difficulties and challenges. For example, when scheduling resources among diverse services, the overlay/perforation framework has a huge influence on the QoS of eMBB in order to enhance the performance requirements of URLLC, and the Lyapunov dual time-scale algorithm improves the continuous QoS of eMBB, but its scheduling time slot is long and the optimization speed is slow. In this paper, a new Dueling DQN-based resource allocation and scheduling algorithm that satisfies the slice requirements is proposed. For the different demand characteristics of eMBB and URLLC services, especially for the ultra-low latency constraint of URLLC services, part of the bandwidth resources occupied by users of eMBB services are scheduled to URLLC users. With spectrum efficiency (SE) and quality of experience (QoE) of the two services as the optimization objectives, we have formed an optimization problem restricted by the rate and delay constraints of the two services and innovatively used Dueling DQN to solve the non-convex optimization problem of slicing resource allocation. Meanwhile, we use the resource scheduling mechanism and ϵ-greedy strategy to select the optimal resource allocation action, adopt the reward-clipping mechanism to enhance the optimization goal and select a reasonable bandwidth resolution (b) to improve the flexibility of bandwidth resource allocation. The main work can be summarized in three aspects.

(1) First, a scenario in which multiple wireless access network slices exist and BSs share bandwidth resources is considered. In this scenario, the resources are allocated and scheduled by BS for users with two different services. For the different demand characteristics of eMBB and URLLC services, especially for the ultra-low latency constraint of URLLC services, some of the bandwidth resources occupied by users of eMBB services are scheduled to URLLC users.

(2) Second, a novelty Dueling DQN-based algorithm aimed at allocating and scheduling of bandwidth resources is proposed. The problem regarding resource allocation and scheduling for eMBB and URLLC is modeled as an optimization problem and plotted as a Markov process, which is addressed through Dueling DQN training. Dueling DQN divides

the action–value function output from the neural network into a state–value function and a dominance function, which reduces the correlation between the current state-action and action selection, and this network architecture is suitable for solving the proposed slicing resource allocation problem in discrete action space. More importantly, we generate the state using the number of packets received by two different service users and define the size of the bandwidth resources allocated to the two slices as actions. Since both the system SE and the QoE of eMBB and URLLC are optimization objectives, it is necessary to consider both the SE and the QoE. Therefore, the reward-clipping mechanism is proposed to encourage the agent to choose the best resource allocation action. Meanwhile, we choose the appropriate bandwidth allocation resolution to ensure the appropriate action space size and increase the flexibility of resource allocation.

(3) Third, the simulations are performed and reasonable data are obtained. It can be seen from the obtained data that the proposed algorithm ensures that the QoEs of eMBB and URLLC are stable at 1.0 with a high probability, meeting the service requirements in this scenario. Moreover, the QoE, SE and network utility show convergence trends. In contrast with Q-learning, DQN as well as Double DQN, the proposed Dueling DQN algorithm improves the network utility by 11%, 8% and 2%, respectively. Furthermore, the resource scheduling mechanism improves the QoE of URLLC, SE and network utility by 3%, 4.5% and 7%, respectively.

The organization of the following section in this paper is as follows. Section 2 builds the system model and formulates the optimization problem. In Section 3, the theoretical basis of Dueling DQN is introduced and the proposed slice resource allocation algorithm based on Dueling DQN is discussed in detail. Section 4 displays the simulation parameters and results. Finally, Section 5 concludes this work.

2. Problem Statements

2.1. System Model

Under the 3GPP standard, the main application scenarios of 5G networks include eMBB, URLLC and mMTC services, but they are widely different. The eMBB is a continuous wide-area coverage and high-capacity hot spot scenario. Its high-capacity hot spots mainly target at local hot spots, providing users with a high-speed data transmission rate and meeting users' high traffic requirements. The URLLC can support ultra-high reliability connection under high-speed mobile conditions with extremely low latency. The mMTC has a large number of connected devices, but it typically sends relatively low amounts of non-latency-sensitive data. Obviously, eMBB and URLLC have higher bandwidth requirements and greater fluctuation in their resource requirements, while mMTC has fewer resource requirements and little fluctuation. Therefore, we do not consider the mMTC service of 5G network applications. This paper focuses on the problem of allocating limited bandwidth resources to URLLC and eMBB hybrid services and scheduling resources between slices. When the bandwidth resources provided by the BS to users are insufficient, the packet dropping probability increases for both eMBB and URLLC services, leading to a decrease in QoE. In order to improve the QoE of URLLC, SE and network utility on basis of ensuring the QoE of eMBB, the bandwidth allocation resolution (b) is chosen for the resource allocation and scheduling.

As presented in Figure 1, we take a scenario of heterogeneous integrated network slicing into account, which is composed of multiple BSs and users. The physical network is divided into N slices $\{1, 2, \ldots, N\}$. The user set U contains M users $\{u_1, u_2, \ldots, u_M\}$, including J eMBB users and K URLLC users. In Figure 1, the network entity represented by the agent in the actual 5G network environment is the software-defined network (SDN), which obtains the information of the BS and users in the environment, controls the BS to slice and allocates bandwidth resources to users as required. The relationship between the agent and the whole environment is as follows. The agent can obtain the changes of eMBB and URLLC requirements and the resource allocation information of the BS in the environment in a timely manner. In the current time slot, when users of different services request resources from the BS, the agent uses the number of packets received by users of two different services to generate a state and defines the size of bandwidth resources allocated to the two slices as actions. Then, they traverse all actions in each state and select the best action according to the ϵ-greedy strategy, which corresponds to the bandwidth allocation scheme of BS. Meanwhile, agents form rewards based on the reward-clipping mechanism and obtain new states according to environmental changes.

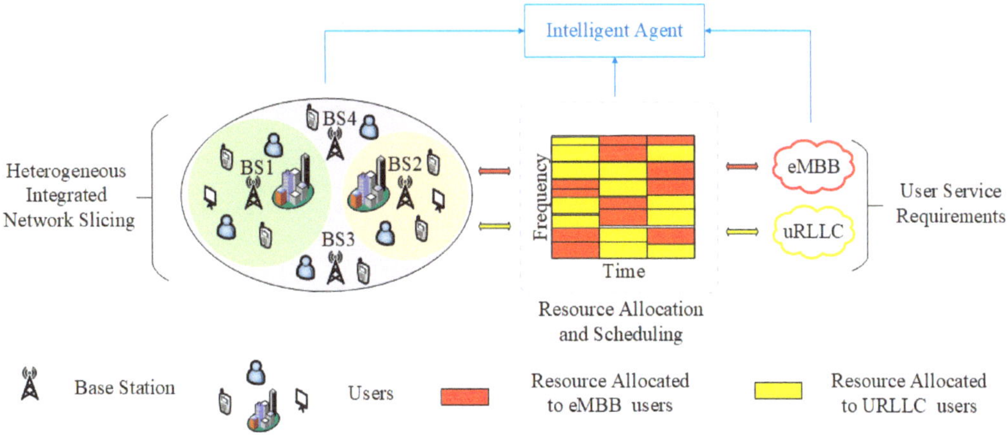

Figure 1. Network slicing scenario with multiple BSs and eMBB and URLLC users.

2.1.1. Resource Allocation of eMBB and URLLC Slices

In a heterogeneous integrated network slicing scenario with hybrid eMBB and URLLC services, both slices share all bandwidth resources. In order to denote the allocation of bandwidth resources, the binary variable $\lambda_e \in (0,1)$ and $\lambda_l \in (0,1)$ are defined, where $\lambda_e = 1$ and $\lambda_l = 1$ indicate that the bandwidth resources are allocated to users of eMBB and URLLC, respectively. The bandwidth allocated by the BS to eMBB users can be denoted by

$$B_{u_e} = \lambda_e \cdot b \cdot d_{u_e} \tag{1}$$

where b denotes the bandwidth allocation resolution and d_{u_e} indicates the amount of bandwidth allocated from the BS to eMBB users. The bandwidth obtained by the URLLC user from the BS is represented as

$$B_{u_l} = \lambda_l \cdot b \cdot d_{u_l} \tag{2}$$

where d_{u_l} is the amount of bandwidth allocated from the BS to URLLC users.

2.1.2. Resource Scheduling between URLLC and eMBB Slices

When URLLC users request bandwidth resources from the BS and the bandwidth resources in the BS are already fully occupied by the eMBB slice and other URLLC users, part of the bandwidth resources occupied by the eMBB users will be scheduled by the BS to the URLLC users. The purpose of resource scheduling is to guarantee the QoE of URLLC. Define the total bandwidth requested by $L(L \leq K)$ URLLC users as $B_{e,l}$. To prevent the termination of partial eMBB service and reduce the impact of bandwidth resource scheduling on the QoE of eMBB due to bandwidth resource scheduling, the bandwidth occupied by any eMBB user will not be fully scheduled for the URLLC users. The bandwidth resources being scheduled for each eMBB user can be given as

$$B_{u_e,u_l} = b \cdot d_{e,l} \tag{3}$$

where $d_{e,l}$ denotes the number of bandwidth resources lost for each eMBB user. However, the resource scheduling must satisfy the following condition

$$B_{u_e} - B_{u_e,u_l} \geq B_0 \tag{4}$$

where B_0 denotes the minimum bandwidth required to guarantee the minimum transmission rate for eMBB service. If L eMBB users are unable to provide bandwidth resources for the URLLC users, other L eMBB users will continue to provide bandwidth resources for the URLLC users.

The bandwidth resources obtained from the BS for eMBB and URLLC slices can be denoted as

$$B_e = \sum\nolimits_{u_e \in \mathrm{U}} B_{u_e} - B_{e,l} \tag{5}$$

$$B_l = \sum\nolimits_{u_l \in \mathrm{U}} B_{u_l} + B_{e,l} \tag{6}$$

In summary, the situation of the bandwidth resources allocated by the BS to the users is completed.

2.2. Problem Formulation

2.2.1. Calculation of the SE

In the system with hybrid eMBB and URLLC services, improving resource utilization becomes a problem to be solved. Let SE denotes a symbolic evaluation index.

The noise in the Rayleigh fading channel cannot be ignored when the base station and the user are connected. However, in our environment, co-channel interference is avoided because each base station operates on a different frequency band to allocate bandwidth to users with different service types. Therefore, we use signal-to-noise ratio (SNR) when calculating the transmission rate. For the eMBB user, the transmission rate is related to the allocated bandwidth resources and the SNR between the BS and the user. It is calculated by

$$r_{u_e} = B_{e,o} \log_2 \frac{h_e \cdot p_e}{B_{e,o} \cdot N_0} \tag{7}$$

where h_e denotes the channel gain between the BS and the eMBB user, p_e represents the transmitted power of the BS connected to the eMBB user and N_0 is the noise power spectral density. However, $B_{e,o}$ has two cases, as follows. The bandwidth resources of eMBB users who lose resources are $B_{u_e} - B_{u_e,u_l}$, and the bandwidth resources of the other eMBB users remain B_{u_e}. The sum of downlink transmission rate for eMBB users can be calculated as

$$R_e = \sum\nolimits_{u_e \in \mathrm{U}} r_{u_e} \tag{8}$$

Similarly, we have the downlink transmission rate of URLLC user

$$r_{u_l} = B_{l,o} \log_2 \frac{h_l \cdot p_l}{B_{l,o} \cdot N_0} \tag{9}$$

where h_l denotes the channel gain between the BS and the URLLC user and p_l represents the transmitted power of the BS connected to the URLLC user. Moreover, $B_{l,o}$ also has two cases, as follows. For URLLC users who obtain resources through scheduling, their bandwidth resources are $B_{u_l} + B_{u_e,u_l}$, while other URLLC users maintain their bandwidth resources at B_{u_l}. We can calculate the sum of the downlink transmission rate for URLLC users as

$$R_l = \sum\nolimits_{u_l \in \mathrm{U}} r_{u_l} \tag{10}$$

since the SE is considered the sum of the downlink transmission rate divided by the total bandwidth (W) allocated from the BS to the users. The SE can be denoted by a variable Y and formulated as

$$Y = \frac{R_e + R_l}{W} \tag{11}$$

2.2.2. Calculation of the QoE

Due to its requirements for low latency and ultra-reliability, we will prioritize the reduction of packet dropping probability for URLLC services. When the bandwidth resources in the current time slot are insufficient, the BS will partially schedule the bandwidth resources occupied by the transmission of eMBB packets to the URLLC users until the resources required for the transmission of URLLC packets are satisfied. The major objective of both services is to obtain low packet dropping probability and high transmission rates. Before quantifying the QoE, we define a binary variable $\rho \in (0,1)$. When the packet transmission is successful, ρ is taken as 1, and when the packet transmission fails, ρ is taken as 0.

Let the number of packets transmitted (pkt) of eMBB users expressed as p_{u_e}. So, the total number of packets transmitted of eMBB users can be calculated as

$$p_e = \sum\nolimits_{u_e \in \mathrm{U}} p_{u_e} \tag{12}$$

As the QoE is defined as the packet dropping probability, we define q_{u_e} as a packet transmitted by an eMBB user. We can formulate the QoE of eMBB as

$$Q_e = \frac{\sum_{u_e \in \mathrm{U}} \sum_{q_{u_e} \in p_{u_e}} \rho}{p_e} \tag{13}$$

Moreover, we denote the pkt of URLLC users as p_{u_l}, and calculate the total number of packets transmitted of URLLC users by

$$p_l = \sum\nolimits_{u_l \in \mathrm{U}} p_{u_l} \tag{14}$$

The q_{u_e} is defined as a packet transmitted by an URLLC user. The QoE of URLLC is formulated as follows

$$Q_l = \frac{\sum_{u_l \in \mathrm{U}} \sum_{q_{u_l} \in p_{u_l}} \rho}{p_l} \tag{15}$$

2.2.3. Calculation of the Network Utility

To address the resource allocation and scheduling problem in a hybrid services system, we achieve a reasonable allocation of resources in a diverse services system by dynamically adjusting the allocation of bandwidth resources for each slice. The optimization objective is the weighted sum of the SE and QoE of the two services, which we define as the network utility function F.

Mathematically, the formulated problem concerning the allocation and scheduling of resources is presented by

$$\max_{B_{u_e}, B_{u_l}} F = \alpha Y + \beta Q_e + \eta Q_l \tag{16}$$

$$\text{S. t. } B_e + B_l \leq \mathrm{W} \tag{16a}$$

$$B_{u_e} \leq C_e \tag{16b}$$

$$B_{u_l} \leq C_l \tag{16c}$$

$$r_{u_e} \geq \overline{r_{u_e}} \tag{16d}$$

$$r_{u_l} \geq \overline{r_{u_l}} \tag{16e}$$

$$t_{u_e} \leq \overline{t_{u_e}} \tag{16f}$$

$$t_{u_l} \leq \overline{t_{u_l}} \tag{16g}$$

where α, β, η denote the importance weight values of the SE, the QoE of eMBB Q_e and the QoE of URLLC Q_l, respectively. The total number of bandwidth resources obtained from the BS for both eMBB and URLLC slices less than the total bandwidth resources of the system is denoted in Equation (16a). Equations (16b) and (16c) indicate that the bandwidth resources allocated by the BS for both eMBB and URLLC slices must not exceed the eMBB and URLLC downlink bandwidth capacity C_e and C_l, respectively. The transmission rate of eMBB and URLLC users must be higher than the transmission rate specifications of both services in 5G scenarios $\overline{r_{u_e}}$ and $\overline{r_{u_l}}$, which are expressed in Equations (16d) and (16e). Additionally, Equations (16f) and (16g) denote that the eMBB and URLLC services transmission latency must be lower than the maximum latency requirement $\overline{t_{u_e}}$ and $\overline{t_{u_l}}$ in 5G scenario, respectively.

In the heterogeneously integrated network slicing scenario in which eMBB and URLLC service requirements coexist, the resource scheduling and allocation processes of front and rear timeslots interact. The resource allocation in each time slot should meet the current fluctuating demand. However, with the time-varying nature of both the remaining resources and user demand, the BS needs to continuously change the resource allocation scheme to ensure the QoE for both services and improve SE and network utility. To illustrate that the optimization problem formed is non-convex and NP-Hard, it is mapped to the 0–1 backpack problem. Assume a backpack of capacity C_n and T_n items and define the value of each item to be p_n and the weight to be w_n. The purpose is to search for a subset $T'_n \in T_n$ that obtains the maximum $\sum_{T'_n \in T_n} p_n$ under the condition that $\sum_{T'_n \in T_n} w_n \leq C_n$. Meanwhile, a simplified form of the optimization problem in this paper is considered, i.e., the case in which only one slice of URLLC is available. Then, the optimization objective of this simplified problem can be denoted as

$$\max_{B_{u_l}} F_l = \alpha Y + \eta Q_l = \alpha \frac{R_l}{B_l} + \eta \frac{\sum_{u_l \in U} \sum_{q_{u_l} \in p_{u_l}} \rho}{p_l} \tag{17}$$

$$\text{S. t.} \quad B_l \leq W \tag{17a}$$

$$B_{u_l} \leq C_l \tag{17b}$$

$$r_{u_l} \geq \overline{r_{u_l}} \tag{17c}$$

$$t_{u_l} \leq \overline{t_{u_l}} \tag{17d}$$

Mapping the 0–1 backpack problem to this optimization objective, the quantity of items T_n corresponds to the users in URLLC slice, the value p_n corresponds to the weighted sum of QoE and SE achieved by the slice and the weight w_n corresponds to the limit of downlink capacity, transmission rate, and latency. Obviously, the optimization problem can be completed in one polynomial time, while the 0–1 backpack problem has NP-Hard characteristics, so the simplified problem can be classified as an NP-Hard problem. It can be concluded that the optimization problem formed in this paper is a non-convex optimization and is NP Hard. Since the prior transfer probability is unknown, it is very difficult to obtain the closed optimal solution of the formulaic problem. However, RL is more suitable for solving such problems in which the probability of prior transfer is unknown. Therefore, we use RL to find the best scheme for the allocation and scheduling of bandwidth resources in heterogeneous integrated networks.

3. Proposed Algorithm

3.1. Foundation of Dueling DQN

It is worth mentioning that DQN, as a branch of DRL, uses two key technologies for improvement and has outstanding advantages in decision making. Firstly, the experience replay mechanism breaks the inherent correlation among samples, making them independent of each other. Secondly, the target value network can enhance the convergence and stability of training by lessening the correlation between the current and target Q value, correspondingly. An intelligent agent obtains information about the environment through trial and error and uses the data obtained during the interaction as observations. Then, the agent traverses the actions in a given state and finds the corresponding action with the largest Q value according to its ϵ-greedy policy. However, DQN has a disadvantage in that the Q value output by its neural network denotes the selected action value in the state, which is dependent on the action and state. This means that the DQN fails to reflect the different effects of state and action on the Q value. Furthermore, DQN is vulnerable to the overestimation problem, resulting in poor training stability. Among the improvements of DQN in recent years, Dueling DQN has outstanding advantages, and its network stability and convergence speed have been significantly improved. In particular, Dueling DQN maintains the advantages of the DQN while improving on the DQN in terms of network structure by dividing the action–value function from the output of the neural network into the state–value function and advantage function. This allows Dueling DQN to learn the value function for each state without considering what action to take in that state. Therefore, the Dueling DQN converges better when the current action is less relevant to the successive state and the current state–action function is also less relevant to the current action selection. Based on the improvements and advantages of Dueling DQN over DQN, we prefer to choose Dueling DQN for iterative optimization of proposed nonconvex optimization problems.

The process of interaction between the agent of Dueling DQN and the environment can be cast into a Markov decision process (S, A, R, P, γ), where S presents the state space, A is the action space. The current state s and the next state s' are stored in the state space, while the current action a and the next action a' are stored in the action space. R denotes the reward function, which is the goal that the agent maximizes during action selection and is the key factor that makes the training process more stable. P is the transfer probability, which represents the probability that the current state will be transferred to another state when an action is performed. γ is a discount factor greater than 0 and less than 1 that moderates near and far-term effects in reinforcement learning. The action–value function can be formulated as

$$Q^{\pi}(s,a) = V^{\pi}(s) + A^{\pi}(s,a) \tag{18}$$

Here, the policy π denotes the distribution that maps state to action. Then, the two functions $A^{\pi}(s,a)$ and $V^{\pi}(s)$ are approximated using the neural network. It can be seen that $V^{\pi}(s)$ relates only to states, while $A^{\pi}(s,a)$ relates to both states and actions. In fact, there are two neural networks with parameters θ in the Dueling DQN: the target Q network and the evaluation Q network, respectively. Let $Q(s,a;\theta,\phi,\varphi)$ denote the value function with parameters θ, which is expressed as

$$Q(s,a;\theta,\phi,\varphi) = V(s;\theta,\phi) + A(s,a;\theta,\varphi) \tag{19}$$

where θ is a shared parameter, ϕ denotes a dominant function parameter and φ is used to indicate a parameter of the action–value function. However, there exists a serious problem in the above equation, which is that the unique $V(s;\theta,\phi)$ and $A(s,a;\theta,\varphi)$ cannot be obtained from $Q(s,a;\theta,\phi,\varphi)$ in Equation (19). Thus, a centralization processing of the advantage function is performed to guarantee that zero dominance will occur for a given action. Further, $Q(s,a;\theta,\phi,\varphi)$ can be reformulated as

$$Q(s,a;\theta,\phi,\varphi) = V(s;\theta,\phi) + [A(s,a;\theta,\varphi) - \frac{1}{|A|}\sum_{a'} A(s,a';\theta,\varphi)] \quad (20)$$

The agent of Dueling DQN makes an observation as it interacts with the environment. The Q-network calculates all Q values for each action when observation is used as state inputs. Then, the agent selects the action that maximizes the Q value relying on a ϵ-greedy strategy and provides the reward value. In Dueling DQN, the target Q value of the target Q-network is updated by copying the current Q value every C iterations. However, the current Q value is reset with real-time updates in each iteration. The target Q value (Q_t) of the target Q-network is denoted by

$$Q_t = r + \gamma \max_{a'} \hat{Q}(s',a';\theta,\phi,\varphi) \quad (21)$$

Then, the loss function $L(\theta)$ in Dueling DQN is defined by

$$L(\theta) = \mathbf{E}\left[(Q_t - Q(s,a;\theta,\phi,\varphi))^2\right] \quad (22)$$

where $\mathbf{E}$ denotes the expected value. Meanwhile, the optimal parameter θ is obtained through the minimization of the square of TD error; that is,

$$\varsigma^2 = [Q_t - Q(s,a;\theta,\phi,\varphi)]^2 \quad (23)$$

Finally, the action–value function $Q(s,a;\theta,\phi,\varphi)$ is updated by

$$Q(s,a;\theta,\phi,\varphi) = Q(s,a;\theta,\phi,\varphi) + \varepsilon[Q_t - Q(s,a;\theta,\phi,\varphi)] \quad (24)$$

The iterative training of Dueling DQN requires that a fixed number of iterations be set. When the iteration is ended, Dueling DQN can utilize the trained neural network for optimal action selection.

3.2. The Dueling DQN Based Slicing Resource Allocation and Scheduling Algorithm

The proposed Dueling DQN-based algorithm is used for resource allocation and scheduling in eMBB and URLLC hybrid traffic. Bandwidth resources are dynamically allocated and scheduled so that the requirements of users are better met and network utility is maximized. In cases in which the bandwidth resources are insufficient, the BS schedules some of the bandwidth resources occupied by eMBB service to the URLLC service, improving the QoE of URLLC and network utility with the premise of ensuring the QoE of the eMBB.

To achieve resource scheduling between eMBB and URLLC users, the following resource scheduling mechanism is set up. When randomly distributed users request resources from the BS, the BS counts the number of users requesting resources and slices the bandwidth by service types. In each iteration, the BS allocates resources to the users of both services while counting the number of users that request resources. When there are insufficient bandwidth resources and URLLC users still request resources, the BS schedules bandwidth resources occupied by the corresponding number of the users of eMBB to URLLC based on the number of URLLC users requesting resources. Furthermore, the bandwidth resources occupied by eMBB users are not all scheduled to avoid service interruption for this user. After several iterations, a resource scheduling scheme can be obtained to provide the best network utility.

The goal of this paper is to improve the SE of the network system while guaranteeing the QoE of both services, so we use a reward-clipping mechanism that allows the agent to optimize both metrics through the algorithm. Since the QoE and the SE of the system are of different orders of magnitude, we use different coefficients in each segment of the reward parameter to make the reward value reach a value that is easy for the agent to simulate and learn. We expect users of both services to achieve satisfactory QoE. In order to ensure that the QoE of slices meet the 5G standard and reach 1.0 as often as possible, we set the QoE threshold of 0.98 in the reward function. If the Q_l cannot satisfy the requirement we mentioned above, a more unfavorable negative reward value in Equation (25) will be calculated by

$$r = -3 + [(Q_l - 1) \cdot 10] \tag{25}$$

When the Q_l satisfies the requirement but the Q_e cannot, we will have a negative reward value as follows

$$r = (Q_l - 1) \cdot 10 \tag{26}$$

Similarly, we would like to see an improvement in SE. We compared the highest value and lowest value of SE in the algorithm training process, and in order to ensure a higher system SE is achieved as often as possible, so that the agent presents a stable training trend in the training process, we set SE as 380 between the highest value and the lowest value in the reward function. If the QoE of both services can be achieved but the SE does not satisfy this condition, a poorer positive reward value can be given as

$$r = Y \cdot 0.01 \tag{27}$$

Conversely, the QoE of both services can be achieved and the SE satisfies this condition; thus, we will calculate a better positive reward value as follows:

$$r = 5 + [(Y - 380) \cdot 0.1] \tag{28}$$

The procedures of resource allocation and scheduling for hybrid eMBB and URLLC services using the proposed Dueling DQN based algorithm with resource allocation and scheduling are as follows. In order to helps readers to understand our process more clearly, the algorithm flowchart is shown in Figure 2.

Before starting the iterative training, the parameters are initialized and a randomly selected policy is required to produce an original state. Moreover, the BS randomly selects an allocation scheme to first allocate bandwidth resources for eMBB and URLLC users, and then schedule the bandwidth resources according to the resource scheduling mechanism. After the end of scheduling, the intelligent agent of the Dueling DQN obtains information during its interaction with the environment and calculates the *pkt* of eMBB and URLLC users as an observation. Afterward, the observation is entered into the Q-network to form the initial state.

Each iteration performs the operations as follows. The BS selects a resource allocation action based on the policy in the Dueling DQN, after which scheduling is performed. Then, the user receives the resource allocated by the BS, and S and A are updated in the Dueling DQN. Each state in the state space is the number of eMBB and URLLC packets successfully transferred. Each action in the action space is the bandwidth resource allocated to users by BS based on the network utility and the state feedback from users. The SE and the packet dropping probabilities are calculated according to Equations (11), (13) and (15). Thereby, the network utility can be calculated as shown in Equation (16). It is worth mentioning that the reward calculation formula is one of Equations (25)–(28). Once again, the *pkt* is calculated as the next state. Then, the s, a, s', r is imported to Dueling DQN for training.

For each iteration, the training process is as follows. Firstly, the agent obtains s, a, s', r in the response of the environment, which is saved in the replay memory as a transfer sample. After enough data are deposited in the sample pool, a minibatch-sized transaction is randomly selected in the sample pool for training. Secondly, the evaluation Q-network of the agent adds the advantage function of centralized processing to the state-value function to obtain the current Q value, as illustrated in Equation (20). Meanwhile, Equation (21) is the formula used by the intelligent agent to calculate the target Q value. Moreover, the action that maximizes the current Q value in a given state is selected on the basis of the

ϵ-greedy strategy. Finally, the current update of Q network parameters is based on the loss function in Equation (22) and the gradient descent method in Equation (23). Consistent with Equation (24), the current Q-network parameters are cloned into the target Q-network by resetting to complete the parameter update of the target Q-network after C iterations.

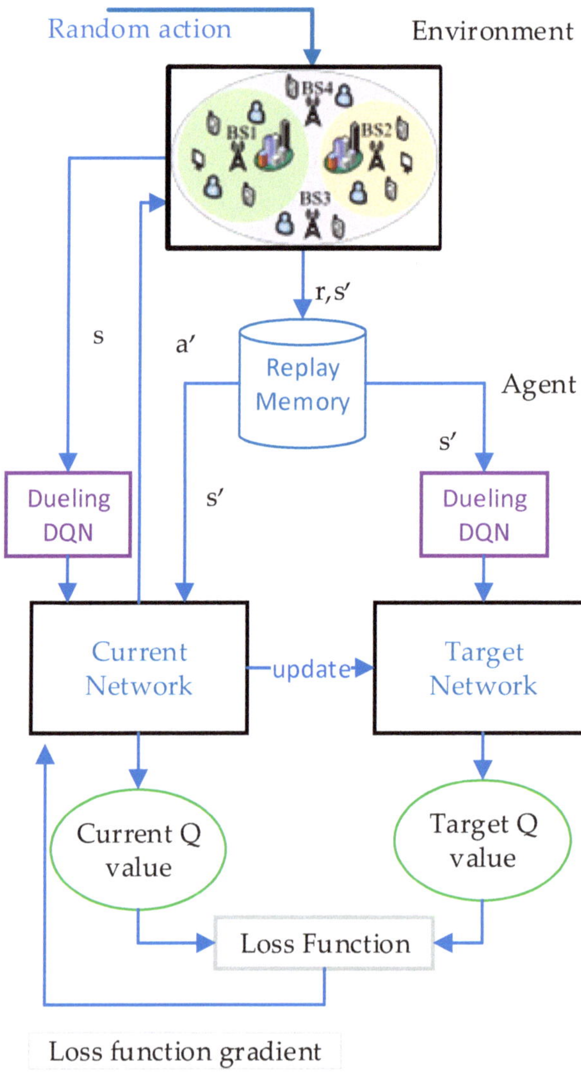

Figure 2. Algorithm flow diagram.

Using the predetermined number of iterations, the value function network with great performance is trained. The Dueling DQN is capable of obtaining an action under the ϵ-greedy strategy for a given state to reduce the loss function and improve the cumulative expected reward. Therefore, the best scheme of resource allocation and scheduling can be obtained in the eMBB and URLLC hybrid service system, which improves the QoE of URLLC, SE and network utility while ensuring the QoE of eMBB. The pseudocode of the proposed algorithm is presented in Algorithm 1.

Algorithm 1. The Dueling DQN based slicing resource allocation and scheduling

Initialize the replay memory D, the capacity P, the current and target action-value function Q and $\hat{Q}$ with random weights θ, the parameter ϕ and φ;
Choose random action a_0 to allocate bandwidth for eMBB and URLLC users;
Scheduling:
    User $\leftarrow$ The bandwidth resources;
  The URLLC users continue to request resources;
  The eMBB users $\xrightarrow{b}$ The URLLC users;
The *pkt* calculated $\rightarrow s$;
Repeat
  For iteration = 1, to T, **do**
    Policy $\rightarrow a$ choosed;
    Execution **Scheduling**;
    The SE is calculated as shown in Equation (11);
    The Q_e and Q_l are calculated on the basis of Equations (13) and (15);
    Calculate the network utility based on Equation (16);
    Calculate the reward based on one of Equations (25)–(28);
    The *pkt* calculated $\rightarrow s'$;
      # Train Dueling DQN;
      The $\{s, a, s', r\}$ is stored in D of Dueling DQN;
The agent samples $\{s_i, a_i, s_{i+1}, r_i\}$ from D;
      Define $Q(s_i, a_i; \theta, \phi, \varphi)$ according to Equation (20);
      Set

$$y_i = \begin{cases} r_i & \text{if terminal step } i+1 \\ r_i + \gamma max_{a'} \hat{Q}(s_{i+1}, a'; \theta^*, \phi, \varphi) & \text{otherwise} \end{cases}$$

      The agent updates the network parameters θ by $[y_i - \hat{Q}(s_i, a_i; \theta, \phi, \varphi)]^2$;
      Executed $\hat{Q} \leftarrow Q$ every C iterations;
  End for
Until The end of the iterations.

3.3. Time Complexity Analysis of Algorithm

The time complexity of the training phase needs to consider the time complexity of training the Q network and the number of attempts needed to train the Q network. In the process of training the Q network, the connection weights between every two adjacent layers of neurons need to be updated. We set the number of layers of the Q network as x_i, the number of neurons in the ith layer to be n_i, and the number of iterations in each training to be t_{train}, then the time complexity c_{train} of training a Q network once can be calculated as.

$$c_{train} = t_{train} D(\sum_{i=1}^{x=1} x_i x_{i+1}) \tag{29}$$

We denote the total number of iterations in the algorithm as t_{total}, and the number of steps in each iteration as t_{step}, then the number of times to train the Q network is $t_{total} \cdot t_{step}$, so the time complexity of the proposed algorithm training phase can be calculated as

$$c_{train} = t_{total} \cdot t_{step} \cdot t_{train} D(\sum_{i=1}^{x=1} x_i x_{i+1}) \tag{30}$$

The time complexity of the online training phase of the deep reinforcement learning algorithm is high, but after the Q network is trained, the Q network does not need to be updated in the running phase, and the time complexity is low, which can meet the requirements of online decision-making time under real-time network conditions. Since the algorithms we compared in the simulation are all deep reinforcement learning algorithms and set the same parameters, they are roughly the same in terms of algorithm complexity.

4. Simulation Results and Analysis

4.1. Simulation Parameters

In this part, we conduct extensive simulations to verify the performance of the algorithm we proposed in heterogeneous integrated networks. The simulation runs on a PC with an Intel Core i7-10750 H CPU at 2.6 GHz. The graphics card we use is NVIDIA GeForce GTX1650 Ti with 4G memory. We use TensorFlow 1.15 deep learning framework and Python 3.8 to implement our algorithm in the simulation. The scenario contains two kinds of services and two types of slices, correspondingly. In order to fit the actual situation and reflect the advantages of the algorithm, we set the available bandwidth W of BS to 20 MHz, the radius of BS to 50 m, and the number of users as 500. We set the rate and delay thresholds of eMBB and URLLC according to the 5G service level agreement. In order to ensure that the action space is not too large and better actions can be obtained, the bandwidth resolution is set to 0.1 MHz. In the actual scenario, the number of users of the eMBB service is often greater than the number of URLLC services, so the proportion of the number of eMBB and URLLC users is set to 3:2. For Rayleigh fading, we set the noise power spectral density to −174 dBm/Hz. The details of distribution of users and *pkt* standards according to [22] are listed in Table 1.

Table 1. Simulation Parameters.

	eMBB	**URLLC**
Channel	Rayleigh fading	
Scheduling	1 s (2000 scheduling slots)	
Noise Spectral Density (σ)	−174 dBm/Hz	
Total Bandwidth (B)	20 MHz	
Bandwidth Allocation Resolution (b)	0.1 MHz	
Number of Users (M)	500	
	300	200
Minimum Rate Constraint (r)	100 Mbps	10 Mbps
Maximum Delay Constraints (t)	10 ms	1 ms
Distribution of Users	Truncated Pareto [=Exponential Parameter: 1.2, Average: 6 ms, Maximum: 12.5 ms]	Constant: 0.3 MByte
Distribution of *pkt*	Truncated Pareto [Exponential Parameter: 1.2, Average: 100 Byte, Maximum: 250 Byte]	Exponential [Mean: 180 ms]

4.2. Performance Evaluation

Next, we illustrate the performance simulation results of the proposed Dueling DQN-based algorithm in detail. Furthermore, it is compared with different environmental parameters and with algorithms based on Q-learning, DQN as well as Double DQN. The Ref. [12] applies Q-learning to resource allocation in a network slicing scenario. In Refs. [21,22], the resource allocation algorithm based on DQN is mentioned and used as a comparison of resource allocation schemes in network slicing scenario. The algorithm based on Double DQN is used to solve the management and allocation of wireless network resources in Refs. [23,25]. In particular, the same learning ratios are set for Q-learning, DQN, Double DQN and Dueling DQN. For the common parameters of the algorithm, we set the learning rate to 0.01, the discount factor to 0.9, and the minimum batch size to 32 after our experiment. To unify the SE and QoE orders of magnitude and thus obtain easily comparable system utilities, the weights of the optimization target SE and QoE of both services are set as $\alpha = 0.01$, $\beta = 1$, $\eta = 3$, respectively. The performance simulation results and brief analysis in terms of the QoE, SE and network utility are as follows.

Figures 3–6 present the performance of the QoE, SE and network utility with the slicing resource allocation algorithms based on Q-learning, DQN, Double DQN and Dueling DQN, respectively. Figures 3 and 4 depict the tendency of QoE for both services with the increasing number of iterations. After about 1000 iterations, although the latency requirement of URLLC is higher than that of eMBB, the Dueling DQN can achieve almost 100% QoE of both services. Compared with Dueling DQN, Double DQN shows slightly less stability for both services, while both DQN and Q-learning show extreme instability for both services. The reason for this is that the Dueling DQN makes improvements compared with the other three algorithms, thereby allocating bandwidth resources more rationally and achieving the desired QoE.

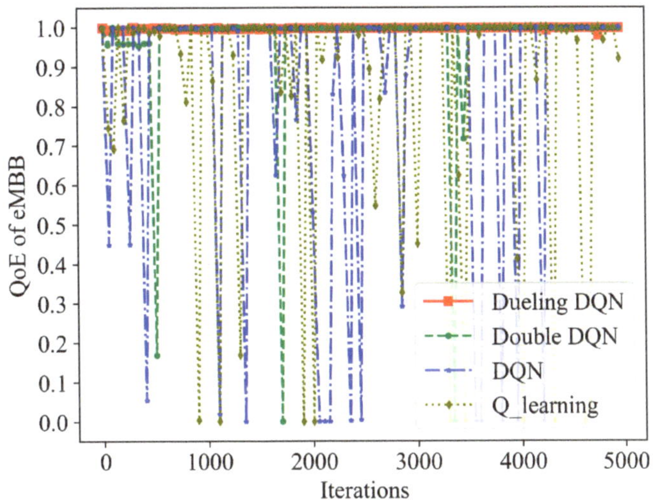

Figure 3. A comparative result of QoE for eMBB service with different slicing algorithms.

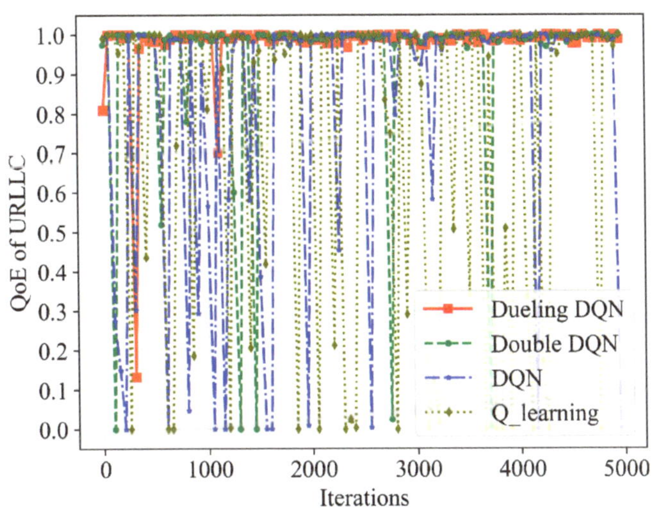

Figure 4. A comparative result of QoE for URLLC service with different slicing algorithms.

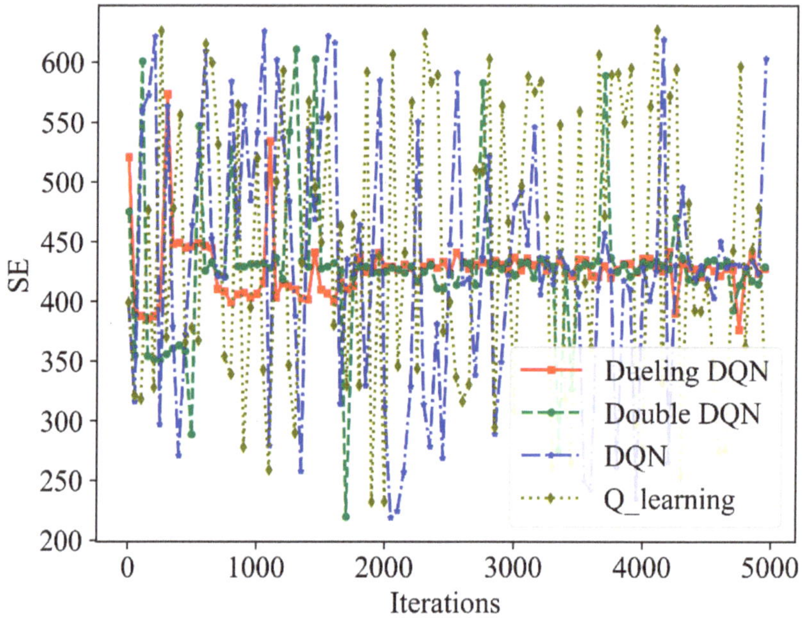

Figure 5. A comparative result of SE with different slicing algorithms.

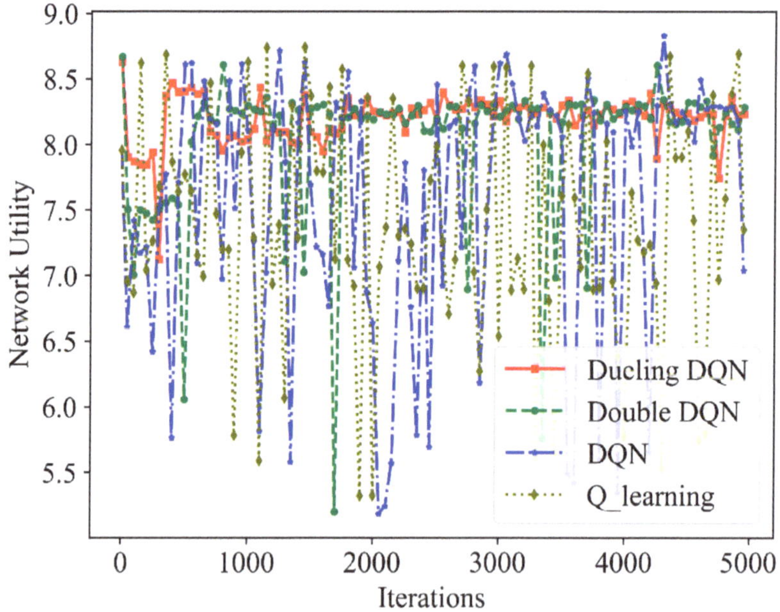

Figure 6. A comparative result of network utility with different slicing algorithms.

Figure 5 presents the trend of SE with the increasing number of iterations. It is indicated that the SE obtained through Dueling DQN can be divided into three stages. The SE is high but unstable in the first 600 iterations, low from 600 iterations to 2000 iterations and basically stable at 430 after around 2000 iterations. This is due to the fact that the number of eMBB users is larger than the number of URLLC users and the BS allocates more bandwidth resources to eMBB users at the beginning of the iteration. In addition, the transmission rate achieved by eMBB users with the same bandwidth is much higher than the transmission rate achieved by URLLC users. Therefore, a large SE is calculated according to Equation (15). Since the proposed algorithm focuses on improving the QoE of URLLC, BS allocates more bandwidth resources to URLLC users after 600 iterations, reducing SE to about 400. After around 2000 iterations, the neural network parameters of the Dueling DQN are updated to a basically stable state, thereby obtaining the high and convergent SE. During the optimization process, SE has some values exceeding 450, which obtain very low network utility and reward because SE cannot guarantee the corresponding service quality. Therefore, the proposed algorithm does not allow the SE to converge to the anomaly height value described above. Through these too high and low abnormal values, we can see the advantages of the proposed Dueling DQN-based algorithm, which improves the SE while prioritizing the service quality. However, the unusual and sudden drop in performance late in the training period is due to the tiny non-zero ϵ-greedy exploration rate. Moreover, it is also possible to obtain the information that the Double DQN obtains the optimal curve of SE close to Dueling DQN but with greater fluctuation, while the other two algorithms have no convergence tendency. The Dueling DQN achieves the best convergence SE among the four algorithms, indicating that Dueling DQN can achieve the purpose of guaranteeing QoE and improving SE for different services by flexibly adjusting the resource allocation and scheduling.

Figure 6 further presents the tendency of the network utility as the increasing number of iterations. We define network utility as the weighted sum of SE and QoE, and QoE converges to 1.0 with high probability. Thus, the network utility has a strong correlation with the optimization of SE, which can also be divided into three stages. Under the influence of unstable SE and QoE, the network utility fluctuates widely at the beginning of the iteration. Although 100% QoE is achieved, the network utility shows a downward trend due to resource scheduling. Under the effect of reward, QoE and SE are improved and remain convergent, so the network utility can basically remain above 8.3 after 2000 iterations. The abnormally high SE in Figure 5 corresponds to the low network utility in Figure 6, which further explains that just a high SE cannot guarantee the service quality of eMBB and URLLC users. However, the Double DQN eventually converges to a value similar to the Dueling DQN but shows less stability, and both Q-learning and DQN show no apparent signs of convergence. Furthermore, in contrast with Q-learning, DQN as well as Double DQN, the proposed algorithm enhances the network utility by 11%, 8% and 2%, respectively.

Afterward, the impact of resource scheduling mechanisms on performance is further investigated. Figures 7 and 8 show the differences in QoE of eMBB and URLLC during the iterative learning of Dueling DQN with and without the resource scheduling mechanism. For the eMBB users, when there is a resource scheduling mechanism, the QoE demonstrates trivial improvement. With the bandwidth resources which are allocated to eMBB users are much greater, the QoE of URLLC occasionally has values that fail to satisfy service requirements in the absence of the mechanism. Moreover, the QoE of URLLC is raised by 3% with the mechanism and there are no low outliers that do not satisfy URLLC feature requirements after optimization. This phenomenon occurs because some of the bandwidth resources occupied by eMBB are scheduled to URLLC in order to increase the bandwidth resources available for the transmission of URLLC packets without interrupting eMBB user services. Therefore, it ensures a low packet dropping probability of eMBB and reduces the packet dropping probability of URLLC. In other words, it improves the convergence stability of the QoE of the URLLC without reducing the QoE of the eMBB.

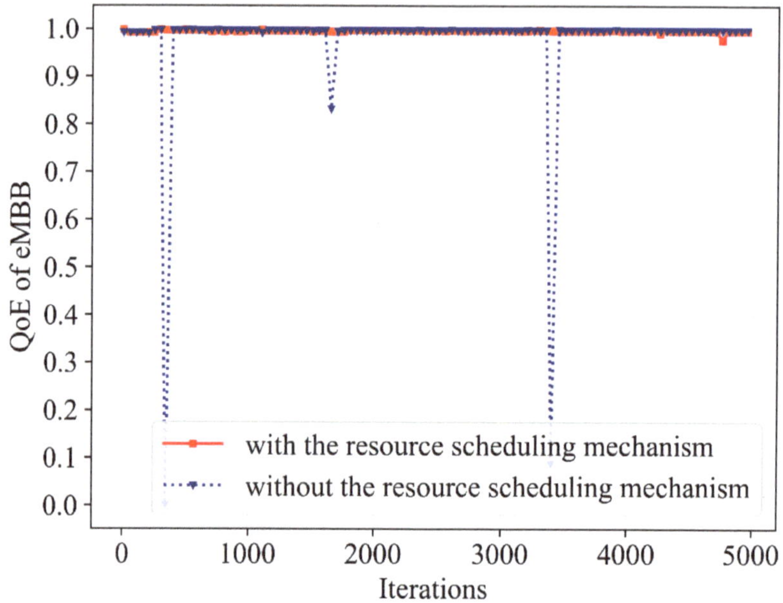

Figure 7. A comparative result of QoE for eMBB service with and without the resource scheduling mechanism.

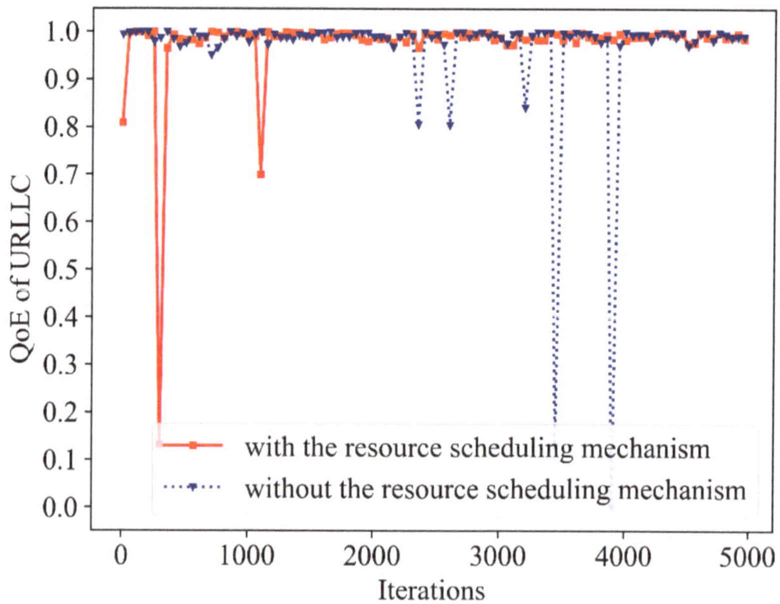

Figure 8. A comparative result of QoE for URLLC service with and without the resource scheduling mechanism.

Figure 9 presents the impact on SE with and without the resource scheduling mechanism. The SE curve without a resource scheduling mechanism shows a trend of increasing first and then decreasing. At the beginning of iterative learning, the agent explores actions beneficial to SE, making SE improve to a better value. However, after iterative learning, it drops and remains at a low value, and occasionally shows a high outlier. When this mechanism is not considered, BS tends to allocate bandwidth in proportion to users, and the demand rate of eMB service is much higher than that of the URLLC service, resulting in high SE. Then, it learns to allocate bandwidth in a more rational manner after iterative training; that is, the bandwidth allocated to URLLC increases. The above factors cause the SE to drop slightly, and thus it fails to converge at higher values. Additionally, we can see that since the bandwidth resources allocated to eMBB are increased at this time, excessive SE values are obtained late in the iteration, corresponding to the low abnormal QoE of URLLC. However, the SE with the mechanism only obtains an excessive SE value at the beginning of the iteration because the QoE of URLLC is not satisfied, and tends to be stable after iterative learning. Meanwhile, the algorithm with the resource scheduling mechanism has a slightly faster convergence speed and better stability and can converge to a better value without outliers. Therefore, we can conclude that through the resource scheduling mechanism, the Dueling DQN can learn more rational bandwidth allocation to improve the stability of SE under the premise of guaranteeing the QoE of different services.

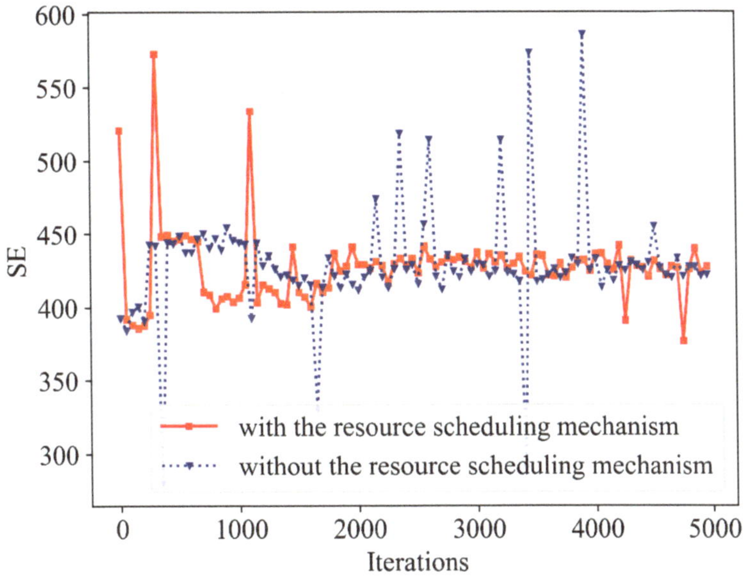

Figure 9. A comparative result of SE with and without the resource scheduling mechanism.

Furthermore, Figure 10 continues to present the differences in network utility with and without the resource scheduling mechanism. Since QoE achieves the optimal value of 1 with a high probability, there is a strong correlation between network utility and SE. The network utility without a resource scheduling mechanism showed a high value in the first 2000 iterations but decreased slightly and remained at a low value in the subsequent iterations. During the iteration, the outlier in network utility is due to the high SE obtained by allocating more bandwidth resources to eMBB users, but at this time, the QoE of URLLC is not satisfied and the reward-clipping mechanism presents a worse reward value. However, the next bandwidth allocation action requires a great reward; thus, the resource allocation action is improved to satisfy the QoE of the URLLC and restore stability to the SE and network utility. Figures 7–10 show that the utilization of the resource scheduling mechanism improves the SE and network utility by 4.5% and 7%, respectively. The simulation curves prove the effect of Dueling DQN with the resource scheduling mechanism.

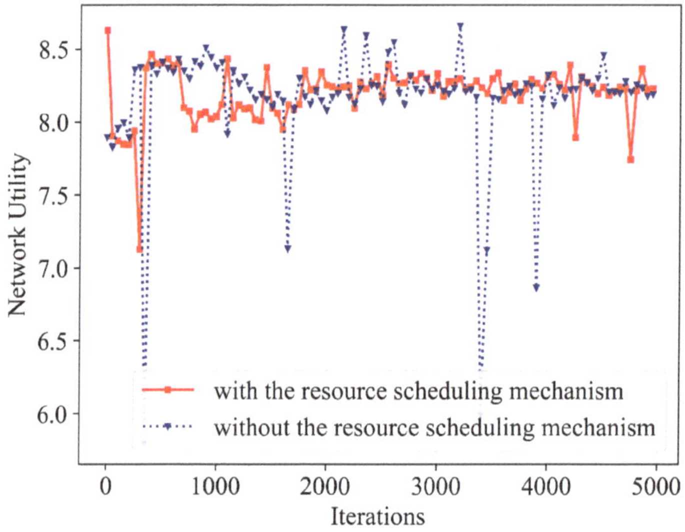

Figure 10. A comparative result of network utility with and without the resource scheduling mechanism.

Figures 11 and 12 compare the QoE tendency of eMBB and URLLC in the different cases in which the b is 0.1 MHz, 0.2 MHz and 0.4 MHz. It shows that the QoE converges almost to the optimum at the condition that b is 0.1 MHz, but the convergence is slightly worse when b is 0.09 MHz and 0.2 MHz. It is worth noting that when b is 0.4 MHz, the QoE of eMBB struggles to satisfy the demand, and the QoE of URLLC also has worse stability. Although we set up a mechanism to only schedule bandwidth beyond the minimum required to satisfy eMBB when performing bandwidth resource scheduling, the QoE of eMBB is severely affected by the large number of bandwidth resources being scheduled each time.

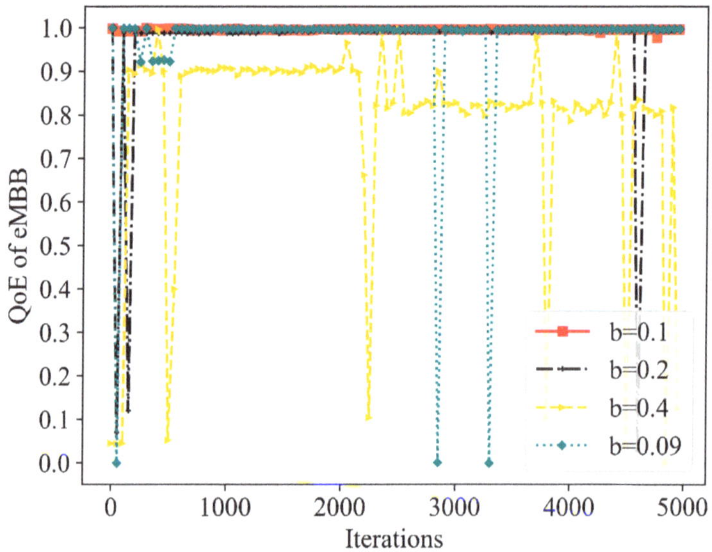

Figure 11. A comparative result of QoE for eMBB service at various b.

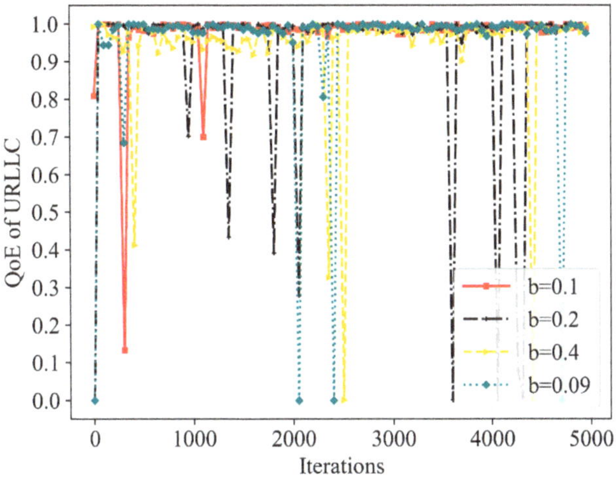

Figure 12. A comparative result of QoE for URLLC service at various *b*.

Figures 13 and 14 further compare the performance of the SE and network utility in the different cases. From the perspective of SE, the convergence of the Dueling DQN is significantly superior to other cases when *b* is 0.1 MHz. Although the SE converges to a higher value at *b* of 0.4 MHz, the QoE fails to satisfy the requirements at this moment. Meanwhile, the scheduling of bandwidth resources is more flexible when *b* is 0.09 MHz, but the stability of SE is not improved. From the perspective of network utility, among the mentioned *b*, the convergence speed is the slowest when *b* is 0.09 MHz and shows a slight difference in other cases. More bandwidth resources will be scheduled without interrupting eMBB user services when we slice the bandwidth more finely. Hence, the delicate *b* can better improve the QoE of URLLC without reducing the QoE of eMBB and can obtain better stability of SE and network utility, due to the increased flexibility in bandwidth resource allocation and scheduling. Moreover, the *b* we set has a great impact on the action space; the excessively delicate *b* can also make the action space too large, leading to a more complex resource allocation process, which influences the effectiveness of the latency time of service and the efficiency of bandwidth allocation and scheduling. The choice of 0.1 MHz for *b* in this paper makes the proposed algorithm perform better when solving the resource allocation and scheduling problem.

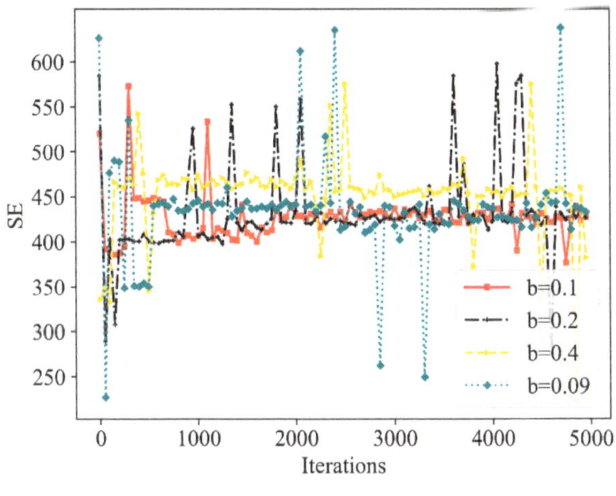

Figure 13. A comparative result of SE at various *b*.

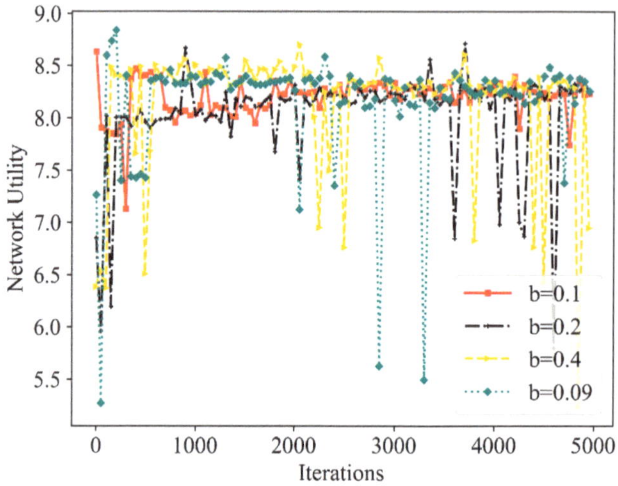

Figure 14. A comparative result of network utility at various b.

After the display and analysis of the above simulation results, the following conclusions can be obtained. Dueling DQN and the resource scheduling mechanism contribute to the stability of QoE and the improvement of SE and network utility. Moreover, the selection of b obtains the appropriate size of action space and improves the flexibility of resource allocation, and the use of the reward-clipping mechanism enables the proposed algorithm to obtain a more stable performance.

5. Conclusions

In this paper, the optimization problem of resource allocation and scheduling in heterogeneous integrated networks is proposed, and the dueling deep Q network (Dueling DQN)-based algorithm for eMBB and URLLC hybrid services is proposed to solve this problem. To prioritize URLLC service requirements, a resource scheduling mechanism between eMBB and URLLC services is proposed. To solve this formulated optimization problem with non-convex properties in relation to the allocation and scheduling of bandwidth resources, an iterative Dueling DQN-based algorithm is proposed. In addition, to enhance the training stability of Dueling DQN, the reward-clipping mechanism is adopted. Moreover, to increase flexibility in resource allocation, a suitable bandwidth allocation resolution (b) is chosen. We verify through simulations that the algorithm based on Dueling DQN for resource allocation and scheduling has excellent performances for quality of experience (QoE), spectrum efficiency (SE) and network utility. We also verify that the Dueling DQN-based algorithm is much better suited to tackling this problem than the Q-learning, DQN and Double DQN, and the resource scheduling mechanism significantly increases the stability of performances.

Author Contributions: Conceptualization, G.C.; data curation, R.S.; formal analysis, G.C. and R.S.; methodology, Q.Z.; validation, F.S.; writing—original draft, G.C.; writing—review and editing, R.S. All authors have read and agreed to the published version of the manuscript.

Funding: This work was supported by the National Natural Science Foundation of China under Grant No. 61701284, 61871370, the Natural Science Foundation of Shandong Province of China under Grant No. ZR2022MF226, the Talented Young Teachers Training Program of Shandong University of Science and Technology under Grant No. BJ20221101, the Innovative Research Foundation of Qingdao under Grant No. 19-6-2-1-cg, the Elite Plan Project of Shandong University of Science and Technology under Grant No. skr21-3-B-048, the National Key R&D Program of China under Grant No. 2019YFE0120700, 2019YFB1803101, the Hundred Talent Program of Chinese Academy of Sciences under Grant No. E06BRA1001, the Sci. & Tech. Development Fund of Shandong Province of China under Grant No. ZR202102230289, ZR202102250695 and ZR2019LZH001, the Humanities and Social Science Research Project of the Ministry of Education under Grant No. 18YJAZH017, the Taishan Scholar Program of Shandong Province under Grant No. ts20190936, the Shandong Chongqing Science and technology cooperation project under Grant No. cstc2020jscx-lyjsAX0008, the Sci. & Tech. Development Fund of Qingdao under Grant No. 21-1-5-zlyj-1-zc, the SDUST Research Fund under Grant No. 2015TDJH102, and the Science and Technology Support Plan of Youth Innovation Team of Shandong higher School under Grant No. 2019KJN024.

Institutional Review Board Statement: Not applicable.

Informed Consent Statement: Not applicable.

Data Availability Statement: Data sharing not applicable.

Acknowledgments: The authors would like to extend their gratitude to the anonymous reviewers and the editors for their valuable and constructive comments, which have greatly improved the quality of this paper.

Conflicts of Interest: The authors declare no conflict of interest.

References

1. Elayoubi, S.E.; Jemaa, S.B.; Altman, Z.; Galindo-Serrano, A. 5g ran slicing for verticals: Enablers and challenges. *IEEE Commun. Mag.* **2019**, *57*, 28–34. [CrossRef]
2. Chahbar, M.; Diaz, G.; Dandoush, A.; Cérin, C.; Ghoumid, K. A comprehensive survey on the e2e 5g network slicing model. *IEEE Trans. Netw. Serv. Manag.* **2021**, *18*, 49–62. [CrossRef]
3. Fossati, F.; Moretti, S.; Perny, P.; Secci, S. Multi-resource allocation for network slicing. *IEEE/ACM Trans. Netw.* **2020**, *28*, 1311–1324. [CrossRef]
4. Richart, M.; Baliosian, J.; Serrat, J.; Gorricho, J. Resource Slicing in Virtual Wireless Networks: A Survey. *IEEE Trans. Netw. Serv. Manag.* **2016**, *13*, 462–476. [CrossRef]
5. Foukas, X.; Patounas, G.; Elmokashfi, A.; Marina, M.K. Network Slicing in 5G: Survey and Chal-lenges. *IEEE Commun. Mag.* **2017**, *55*, 94–100. [CrossRef]
6. Mei, J.; Wang, X.; Zheng, K. An intelligent self-sustained RAN slicing framework for diverse service provisioning in 5G-beyond and 6G networks. *Intell. Converg. Netw.* **2020**, *1*, 281–294. [CrossRef]
7. Dai, P.; Liu, K.; Wu, X.; Liao, Y.; Lee, V.C.S.; Son, S.H. Bandwidth Efficiency and Service Adaptiveness Oriented Data Dissemination in Heterogeneous Vehicular Networks. *IEEE Trans. Veh. Technol.* **2018**, *67*, 6585–6598. [CrossRef]
8. Guo, Y.; Yang, Q.; Kwak, K.S. Quality-oriented Rate Control and Resource Allocation in Time-Varying OFDMA Networks. *IEEE Trans. Veh. Technol.* **2017**, *66*, 2324–2338. [CrossRef]
9. Ko, H.; Lee, J.; Pack, S. Priority-Based Dynamic Resource Allocation Scheme in Network Slicing. In Proceedings of the 2021 International Conference on Information Net-Working (ICOIN), Jeju Island, Republic of Korea, 13–16 January 2021.
10. Alfoudi, A.S.D.; Newaz, S.H.S.; Otebolaku, A.; Lee, G.M.; Pereira, R. An Efficient Resource Management Mechanism for Network Slicing in a LTE Network. *IEEE Access* **2019**, *7*, 89441–89457. [CrossRef]
11. Abdelsadek, M.Y.; Gadallah, Y.; Ahmed, M.H. Resource Allocation of URLLC and eMBB Mixed Traffic in 5G Networks: A Deep Learning Approach. In Proceedings of the GLOBECOM 2020–2020 IEEE Global Communications Conference, Taipei, Taiwan, 7–11 December 2020; pp. 1–6. [CrossRef]
12. Iqbal, M.U.; Ansari, E.A.; Akhtar, S. Interference Mitigation in HetNets to Improve the QoS Using Q-Learning. *IEEE Access* **2021**, *9*, 32405–32424. [CrossRef]
13. Mao, Q.; Hu, F.; Hao, Q. Deep Learning for Intelligent Wireless Networks: A Comprehensive Survey. *IEEE Commun. Surv. Tutor.* **2018**, *20*, 2595–2621. [CrossRef]
14. Zhang, C.; Patras, P.; Haddadi, H. Deep Learning in Mobile and Wireless Networking: A Survey. *IEEE Commun. Surv. Tutor.* **2019**, *21*, 2224–2287. [CrossRef]
15. Li, R.; Zhao, Z.; Sun, Q.; Chih-Lin, I.; Yang, C.; Chen, X.; Zhao, M.; Zhang, H. Deep reinforcement learning for resource management in network slicing. *IEEE Access* **2018**, *6*, 74429–74441. [CrossRef]
16. Chen, X.; Zhao, Z.; Wu, C.; Bennis, M.; Liu, H.; Ji, Y.; Zhang, H. Multi-Tenant Cross-Slice Resource Orchestration: A Deep Reinforcement Learning Approach. *IEEE J. Sel. Areas Commun.* **2019**, *37*, 2377–2392. [CrossRef]
17. Chen, X.; Wu, C.; Chen, T.; Zhang, H.; Liu, Z.; Zhang, Y.; Bennis, M. Age of Information Aware Radio Resource Management in Vehicular Networks: A Proactive Deep Reinforcement Learning Perspective. *IEEE Trans. Wirel. Commun.* **2020**, *19*, 2268–2281. [CrossRef]
18. Qi, C.; Hua, Y.; Li, R.; Zhao, Z.; Zhang, H. Deep Reinforcement Learning with Discrete Normalized Advantage Functions for Resource Management in Network Slicing. *IEEE Commun. Lett.* **2019**, *23*, 1337–1341. [CrossRef]
19. Sciancalepore, V.; Costa-Perez, X.; Banchs, A. RL-NSB: Reinforcement Learning-Based 5G Network Slice Broker. *IEEE/ACM Trans. Netw.* **2019**, *27*, 1543–1557. [CrossRef]

20. Zhang, Q.; Saad, W.; Bennis, M. Distributional Reinforcement Learning for mmWave Communications with Intelligent Reflectors on a UAV. In Proceedings of the GLOBECOM 2020-2020 IEEE Global Communications Conference, Taipei, Taiwan, 7–11 December 2020; pp. 1–6.
21. Hua, Y.; Li, R.; Zhao, Z.; Chen, X.; Zhang, H. GAN-Powered Deep Distributional Reinforcement Learning for Resource Management in Network Slicing. *IEEE J. Sel. Areas Commun.* **2020**, *38*, 334–349. [CrossRef]
22. Li, R.; Wang, C.; Zhao, Z.; Guo, R.; Zhang, H. The LSTM-Based Advantage Actor-Critic Learning for Resource Management in Network Slicing with User Mobility. *IEEE Commun. Lett.* **2020**, *24*, 2005–2009. [CrossRef]
23. Yuan, Y.; Li, Z.; Liu, Z.; Yang, Y.; Guan, X. Double Deep Q-Network Based Distributed Resource Matching Algorithm for D2D Communication. *IEEE Trans. Veh. Technol.* **2022**, *71*, 984–993. [CrossRef]
24. Sun, G.; Xiong, K.; Boateng, G.O.; Liu, G.; Jiang, W. Resource slicing and customization in RAN with dueling deep Q-Network. *J. Netw. Comput. Appl.* **2020**, *157*, 102573. [CrossRef]
25. Chen, G.; Zhang, X.; Shen, F.; Zeng, Q. Two tier slicing resource allocation algorithm based on deep reinforcement learning and joint bidding in wireless access networks. *Sensors* **2022**, *22*, 1424–8220. [CrossRef] [PubMed]
26. Boateng, G.O.; Ayepah-Mensah, D.; Doe, D.M.; Mohammed, A.; Sun, G.; Liu, G. Blockchain-Enabled Resource Trading and Deep Reinforcement Learning-Based Autonomous RAN Slicing in 5G. *IEEE Trans. Netw. Serv. Manag.* **2022**, *19*, 216–227. [CrossRef]
27. Zhao, N.; Liang, Y.; Niyato, D.; Pei, Y.; Wu, M.; Jiang, Y. Deep Reinforcement Learning for User Asso-ciation and Resource Allocation in Heterogeneous Cellular Networks. *IEEE Trans. Wirel. Commun.* **2019**, *18*, 5141–5152. [CrossRef]
28. Esswie, A.A.; Pedersen, K.I. Multi-User Preemptive Scheduling for Critical Low Latency Communications in 5G Networks. In Proceedings of the 2018 IEEE Symposium on Computers and Communications (ISCC), Natal, Brazil, 25–28 June 2018; pp. 00136–00141.
29. Alsenwi, M.; Tran, N.H.; Bennis, M.; Bairagi, A.K.; Hong, C.S. eMBB-URLLC Resource Slicing: A Risk-Sensitive Approach. *IEEE Commun. Lett.* **2019**, *23*, 740–743. [CrossRef]
30. Feng, L.; Zi, Y.; Li, W.; Zhou, F.; Yu, P.; Kadoch, M. Dynamic Resource Allocation with RAN Slicing and Scheduling for uRLLC and eMBB Hybrid Services. *IEEE Access* **2020**, *8*, 34538–34551. [CrossRef]
31. Han, Y.; Tao, X.; Zhang, X.; Jia, S. Hierarchical Resource Allocation in Multi-Service Wireless Networks with Wireless Network Virtualization. *IEEE Trans. Veh. Technol.* **2020**, *69*, 11811–11827. [CrossRef]
32. Wang, C.; Duan, X.; Jiao, B. Q-learning based Resource Allocation for hybrid Services with Self-similar Traffic. In Proceedings of the 2020 IEEE 5th International Conference on Signal and Image Processing (ICSIP), Nanjing, China, 23–25 October 2020; pp. 1117–1121.
33. Li, J.; Zhang, X. Deep Reinforcement Learning-Based Joint Scheduling of eMBB and URLLC in 5G Networks. *IEEE Wirel. Commun. Lett.* **2020**, *9*, 1543–1546. [CrossRef]
34. Alsenwi, M.; Tran, N.H.; Bennis, M.; Pandey, S.R.; Bairagi, A.K.; Hong, C.S. Intelligent Resource Slicing for eMBB and URLLC Coexistence in 5G and Beyond: A Deep Reinforcement Learning Based Approach. *IEEE Trans. Wirel. Commun.* **2021**, *20*, 4585–4600. [CrossRef]
35. Yin, H.; Zhang, L.; Roy, S. Multiplexing URLLC Traffic Within eMBB Services in 5G NR: Fair Scheduling. *IEEE Trans. Commun.* **2021**, *69*, 1080–1093. [CrossRef]
36. Almekhlafi, M.; Chraiti, M.; Arfaoui, M.A.; Assi, C.; Ghrayeb, A.; Alloum, A. A Downlink Puncturing Scheme for Simultaneous Transmission of URLLC and eMBB Traffic by Exploiting Data Similarity. *IEEE Trans. Veh. Technol.* **2021**, *70*, 13087–13100. [CrossRef]
37. Darabi, M.; Jamali, V.; Lampe, L.; Schober, R. Hybrid puncturing and superposition scheme for joint scheduling of urllc and embb traffic. *IEEE Commun. Lett.* **2022**, *26*, 1081–1085. [CrossRef]

sensors

MDPI

Article

Movie Scene Event Extraction with Graph Attention Network Based on Argument Correlation Information

Qian Yi [1,2], Guixuan Zhang [1], Jie Liu [1,*] and Shuwu Zhang [1]

[1] Beijing Engineering Research Center of Digital Content Technology, Institute of Automation, Chinese Academy of Sciences, Beijing 100038, China
[2] School of Computer and Control Engineering, University of Chinese Academy of Sciences, Beijing 100038, China
* Correspondence: jie.liu@ia.ac.cn

Abstract: Movie scene event extraction is a practical task in media analysis, which aims at extracting structured events from unstructured movie scripts. However, although there have been many studies regarding open domain event extraction, there have only been a few studies focusing on movie scene event extraction. Specifically aimed at instances where different argument roles have the same characteristics in a movie scene, we propose the utilization of the correlation between different argument roles, which is beneficial for both movie scene trigger extraction (trigger identification and classification) and movie scene argument extraction (argument identification and classification) in event extraction. To model the correlation between different argument roles, we propose the superior role concept (SRC), a high-level role concept based upon the ordinary argument role. In this paper, we introduce a new movie scene event extraction model with two main features: (1) an attentive high-level argument role module to capture SRC information and (2) an SRC-based graph attention network (GAT) to fuse the argument role correlation information into semantic embeddings. To evaluate the performance of our model, we constructed a movie scene event extraction dataset named MovieSceneEvent and also conducted experiments on a widely used dataset to compare the results with other models. The experimental results show that our model outperforms competitive models, and the correlation information of argument roles helps to improve the performance of movie scene event extraction.

Keywords: event extraction; graph attention network; argument correlation information

Citation: Yi, Q.; Zhang, G.; Liu, J.; Zhang, S. Movie Scene Event Extraction with Graph Attention Network Based on Argument Correlation Information. *Sensors* **2023**, *23*, 2285. https://doi.org/10.3390/s23042285

Academic Editor: Sung-Bae Cho

Received: 6 February 2023
Revised: 10 February 2023
Accepted: 14 February 2023
Published: 17 February 2023

1. Introduction

Event extraction (EE) is an essential task in information extraction. Movie scene event extraction is a practical task in media analysis, which can help viewers to understand the movie plot. In movie scene event extraction, after the event trigger and its argument are obtained from unstructured text, a predefined event type and a role are then assigned to the trigger and the argument, respectively. For example, in the sentence "Peter picks up the pistol and shoots Ruth.", the predefined event type "Exchange of Fire" and its corresponding trigger "shoots" are first extracted by the trigger extraction module. The argument extraction module needs to further extract the argument subjects, "Peter" and "Rose", and their argument roles, "Attacker" and "Victim".

Since event extraction plays a vital role in many downstream tasks, great efforts have been made to improve the performance of event extraction. Conventional event extraction models utilized handcraft features and adopted kernel-based methods [1–5]. However, with more and more attention being paid to deep learning, the application of distributional representation-based methods became more and more popular and achieved better performance [6–9]. Some recent works have proposed advanced methods to further improve event extraction, for instance, the question answering (QA) framework, with more external information being adopted [10,11].

However, open domain event extraction has attracted a lot of attention, and there is not much research that specifically focuses on movie scene event extraction. In movie scene event extraction, it is not uncommon for different argument roles to have some similar attributes. Take the sentence "Peter picks up the pistol and shoots Ruth.", for example. The argument roles of "Peter" and "Ruth" are "Attacker" and "Victim", respectively, and both of these argument roles stem from the SRC "person". This correlating information between argument roles not only benefits the argument extraction, but also the trigger extraction. Intuitively, the SRC also makes it easier to classify the argument role. For instance, if we know the "Attacker Job" belongs to the SRC "person", it is easier to classify it into the role of "Attacker". SRC information also helps to identify the trigger and its event type. For example, for the trigger type "Exchange of Fire", its arguments usually involve people. So, if we know that the respective candidate arguments are both an SRC "person", as given in the sentence, the trigger type of the model will tend to be classified as "Exchange of Fire".

In order to obtain the correlation information between argument roles and further improve movie scene event extraction, we proposed a novel, argument correlation, information-based GAT. To capture the correlation information between argument roles, an attentive high-level argument role module is applied to obtain SRC features. Then, the GAT [12] is employed to extract semantic features via a dependency tree, which is able to link related mentions and shorten the distance between the trigger and its arguments. In addition, the attention unit in the GAT is utilized to integrate SRC information into the semantic features.

In addition, because there is no special dataset for movie scene event extraction, the lack of professional datasets has also become an urgent problem that requires a solution. In order to further explore movie scene event extraction and solve this problem, we constructed an event extraction dataset for movie scenes. We choose movie scene sentences from the film scripts and labeled them manually. The dataset contains 5852 training samples and 486 testing samples, with 12 event types and 18 argument roles. This dataset helps us further verify the effectiveness of our algorithm in the task of movie scene event extraction.

In this paper, our main contributions are as follows:

- We introduce the correlation information of argument roles to further improve joint movie scene event extraction.
- We propose an SRC-based GAT to capture the semantic features and integrate the correlation information of argument roles into the semantic features.
- We constructed a movie scene extraction dataset to verify the effectiveness of our model. The experimental results show that our model outperforms competitive models, and the correlation information between argument roles can help to improve the performance of movie scene event extraction.

The remainder of this paper is organized as follows. In Section 2, we present related work concerning event extraction. In Section 3, we outline our proposed relation extraction model. Next, in Section 4, we present the experimental results of our model and then analyze the results. Finally, in Section 5, we give the conclusions of our paper and introduce our future work.

2. Related Work

Natural language processing (NLP) technology is widely used in media analysis [13–16]. As an essential task of NLP, event extraction can be divided into two subtasks: (1) event trigger extraction, in which the "trigger" (that represents the occurrence of an event) is extracted and then assigned an event type, and (2) argument extraction, in which the arguments of the trigger are detected, and then each argument is assigned an argument role with respect to the event type.

Earlier works paid more attention to the pipeline methods or one of the subtasks. Two types of efforts can be identified. One relies on hand-crafted features [1–5,17]. For instance, Mcclosky [1] et al. utilized the tree of event–argument relations as features for event extraction. Liao et al. [2] and Ji et al. [4] used the document-level features and Li et al. [3] used global features to improve performance, while Huang et al. [5] modeled the textual cohesion of the text as features. The other is the neural network method [7,8]. Nguyen et al. [7], for example, utilized graph convolutional networks (GCNN) for event detection, and in [8], they adopted domain adaptation convolutional neural networks (CNN) to improve event detection.

However, as error propagation problems are present in these works, the accuracy of downstream subtasks will be impacted. Thus, more and more joint models have been proposed to solve this problem [18–20]. Chen et al. proposed the dynamic multi-pooling CNN model to extract semantic features [9]. Sha et al. [6] utilized recurrent neural networks (RNN) to embed a dependency bridge, which achieved good performance. Li et al. [21] introduced external knowledge to improve domain-specific event extraction. Recently, some works have adopted a generative method [22,23] to solve event extraction problems. QA frameworks have been adapted for event extraction and can also effectively solve the problem of few-shot event extraction [10,11].

However, the above methods are mainly aimed at open domain event extraction tasks and do not take the characteristics of movie scene event extraction into consideration. In this paper, considering that argument roles in movie scene event extraction usually belong to several specific categories, we define several SRCs to model the correlation between different argument roles when proceeding with joint event extraction. Furthermore, we designed two specific modules to utilize the correlation information to improve the performance of the two subtasks of event extraction.

3. Model

In this section, we introduce our model for movie scene event extraction in detail. The whole process of movie scene event extraction is shown schematically in Figure 1. As Figure 1 shows, the whole process can be divided into three steps: (1) obtaining the argument-oriented SRC embedding through the attentive high-level role module; (2) incorporating the argument-oriented SRC embedding into GAT for trigger extraction; and (3) incorporating the argument-oriented SRC embedding into GAT for argument extraction. In the remainder of this section, we first present how to obtain argument-oriented SRC embedding through an attentive high-level role module, then we introduce the framework of GAT. Finally, we demonstrate how to incorporate the argument-oriented SRC embedding into GAT and apply it for trigger extraction and argument extraction.

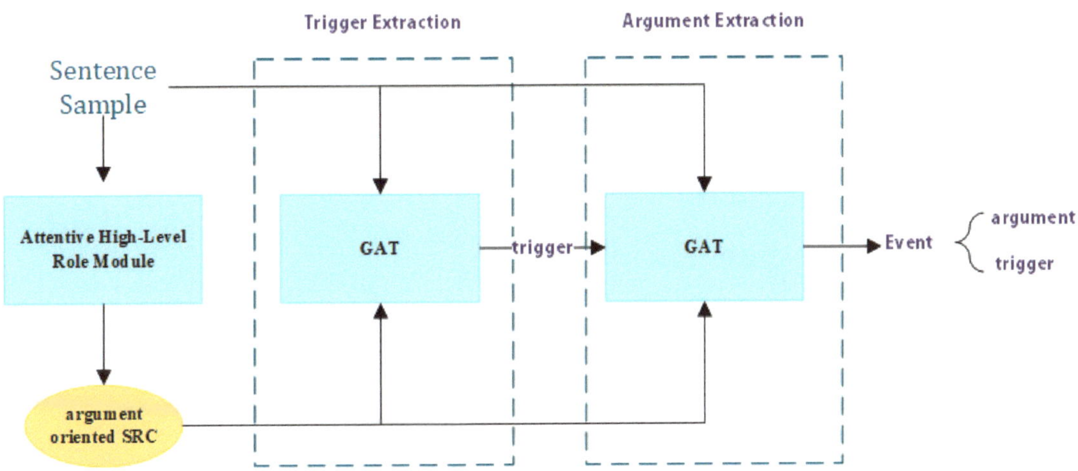

Figure 1. The process of movie scene event extraction.

3.1. Attentive High-Level Role Module

Figure 2 shows the structure of the attentive high-level role module, and we introduce it in detail here.

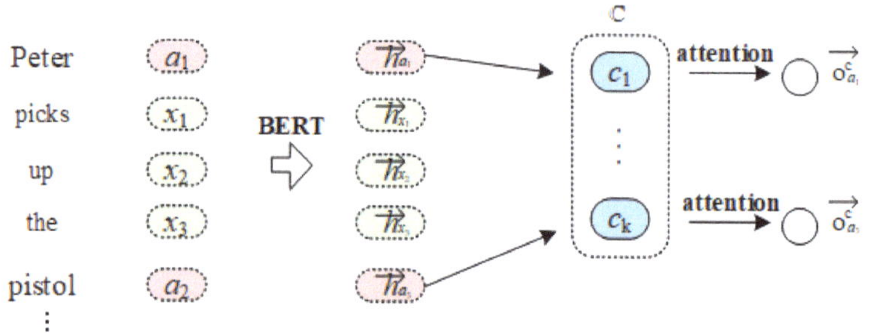

Figure 2. Framework of attentive high-level role modules.

The input sentence can be denoted as a n-word sequence *s* = *{x_1, x_2, . . . , a_1, . . . , a_k, . . . , x_(n − k)}*, in which *a_i* denotes the candidate argument. The candidate arguments are the named entities extracted by StanfordNER [24]. To obtain the hidden embedding of each word, we employed BERT [25] as the encoder, which achieved the STOA performance on a wide range of NLP tasks. So, we can embed the members of the word sequence into their hidden representations:

$$\{h_1, h_2, \ldots, h_{a1}, \ldots, h_{ak}, h_{n-k}\} = BERT(x_1, x_2, \ldots, a_1, \ldots, a_k, \ldots, x_{n-k}) \tag{1}$$

where $h_i \in R^d$ and d are the word embedding sizes.

In terms of candidate argument a_i, different SRCs have different degrees of correlation with it. Firstly, for each SRC c, we assign a trainable embedding vector $u_c \in R^d$. Then, with respect to the given candidate argument a_i, we calculate the attention score $s^c_{a_i}$ of each SRC c:

$$h^c_{a_i} = ReLU(\mathbf{W}_a \cdot [u_c || h_{ai}]); \tag{2}$$

$$s^{c_j}_{a_i} = \frac{exp\left(\mathbf{W}_b \cdot h^{c_j}_{a_i}\right)}{\sum_C exp\left(\mathbf{W}_b \cdot h^{c_k}_{a_i}\right)}, \tag{3}$$

where $\mathbf{W}_a \in R^{d \times 2d}$ and $\mathbf{W}_b \in R^d$ are the weighted matrices, | | denotes the concatenation operation through the dimensions, and C is the set of candidate superior role concepts. Finally, we are able to obtain the argument-oriented SRC embedding vector o^C by calculating the weighted sum of the embedding vectors of superior role concepts:

$$o^C_{a_i} = \sum\nolimits_{c \in C} s^c_{a_i} \cdot u_c. \tag{4}$$

Intuitively, $o^C_{a_i}$ contains the semantic information of the SRC that has a stronger correlation with the candidate argument a_i.

3.2. Event Trigger Extraction

Figure 3 shows an example of the whole event extraction process and the details of trigger extraction. In this section, we introduce the details of the latter process.

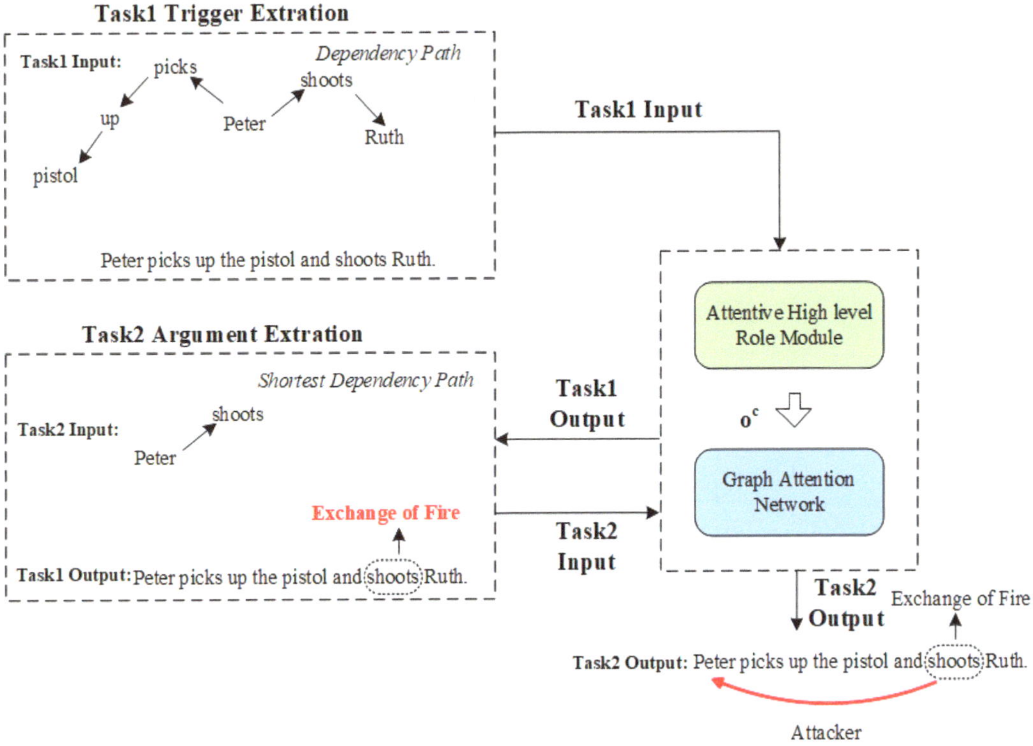

(a)

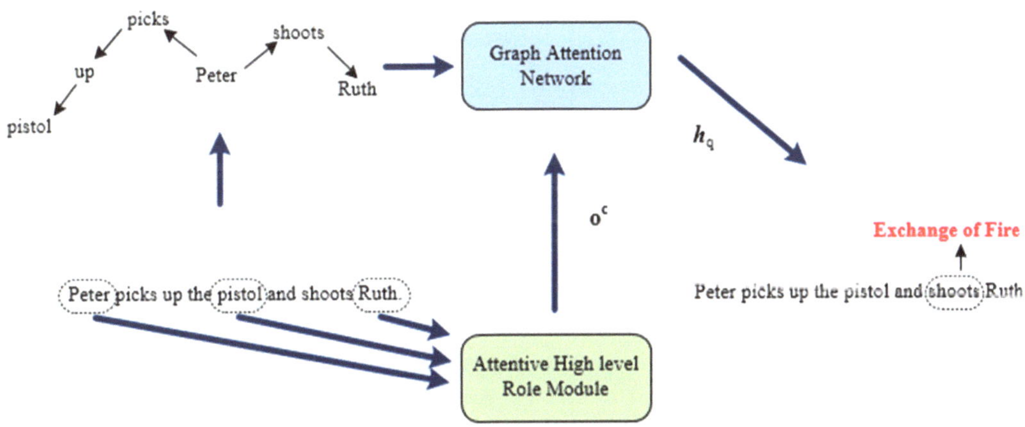

(b)

Figure 3. (**a**) The overall framework of the whole event extraction model; (**b**) the details of trigger extraction.

The Graph Attention Network The GAT employs the attention mechanism to embed the tree-structured topology, as its name suggests. It shows improvements in capturing semantic features when compared to conventional sequential models. GAT can be seen as an extension of memory networks [26].

For each node in GAT, their hidden states are obtained by using the attention unit to calculate the weighted sum of the hidden states of their children nodes in the graph structure. Figure 4a shows the operation of one node in GAT. For each node on the dependency tree, $S(q)$ denotes the set of children nodes of node q. We can obtain the corresponding attention score using the following equation:

$$\alpha_{qj} = \frac{exp(LeakyReLU(h_j \cdot \mathbf{W}_g \cdot h_q))}{\sum_{k \in S(q)} exp(LeakyReLU(h_k \cdot \mathbf{W}_g \cdot h_q))}, \quad (5)$$

where node j is the child node of q, and $\mathbf{W}_q \in R^{d \times d}$ is the weighted matrix.

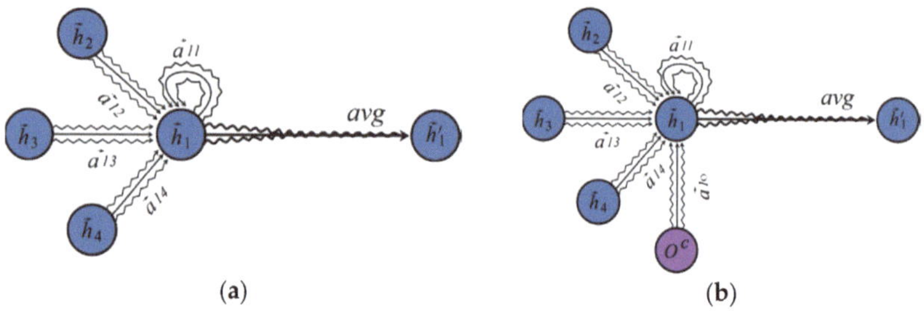

Figure 4. (**a**) An original graph attention unit; (**b**) an SRC-based graph attention unit. The purple circle labelled o^c represents the SRC features.

Then, we obtain the embedding h_q of q by calculating the weighted sum of the hidden embedding of $S(q)$:

$$h_q = \sigma\left(\sum_{k \in S(q)} \alpha_{qj} \cdot h_k\right). \quad (6)$$

where σ represents the sigmoid function, and h_k is the hidden embedding of node k.

In order to obtain the hidden embedding of node j, the graph attention unit calculates all of its children nodes' hidden states through depth-first traversal [27].

Superior Role Concept-Based Graph Attention Network Utilizing Formulas (1)–(4), we are able to obtain the argument-oriented SRC embedding o_a^C. Specifically, for each candidate argument, we obtain the correlation between each argument word by calculating the attention score of each SRC embedding u_c. Finally, we can obtain an argument-oriented SRC embedding vector o_a^C by calculating the weighted sum of u_c. It is worth noting that we use the full set of all superior role concepts C to calculate the argument-oriented SRC embedding vector during trigger extraction, since, in trigger extraction, we do not know the trigger and the event type and, therefore, have not obtained their corresponding argument roles.

After the argument-oriented SRC embedding o_a^C is obtained, we then further incorporate it into the GAT to use the SRC information to improve the trigger extraction. Figure 4b demonstrates how to fuse the argument-oriented SRC embedding into GAT.

For a given sentence, we first utilize the Stanford Parser [24] to obtain its dependency tree, and the tree structure will then be fed to the GAT. For each node q on the dependency tree, $S(q)$ denotes the set of children nodes of node q, and o_q^C is the SRC information embedding of q. o_q^C is set to 0 if q is not a candidate argument. Normally, in the original GAT, we calculate the attention score of each node j in *S(q)* directly. However, to merge the extra-high-level argument role information, we treat the embedding vector o_q^C as a child node of q, so we are able to obtain the corresponding attention score:

$$\alpha_{qj} = \frac{exp(LeakyReLU(h_j \cdot \mathbf{W}_g \cdot h_q))}{\sum_{k \in S(q) \cup \{o\}} exp(LeakyReLU(h_k \cdot \mathbf{W}_g \cdot h_q))}, \quad (7)$$

where node j can be either the child node of q or the argument-oriented SRC embedding vector of q. When q is not a candidate argument node, the attention score of o is 0.

Then, we obtain the embedding of q by calculating the weighted sum of the hidden embedding of $S(q)$ and the argument-oriented SRC embedding vector o^c:

$$h'_q = \sigma\left(\sum_{k \in S(q) \cup \{o\}} \alpha_{qj} \cdot h_k\right). \tag{8}$$

where σ represents the sigmoid function. To capture more affluent features, a multi-head attention is applied. So, the final embedding of q can be obtained from the expression below:

$$h'_q = \sigma\left(\frac{1}{M} \sum_{m=1}^{M} \sum_{k \in S(q) \cup \{o\}} \alpha_{qj} \cdot h_k\right) \tag{9}$$

where M is the number of attention heads; here, we adopt a three-head attention. After the hidden state of each node is obtained, we send the embedding to a feed forward layer with a SoftMax classifier in order to predict the trigger type for each node and optimize the parameters by minimizing a negative log-likelihood loss.

3.3. Event Argument Extraction

After all candidate event triggers are obtained after trigger extraction, we begin extracting event arguments with respect to the given triggers. Figure 5 shows the details of the event argument extraction module.

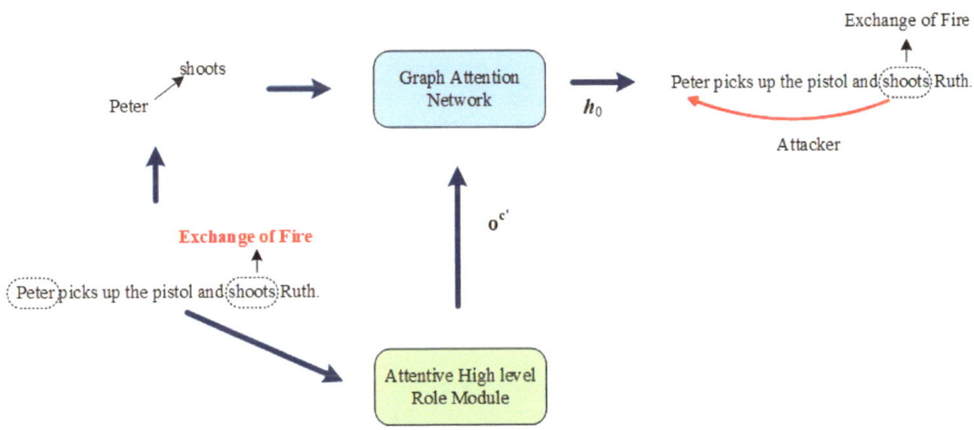

Figure 5. The details of the event argument extraction.

Unlike trigger extraction, we adopt the shortest dependency path (SDP) as the input for SRC information-based GAT in argument extraction, as SDP can better capture the correlation between the trigger and argument. Figure 4 shows an example of SDP and its corresponding dependency tree—it is easy to see that SDP is more concise, as redundant information can be eliminated.

Another difference between argument extraction and trigger extraction is that, in argument extraction, we know the trigger and the event type. So, when calculating the argument-oriented SRC embedding o^c in the argument extraction procedure, we only utilize the corresponding SRC of the given event type to calculate o^c. For instance, for the event type "Exchange of Fire", its argument roles stem from three role concepts: person, place, and item. Thus, when calculating the argument-oriented SRC embedding for the arguments of the event type "Exchange of Fire", following Formulas (2)–(4), the set of candidate superior role concepts is C = {Person, Place, Item}.

Specifically, given an event type and its candidate argument, its corresponding SRC set is $C' = \{c_1, c_2, \ldots, c_k\}$, where k is the number of SRCs corresponding to the given event type. Following Formulas (1)–(4), we are able to obtain the argument-oriented SRC embedding $o_a^{C'}$. We then utilize the SRC-based GAT introduced in the event trigger extraction module to encode each node in the SDP into a new hidden state representation and use the hidden embedding of the root node h_0 as the embedding vector of the input. For each candidate argument, we can obtain an argument-role-oriented instance embedding h_0.

Then, we use the argument-role-oriented instance embedding h_0 as the input feature for argument role classification and send the embedding to a feed forward layer with a SoftMax classifier to predict the argument role. We optimize the parameters by minimizing a negative log-likelihood loss.

4. Experiments

4.1. Experiment Setup

Datasets

1. MovieSceneEvent: We constructed a movie scene event extraction dataset named MovieSceneEvent for this research. To construct a movie-scene-specific event extraction dataset, we first summarized 12 common types of events based on the research needs and the suggestions of professionals in the film field. Then, we chose sentences related to these events from movie script texts. These movie scripts were selected from 13 common genres of movies (including romance, comedy, action, war movies, and so on). According to the defined event types, we first used the manually defined template to roughly screen out the texts related to the defined event type from the script text and then manually filter these texts. Finally, these sentences were further labeled manually. We asked two annotators to label each sample. If their labeling was consistent, that result was used for the sample. If not, a third annotator was used to ensure the accuracy of the labeling. The movie scene event extraction dataset contains 5852 training samples and 486 testing samples, with 12 event types and 18 argument roles.
2. ACE2005: Following previous works [3,28], we also adopted ACE2005, the widely used event extraction dataset, to evaluate the effectiveness of our model. It contains 599 documents, with 13,672 labeled sentences in the ACE2005 dataset, and these sentences are labeled with 8 given event types, 33 event subtypes, and 35 argument roles. Following [3,26], we split the ACE2005 dataset into 529, 30, and 40 documents for training, development, and testing, respectively.

Evaluation Measures Following previous work [6–9], we evaluated our model on four metrics: (1) trigger detection: a trigger is considered correctly detected if span offsets correctly match the label; (2) trigger classification: a trigger is considered correctly classified only if both of its span offsets and event subtypes are correct; (3) argument detection: an argument is considered correctly detected if its span offsets match the label and its event subtypes exactly match the label; and (4) argument classification: an argument is considered correctly classified only if its span offsets, event subtypes, and argument roles are correct. For example, in the sentence "Peter picks up the pistol and shoots Ruth.", for the first metric, the trigger word "shoots" is considered correctly detected if span offsets correctly match the label "(6,1)", in which "6" indicates the start index and "1" indicates the span. For Metric 2, the trigger "shoots" is considered correctly classified only if both of its span offsets "(6,1)" and the event subtype are correctly classified as "Exchange of Fire". For Metric 3, the argument "Peter" is considered correctly detected if its span offsets match the label "(0,1)" and its event subtype exactly matches the label "Exchange of Fire". For Metric 4, the argument "Peter" is considered to be correctly classified only if its span offsets ("(0,1)"), event subtype ("Exchange of Fire"), and argument role ("Attacker") are correct. All experimental results are presented in the form of precision (P), recall (R), and F-measure ($F1 = 2 * P * R/(P + R)$) for each metric.

The definition of the superior role concept is based on experience; thus, we define four superior role concepts manually: person, place, item, and time. The results of superior role concepts in this paper cannot be directly extended to other datasets with different label definitions, but the definition method is simple.

Hyperparameters We used BERT for sequence encoding to obtain the hidden embeddings and adopt the BERT–BASE–CASED [25] model. As for the hyperparameters, we tried the following parameter settings: learning rate = {0.0025, 0.005, 0.01}; batch size = {15, 25, 50}. The experiment results of different settings of parameters on the MovieSceneEvent dataset are shown in Figures 6 and 7. We ultimately chose 0.005 and 25 as the learning rate and batch size, respectively. The hyperparameters are listed in Table 1. We utilized two NVIDIA K40 as the running environment, and the details of the model training are also listed in Table 1.

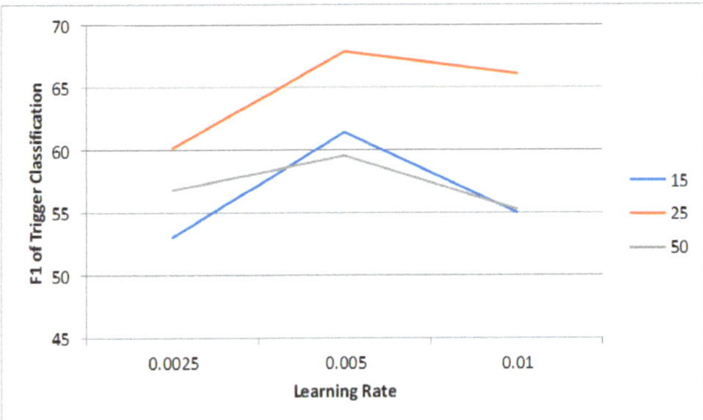

Figure 6. The influence of learning rate and batch size on trigger classification.

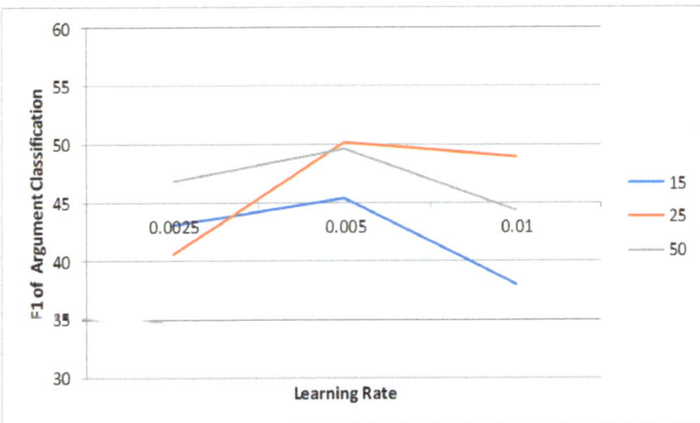

Figure 7. The influence of learning rate and batch size on argument classification.

Table 1. Hyperparameters.

Parameter	Value
Word embedding size d	768
Batch size	25
Epoch size	20
Dropout rate	0.5
Learning rate	0.005
Optimizer	AdaGrad

4.2. Overall Performance

We compared the performance of our model with several representative models. **JOINTFEATURE** [28] considered the relationship between triggers, arguments, and their correlations and conducted joint inference of these variables across a document. **dbRNN** [6] adopted LSTM as encoder to embed the dependency tree structure in order to extract the event trigger and argument role. **Joint3EE** [7] conducted the event extraction in a multi-task module with shared BiGRU hidden representations. **BS** [10] is a representative model-adapted QA framework, which used bleached statements (BSs) to give a model access to information contained in annotation manuals. **Text2Event** [29] is a sequence-to-structure generation model that can directly extract events from the text in an end-to-end manner.

Table 2 shows the comparison between our model and other models using the movie scene event extraction dataset. Table 3 shows the comparison between our model and the above models on an open domain event extraction dataset. Tables 2 and 3 demonstrate that our model achieves the best performance on most of the evaluation criteria, especially on the F1 scores in the classification results. Specifically, in the MovieSceneEvent dataset, our model achieves F1 scores of 70.3% on trigger identification, which is 8.2%, 7.8%, 1.2%, 2.1%, and 1.1% higher than that of JOINTFEATURE, dbRNN, Joint3EE, BS, and Text2Event, respectively. Our model achieves F1 scores of 67.3% on trigger classification, which is 7.9%, 11.9%, 3.9%, 2.6%, and 3.2% superior to that of JOINTFEATURE, dbRNN, Joint3EE, BS, and Text2Event, respectively. Our model achieves F1 scores of 53.7% on argument identification, which is 8.3%, 8.5%, 3.8%, 11.1%, and 7.5% higher than that of JOINTFEATURE, dbRNN, Joint3EE, BS, and Text2Event, respectively. Our model achieves F1 scores of 53.7% on argument classification, which is 7.2%, 5.6%, 3.7%, 12.2%, and 2.3% better than that of JOINTFEATURE, dbRNN, Joint3EE, BS, and Text2Event, respectively. On the ACE2005 dataset, our model achieves F1 scores of 73.3% on trigger identification, which is 3.2%, 0.8%, and 0.4% higher than that of JOINTFEATURE, Joint3EE, and BS, respectively. Our model achieves F1 scores of 72.6% on trigger classification, which is 3.9%, 0.7%, 2.8%, 2.1%, and 0.8% higher than that of JOINTFEATURE, dbRNN, Joint3EE, BS, and Text2Event, respectively. Our model achieves F1 scores of 55.7% on argument identification, which is 5.1% and 12.7% higher than that of JOINTFEATURE and BS, respectively, and 1.5% and 4.2% lower than that of dbRNN and Joint3EE, respectively. Our model achieves F1 scores of 54.7% on argument classification, which is 6.3%, 4.6%, 2.6%, 12.3%, and 0.3% better than that of JOINTFEATURE, dbRNN, Joint3EE, BS, and Text2Event, respectively. From the experimental results, we can see the following:

(1) Our model steadily outperforms all other competitive models in both the trigger extraction and argument extraction of movie scene event extraction and open domain event extraction, which indicates that the SRC information can benefit both trigger and argument extraction in event extraction.
(2) In argument extraction, our model significantly outperforms prior work, which may be due to the fact that the SRC information has a more direct correlation with the argument role.
(3) It worth noting that the drop of F1 between both argument identification and classification, as well as trigger identification and classification, is smaller than in previous works, which means the SRC information is able to benefit the classification of both argument role and trigger event types. SRC information helps to maintain more semantic information between identification and classification.
(4) When concerning the performance on open domain datasets, the improvement of our model is much smaller. This is probably due to the composition of argument roles in movie scene event extraction being much easier to generalize into several superior role concepts. Thus, the influence of SRC information is more significant.

Table 2. Results from the MovieSceneEvent dataset.

Model	Trigger						Argument					
	Identification			Classification			Identification			Classification		
	P	R	F1	P	R	F1	P	R	F1	P	R	F1
JOINTFEATURE	61.0	63.2	62.1	70.1	51.6	59.4	51.0	40.9	45.4	44.3	41.6	42.9
DbRNN	63.3	61.8	62.5	61.1	50.7	55.4	41.7	49.5	45.2	43.5	45.6	44.5
Joint3EE	65.8	72.9	69.1	60.5	66.7	63.4	48.9	51.1	49.9	50.7	42.8	46.4
BS	66.4	70.8	68.2	61.7	68.1	64.7	42.0	43.3	42.6	40.1	35.9	37.9
Text2Event	68.2	70.3	69.2	62.1	66.2	64.1	45.3	47.2	46.2	47.3	48.4	47.8
Ours	69.1	71.6	70.3	65.6	69.1	67.3	50.6	57.3	53.7	53.3	47.3	50.1

Table 3. Results from the ACE2005 dataset.

Model	Trigger						Argument					
	Identification			Classification			Identification			Classification		
	P	R	F1	P	R	F1	P	R	F1	P	R	F1
JOINTFEATURE	77.6	65.4	70.1	75.1	63.3	68.7	73.7	38.5	50.6	70.6	36.9	48.4
dbRNN	-	-	-	70.1	69.8	71.9	-	-	57.2	-	-	50.1
Joint3EE	70.5	74.5	72.5	68.0	71.8	69.8	59.9	59.8	59.9	52.1	52.1	52.1
BS	68.9	77.3	72.9	66.7	74.7	70.5	44.9	41.2	43.0	44.3	40.7	42.4
Text2Event	-	-	-	71.2	72.5	71.8	-	-	-	54.0	54.8	54.4
Ours	70.4	76.6	73.3	70.2	75.1	72.6	58.4	53.3	55.7	56.7	52.8	54.7

4.3. Effect of Superior Role Concept

To better understand how the SRC influences our model's performance, in this section, we conducted an ablation study by adding the SRC into different procedures of the whole model, and Table 4 presents the results of the ablation experiments.

Table 4. Effect of SRC.

Model	Trigger						Argument					
	Identification			Classification			Identification			Classification		
	P	R	F1	P	R	F1	P	R	F1	P	R	F1
GAT	65.8	68.3	67.0	66.7	64.8	65.7	50.9	52.1	51.5	45.1	42.1	43.5
GAT-TRI+SRC	66.1	65.2	65.6	66.0	62.2	64.0	55.6	51.1	53.2	48.5	40.6	44.2
GAT-ARG+SRC	65.3	72.4	68.6	65.5	70.0	67.7	52.8	50.9	51.8	50.9	47.1	48.9
Ours	69.1	71.6	70.3	65.6	69.1	67.3	50.6	57.3	53.7	53.3	47.3	50.1

In Table 4, the GAT model shows the removal of the SRC information from both the trigger extraction step and the argument extraction step. It uses GAT to embed the tree structure directly, without any other information. The GAT-TRI+SRC and the GAT-ARG+SRC mean that only the SRC information is introduced into the trigger extraction module and the argument extraction module.

As Table 4 shows, when we adopt the GAT model without any extra information, the performances of the four subtasks drop significantly when compared to the whole model. The F1 scores of the 4 subtasks drop by 3.3%, 1.6%, 2.2%, and 6.6%, respectively. This indicates that the correlation information between argument roles is beneficial to both the trigger extraction and the argument extraction.

However, the situations are different when the SRC information is removed from trigger extraction and argument extraction. The improvement was not obvious when only SRC information was added to the trigger extraction module, although the argument extraction performance was improved when only SRC information was added to the argument extraction module. We think this situation arises not because the high-level role information is not useful for trigger extraction, but because the hierarchical relationship between the SRC and the argument role can only be updated through the argument extraction module. In other words, the SRC embedding vectors are mainly trained in the argument extraction module.

4.4. Influence of Dataset Size

We also explored how the size of the dataset influences the performance of our model. We selected 75%, 50%, and 25% samples from the training data to train our model. These samples were selected randomly and uniformly. The results are given in Table 5.

Table 5. Influence of dataset size.

Size	Trigger						Argument					
	Identification			Classification			Identification			Classification		
	P	R	F1	P	R	F1	P	R	F1	P	R	F1
25%	45.8	32.3	37.9	29.6	35.4	32.2	25.6	21.1	23.1	21.1	28.1	24.1
50%	49.1	47.2	48.1	46.0	52.2	48.9	45.3	49.1	47.1	38.5	42.6	40.4
75%	65.5	68.4	66.9	64.5	70.0	67.1	50.8	52.9	51.8	49.9	46.1	47.9
100%	69.1	71.6	70.3	65.6	69.1	67.3	50.6	57.3	53.7	53.3	47.3	50.1

From the experimental results, we can see that when the size of the dataset is reduced to 75%, the performance of the model only decreases slightly. However, when the size of the dataset is reduced to 50%, or even 25%, the performance of the model decreases significantly. When the size of the dataset is reduced to 25%, the F1 scores of the trigger and argument classification drop by 35.1% and 26.0%, respectively. When the size of the dataset is reduced to 75%, the F1 scores of the trigger and argument classification only drop by 2.4% and 2.2%, respectively. Therefore, the size of the dataset has a significant impact on the effect of the model. However, when the dataset reaches a certain size, the impact of increasing the size of the training set on the results gradually decreases.

5. Discussion

In this paper, we proposed an argument correlation, information-based, graph attention network for movie scene event extraction. Specifically, in order to verify the effectiveness of the proposed model, we compared and discussed the performance of the five models JOINTFEATURE, dbRNN, Joint3EE, BS, and Text2Event on two datasets, MovieSceneEvent and ACE2005. The experimental results demonstrate that our models steadily outperform all other competitive models, regarding both trigger extraction and argument extraction, on movie scene event extraction and open domain event extraction. However, when compared with the performance in the open domain dataset, the improvement of our model in the movie scene event extraction dataset is much more significant. This is probably due to the composition of argument roles in movie scene event extraction being much easier to generalize into several superior role concepts. Moreover, the ablation study in Section 4.3 verifies that that the correlation information between argument roles is beneficial to both the trigger extraction and the argument extraction. Furthermore, the study regarding the influence of dataset size indicates that this factor has a significant influence on the performance when the size of the dataset decreases by a significant amount. However, when the dataset reaches a certain size, the impact of increasing the size of the training set on the results is not significant.

6. Conclusions

In this paper, we propose an argument correlation, information-based movie scene event extraction model because existing open domain event extraction methods have not paid attention to the specific information in a certain field nor made full use of the correlation information of argument roles, which is important implicit semantic information in movie scene event extraction. In order to fully utilize this important implicit semantic information to improve movie scene event extraction, we design an SRC-based GAT to capture this implicit information and integrate correlation information of argument roles into the semantic features. The GAT module can capture the semantic features through the dependency tree structure, while fusing the SRC information into the nodes' hidden embedding. In order to verify the effectiveness of our model, we constructed a movie scene event extraction dataset. Experimental results show that the SRC helps to improve the performance of both event trigger extraction and argument extraction. Our model can significantly improve the performance of movie scene event extraction. Meanwhile, it also has good performance in open domain event extraction.

In the future, we will further exploit the influence of external information on event extraction. We will also attempt to integrate our model with recently popular structures, such as a QA framework.

Author Contributions: Conceptualization, Q.Y. and G.Z.; methodology, Q.Y.; software, Q.Y.; validation, G.Z.; formal analysis, Q.Y.; investigation, Q.Y.; resources, J.L.; writing—original draft preparation, Q.Y.; writing—review and editing, S.Z.; visualization, Q.Y.; supervision, S.Z.; project administration, J.L.; funding acquisition, G.Z. All authors have read and agreed to the published version of the manuscript.

Funding: This work was supported by the National Key R&D Program of China (2021YFF0900600).

Institutional Review Board Statement: Not applicable.

Informed Consent Statement: Not applicable.

Data Availability Statement: The data presented in this study are available from the corresponding authors. The data cannot be made public, as they relate to ongoing projects.

Acknowledgments: This paper is based on our previous work [30], presented at the 2nd International Conference on Culture-oriented Science & Technology (ICCST).

Conflicts of Interest: The authors declare no conflict of interest regarding the publication of this article.

References

1. Mcclosky, D.; Surdeanu, M.; Manning, C.D. Event Extraction as Dependency Parsing. In Proceedings of the 49th Annual Meeting of the Association for Computational Linguistics: Human Language Technologies, Portland, OR, USA, 19–24 June 2011; pp. 1626–1635.
2. Liao, R. Using Document Level Cross-Event Inference to Improve Event Extraction. In Proceedings of the 48th Annual Meeting of the Association for Computational Linguistics, Uppsala, Sweden, 11–16 July 2010; pp. 789–797.
3. Li, Q.; Ji, H.; Huang, L. Joint Event Extraction via Structured Prediction with Global Features. In Proceedings of the 51st Annual Meeting of the Association for Computational Linguistics, Sofia, Bulgaria, 4–9 August 2013; pp. 73–82.
4. Ji, H.; Grishman, R. Refining Event Extraction through Cross-Document Inference. In Proceedings of the ACL-08: HLT, Columbus, OH, USA, 19–20 June 2008; pp. 254–262.
5. Huang, R.; Riloff, E. Modeling Textual Cohesion for Event Extraction. *Proc. AAAI Conf. Artif. Intell.* **2021**, *26*, 1664–1670. [CrossRef]
6. Sha, L.; Qian, F.; Chang, B.; Chang, B.; Sui, Z. Jointly Extracting Event Triggers and Arguments by Dependency-Bridge Rnn and Tensor-Based Argument Interaction. *Proc. AAAI Conf. Artif. Intell.* **2018**, *32*, 5916–5923. [CrossRef]
7. Nguyen, T.H.; Grishman, R. Graph Convolutional Networks with Argument-Aware Pooling for Event Detection. *Proc. AAAI Conf. Artif. Intell.* **2018**, *32*, 5900–5907. [CrossRef]
8. Nguyen, T.H.; Grishman, R. Event Detection and Domain Adaptation with Convolutional Neural Networks. In Proceedings of the 53rd Annual Meeting of the Association for Computational Linguistics and the 7th International Joint Conference on Natural Language Processing (Volume 2: Short Papers), Beijing, China, 26–31 July 2015; Association for Computational Linguistics: Stroudsburg, PA, USA, 2015.
9. Chen, Y.; Xu, L.; Liu, K.; Zeng, D.; Zhao, J. Event Extraction via Dynamic Multi-Pooling Convolutional Neural Networks. In Proceedings of the 53rd Annual Meeting of the Association for Computational Linguistics and the 7th International Joint Conference on Natural Language Processing (Volume 1: Long Papers), Beijing, China, 26–31 July 2015; Association for Computational Linguistics: Stroudsburg, PA, USA, 2015.
10. Chen, Y.; Chen, T.; Ebner, S.; White, A.S.; Van Durme, B. Reading the Manual: Event Extraction as Definition Comprehension. *arXiv* **2019**, arXiv:1912.01586.
11. Du, X.; Cardie, C. Event Extraction by Answering (Almost) Natural Questions. *arXiv* **2020**, arXiv:2004.13625.
12. Petar, V.; Cucurull, G.; Casanova, A. Graph Attention Networks. *arXiv* **2017**, arXiv:1710.10903.

13. Yan, M.; Lou, X.; Chan, C.A.; Wang, Y.; Jiang, W. A semantic and emotion-based dual latent variable generation model for a dialogue system. *CAAI Trans. Intell. Technol.* **2023**, 1–12. [CrossRef]
14. Yi, Q.; Zhang, G.; Zhang, S. Utilizing Entity-Based Gated Convolution and Multilevel Sentence Attention to Improve Distantly Supervised Relation Extraction. *Comput. Intell. Neurosci.* **2021**, *2021*, 6110885. [CrossRef] [PubMed]
15. Liu, W.; Pang, J.; Du, Q.; Li, N.; Yang, S. A Method of Short Text Representation Fusion with Weighted Word Embeddings and Extended Topic Information. *Sensors* **2022**, *22*, 1066. [CrossRef] [PubMed]
16. Pota, M.; Ventura, M.; Catelli, R.; Esposito, M. An Effective BERT-Based Pipeline for Twitter Sentiment Analysis: A Case Study in Italian. *Sensors* **2021**, *21*, 133. [CrossRef] [PubMed]
17. Yan, M.; Li, S.; Chan, C.A.; Shen, Y.; Yu, Y. Mobility Prediction Using a Weighted Markov Model Based on Mobile User Classification. *Sensors* **2021**, *21*, 1740. [CrossRef] [PubMed]
18. Kriman, S.; Ji, H. Joint Detection and Coreference Resolution of Entities and Events with Document-Level Context Aggregation. In Proceedings of the 59th Annual Meeting of the Association for Computational Linguistics and the 11th International Joint Conference on Natural Language Processing: Student Research Workshop, Bangkok, Thailand, 5–6 August 2021; Association for Computational Linguistics: Stroudsburg, PA, USA, 2021.
19. Lyu, Q.; Zhang, H.; Sulem, E.; Roth, D. Zero-Shot Event Extraction via Transfer Learning: Challenges and Insights. In Proceedings of the 59th Annual Meeting of the Association for Computational Linguistics and the 11th International Joint Conference on Natural Language Processing (Volume 2: Short Papers), Online, 1–6 August 2021; Association for Computational Linguistics: Stroudsburg, PA, USA, 2021.
20. Lin, Y.; Ji, H.; Huang, F.; Wu, L. A Joint Neural Model for Information Extraction with Global Features. In Proceedings of the 58th Annual Meeting of the Association for Computational Linguistics, Online, 5–10 July 2020; pp. 7999–8009.
21. Li, D.; Huang, L.; Ji, H.; Han, J. Biomedical Event Extraction Based on Knowledge-Driven Tree-LSTM. In Proceedings of the NAACL2019, Minneapolis, MN, USA, 2–7 June 2019; pp. 1421–1430.
22. He, H.; Ning, Q.; Roth, D. QuASE: Question-Answer Driven Sentence Encoding. *arXiv* **2019**, arXiv:1909.00333.
23. Liu, J.; Chen, Y.; Liu, K.; Bi, W.; Liu, X. Event Extraction as Machine Reading Comprehension. In Proceedings of the 2020 Conference on Empirical Methods in Natural Language Processing (EMNLP), Online, 16–20 November 2020; Association for Computational Linguistics: Stroudsburg, PA, USA, 2020.
24. Finkel, J.R.; Grenager, T.; Manning, C. Incorporating Non-Local Information into Information Extraction Systems by Gibbs Sampling. In Proceedings of the 43rd Annual Meeting on Association for Computational Linguistics—ACL'05, Ann Arbor, MI, USA, 25–30 June 2005; Association for Computational Linguistics: Morristown, NJ, USA, 2005.
25. Devlin, J.; Chang, M.-W.; Lee, K.; Toutanova, K. BERT: Pre-Training of Deep Bidirectional Transformers for Language Understanding. *arXiv* **2018**, arXiv:1810.04805.
26. Weston, J.; Chopra, S.; Bordes, A. Memory Networks. *arXiv* **2014**, arXiv:1410.3916.
27. Tarjan, R. Depth-first search and linear graph algorithms. *SIAM J. Comput.* **1972**, *1*, 146–160. [CrossRef]
28. Yang, B.; Mitchell, T.M. Joint Extraction of Events and Entities within a Document Context. *arXiv* **2016**, arXiv:1609.03632.
29. Lu, Y.; Lin, H.; Xu, J.; Han, X.; Tang, J.; Li, A.; Sun, L.; Liao, M.; Chen, S. Text2Event: Controllable Sequence-Tostructure Generation for End-to-End Event Extraction. *arXiv* **2021**, arXiv:2106.09232.
30. Yi, Q.; Zhang, G.; Liu, J.; Zhang, S. Movie Scene Argument Extraction with Trigger Action Information. In Proceedings of the 021 International Conference on Culture-oriented Science & Technology (ICCST), Beijing, China, 18–21 November 2021.

sensors

MDPI

Article

A Novel Swarm Intelligence Algorithm with a Parasitism-Relation-Based Structure for Mobile Robot Path Planning

Hui Ren [1,2,3], Luli Gao [1,2,3,*], Xiaochen Shen [1,2,3,*], Mengnan Li [1,2,3] and Wei Jiang [1,2,3]

1 School of Information and Communication Engineering, Communication University of China, No. 1 Dingfuzhuang East Street, Chaoyang District, Beijing 100024, China
2 State Key Laboratory of Media Convergence of Communication, Communication University of China, Beijing 100024, China
3 Key Laboratory of Acoustic Visual Technology and Intelligent Control System, Ministry of Culture and Tourism, Beijing 100024, China
* Correspondence: gaoluli@cuc.edu.cn (L.G.); shen2chen2@126.com (X.S.)

Abstract: A multi-swarm-evolutionary structure based on the parasitic relationship in the biosphere is proposed in this paper and, according to the conception, the Para-PSO-ABC algorithm (ParaPA), combined with merits of the modified particle swarm optimization (MPSO) and artificial bee colony algorithm (ABC), is conducted with the multimodal routing strategy to enhance the safety and the cost issue for the mobile robot path planning problem. The evolution is divided into three stages, where the first is the independent evolutionary stage, with the same evolution strategies for each swarm. The second is the fusion stage, in which individuals are evolved hierarchically in the parasitism structure. Finally, in the interaction stage, a multi-swarm-elite strategy is used to filter the information through a predefined cross function among swarms. Meanwhile, the segment obstacle-avoiding strategy is proposed to accelerate the searching speed with two fitness functions. The best path is selected according to the performance on the safety and consumption issues. The introduced algorithm is examined with different obstacle allocations and simulated in the real routing environment compared with some typical algorithms. The results verify the productiveness of the parasitism-relation-based structure and the stage-based evolution strategy in path planning.

Keywords: Para-PSO-ABC algorithm; dual-community-evolutionary structure; parasitic relationship; path planning; multi-swarm evolution

Citation: Ren, H.; Gao, L.; Shen, X.; Li, M.; Jiang, W. A Novel Swarm Intelligence Algorithm with a Parasitism-Relation-Based Structure for Mobile Robot Path Planning. *Sensors* **2023**, *23*, 1751. https://doi.org/10.3390/s23041751

Academic Editor: Gregor Klancar

Received: 29 December 2022
Revised: 19 January 2023
Accepted: 2 February 2023
Published: 4 February 2023

1. Introduction

With the development of control technology, mobile robotics has been developed in multiple fields, such as rescue, military, industry, etc. As one of the essential parts of the mobile robotics field, the path planning problem establishes an effective path for the robot to reach the target and finish the task without any collisions based on the specific environment. According to the preknowledge of the working environment, the path planning approach can be classified into two parts, which are the global path planning with static information and the local planning based on the sensor information. The former one needs to find a suitable route according to previous map information. While the local aspect should have the decision capacity upon real-time information and find where the problem is, such as the local obstacle distribution, so as to optimize a solution from the current node to a sub-target until the mission is completed. Successful planning should meet both optimization criteria in terms of time and traveling distance, etc.

Recently, various methods of intelligent planning have been studied to find the most productive solution. All the methods can be classified into traditional path-planning methods based on environment modeling, search-based method, and artificial intelligence algorithm (seen in Figure 1). The traditional path planning algorithms are built with the previously defined map, and most of them require environment information in advance to guide robot movement [1] or make a mobility prediction [2]. It is normally utilized to solve a global problem, such as the potential field category, which produces an artificial field based on the motion environment. The movement of the robot can be guided by descent direction, such as gravity, to avoid the repulsive fields (obstacle) from the start to the target. However, it cannot guarantee the path is the global best, even for a certain search range. Hence, the optimization algorithm is normally utilized to optimize the path generated by the artificial potential field (APF), so as to increase the effectiveness of the hybrid algorithm [1]. Another category is the search-based method, such as Dijkstra and A* algorithm, whose complexities increase with the dimensions of problems, resulting in lower effectiveness [3]. Further, a dynamic environmental problem is hard to deal with for those methods that may increase the cost of the planning. Moreover, the environment-modeling-based method, such as the Voronoi Diagram [4], Visibility Graph [5], and Cell Decomposition [6], decomposes the environment into several regions and transforms the complex workspace into a simple map search problem. This type of algorithm has a strong ability to guarantee safety, but the local optima cannot be avoided.

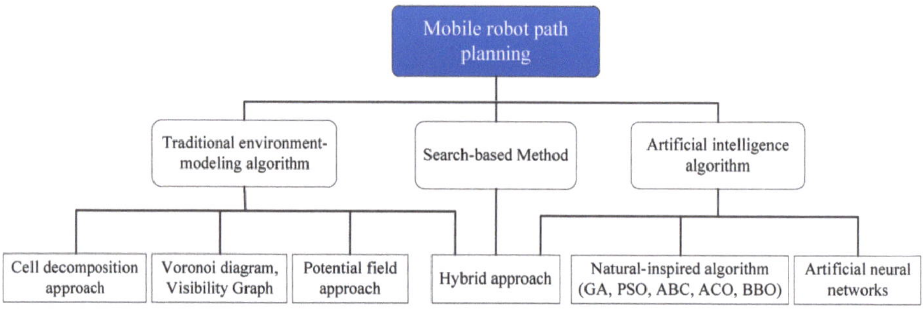

Figure 1. Classification of mobile robot path planning approaches.

The artificial intelligence algorithm is probably the most commonly adopted approach for mobile robot path planning, which transforms the best path planning into a constrained optimization with one or several objective functions. The optimization algorithm is utilized to solve the nonlinear and multi-constrained models for an optimum or approximate solution, such as particle swarm optimization (PSO), ant colony optimization (ACO), artificial bee colony algorithm (ABC), genetic algorithm (GA), artificial neural network [7–12], etc. They have a great ability to handle the uncertainty condition in the complex environment and are flexible for global or local routing problems. However, the performance of this kind of method is unstable due to defects in the algorithm, such as the premature problem and the balance between exploration and exploitation, which may cause inferior work efficiency with a redundant computation. Therefore, the algorithm should be modified according to its merits. For instance, to address the slow convergence problem in the GA, Hong et al. introduced a co-evaluation strategy to provide some margin of error during evolution. Similarly, Ref. [13] proposed a parallel strategy in ACO to address the premature problem. The hybrid algorithm is another productive and flexible method that can merge the advantages of each algorithm to find the best path. Ref. [14] introduced a hybrid algorithm that combined biogeography-based optimization (BBO) and PSO based on the Voronoi diagram to deal with static path planning. Normally, a traditional path planning algorithm could generate a more stable static path. Ref. [15] utilized the merit of the potential field method, combined with a bacterial evolutionary algorithm, to reduce the disadvantages of the intelligence algorithm when it is applied in an environment with a dynamic condition. The traditional map-based method is applied with the evolution algorithm is also an effective method. Ref. [16] used fuzzy logic to enhance the searchability for a dynamic path optimization based on the A* and GA. In addition, the hybrid method can extract some special evolution mechanisms to compensate for the defect in the iteration process. Ref. [17] pointed out that the parameters in the bat algorithm (BA) can be optimized by PSO in multi-objective optimization. In this category, however, the robot always moves precisely along the predetermined path [18]. In the case of an accident, the original path would be affected and redesigned with a local search method. Hence, the falling into a local optimum and computational complexities problems of the intelligence algorithm is trickier, so as to reduce the robustness and working safety. The convergence recourse should be regrouped and allocated to balance exploration and exploitation. Most swarm intelligence algorithms are improved by the evolutionary approach, such as the whale algorithm [19] and the bat algorithm [20], which changes the form of evolution. However, such modification cannot solve the search balance problem. The other method is to construct evolutionary structures, such as ABC [21], artificial fish algorithm [22], or wolf colony algorithm [23]. Although these algorithms can effectively improve the efficiency of population utilization by allocating responsibilities to the populations, they still have inevitable problems, such as the weak local exploitation ability in the ABC. The reasons are that it cannot control the evolutionary direction and the nectar replacement mechanism, resulting in a lower convergence accuracy.

In this paper, a conception of the dual-community-evolutionary structure inspired by the parasitism relation is proposed to balance the exploration and exploitation in optimization. Then, the parasitism-relation-based algorithm, composed of the PSO and ABC algorithm (ParaPA), is applied tp the path planning problem. In the paradigm algorithm, the memory swarm in PSO is utilized to prevent the loss of optima at the superior level, while the evolution approach in ABC is tailored to maximize the convergence-resource usage at the bottom level. The swarm intelligence algorithm is a search optimization based on probability where a more uniform distribution at an earlier stage obtains a better optimal solution easier. Hence, during the evolution, the initialization phase uses a chaos-based logistic map to create a chaotic status in the first stage. Then, the personal best particles in PSO are selected in the superior population, whereas the global best is produced by the nectar selection method in the ABC process. Meanwhile, the environment is divided into several segments to decrease the dimension of the variable, which improves the adaptability of the algorithm. In addition, multi-swarm evolution with the elitist-based information changing strategy is conducted to guarantee the algorithm diversity directly in the upper layer through the cross function. Finally, the proposed algorithm is examined in some path-planning environments and compared with other path-planning for verifying its effectiveness and safety.

Notations: $\|\cdot\|$ represents Euclidean norm. $\mathbb{R}^n$ is the n-dimensional real number space. $\subsetneq$ is the non-true subset operator.

2. Preliminary Knowledge and Analysis of ABC Algorithm

2.1. Basic Conception of Artificial Bee Colony Algorithm

The artificial bee colony algorithm is derived from the honey-harvesting behavior of honey bees. The bee colony is divided into several groups with different tasks and shares the group information to find the optimal solution. The major assignments can be divided into two swarms, namely employed bees and onlookers [21]. The former forages for food sources and searches only in the local region. The number of food sources equals the onlooker bees population. Once the nectar collection is completed, associated bees can become scout bees and repeat the food search within the entire space and bring the new location into the colony. The procedure for mass fundamentals is described below:

A. Employed bee

In charge of the food exploitation by the defined equation as follows:

$$v_{ij} = x_{ij} + \phi\left(x_{ij} - x_{kj}\right) \tag{1}$$

where $i \neq k$, j indicates the dimension index from $\{1, 2, \ldots D\}$, and ϕ is a random factor selected from $[-1, 1]$. $X_i = \{x_{i1}, x_{i2}, \ldots, x_{iD}\}$ represents the current food source location, D is the dimensions number of X_i, and $v_i = \{v_{i1}, v_{i2}, \ldots, v_{iD}\}$ is the location of a new food source searched for by employed bee. Note that if the fitness of the new sources is higher, the memories of employed bees will keep the new position. Otherwise, the previous one is kept to the next iteration until the nectar is substituted.

B. On looker

A bee is looking for suitable food sources by roulette wheel selection, as shown in Equation (2):

$$p_i = \frac{fit_i}{\sum_{i=1}^{SN} fit_i} \tag{2}$$

where fit_i is the fitness value of X_i, and p_i the is the probability of selection. Once the target is locked, onlookers are transformed into employed bees and exploit food sources.

C. Scout bee

If a food source cannot be further improved, new nectar would be produced randomly in the searching space, using Equation (3):

$$x_{ij} = x_{\min,j} + random(0,1)\left(x_{\max,j} - x_{\min,j}\right) \tag{3}$$

where $x_{\max,j}$ and $x_{\min,j}$ are the maximum and minimum search bounds.

2.2. The Pros and Cons Analysis of ABC

In ABC, the group evolution strategy can maximize the utilization of convergence resources. For instance, the setting of the nectar harvesting mechanism allows the algorithm to retain a dynamic search ability even in the later period. However, because of this kind of mechanism, a potential optimum may be disturbed, resulting in the algorithm being unable to achieve better local convergence accuracy in the limited number of iterations. To address the question, Li et al. pointed out that the successful experience generated in the evolution can be used to guide the next foraging behavior [24]. The best value is kept throughout the whole process. While Ref. [25] found that the nectar collection is a random process, if different nectars have the same objective value, the search phase of onlookers might be invalid in the evolution. Therefore, they proposed a neighborhood selection method to improve the updated format. In addition, Zhou et al. introduced a multi-elite strategy to increase the guidance in position updating. Moreover, the employed bee and onlooker bee adapt two different search equations [26]. Nevertheless, this approach does not solve the problem of search range overlapping among elites. Regardless of either approach, the fundament is to prevent the loss of optimal solutions with iterations at the cost of the exploration capacity of the algorithm.

The diversity and local convergence accuracy in the later stage is a pair of contradictory factors. A better solution is to explore the search space as much as possible in the initialization and establish an appropriate promotion method in the middle stages of evolution so as to avoid the limitations on diversity caused by the homogeneous potential optimum in the later stages. While the ABC algorithm can deal with the problem by collecting and reusing the converging resources through the division of functional responsibilities into different swarms. These advantages can be utilized, and this article conducts a novel approach incorporated with PSO to address the shortcomings of the ABC algorithm in a dual-community structure. The community can be regarded as a swarm with a specific evolution pattern.

3. The Proposed Parasitism-Relation Structure and ParaPA Algorithm

It is worth noting that the improvement on certain parts of the algorithm always comes at the expense of other performance, and this approach can only achieve better application results with specific requirements. However, in most of the modifications, the efficiency of the algorithm does not increase, and the extra parameter setting in the mechanism also asks for higher requirements during the initialization process. While the structure of dual communities can compensate for the defects of each algorithm by establishing the parasitic relationship to integrate the advantages of algorithms. The problem is how to ensure the information interchange and rebuild the evolutionary mechanism based on the relationship to balance the exploration and exploitation ability during the whole process. Hence, the major purpose of the structure is to address the search balance problem.

3.1. The Parasitism-Relation Structure

The superiority of the PSO is that it can utilize the best personal and global memories to increase evolution efficiency. This paper establishes the relationship between particle swarms in PSO and bee colonies by the parasitism phenomenon in the biosphere. In previous studies, some researchers have used the symbiosis phenomenon to construct the evolutionary process (symbiotic organisms search algorithm, called the SOS algorithm) [27,28]. Although the same conception of symbiosis is utilized, the understanding of symbiosis is quite different. In essence, the SOS algorithm divides the evolution process into three stages, named mutualism, commensalism, and parasitism, which correspond to the initialization stages for producing diversity, the middle stage for evolving toward the best position, and the last stage for preventing populations from stagnating by generating a random sample, respectively. In terms of modes and the formulation established in SOS, the essence is a DE algorithm with a phased evolution. As the concept of symbiosis described in Ref. [29], to mimic the symbiosis phenomenon, a more appropriate approach is to focus on the swarm relationship rather than individuals. The survival of any swarm is determined by whether it can converge to the final evolution. While in a parasitic relationship, the host has a superior resource, and the inferior party can only attach to the dominant position to keep evolution and competition for a chance to survive. The organism, however, has limited nutrition, which means the exploited position should keep changing throughout the whole process. When the convergence resources are exhausted, the attached particle also loses the right to survive and thus enters a chaotic state. The interpretation of this relation in mathematics is (Note that the multi-path planning is related to the multimodal optimization, which is adapted in this paper.):

Definition 1. *Global best. If* $\exists x^* \in S$, *for* $\forall x \in S$ *where* S *is the search space with* $S \subset \mathbb{R}^n$, *can have* $f(x) \leq f(x^*)$, *then the* x^* *is regarded as the global best decision variable in* S *and* $f(x^*)$ *is named the global best value.*

Definition 2. *Multimodal Function.* f^* *is the best value on* S *based on* f, *if there are different decision variables* $x_1, x_2, \cdots, x_m \in S$ *where* $f(x_i)(i = 1, 2, \cdots m)$ *are the global best or local best value, the function* $f(x)$ *can be named as the multi-peaks function or multimodal function.*

Based on Definition 2, the process of finding the best decision variables in S based on the multimodal function f can be named multimodal optimization.

Definition 3. *Parasitism relationship in multimodal optimization.* S *is regarded as the living space for the decision variable* x. *Suppose there are* m *optima in* S, *recorded as* $\Phi_1, \Phi_2, \cdots, \Phi_m$, $\exists S_1 \subsetneq S$, *when the iteration is* t *and the* S_1 *can be recorded as* $S_1(t)$, *then* $\exists P, 0 < P < 1$, *satisfy*

$$\lim_{t\to\infty} \prod_{i=1}^{m} P(\Phi_i \in S_1(t)) \geqslant P$$

P is the convergence probability, which is defined in fuzzy logic. Further, for $\forall S_2 \subsetneq S$, $S_1 \neq S_2$, *for* $\forall 0 < \delta < 1$, *can satisfy*

$$\lim_{t\to\infty} \prod_{i=1}^{m} P(\Phi_i \in S_2(t)) < \delta$$

Then called the S_1 and S_2 is the parasitism relationship in the living space S.

Definition 4. *Suppose Φ_i is the i-th optimum in the living space S, if $\exists\beta \in S$, $\forall\delta > 0$, if $\|\Phi_i - \beta\| < \delta$, then*

$$P(\Phi_i \in S) = 1.$$

If a threshold value is determined, let $0 < \theta_1 < 1$, $\exists\theta_2$ and $0 < \theta_2 < \theta_1$, when $\exists\beta \in S$ and $\|\Phi_i - \beta\| < \theta_1$, then

$$P(\Phi_i \in S) = \frac{\|\Phi_i - \beta\|}{|\theta_1 - \theta_2|}.$$

Further, for $\forall\beta \in S$, it always has $\|\Phi_i - \beta\| > \theta_1$, then

$$P(\Phi_i \in S) = 0$$

To summarize, as

$$P(\Phi_i \in S) = \begin{cases} 1, & \textit{Totally convergence} \\ 0, & \textit{Totally not convergence} \\ \dfrac{\|\Phi_i - \beta\|}{|\theta_1 - \theta_2|}, & \textit{Fuzzy convergence} \end{cases}$$

The commensalism and mutualism relationship is described in Appendix A.

In the ParaPA algorithm, the bee colony is regarded as the host part to control the better convergence resources, while the survival of particles in PSO is living in the bottom population, such as the structure, as shown in Figure 2. The life of particles in PSO swarm leeches onto the onlookers and personal best swarms. In other words, the bottom particles cannot find the evolution direction without the instruction from the pbest and bees swarms. If the populations are divided into different hierarchies according to their objective function fitness, the result can be illustrated as in Figure 3. The S_4 level represents the bottom swarm whose evolution is dependent on S_3. Meanwhile, the S_3 level is constructed from S_4, and they hold exclusive evolution strategies, such as the multi-swarm-elites strategy in pbest swarm and look limitation strategy in the bee colony, which also aims to enhance the algorithm performance in exploitation and diversity, respectively. S_1 and S_2 belong to the superior level, which is mainly responsible for exchanging the best information in its population with other populations and also conveying the feedback to S_3. Hence, the major function of a superior level is to guarantee the information changes among the different populations to avoid evolutionary problems from a monotonous population.

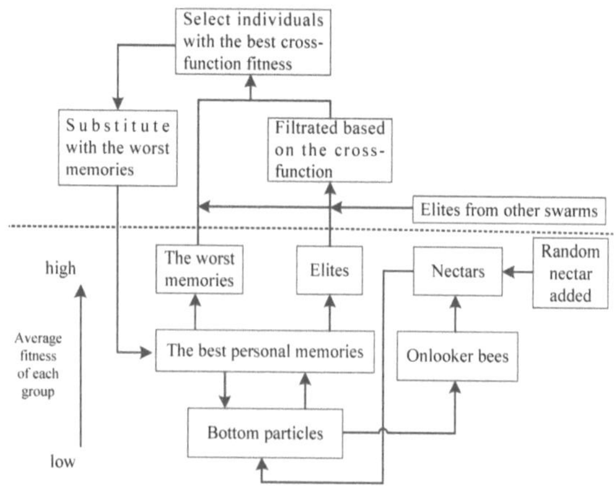

Figure 2. Evolutionary structure of the ParaPA algorithm.

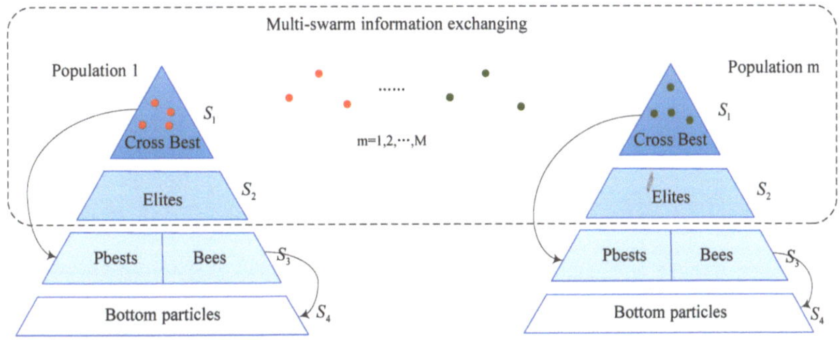

Figure 3. The hierarchical illustration of the ParaPA algorithm with multi-swarm strategy.

3.2. Evolutionary Process of ParaPA

3.2.1. Independent Evolution Stage

A chaotic distribution has been proven to achieve a better statistical property and have faster convergence to the algorithm [30]. To prosper the diversity of the food sources initialization, in this paper, the Logistic map, which is widely used in the chaos-based initialization, is given as follows:

$$x_{n+1} = \mu x_n(1 - x_n) \tag{4}$$

where n is the number of chaotic variables and $x_0 \notin \{0, 0.25, 0.5, 0.75, 1\}$. The chaotic control parameter $\mu \in [3.57, 4]$. We use $\mu = 4$ to produce the chaotic system.

During the initiation, the food resources are selected based on objective fitness, and then all the particles in PSO will transform to onlooker bees to choose the nectar as the global best. Moreover, personal memory is also kept to create more possibilities during the iteration.

3.2.2. The Fusion Stage

Based on the independent evolution, the novel position update incorporated with the bee colony is introduced as follows:

$$V_{id}^{t+1} = \omega * V_{id}^{t} + r_1 * (pbest_{id}^{t} - X_{id}^{t}) + r_2 * (onlooker_{id}^{t} - X_{id}^{t}) \tag{5}$$

$$X_{id}^{t+1} = X_{id}^{t} + V_{id}^{t+1} \tag{6}$$

where V_{id} is the d dimensional velocity of the current particle after t iterations, X_{id} is the d dimensional location of the current particle after t iterations, and the inertial weight w is utilized for the exploration ability, changing from $[0.1, 0.9]$ (as shown in Equation (7)) through the evolutionary process, which is a self-regulating method.

$$\omega = \omega_{\max} - (\omega_{\max} - \omega_{\min}) * \frac{g}{Maxgen} \tag{7}$$

where g is the current iterative time, and *Maxgen* is the total number of iterations. r_1 and r_2 are two constants, which are taken randomly from 0 to 2 in this paper. The global best is replaced by the best onlooker bee, which is selected by the roulette wheel, as shown in Equation (2).

The limitation of nectar collection is set to three times the number of the population, which means that when all individuals visit the current solution more than three times, the nectar should be abandoned and the onlooker bee on the current nectar would be respawned inside the solution space according to Equation (3). Meanwhile, the original onlooker bees, regarded as the host, compete with the parasitic individuals during the evolution. The host would be directly replaced once a better position appears and the visited number is also restarted. It is worth noting that there is no memory preservation for the onlooker. Originally, the attached individuals would fall into a chaos status when the onlooker is regenerated. However, due to the record of the personal best position, the parasitic individual can still move toward the best position in its memory after losing the global guidance so that the algorithm can keep the exploitation capacity in the middle stage of evolution and avoid the waste of iterations.

3.2.3. Interaction Stage with the Multi-Swarm Elite Strategy

In order to enhance the diversity of the algorithm, this paper adopts a multiple swarm parallel strategy. Each swarm is an independent biosphere, and its parasitic relationships are bound in the biosphere, which is not influenced by other environments. To avoid overlapping regions of the host during the convergence, each population takes out the best memory to join the mixing pool after each evolution (as illustrated in Figure 2). The quality of all elites in the mixing pool is judged by the cross function instead of the previous objective function. Furthermore, the worst individuals are usually discarded via evolution, but the amount of information carried by them can create substantial value in multimodal situations. For the consideration of diversity, the worst individuals in the personal best memory are also mixed with the individuals selected from the mixed pool who have better cross function values and then re-screened by the related objective function. Finally, the individuals with the worst fitness in each swarm would be replaced by those chosen memories in the next iteration (the pseudocode of multi-swarm elites selection is shown in Algorithm 1).

Algorithm 1 Multi-swarm-elite selection strategy.

Input: Select elites from the best personal memories in each sub-swarm based on the objective functions and add them to the mixed pool.
Steps: **for** each sub-swarm
1. Select several worst personal memories in each sub-swarm (suppose the number is w).
2. Choose the best individuals from the mixed pool according to the cross fitness. Then, mix the worst w individuals into with them and remove other elites in the pool.
3. **for** each individual in the mixed pool
   Evaluated it by the cross function
   if current individual is better
   Record it in the pool-output group (In this paper, the population of pool-output group is equal to 1)
   end
   end
4. Replace the worst individuals in sub-swarm by the pool-output group into next iterations

end
Output: Each sub-swarm with elites who has the best cross fitness. To note that the replaced individuals always keep the best memories.

4. Problem Formulation and the Strategies for Path Planning

In this paper, several static obstacles are listed in the environment along with some premises and assumptions, shown as follows:

(1) Global path planning is the main target in this paper, which means all obstacles are known before algorithm execution.
(2) The environment of path planning is built in a 2D workspace. The path planning will consider obstacle avoidance without height.
(3) If the planning path can avoid the obstacles successfully under environmental constraints, it means the algorithm has the ability to build a safe path in the static condition. Hence, the physical characteristics of the robot are not considered in this paper.
(4) In the static condition, the robot speed is constant.

4.1. Workspace Formulation

Normally, the path planning question has two or three decision variables. Although it can enrich the diversity in the evolution process, the convergence efficiency of the algorithm will be significantly reduced in the face of complex obstacle situations. Therefore, a new coordinate is applied to connect the start and target positions, as shown in Figure 4. Then the new x'-axis is divided into N segments averagely, which means the x dimension is fixed in the new coordinate. While the y dimension is optimized along the parallel line L_i $(i = 1, 2, \ldots, N-1)$, which is vertical to the new x'-axis. Hence, the planning path can be presented as $(S_1, S_2, S_3, \ldots, S_n)$, where $n = 1, 2, \ldots, N$ and the start and target point are fixed. The corresponding transformation formula is shown in Equation (8).

$$\begin{bmatrix} x' \\ y' \end{bmatrix} = \begin{bmatrix} \cos\phi & -\sin\phi \\ \sin\phi & \cos\phi \end{bmatrix} \begin{bmatrix} x \\ y \end{bmatrix} + \begin{bmatrix} Start_x \\ Start_y \end{bmatrix} \quad (8)$$

where the (x, y) and $(Start_x, Start_y)$ belong to the original cartesian coordinate frame. ϕ is the rotation angle to the new frame. Moreover, the map boundary is changed. For instance, in Figure 4, the new y'-axis intersects with the original y-axis. For any point on the parallel line L_i, the boundary of y' in the new frame is determined as follows:

$$y'_{\max} = \begin{cases} \frac{x_i}{\sin\phi}, x_i < x_m \text{ and } y_i < y_m \\ \frac{ul_y - y_i}{\cos\phi}, x_i > x_m \text{ and } y_i > y_m \end{cases} \quad (9)$$

$$y'_{\min} = \begin{cases} -\frac{y_i}{\cos\phi}, x_i < x_m \text{ and } y_i < y_m \\ -\frac{ul_x - x_i}{\sin\phi}, x_i > x_m \text{ and } y_i > y_m \end{cases} \quad (10)$$

where ul_y and ul_x are the upper limits of the y-axis and x-axis, respectively. Further, the segment points are located on the x'-axis, such as $P_i(x_i, y_i)$, with respect to the original frame. For any points that go over the boundary during the evolution, the whole path $(S_1, S_2, S_3, \ldots, S_n)$ should be regenerated in the solution space. To satisfy the safety issue, each segment S_n should not intersect with any obstacles, and all the objective functions are calculated on the original cartesian coordinate frame.

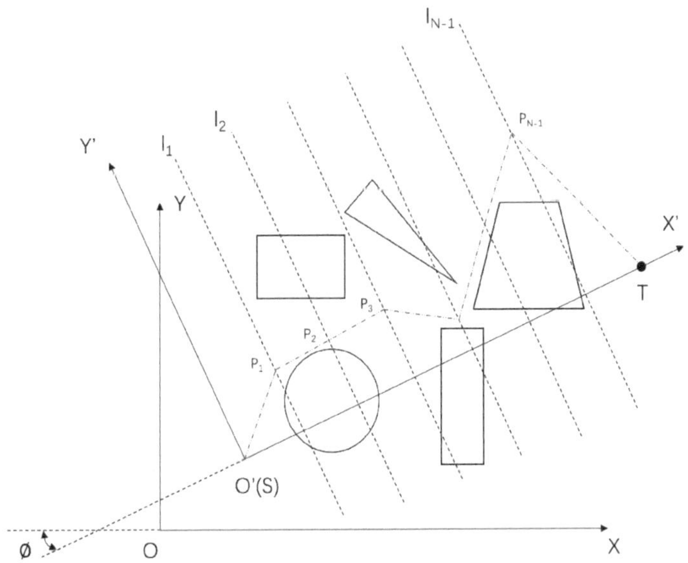

Figure 4. The coordinate transformation for path planning.

4.2. Design of Objective Function

The quality of a path is examined by the fitness value. In the previous study, only feasible paths were evaluated, and information on infeasible paths was neglected [31], which makes it difficult to take the valid yet easily ignored information into account for global planning. Therefore, in this paper, different objective functions are used for the assessment of feasible and infeasible paths, respectively. According to the previous definition, when any segments are overlapped with obstacles, the current path is regarded as an infeasible solution. For the evaluation of infeasible paths, the penalty for worse individuals should be increased so that the evolution of the infeasible solutions can move toward the local best, which has the least intersecting segments. While for the feasible path, when the speed is constant, the total path length should be considered primarily. Since the idealized robot model is adopted in this paper, the kinetics effects are not considered, but the path should be smoothed as much as possible during planning. Based on the aforementioned hypothesis, a path is composed of N segments and $N+1$ nodes; then the evolutionary variable can be represented as $\text{path}_i\ (p_1, p_2, \ldots, p_{N+1})$ where p_1 should be the fixed start point, and p_{N+1} is the target point. Hence, the length of the path can be calculated as

$$\begin{aligned} f_1 &= \sum_{n=1}^{N} \sqrt{(x_{n+1}-x_n)^2 + (y_{n+1}-y_n)^2} \\ &= \sum_{n=1}^{N} \sqrt{\left(\frac{d(x'_1, x'_{N+1})}{N}\right)^2 + (y'_{n+1}-y'_n)^2} \end{aligned} \quad (11)$$

where x_n and y_n are the original coordinates of node $p_n(x_n, y_n) \cdot d(x'_1, x'_{N+1})$ is the distance from the start point to the target along the x'-axis and y'_n is the transformed coordinates.

The smooth function can be defined as follows:

$$f_2 = \pi - \sum_{j=2}^{N} \cos^{-1}\left[\frac{(x_j - x_{j-1})(x_{j+1} - x_j) + (y_j - y_{j-1})(y_{j+1} - y_j)}{\sqrt{(x_j - x_{j-1})^2 + (y_j - y_{j-1})^2} \times \sqrt{(x_{j+1} - x_j)^2 + (y_{j+1} - y_j)^2}}\right] \quad (12)$$

which represents the summation of the angles between every connected segment. It starts from the second point and calculates the angle between the first and second segments, et cetera. The small f_2 value means a small direction change in each turn, representing a better path.

The overall objective function for the feasible function is expressed as follows:

$$F_{feasible} = \frac{1}{f_1} + kf_2 + C \quad (13)$$

where k is the weight of smoothing and a higher k can obtain a smoother path, but the diversity might be affected. C is the feasible and practical reward parameter which is a positive number that is not greater than the maximum path length in the current environment.

For the infeasible condition, except for the length of the path, the ratio of infeasible segments and the ratio of the infeasible path over the total length are involved in the estimation as follows:

$$f_3 = \frac{N_{\text{inf}}}{N} \quad (14)$$

$$f_4 = \sum_{i=1}^{N_{\text{inf}}} \frac{d^i_{obs}}{f_1} \quad (15)$$

where the N_{inf} is the number of infeasible segments and d^i_{obs} is the overlapping distance between each infeasible segment and the obstacles. The objective function for the infeasible function is expressed as follows:

$$F_{infeasible} = \frac{1}{wf_1} + \frac{1}{f_3 + f_4} \tag{16}$$

where w is the weight to adjust the influence of total infeasible path length.

4.3. Design of the Cross Function

The main purpose of the multimodal strategy is to increase the efficiency of path planning, which can provide several collision-free paths from the start position to the target. Once a path falls into an emergency condition, the potential solution can change to another option immediately. The target and initial point for the robot are already known before the real application. The motion of the robot is from its current position to reach the next subsequent segment, and the process will continue until it reaches its goal position. Hence, segments of the robot should not intersect with obstacles or any other potential solution. For the design of the cross function, the objective is to generate a constraint that minimizes the arrival time for each planned path, and meanwhile, each segment cannot overlap with other paths as much as possible. Based on the above analysis of the algorithm, the best individuals in each sub-swarm are first mixed, and the cross fitness of each population elite is calculated based on the following equation, Equation (17), assuming that the number of elites is E.

$$cross\ function = \sum_{e=1}^{E}\sum_{n=1}^{N} Cp_n^e = \sum_{e=1}^{E}\sum_{n=1}^{N} \begin{pmatrix} Cp_1^1 & \cdots & Cp_N^1 \\ \vdots & \ddots & \vdots \\ Cp_1^E & \cdots & Cp_N^E \end{pmatrix} \tag{17}$$

where the index n is the n-th segment in the current evaluated path, the index e is the e-th elite in the other sub-swarms, and Cp_n^e is the cross value, which represents the cross fitness of individual i and interactive elite e. The cross function should be applied to each sub-swarm, and the cross evaluation should go through the sub-swarms except for the current swarm. Moreover, only the elite with the lowest number of segment intersections will be selected and mixed again with the worst individuals in each group memory. After that, all individuals are examined by the cross function again, and the elite who has the greatest difference from others in the mixed group is selected in the next iteration. It is worth noting that all mixing processes are evaluated only by cross fitness to avoid the influence of other factors.

4.4. Multimodal Path Planning Strategy

Multimodal path planning is designed to find multiple paths in a single run. When an emergency occurs on a path, the potential solution can be immediately changed to another option, thus addressing the inefficiencies of traditional path planning. Different from the scheme of a bug algorithm robot walking around obstacles when encountering obstacles and walking along a straight line without encountering obstacles [32], the multimodal path planning strategy adopts the segmented method as described in Section 4.1. Each node position of the segmented $(P_1, P_2, \ldots, P_N)$ corresponds to a dimension of the variable; then the paraPA algorithm is used to find multiple paths. The optimal value with the best fitness that each swarm finally converges to is a global optimal path, while coevolution between multiple swarms ensures the diversity of the paths.

During the demonstration, we found that normally, there are one or two segments that are infeasible, causing the failure of the whole path, which is especially common among the superior populations at the late stage of convergence. To address this problem, this paper proposed a multi-path-based reverse planning strategy, which can be regarded as a complementary strategy for the replacement strategy to nectar sources in the bee colony. If the replaced nectar source only has one or two infeasible segments, the reverse planning strategy is triggered. Specific details of this strategy is shown in Algorithm 2:

Algorithm 2 Multi-path-based reverse planning strategy.

Input: Paths with segments that are infeasible.
Steps: **for** each infeasible path
    1. Find the index of the non-viable section (suppose S_i is the infeasible segment).
    2. Go through the *pbest* group to check.
        if there is a feasible solution in the same section (S_j)
            Examine the segments from S_j back to the start segment.
                if segments from start to S_j in $pbest_p$ are all feasible
                    Record the corresponding node in this $pbest_p$ and replace the corresponding positions in nectar.
                end
        else
            Trigger the random generation process of nectar.
        end
    end
Output: New feasible paths.

5. Experiments and Analysis

5.1. Environments and Comparison of Algorithms

In recent years, researchers have been using the population intelligence approach for path planning [33]. In order to examine the effectiveness of the ParaPA algorithm on different obstacle situations, three different types of scenarios are conducted in this paper to examine the robustness, efficiency, and convergence performance of the algorithm. The first is 20 × 20 maps, which are used to detect the sensitivity of algorithms to different types of obstacles. Second, rectangles are randomly generated in 50 × 50 maps with various sizes as obstacles, which is applied to test the adapted capacity of the algorithm under the complex environment. The third category, an actual scenario simulation, named Small Cultural Complex, is adopted as the application environment built-in 100 × 100 maps. The Small Cultural Complex is a new cultural industry model, which achieves different cultural function requirements by transforming the inside construction of the building, and during the transformation, mobile robots would take charge of the functional equipment transport, such as stage props or lighting. In this paper, the exhibition, sports activity, and theatrics functions are modeled, respectively. Meanwhile, to further show the advantages of the multi-swarm strategy, this paper uses PSO, CPSO, WPSO, ABC, and ACO algorithms for comparison experiments. Furthermore, MSPO and MABC are applied to test the effect of the proposed structure, also to evaluate the algorithm in terms of planning effectiveness, stability, etc.

5.2. Experimental Settings

For all algorithms, the data are calculated from 1000 runs, and the maximum evolution times are 200. For parameters setting in the PSO algorithm, the population size for each swarm is $NP = 50$, and the maximum and minimum evolution steps are $Vmax = 2$ and $Vmin = -2$, respectively. The acceleration factors are accepted as a constant 1.49445 in this paper. Specifically, for the WPSO, the inertia weight w is linearly changed from 0.1 to 0.9. Otherwise, the w is adapted as a constant 0.7. While for CPSO, the chaos-based initialization is utilized with a logistic chaos map to enrich the algorithm diversity where $\mu = 4$ is adapted. For the parameter setting of the ABC algorithm, the look limitation is adapted as $5 \times NP$, which means all individuals in the swarm should visit the nectar more than five times, then the nectar can be replaced. In ABC and MABC, the size of onlookers is equal to the size of employed bees, but in the ParaPA algorithm, this number is equal to $NP/5$. Moreover, the number of elites in each swarm is taken as 10 while 3 of them will be selected in the cross-verification process. Moreover, the number of elites in each swarm is taken as 10, while 3 of them will be selected in the cross-verification process. The start points for 20 × 20 and 50 × 50 maps are set at point $S(0.5, 0.5)$ and the target points are

$E_{20}(19.5, 19.5)$ and $E_{50}(49.5, 49.5)$, respectively. While for 100 × 100 maps, the start points are set to $S_{100}(3.5, 3.5)$, and the target points are $E_{100}(99.5, 99.5)$.

5.3. Results Representation and Analysis

5.3.1. Scenario 1: Path Planning on 20 × 20 Maps with Different Types of Obstacles

In the first case, different obstacles are constructed in a 20 × 20 map, as shown in Figure 5a–g to examine the sensitivity of the algorithm on obstacle shapes. The length of each chromosome is set to 12 for all algorithms. Results are shown in Table 1, and all the best values are highlighted in bold. First, by comparing with the particle swarm algorithm in the original coordinate, it can be clearly verified that the performance of the algorithm with the new coordinate system is significantly improved, but the problem is also ParaPA. As can be seen in the test of Map 4, the mean and variance under the original coordinates are the best, which means that the new coordinate system is inferior to dealing with the S-shaped trajectory planning. Because one dimension is fixed in the new coordinate system, the changeable range for each individual is less than the original coordinate. Hence, when in an environment with several continuous transverse obstacles, such as Map 4, the path generated by the algorithm is relatively monotonous, resulting in a worse convergence effect. It is worth noting that the best value of optima, as well as the worst individual in Table 1, are not in the original coordinate system, indicating that the algorithm in the new coordinate system could be superior during the evolution once the algorithm can find a feasible sample as soon as possible. From the results of other maps, the ParaPA algorithm has the best performance in 1000 runs, and the final convergence achieved the shortest path length in most scenarios. While the algorithm with the best stability is ACO, its evolutionary process is easily stagnant, resulting in unsatisfactory final convergence results. In comparison, the multi-swarm strategies, such as MABC and ParaPA algorithm, can increase the path diversity in a complex environment, and the proposed structure can be better exploited locally for better convergence results.

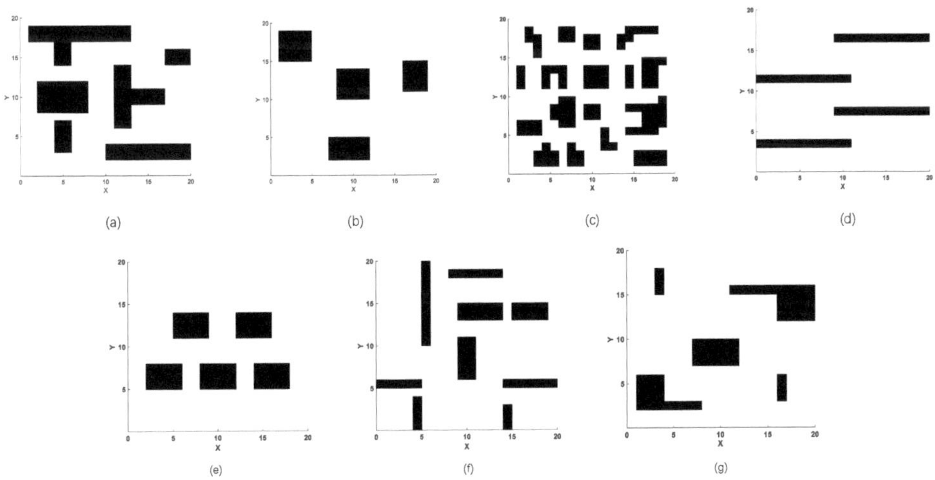

Figure 5. Set of test 20 × 20 maps for the first scenario in which (**a–g**) show the different distributions of obstacles on 20 × 20 maps.

Table 2 shows the success rate. Note that the multiple population strategy proposed in this paper, combined with the parasitic relationship of two communities, can increase the probability of a successful search for the special continuous transverse obstacles in Map 4. In most cases, the algorithm can successfully find a suitable path. The results, however, are different in terms of the algorithm's efficiencies combined with data from Table 1. The structure of ParaPA can utilize the population differences in multiple swarms and transform them into the diversity of path samples so as to improve the effectiveness of the algorithm, and the diversity of the algorithm is expressed by the individual disparity in the population in the following:

$$DI^G = \frac{1}{NP}\sqrt{\sum_{i=1}^{NP}\left\|X_i^G - \frac{1}{NP}\sum_{j=1}^{NP}X_j^G\right\|^2} \quad (18)$$

where DI is the individual diversity [34], G means generation, and NP is the number of populations.

Table 1. Convergence performance comparison for scenario 1 on 20 ∗ 20 maps.

MAPs		ParaPA	PSO(ori)	PSO	CPSO	WPSO	MPSO	ACO	ABC	MABC
Map1	Mean	**28.531** [1]	32.311	30.059	29.919	30.079	32.644	38.002	28.667	28.583
	Std	0.143	4.245	2.767	1.851	1.719	8.640	0.063	0.170	**0.042**
	Worst	29.825	58.069	89.130	40.225	38.921	212.317	40.000	31.328	**28.802**
	Optima	**28.455**	29.228	28.459	28.459	28.455	28.475	38.000	28.480	28.480
Map2	Mean	**27.024**	32.526	28.041	28.007	28.009	27.496	38.032	27.044	27.033
	Std	**0.005**	5.596	1.116	1.077	1.015	0.492	0.251	0.018	0.006
	Worst	27.098	53.125	36.749	34.031	32.926	30.317	41.000	27.173	**27.060**
	Optima	**27.022**	27.295	27.022	27.022	27.022	27.025	37.000	27.022	27.022
Map3	Mean	**28.265**	39.905	31.822	30.668	30.541	30.171	38.000	29.777	28.670
	Std	0.419	7.300	6.352	2.084	1.885	2.056	**0.000**	1.542	0.571
	Worst	**30.495**	84.383	99.917	41.312	39.035	42.676	38.000	37.597	33.039
	Optima	**27.848**	29.482	27.882	27.900	27.914	27.994	38.000	28.102	28.170
Map4	Mean	57.444	**47.617**	72.308	52.553	NaN	82.221	81.612	84.206	73.934
	Std	14.674	**3.723**	14.388	5.138	NaN	16.069	17.741	15.514	13.658
	Worst	96.037	69.024	108.846	**60.659**	NaN	114.122	150.000	113.900	114.461
	Optima	41.738	40.665	46.061	**38.334**	NaN	55.814	50.000	56.989	48.293
Map5	Mean	**28.922**	36.411	31.515	31.430	31.538	29.455	38.004	30.795	29.431
	Std	1.795	7.086	2.341	2.370	2.238	1.845	**0.089**	2.635	1.627
	Worst	33.343	69.188	41.791	42.173	39.855	36.349	40.000	39.659	**32.952**
	Optima	27.290	27.828	27.297	27.302	27.295	**27.275**	38.000	27.376	27.395
Map6	Mean	**28.482**	37.455	29.827	29.868	29.787	29.600	38.000	29.899	29.117
	Std	0.379	6.885	1.434	1.276	1.017	1.598	**0.000**	0.532	0.119
	Worst	**29.487**	63.969	53.169	37.791	37.399	73.792	38.000	32.878	29.724
	Optima	**27.982**	29.696	28.046	28.001	28.041	28.421	38.000	28.936	29.111
Map7	Mean	30.918	37.452	33.140	33.169	32.956	32.700	38.038	33.414	**30.018**
	Std	1.916	3.087	3.128	3.238	3.036	2.730	**0.273**	2.795	0.388
	Worst	37.353	54.301	47.069	46.973	42.344	**29.084**	40.000	40.827	34.502
	Optima	29.139	29.594	29.151	**29.127**	29.136	42.175	38.000	29.385	29.285

[1] The bold represents the best value.

Table 2. Success rate on various maps.

	Maps	ParaPA	PSO(ori)	PSO	CPSO	WPSO	MPSO	ACO	ABC	MABC
Maps 20∗20	Map1	1	0.988	0.998	0.998	0.999	0.998	1	1	1
	Map2	1	1	1	1	1	1	1	1	1
	Map3	1	0.493	0.871	0.94	0.944	0.966	1	1	1
	Map4	0.499	1	0.07	0.041	0	0.04	0.83	0.016	0.237
	Map5	1	1	1	1	1	1	1	1	1
	Map6	1	0.361	0.944	0.928	0.926	0.995	1	1	1
	Map7	1	1	0.966	0.969	0.936	1	1	1	1
Maps 50∗50	Map1	1	0.131	0.616	0.794	0.617	0.98	1	1	1
	Map2	1	0.352	0.958	0.793	0.947	0.997	0.836	1	1

5.3.2. Scenario 2: Path Planning on 50 × 50 Maps with Randomly Generated Rectangles as Obstacles

While the merits of diversity are not obvious in the simple condition. To verify the convergence and searching capacity of different algorithms, more complex environments are built with randomly generated rectangles in 50 × 50 maps, as shown in Figure 6. In terms of the success rate from Table 2, ParaPA, ABC, and MABC are the only algorithms that can guarantee the finding of a feasible path, while Table 3 shows the convergence performance of the algorithm in the 50 × 50 environments. It is worth noting that the most stable algorithm is MABC, while the ParaPA algorithm makes some stability sacrifices in order to obtain a better search breadth in return for more accurate local convergence. This can be seen from the best value of optima in each map, indicating that the ParaPA algorithm has a stronger local searching ability. The standard deviation reflects the stability of the algorithm's performance in a scenario, while the value of DI is an indicator of the diversity of the algorithm. In Table 4, the above algorithms with better performance are selected to compare the diversity. The standard deviations of CPSO and WPSO are larger compared with others. Meanwhile, from the performance of DI, the algorithm does not construct more feasible paths during chaos search, resulting in poorer sample diversity. The proposed structure performs chaos at the bottom to create more feasible paths, and the superior individuals are responsible for fine-tuning the optima. Combining the optima value in Table 3 and the DI performance in Table 4 shows the capacity of the ParaPA algorithm both in local convergence accuracy and global exploration.

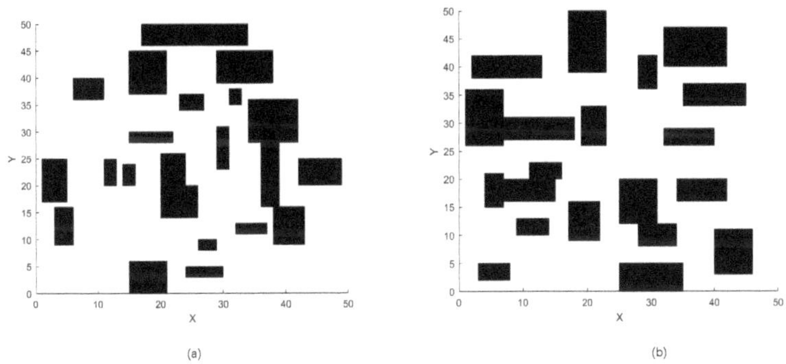

Figure 6. Set of test 50 × 50 maps for the second scenario in which (**a**,**b**) show the different distributions of obstacles on 50 × 50 maps.

Table 3. Convergence performance on 50 × 50 maps.

MAPs		ParaPA	PSO(ori)	PSO	CPSO	WPSO	MPSO	ACO	ABC	MABC
Map1	Mean	**74.776**	94.889	87.125	96.964	86.833	80.842	160.233	82.594	79.882
	Std	3.005	21.593	24.458	39.288	20.584	7.851	22.051	3.939	**1.778**
	Worst	103.862	200.684	295.924	315.496	219.431	223.645	266.000	106.806	**86.220**
	Optima	**70.107**	72.954	72.299	73.702	72.403	70.300	114.000	74.303	74.052
Map2	Mean	**76.558**	91.572	83.963	97.128	85.062	78.558	167.916	80.015	78.372
	Std	2.671	14.287	13.566	38.198	16.207	4.347	25.126	2.818	**0.981**
	Worst	**83.936**	152.965	255.782	299.138	238.817	105.914	320.000	96.480	87.981
	Optima	71.287	71.959	72.946	74.968	73.867	**71.361**	116.000	75.309	75.876

Table 4. Diversity of individuals (DI) on 50 × 50 maps.

50 * 50 Maps	ParaPA	CPSO	WPSO	MPSO	MABC
Map1	**8.897**	5.073	4.279	6.776	7.021
Map2	**9.338**	7.523	5.618	7.013	5.077

The convergence curves in 50 × 50 maps for the compared algorithms are shown in Figure 7 to observe the convergence variation during an evolution where the maximum number of evolutions is set to 4000. The frequency and magnitude of the changes in the path length are small enough to be ignored after approximately 300 iterations. Hence only the first 400 iterations are taken into account, as the results shown in Figure 7a. It can be seen that all algorithms complete the planning for a feasible path within 60 iterations, after which the best value is planned locally around it, and the changing magnitude is reduced. Note that the best path is recorded based on the parasitism swarm, indicating that the best position has been locked by the inferior population at around 400 iterations. After that, what needs to be performed for the algorithm is to maintain the diversity to create more possibilities for path convergence. The DI is measured in 4000 generations to observe the tendency, as shown in Figure 7a. It can be seen that it is difficult to maintain a stable diversity for CPSO as well as WPSO, which means that their convergence tendency will fall into a unified position. ParaPA and MABC are more stable in terms of diversity performance.

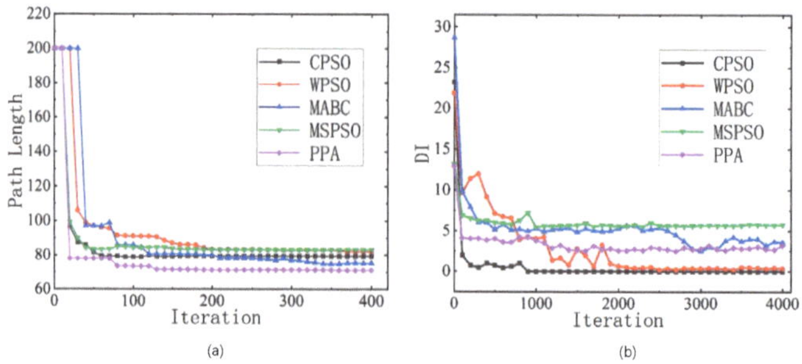

Figure 7. Convergence comparison on 50 × 50 maps where (**a**) is the path length change curve during iteration and (**b**) shows the curves of the diversity of individuals (DI) within 4000 iterations.

5.3.3. Scenario 3: Path Planning for Real Application on 100 × 100 Maps

In practical map applications, as illustrated in Figure 8 [35], this paper chooses algorithms with a better performance in the above test to make a comparison.

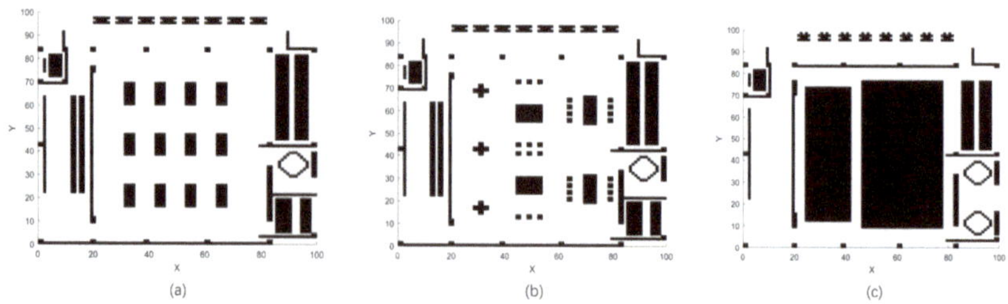

Figure 8. Real application on 100 × 100 maps of the Small Cultural Complex where (**a–c**) are the layouts corresponding to the function of the exhibition, sports, and performance, respectively.

The results in Tables 5 and 6 show that the ABC series algorithm is able to create more sample diversity but hardly converts the recourses to the final convergence accuracy. For the ParaPA algorithm with a parasitic relationship, the bottom particles can guarantee the search ability, but the "nutrients" they can pass to the superior layer would be reduced. It can be solved by building multiple population interactions to increase the diversity in the upper layer population directly. From the test on Maps 2 and 3, MABC has a better performance in terms of stability. In other words, MABC does not have any internal mechanism to refine the local search, so the higher stability can be regarded as the outcome at the sacrifice of the exploitation ability as the optima result in Table 5. Consequently, the proposed algorithm has a better performance on most of the measurements, but how to choose an algorithm should consider the specific requirements of the real application. Furthermore, in order to show the results of multimodal path planning more intuitively, we take map (a) in Figure 8 as an example and show the multiple paths planned by the single-run algorithm in Figure 9. In a single run, it generates the three paths with the best fitness values, as shown in Figure 9.

Table 5. Convergence performance on 100 × 100 maps.

100 ∗ 100 Maps		ParaPA	CPSO	MPSO	ABC	MABC
Map1	Mean	**147.727**	157.103	155.820	167.317	161.270
	Std	5.702	8.130	7.914	7.717	**4.940**
	Worst	**168.755**	203.790	198.764	206.522	177.713
	Optima	**141.006**	142.508	142.262	148.932	148.287
Map2	Mean	**145.878**	154.001	153.771	161.662	157.390
	Std	9.356	10.786	10.398	5.013	**4.021**
	Worst	347.053	263.723	347.306	188.500	**169.377**
	Optima	**139.750**	141.253	141.306	145.790	145.765
Map3	Mean	**166.445**	171.323	176.595	171.036	169.858
	Std	18.047	2.486	12.6525	4.1758	**0.481**
	Worst	554.350	175.755	335.206	255.509	**170.056**
	Optima	**165.459**	168.992	165.795	168.958	169.838

Table 6. Diversity of individuals (DI) on 100 × 100 maps.

100 ∗ 100 Maps	ParaPA	CPSO	MPSO	ABC	MABC
Map1	7.171	6.973	9.249	12.728	**13.409**
Map2	5.601	6.034	8.104	13.430	**14.664**
Map3	4.078	8.424	9.129	**17.583**	10.043

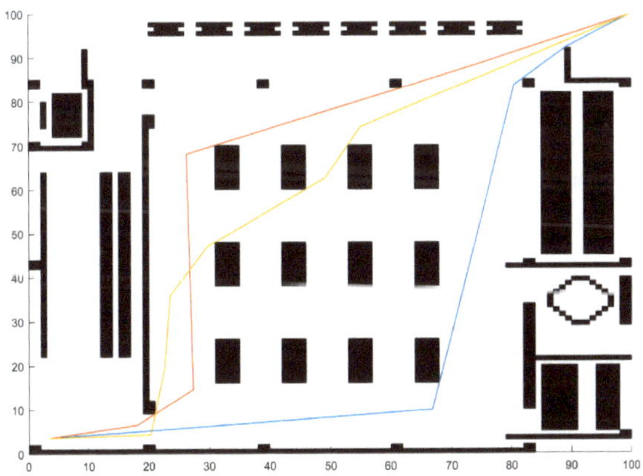

Figure 9. Path diagram in which colorful lines represent three routes obtained by a signle run on a 100 × 100 map.

6. Conclusions

Primarily, this paper introduces a dual-community-based evolutionary algorithm model that mimicked parasitic relationships in the biosphere. In the established structure, the bottom particles guarantee the algorithm diversity, while the superior community controls the better convergence resources and is responsible for completing the local convergence. Meanwhile, a multiple population strategy is conducted that is utilized to build information interaction channels among superior populations and directly shares the best information contained in superior populations. Finally, based on the above algorithm structure, ParaPA is proposed to solve the path planning problem. Meanwhile, we design a cross function to filter the high-level interactive information through the proposed multiple swarm framework so as to ensure path diversity in the superior population. From the comparison of the algorithm diversity and the average length of paths, the proposed approach is able to produce more path possibilities by using the structure, achieving a better solution in regard to path length.

Compared with traditional path planning strategies, such as A* algorithm, because of the characteristics of ABC and PSO algorithms, the ParaABC algorithm can effectively solve high-dimensional path planning problems. In the process of algorithm operation, once the target point is unreachable, the proposed algorithm can discard the current path directly or re-plan the infeasible section according to prior knowledge, thus saving performance loss. In addition, the ParaABC algorithm can plan multiple optimal paths through a single operation. Compared with the popular method of neural networks in recent years, the proposed method only needs to adjust a few parameters and has low equipment requirements. However, although the performance of ParaABC is improved compared with similar algorithms, it still faces the problems of diversity disappearance and premature convergence.

In the future, the three symbiosis-relation-based evolutionary structures introduced in this paper could be studied for different applications. Under the mutualism relationship, the convergence probabilities of multiple populations are equivalent, similar to the ring topology, which is suitable for multimodal or multi-objective optimization problems. For the commensal relationship, the interspecies relationship is inclined to a competition that has a continuous dynamic optimization capability. While in the parasitic relationship, where the convergence will tend to be homogeneous during the evolution, the avoidance of the premature problem is required. Note that the symbiosis relationship construction establishes an information bridge among multiple populations rather than a specific individual. Meanwhile, the development of symbiosis-relation-based evolutionary conception for different applications through the adapted framework is an interesting field that may be undertaken in the future.

Author Contributions: Conceptualization, H.R. and X.S.; Methodology, L.G. and X.S.; Software, L.G. and X.S.; Validation, H.R. and L.G.; Formal analysis, H.R. and X.S.; Investigation, H.R., L.G., X.S. and M.L.; Resources, H.R. and W.J.; Data curation, H.R., L.G., X.S. and M.L.; Writing—original draft, H.R., L.G. and X.S.; Writing—review and editing, H.R. and L.G.; Visualization, L.G., X.S. and M.L.; Supervision, H.R. and W.J.; Project administration, H.R.; Funding acquisition, H.R. and W.J. All authors have read and agreed to the published version of the manuscript.

Funding: This research was funded by the National Key Research and Development Funding grant number 2018YFB1403703.

Institutional Review Board Statement: Not applicable.

Informed Consent Statement: Not applicable.

Data Availability Statement: Not applicable.

Conflicts of Interest: This manuscript has not been published and is not under consideration for publication elsewhere. The authors have no conflicts of interest to disclose.

Appendix A

According to the description in [29], the commensalism and mutualism relationship should be defined as follows. All the relationships are defined under the multimodal optimization condition.

Commensalism relationship: The weaker population should give resources to the superior population while it is not affected by the interaction.

Mutualism relationship: The evolutional process will not be interrupted by others among two or more populations and can obtain benefits from group interaction. In other words, each evolution is independent while the information changing among different populations can contribute to their convergence.

We define the relationships in mathematics as:

Definition A1. *Suppose there are m optima in S, recorded as* $\Phi_1, \Phi_2, \ldots, \Phi_m$*. There exists* $S_1 \subsetneq S$, $S_2 \subsetneq S$, $S_1 \neq S_2$*. In the t-th iteration,* S_1 *and* S_2 *are recorded as* $S_1(t)$ *and* $S_2(t)$*. If the following function is true*

$$\lim_{t\to\infty}\prod_{i=1}^{m} P(\Phi_i \in S_1(t)) > \lim_{t\to\infty}\prod_{i=1}^{m} P(\Phi_i \in S_2(t))$$

then the S_1 *and* S_2 *is the commensalism relationship in the living space S.*

Definition A2. *Suppose there are m optima in S, recorded as* $\Phi_1, \Phi_2, \ldots, \Phi_m$*.* $\exists S_1 \subsetneq S$*, when the iteration is t and* S_1 *can be recorded as* $S_1(t)$*, then* $\exists P_1(t)$, $0 < P_1(t) < 1$*, satisfies*

$$\lim_{t\to\infty}\prod_{i=1}^{m} P(\Phi_i \in S_1(t)) = P_1(t)$$

and for $\exists S_2 \subsetneq S$, $S_1 \neq S_2$, $\exists P_2(t), 0 < p_2(t) < 1$ *satisfies*

$$\lim_{t\to\infty}\prod_{i=1}^{m} P(\Phi_i \in S_2(t)) = P_2(t)$$

For $\forall \delta > 0$*, always have*

$$\lim_{t\to\infty}\|P_1(t) - P_2(t)\| < \delta$$

called the S_1 *and* S_2 *is the mutualism relationship in the living space S.*

References

1. Bounini, F.; Gingras, D.; Pollart, H.; Gruyer, D. Modified artificial potential field method for online path planning applications. In Proceedings of the 2017 IEEE Intelligent Vehicles Symposium (IV), Los Angeles, CA, USA, 11 June 2017.
2. Yan, M.; Li, S.; Chan, C.A.; Shen, Y.; Yu, Y. Mobility Prediction Using a Weighted Markov Model Based on Mobile User Classification. *Sensors*, **2021**, *21*, 1740. [CrossRef] [PubMed]
3. Chang, J.R.; Jheng, Y.H.; Chang, C.H.; Lo, C.H. An Efficient Algorithm for Vehicle Guidance Combining Dijkstra and A* Algorithm with Fuzzy Inference Theory. *J. Internet. Technol.* **2015**, *16*, 189–200.
4. Xiong, C.; Chen, D; Lu, D. Path planning of multiple autonomous marine vehicles for adaptive sampling using Voronoi-based ant colony optimization. *Robot. Auton. Syst.* **2019**, *115*, 90–103. [CrossRef]
5. Rashid, A. T.; Ali, A. A,; Frasca, M. Path planning with obstacle avoidance based on visibility binary tree algorithm. *Robot. Auton. Syst.* **2013**, *61*, 1440–1449. [CrossRef]
6. Gonzalez, R., Kloetzer, M.; Mahulea, D. Comparative study of trajectories resulted from cell decomposition path planning approaches. In Proceedings of the 2017 21st International Conference on System Theory, Control and Computing (ICSTCC), Sinaia, Romania, 19 October 2017.
7. Song, B.; Wang, Z.; Zou, L. On global smooth path planning for mobile robots using a novel multimodal delayed PSO algorithm. *Cogn. Comput.* **2017**, *9*, 5–17. [CrossRef]
8. Yen, C.T; Cheng, M.F. A study of fuzzy control with ant colony algorithm used in mobile robot for shortest path planning and obstacle avoidance. *Microsyst. Technol.* **2018**, *24*, 125–135. [CrossRef]
9. Rajput, U.; Kumari, M. Mobile robot path planning with modified ant colony optimisation. *Int. J. Bio Inspired Comput.* **2017**, *9*, 106–113. [CrossRef]
10. Lamini, C.; Benhlima, S.; Elbekri, A. Genetic algorithm based approach for autonomous mobile robot path planning. *Procedia Comput. Sci.* **2018**, *127*, 180–189. [CrossRef]
11. Zhang, Y.; Li, S.; Guo, H. A type of biased consensus-based distributed neural network for path planning. *Nonlinear Dynam.* **2017**, *89*, 1803–1815. [CrossRef]
12. Yan, M.; Yuan, H.; Xu, J.; Yu, Y.; Jin, L. Task allocation and route planning of multiple UAVs in a marine environment based on an improved particle swarm optimization algorithm. *Eurasip. J. Adv. Sig. Pr.* **2021**, *94*, 2021. [CrossRef]
13. Yan, F. Autonomous vehicle routing problem solution based on artificial potential field with parallel ant colony optimization (ACO) algorithm. *Pattern Recogn. Lett.* **2018**, *116*, 195–199. [CrossRef]
14. Mo, H.; Xu, L. Research of biogeography particle swarm optimization for robot path planning. *Neurocomputing* **2015**, *148*, 91–99. [CrossRef]

15. Montiel, O.; Orozco-Rosas, U.; Sepúlveda, R. Path planning for mobile robots using bacterial potential field for avoiding static and dynamic obstacles. *Expert. Syst. Appl.* **2015**, *42*, 5177–5191. [CrossRef]
16. Oleiwi, B.K.; Al-Jarrah, R.; Roth, H. Multi objective optimization of trajectory planning of non-holonomic mobile robot in dynamic environment using enhanced GA by fuzzy motion control and A*. In Proceedings of the International Conference on Neural Networks and Artificial Intelligence 2014, Brest, Belarus, June 3-6, 2014.
17. Ajeil, F.H.; Ibraheem, I.K.; Sahib, M.A. Multi-objective path planning of an autonomous mobile robot using hybrid PSO-MFB optimization algorithm. *Appl. Soft. Comput.* **2020**, *89*, 1–13. [CrossRef]
18. Zafar, M.N.; Mohanta, J.C. Methodology for path planning and optimization of mobile robots: A review. *Procedia Comput. Sci.* **2018**, *133*, 141–152. [CrossRef]
19. Gharehchopogh, F.S.; Gholizadeh, H. A comprehensive survey: Whale Optimization Algorithm and its applications. *Swarm. Evol. Comput.* **2019**, *48*, 1–24. [CrossRef]
20. Yang, X.S.; He, X. Bat algorithm: Literature review and applications.*Int. J. Bio Inspired Comput.* **2013**, *5*, 141–149. [CrossRef]
21. Contreras-Cruz, M.A.; Ayala-Ramirez, V.; Hernandez-Belmonte, U.H. Mobile robot path planning using artificial bee colony and evolutionary programming. *Appl. Soft. Comput.* **2015**, *30*, 319–328. [CrossRef]
22. Neshat, M.; Adeli, A.; Sepidnam, G. A review of artificial fish swarm optimization methods and applications. *Int. J. Smart. Sens. Int.* **2012**, *5*,107–148. [CrossRef]
23. Qu, C.; Gai, W.; Zhang, J. A novel hybrid grey wolf optimizer algorithm for unmanned aerial vehicle (UAV) path planning. *Knowl-Based. Syst.* **2020**, *194*, 1–14. [CrossRef]
24. Li, X.; Yang, G. Artificial bee colony algorithm with memory. *Appl. Soft. Comput.* **2016**, *41*, 362–372. [CrossRef]
25. Wang, H.; Wang, W.; Xiao, S.; Cui, Z.; Xu, M.; Zhou, X.. Improving artificial Bee colony algorithm using a new neighborhood selection mechanism. *Inform. Sci.* **2020**, *527*, 227–240. [CrossRef]
26. Zhou, X.; Lu, J.; Huang, J.; Zhong, M.; Wang, M. Enhancing Artificial Bee Colony Algorithm with Multi-elite Guidance. *Inform. Sci.* **2020**, *543*, 242–258. [CrossRef]
27. Cheng, M.Y.; Prayogo, D. Symbiotic organisms search: A new metaheuristic optimization algorithm. *Comput. Struct.* **2014**, *139*, 98–112. [CrossRef]
28. Ezugwu, A. E; Prayogo, D. Symbiotic organisms search algorithm: Theory, recent advances and applications. *Expert. Syst. Appl.* **2019**, *119*, 184–209. [CrossRef]
29. Ren, H.; Shen, X.; Jia, X. A novel dual-biological-community swarm intelligence algorithm with a commensal evolution strategy for multimodal problems.*J. Supercomput.* **2021**, *77*, 10850–10895. [CrossRef]
30. Tharwat, A.; Elhoseny, M.; Hassanien, A.E. Intelligent Bézier curve-based path planning model using Chaotic Particle Swarm Optimization algorithm. *Cluster Comput.* **2019**, *22*, 1–22. [CrossRef]
31. Tuncer, A.; Yildirim, M. Dynamic path planning of mobile robots with improved genetic algorithm.*Cluster Comput.* **2012**, *38*, 1564–1572. [CrossRef]
32. Melo W. D. ; Jorge D.; Marques V. Low-cost thermal explorer robot using a hybrid neural networks and intelligent bug algorithm model. *Int. J. Comput. Appl. T.* **2021**, *65*, 245–252. [CrossRef]
33. Nie, Z.; Yang, X.; Gao, S. Research on autonomous moving robot path planning based on improved particle swarm optimization. In Proceedings of the 2016 IEEE Congress on Evolutionary Computation (CEC), Vancouver, BC, Canada, 24 July 2016.
34. Tang, L.; Dong, Y.; Liu, J. Differential evolution with an individual-dependent mechanism. *IEEE. Trans. Evolut. Comput.* **2014**, *19*, 560–574. [CrossRef]
35. Liu, Z.; Tian, H.; Wang, Y.; Wu, Z. Talking about the Construction Requirements of Multi-functional Basic-level Cultural Service Complex. *Eniertainment Technol.* **2020**,*10*, 66–72.

sensors

MDPI

Article

Indoor Positioning Design for Mobile Phones via Integrating a Single Microphone Sensor and an H_2 Estimator

Yung-Hsiang Chen [1], Pei-Yu Chang [2] and Yung-Yue Chen [3,*]

1 Department of Mechanical Engineering, National Pingtung University of Science and Technology, Pingtung 912301, Taiwan
2 National Chung-Shan Institute of Science and Technology, Taoyuan 32546, Taiwan
3 Department of Systems and Naval Mechatronic Engineering, National Cheng Kung University, Tainan 701401, Taiwan
* Correspondence: yungyuchen@mail.ncku.edu.tw; Tel.: +886-2757575 (ext. 63541)

Abstract: An indoor positioning design developed for mobile phones by integrating a single microphone sensor, an H_2 estimator, and tagged sound sources, all with distinct frequencies, is proposed in this investigation. From existing practical experiments, the results summarize a key point for achieving a satisfactory indoor positioning: The estimation accuracy of the instantaneous sound pressure level (SPL) that is inevitably affected by random variations of environmental corruptions dominates the indoor positioning performance. Following this guideline, the proposed H_2 estimation design, accompanied by a sound pressure level model, is developed for effectively mitigating the influences of received signal strength (RSS) variations caused by reverberation, reflection, refraction, etc. From the simulation results and practical tests, the proposed design delivers a highly promising indoor positioning performance: an average positioning RMS error of 0.75 m can be obtained, even under the effects of heavy environmental corruptions.

Keywords: indoor positioning design; sound pressure level; received signal strength; H_2 estimator; energy consumption

Citation: Chen, Y.-H.; Chang, P.-Y.; Chen, Y.-Y. Indoor Positioning Design for Mobile Phones via Integrating a Single Microphone Sensor and an H_2 Estimator. *Sensors* **2023**, *23*, 1508. https://doi.org/10.3390/s23031508

Academic Editors: Chien Aun Chan, Chunguo Li and Ming Yan

Received: 28 December 2022
Revised: 26 January 2023
Accepted: 27 January 2023
Published: 29 January 2023

1. Introduction

Indoor positioning designs for mobile communication systems have attracted increasing attention recently due to emergency and security concerns. With the increase in time spent by mobile phone users in indoor environments, a mandatory bill, called Enhanced 911 (E911) was passed by the USA, providing the specific requirements for the indoor positioning accuracy of mobile communication systems. The indoor positioning accuracy requirements of E911 are less than 100 m (67% calls) and 300 m (95% calls) [1]. However, the well-known outdoor positioning system, GPS, is not useful for guaranteeing the accuracy of indoor positioning because of the shielding effect of the positioning signal transmission of satellites, i.e., the positioning ability of GPS is weak in regards to indoor environments such as shopping malls, hypermarkets, office building, etc. A variety of theoretical and available technologies have been proposed in recent decades for achieving the requirements of the indoor positioning [2–5], based on methods using infrared rays (IR), ultrasound, radio-frequency identification (RFID), wireless local area networks (WLAN), Bluetooth, audible sound, and other technologies. Nevertheless, not all of the above-mentioned methods can be applied to mobile communication systems due to complexity of implementation and the costs of the hardware and software; hence, a new design which is suitable for performing indoor positioning, based on the integration of a single microphone of the mobile communication system and one positioning estimation design. is proposed in this investigation.

Many indoor positioning methods have been proposed in recent years, and an accurate positioning estimation is the common goal of these designs. The earliest design was based on an infrared ray (IR) system [5–9], and until now, it was the most simple and common design for the purpose of indoor positioning. It is possible that the indoor positioning performance of these types of IR-based designs is acceptable in indoor environments which are well-arranged. However, practically, environmental light sources, such as florescent light and sunlight, are always tricky problems which strongly reduce position accuracy [8]. To compensate for the effect of this environment disturbance, several filters have been developed [8,10]. Applications of IR-based indoor positioning designs are constrained due to these interferences, and constructing such a light-based system comes at a high cost. Radio frequency (RF)-based technologies [11,12] are the other designs used for achieving the indoor positioning goal. Categories of RF-based positioning methods are mainly divided into two groups: 1. the radio frequency identification (RFID) method, and 2. the Wireless Local Area Network (WLAN) method. WhereNet is a real-time indoor and outdoor positioning design developed by Zebra Technologies via the RFID method for users who are in intricate environments, such as libraries, offices, etc. [4,13]. Key parts comprising this positioning system include tags, positioning antennas, processors, servers, where ports, and a software algorithm using the differential time of arrival method (TDOA) for calculating the locations of moving tags.

The major disadvantage of this kind of positioning method is that is necessitates an enormous cost for building numerous infrastructures in the working area. As to the WLAN positioning design, most of WLAN-based algorithms are developed via adopting the received signal strength of the WLAN signals. Generally speaking, WLAN-based designs possess a low-cost feature due to the popularity of WLAN infrastructures in indoor environments. By building on this advantage, a RADAR positioning system has been developed by a Microsoft research group, combining received signal strength detection with the triangulation positioning method. Unfortunately, the received signal strength of WLAN is naturally and inevitably affected by various environmental uncertainties, such as the multipath effect, the no line-of-sight effect [14–17], etc.; hence, some auxiliary designs are proposed to reduce the effect of environmental uncertainties [15,18,19], e.g., a radio map using the fingerprinting method, or the approximation design of a specific environment using the fuzzy logic approach. Fingerprinting techniques work well when the stored information of WLAN access points (APs) increase significantly, i.e., more APs are needed [20]. Algorithms based on sound detection provide another potential method for indoor positioning designs. The Active Bat system was developed by AT&T Cambridge by mimicking the navigation behavior of bats [21].

For improving the accuracy of the Active Bat system, a new design called the Cricket system combines the overall design of the Active Bat system with an extra RF method [22,23]. For the above developments, multi-sensors and ultrasound designs are adopted. Daredevil, developed by Microsoft, uses audible sound sources, providing an indoor positioning ability for mobile phones with at least two embedded microphones [4]. In the work in [14], the mobile phone is used as an emitter, and it collocates with the Wi-Fi network to achieve higher accuracy. A contrasting design, in which the handheld devices are arranged as receivers for some predefined emitters, is another popular design because mobile phone users always need the real-time display of the mobile phones' monitors [24,25]. The positioning algorithm based on TOA or TDOA methods requires that multi-sensors be used, i.e., the total costs will be higher than those using the single sensor design; besides, the TOA or TDOA methods are degraded by four main factors: 1. background noise, 2. the multipath effect [26,27], 3. non line-of-sight propagation [28,29], and 4. mobile synchronization recovery [30,31].

Therefore, an indoor positioning algorithm which can be easily implemented in a mobile phone using just one microphone is proposed. Unlike in the above design, in which mobile phones were set up as emitters, in this study, the mobile phone's microphone is arranged as the receiver of tagged sound sources using distinct frequencies. Four tagged sound sources with distinct frequencies are broadcasted from speakers placed in the corners of an indoor space, and for reducing the total cost of the hardware, only one mobile phone microphone is used to collect messages of different tagged sound sources. Regarding the development of the methodology, one novel indoor positioning method combining the received signal strength (RSS) method, the fast Fourier transform (FFT) method, H_2 estimation design, and the intersection of circles method is proposed. In this proposed method, RSS is used to calculate the strength of each tagged sound source, FFT is adopted to analyze the spectrum of the collected data of tagged sound sources, the H_2 estimation design is used to purify the noisy tagged sound source, and output denoised sound pressure level (SPL), and an accurately estimated position of the user in an indoor environment can then be solved by the intersection of circles method.

2. The Proposed Indoor Positioning Algorithm

The overall schematic of the proposed indoor positioning design is shown as Figure 1. In this investigation, this procedure is separated into three stages. A RSS analysis of the first stage is used to verify the pressure of the received signals measured by the single microphone of the mobile phones; additionally, in this stage, one system identification method—recursive least square (RLS)—and the famous power spectrum analysis tool—FFT—are utilized to simulate the walk behavior of users and make SPL selections regarding which two tagged sound sources will be adopted. In the second stage, one novel estimation design possesses an effective reduction ability regarding environmental corruptions. A set of purified SPLs can be obtained by using this proposed estimation design, and the first two of these four SPLs, which exhibit the strongest intensities, will be used as inputs of geometry equations in the third stages. Based on the purified SPLs, four geometry equations, which are functions of the users' indoor position $(x_r(k), y_r(k))$, can be easily determined from the relationship between tagged sound sources and the microphone of the user's mobile phone. The users' indoor position $(x_r(k), y_r(k))$ can be further solved by the intersection of circles method in the third stage.

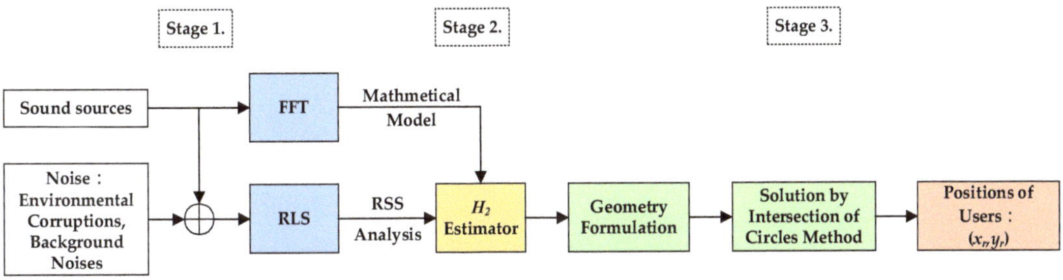

Figure 1. Flow chart of the proposed indoor positioning algorithm.

2.1. Data Acquisition and RSS Analysis

In the following, a detailed mathematical expression for these three stages will be derived. In the first stage, the received raw data of the four tagged sound sources are as follows: S_n, for n =1 to 4, with distinct tagged frequencies, should be identified. As to the choice of these two tagged sound sources, there are two steps arranged previously: calculations of magnitudes and frequencies. For identifying the magnitude and frequency, FFT and RSS analysis are used. The tagged frequency of each sound source cannot be higher than 20 kHz due to the physical limitation of the standard microphone used, and the sampling frequency of the mobile phone is 8 kHz; hence, tag frequencies of the sound sources are selected as 15.5 kHz, 16.5 kHz, 17.5 kHz, and 18.5 kHz, respectively. For separating the four tagged sound sources, a specific frequency tag f_{Sn} is assigned for each of the four tagged sound sources, as follows:

$$O_n : (14000 + 1000n)\text{Hz} < f_{Sn} < (15000 + 1000n)\text{Hz, for } n = 1 \text{ to } 4 \quad (1)$$

Denote the received signals of the tagged sound sources S_n as z_{Sn}, for n =1 to 4, which contain all surrounding sounds. In this stage, two tagged sound candidates will be selected based on the messages of magnitudes and frequencies of these four tagged sound sources. For the purpose of calculating magnitudes (dB) and frequencies (Hz) of the tagged sound sources, the famous power spectrum method FFT is used. According to the definition of FFT, the magnitude for these four distinct frequencies can be expressed as:

$$F_{Sn}(k) = \text{FFT}\{z_{Sn}(k)\}, \text{ for } n = 1 \text{ to } 4 \tag{2}$$

A mean value magnitude adopted for calculation of the intensities of the tagged sound sources is defined as the following:

$$C_{Sn} = \frac{1}{B}\sum_{\kappa=0}^{B} F_{Sn}(k), \text{ for } n = 1 \text{ to } 4 \tag{3}$$

where B is the number of bins. Equations (2) and (3) will be utilized to identify the received tagged sound sources.

2.2. Indoor Positioning Algorithm

The geometry relationship between the carried receiver and distinct tagged sound sources is illustrated in Figure 2. In Figure 2, the carried receiver is set up as the mobile phone and is initialized as $u_r(0) = \{x_r(0), y_r(0)\}$. Suppose at least four tagged sound sources, with distinct frequencies, are placed in four corners of an indoor space, and their coordinates are represented as (x_1, y_1), (x_2, y_2), (x_3, y_3), and (x_4, y_4).

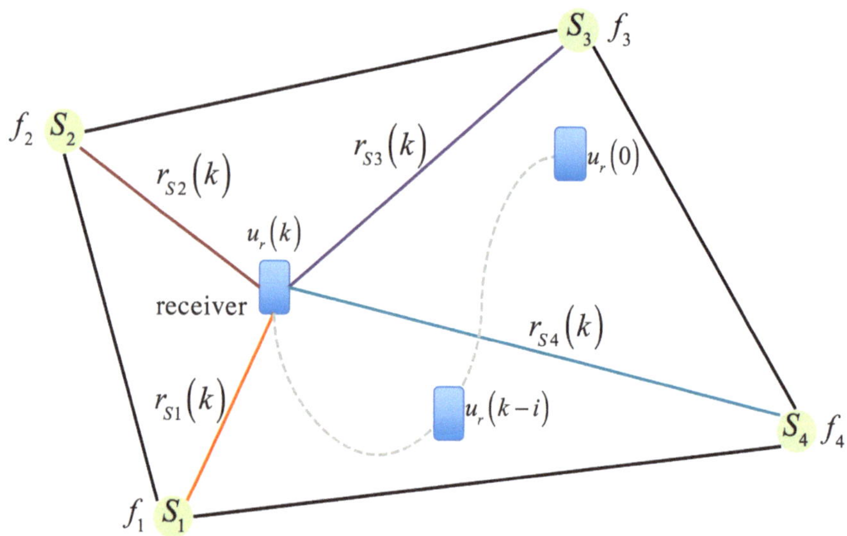

Figure 2. Geometric relationship between the carried receivers with respect to tagged sound sources.

Figure 2 is an irregular-shaped indoor plane, and there are four tagged sound sources placed in four corners. Tagged sound sources are denoted as S_n, for n =1 to 4, and tag frequencies of these sound sources are set up as f_{Sn}. As to the initial locations of the tagged sound sources, they are placed in (x_n, y_n) of the test indoor plane. The moving data of the carried receiver in every time instant is defined as:

$$u_r = \left\{(x_r(k), y_r(k)) \in R^2\right\}, \text{ for } 1 \le k \le N \tag{4}$$

where N is number of indoor positioning iterations.

In the following, the transformation of SPLs and the relative distances will be detailed.

Data collection of SPLs: The carried receiver collects four different SPL values from the tagged sound sources at each sampling time, which can be expressed as a set L_{Sn}:

$$L_{Sn} = \left\{ l_{Sn}(k) \in R^1 \right\}, \text{ for } n = 1 \text{ to } 4, \text{ and } 1 \leq k \leq N \tag{5}$$

where l_{Sn} is the measured SPL of S_n with the tag frequency f_{Sn}.

Relative distances between each tagged sound source and the receiver can be expressed as the following four sets.

$$R_{Sn} = \left\{ r_{Sn}(k) \in R^1 \right\}, \text{ for } n = 1 \text{ to } 4, \text{ and } 1 \leq k \leq N \tag{6}$$

where $r_{Sn}(k)$ is the relative distance from the positions of tagged sound sources S_n to the receiver. Based on Equations (5) and (6), the instantaneous relative distance between each tagged sound source and the receiver can be calculated from the SPL difference $l_{Sn}(k)$ and $l_{Sn}(k-1)$ as

$$r_{Sn}(k) = r_{Sn}(k-1)/10^{\frac{l_{Sn}(k) - l_{Sn}(k-1)}{20}} + \left(\frac{R_C + 16\pi r_2^2}{R_C + 16\pi r_1^2} \right)^{\frac{1}{2}}, \text{ for } n = 1 \text{ to } 4, \text{ and } 1 \leq k \leq N \tag{7}$$

where R_c is the uncertainty room constant. Due to the fact that R_c cannot be measured in prior, it is regarded as a modeling uncertainty. Based on this, the corrupted relative distance $r_{Sn}(k)$ can be further expressed as below:

$$r_{Sn}(k) = r_{Sn}(k-1)/10^{\frac{l_{Sn}(k) - l_{Sn}(k-1)}{20}} + w(k), \text{ for } n = 1 \text{ to } 4, \text{ and } 1 \leq k \leq N \tag{8}$$

where initial values of $r_{Sn}(k)$ and $l_{Sn}(k)$ are $r_{Sn}(0)$ and $l_{Sn}(0)$, respectively, and $w(k) = \left(\frac{R_C + 16\pi r_2^2}{R_C + 16\pi r_1^2} \right)^{\frac{1}{2}}$.

The next section will introduce the method of extracting the noiseless SPLs with distinct tagged frequencies and derive the geometry mathematical formulation for finding the instantons positions of the mobile phone users.

2.2.1. Geometric Mathematical Formulation

In Figure 3, there are four right angle triangles inside a quadrilateral, and each of them describes the relationship between the receiver and each tagged sound sources.

Based on Figure 3, four equations are derived, and these equations express the geometric relationships between the receiver and the tagged sound sources in an indoor plane.

$$(x_1 - x_r(k))^2 + (y_1 - y_r(k))^2 = r_{S1}(k)^2 \tag{9}$$

$$(x_2 - x_r(k))^2 + (y_2 - y_r(k))^2 = r_{S2}(k)^2 \tag{10}$$

$$(x_3 - x_r(k))^2 + (y_3 - y_r(k))^2 = r_{S3}(k)^2 \tag{11}$$

$$(x_4 - x_r(k))^2 + (y_4 - y_r(k))^2 = r_{S4}(k)^2 \tag{12}$$

where (x_1, y_1),(x_2, y_2),(x_3, y_3), and (x_4, y_4) are the coordinates of the tagged sound sources, and these positions are fixed and known. $(x_r(k), y_r(k))$ is the coordinate of the receiver at time instant k. $r_{Sn}(k)$, $n = 1$ to 4 are the relative distances of the receiver to the tagged sound sources S_n at time instant k and can be calculated from Equation (8).

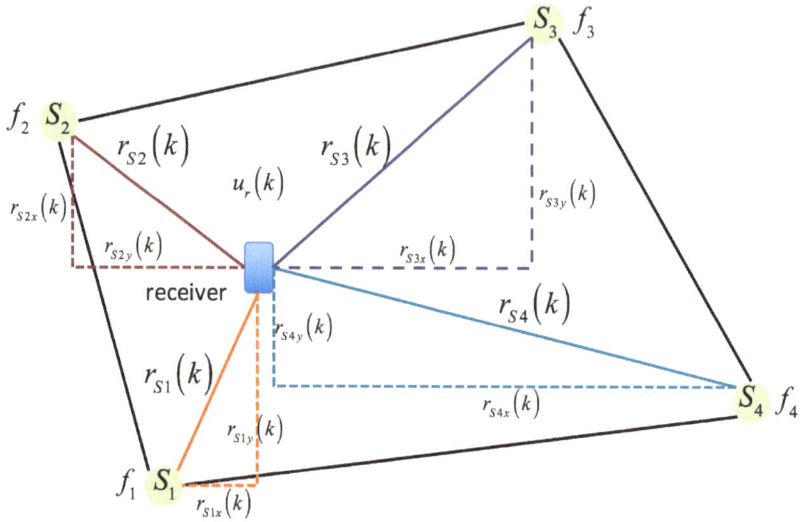

Figure 3. The geometric relationship between the tagged sound sources and the receiver in an indoor plane.

From the above mathematical formulation for the geometric relationship between the tagged sound sources and the receiver, four equations are obtained. However, only two of these four equations will be utilized to solve the two unknowns:$(x_r(k), y_r(k))$, and two of these four equations—Equations (9)–(12)—will be selected according to the measured SPLs by using Equations (2) and (3). The criterion for selecting the two equations is $l_{Max} = Max\ L_{Sn}$ and $l_{\text{Sec}} = Max\{L_{Sn} - Max\ L_{Sn}\}$, e.g., if the intensity sequence of the measured SPLs is in sequence: $l_{S2} > l_{S3} > l_{S1} > l_{S4}$ based on Equations (3), (10) and (11), the following will then be selected and combined as a set of binary quadratic equations as:

$$(x_2 - x_r(k))^2 + (y_2 - y_r(k))^2 = r_{S2}(k)^2 \tag{13}$$

$$(x_3 - x_r(k))^2 + (y_3 - y_r(k))^2 = r_{S3}(k)^2 \tag{14}$$

2.2.2. Solution by Intersection of Two Circles Method

For solving the solution $(x_r(k), y_r(k))$ of the binary quadratic equations, the intersection of two circles method is applied. The reason for using this method is that it offers calculation convenience and a low computational burden for calculators of mobile phones.

In Figure 4, S_1 and S_2 are tagged sound sources which are selected from Figure 3. The coordinates are (x_1, y_1) and (x_2, y_2), respectively. Two overlapped circles can be illustrated when a radius $r_{S1}(k)$ and $r_{S2}(k)$, which are the relative distance between the receiver and the tagged sound source, are assigned for tagged sound sources S_1 and S_2, respectively. Suppose $r_{S1}(k)$ and $r_{S2}(k)$ are large enough to intersect with each other. Two intersect point, $u_{r1}(k)$ and $u_{r2}(k)$, can be obtained, as shown in Figure 4. Theoretically, one of these two points, $(u_{r1}(k)$ and $u_{r2}(k))$, would be the solution of the corresponding binary quadratic equations.

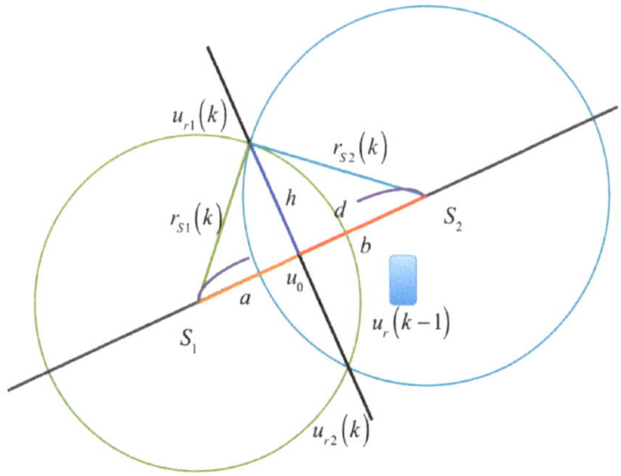

Figure 4. Intersection of two circles.

Denoting the distance $\overline{S_1S_2}$ as d, it can be expressed as:

$$d = a + b = \sqrt{(x_2 - x_1)^2 + (y_2 - y_1)^2} \tag{15}$$

where a is $\overline{S_1u_0}$ and b is $\overline{S_2u_0}$.

There are two equations obtained from triangles $\Delta S_1u_{r1}u_0$ and $\Delta S_2u_{r1}u_0$ as

$$\begin{array}{c} a^2 + h^2 = r_{S1}(k)^2 \\ b^2 + h^2 = r_{S2}(k)^2 \end{array} \tag{16}$$

Using $d = a + b$, the variable a can be solved as:

$$a = \left(r_{S1}(k)^2 - r_{S2}(k)^2 + d^2\right)/(2d) \tag{17}$$

The solution of the point $u_0(x_{u_0}, y_{u_0})$ can be obtained

$$(x_{u_0}, y_{u_0}) = (x_1, y_1) + \frac{a}{d}(x_2 - x_1, y_2 - y_1) \tag{18}$$

Substituting a into Equation (16) to solve h, the coordinate of the intersection point $u_r(k) = (x_r(k), y_r(k))$ can be obtained as below:

$$\begin{array}{c} x_r(k) = x_{u_0} \pm \frac{h}{d}(y_2 - y_1) \\ y_r(k) = y_{u_0} \pm \frac{h}{d}(x_2 - x_1) \end{array} \tag{19}$$

$$\begin{array}{c} x_r(k) = x_1 + \frac{a}{d}(x_2 - x_1) \pm \frac{\sqrt{r_{S1}(k)^2 - a^2}}{d}(y_2 - y_1) \\ y_r(k) = y_1 + \frac{a}{d}(y_2 - y_1) \pm \frac{\sqrt{r_{S1}(k)^2 - a^2}}{d}(x_2 - x_1) \end{array} \tag{20}$$

In Equation (20), two position solutions of the receiver are obtained simultaneously, but only one of them is the correct answer. There is a simple method to determine whether it is the correct one: the correct solution $u_r(k)$ must be inside the indoor plane.

However, there are some special cases in which two solutions of Equation (20) are both inside the quadrilateral $S_1S_2S_3S_4$. One judgment is proposed for the reasonable selection of these special cases by considering the moving velocity of the users. Normally, a moving velocity of 10 m/s is the maximum limitation for most of users. To calculate the moving velocity by the difference of current positioning solution $u_r(k)$ and the previous solution $u_r(k-1)$, one checking condition can found for the selection of the correct solution as:

$$\Delta u_r(k) = \frac{u_r(k) - u_r(k-1)}{t} \leq 10\frac{m}{s} \tag{21}$$

where t is the sampling time of the system.

The solution of Equation (20) based on the accurate measurement of SPLs $l_{Sn}(k)$, is $n = 1$ to 4. Naturally, environmental corruptions, such as reverberation, interference, etc., are always contained in the measurement process of SPLs, hence the SPLs $l_{Sn}(k)$ should be denoised before calculating Equations (8) and (20). For treating this noise reduction problem regarding the corrupted SPLs $l_{Sn}(k)$, an estimation design for effectively removing the environmental corruptions and purifying the SPLs $l_{Sn}(k)$ is proposed. In the following, a systematic estimation design, combining RLS system modeling and a steady-state optimal estimator, is developed for denoising the corrupted SPLs.

2.3. System Modeling

Before estimating the correct SPLs via adopting an optimal estimation method, each of the received SPLs should first be mathematically modeled. Suppose a set of the measured SPLs can be described as:

$$L_{all} = [l_{S1}, \ldots, l_{S4}] \tag{22}$$

where L_{all} is the set of all measured SPL data, and l_{Sn}, for n =1 to 4 is the set of measured SPL data from the tagged sound sources.

The mathematical models of the received SPLs can be expressed as the following white-noise driven difference equation:

$$l_{Sn}(k) = \sum_{i=1}^{m} a_{Sni} l_{Sn}(k-i) + w_{Sn}(k), \text{ for } n = 1, \ldots, 4 \tag{23}$$

$$Z_{Sn}(k) = l_{Sn}(k) + v_{Sn}(k) \tag{24}$$

where $Z_{Sn}(k)$ is the noisy measurement output of the tagged sound source n. Moreover, $w_{Sn}(k)$ and $v_{Sn}(k)$ are Gaussian white noise, with a zero mean and which are uncorrelated with $l_{Sn}(k)$. a_{Sni}, for $i = 1, \ldots, m$ are identifiable system parameters of the tagged sound source n, and m is the system order.

A regressive form is used to express the difference equation in Equation (23):

$$l_{Sn}(k) = \Psi_{Sn}(k)^T \hat{\lambda}_{Sn}(k) + w_{Sn}(k) \tag{25}$$

where $\Psi_{Sn}(k) = [l_{Sn}(k-1) \cdots l_{Sn}(k-m)]^T$ is the regression vector which contains the measured SPL data, and $\hat{\lambda}_{Sn}(k) = [a_{Sn1} \cdots a_{Snm}]$ is the parameter vector.

Equation (25) can be further formulated as a state space form, as follows:

$$L_{Sn}(k+1) = \Psi_{Sn} L_{Sn}(k) + \Lambda w_{Sn}(k) \tag{26}$$

$$Z_{Sn}(k) = \Omega L_{Sn}(k) + \Pi v_{Sn}(k) \tag{27}$$

where $\Psi_{Sn} = \begin{bmatrix} a_{Sn1} & a_{Sn2} & \cdots & a_{Snm-1} & a_{Snm} \\ 1 & 0 & \cdots & \cdots & 0 \\ 0 & \ddots & \ddots & \ddots & \vdots \\ \vdots & \ddots & \ddots & \ddots & \vdots \\ 0 & \cdots & 0 & 1 & 0 \end{bmatrix} \in \Re^{m \times m}, \Lambda = \begin{bmatrix} 1 \\ 0 \\ 0 \\ \vdots \\ 0 \end{bmatrix} \in \Re^{m \times 1}, \Omega = \begin{bmatrix} 1 \\ 0 \\ 0 \\ \vdots \\ 0 \end{bmatrix}^T \in$ $\Re^{1 \times m}$, $\Pi = 1$ and $L_{Sn}(k) = [l_{Sn}(k)\; l_{Sn}(k-1) \cdots l_{Sn}(k-m+2)\; l_{Sn}(k-m+1)]^T$ is the state vector, and $w_{Sn}(k)$ and $v_{Sn}(k)$ are white noises.

Remark 1. *The system order m is a user decided parameter. Theoretically, a more accurate model can be found by increasing this system order. Surely, a high order system needs more calculation time and storage memory. There exists a tradeoff when selecting the system parameter m. System parameters* a_{Sni}, for $i = 1$ *to m can be optimally calculated by several identification methods, such as the recursive least squares method (RLS), the stochastic subspace identification method (SSI), and the system realization using information matrix (SRIM). In this investigation, the system parameters* a_{Sni} *in Equation (23) will be optimally determined by the RLS method.*

The RLS algorithm is often used for searching the optimal parameters of systems with a set of unknown parameters by using the input and output measured raw data. The standard identification process of this algorithm can be expressed as follows:

$$\varepsilon_{Sn}(k) = Z_{Sn}(k) - \Psi_{Sn}(k)^T \hat{\lambda}_{Sn}(k-1) \tag{28}$$

$$\overline{Q}_{Sn}(k) = \frac{1}{\gamma}\left[\overline{Q}_{Sn}(k-1) - \frac{\overline{Q}_{Sn}(k-1)\Psi_{Sn}(k)\hat{\lambda}_{Sn}(k)^T\overline{Q}_{Sn}(k-1)}{\gamma + \Psi_{Sn}(k)^T\overline{Q}_{Sn}(k-1)\hat{\lambda}_{Sn}(k)}\right] \tag{29}$$

$$\hat{\lambda}_{Sn}(k) = \hat{\lambda}_{Sn}(k-1) + \overline{Q}_{Sn}(k)\Psi_{Sn}(k)\varepsilon_{Sn}(k) \tag{30}$$

where $\overline{Q}_{Sn}(k)$ is the estimation of the coefficient covariance at time instant k, $\hat{\lambda}_{Sn}(k)$ is the identified parameter, $\Psi_{Sn}(k)$ is the input data, $\varepsilon_{Sn}(k)$ is the prediction error, $Z_{Sn}(k)$ is the measurement output, and γ is the forgetting factor. The range of forgetting factor γ is usually given within 0.95 to 1.

2.4. H_2 Estimation Design

Based on the identified system parameters $\hat{\lambda}_{Sn}(k)$ in Equation (30), an H_2 estimator is proposed for eliminating the influence of the environmental corruptions.

Equations (26) and (27) represent the state-space system of measured SPLs, and the purified SPLs $h_{Sn}(k)$ can be reconstructed as:

$$h_{Sn}(k) = JL_{Sn}(k) \tag{31}$$

where J is a constant matrix that is set up to draw out the purified SPL $h_{Sn}(k)$ from state vector $L_{Sn}(k)$. The designed target is to hunt for the estimation $\hat{L}_{Sn}(k)$ from the measured SPL $Z_{Sn}(k)$, which is corrupted by environmental noises; hence, the state estimator for purifying the corrupted SPL is formulated as the following:

$$\begin{aligned} &\hat{L}_{Sn}(k+1) = \Psi_{Sn}\hat{L}_{Sn}(k) + G_{Sn}\left[Z_{Sn}(k) - \Omega\hat{L}_{Sn}(k)\right] \\ &\hat{h}_{Sn}(k) = J\hat{L}_{Sn}(k) \end{aligned} \tag{32}$$

where $G_{Sn} \in \Re^{m\times 1}$ is the designed estimation gain in a steady state.

Define the estimation error between purified SPL signal and estimation signal as follows:

$$\begin{aligned} \tilde{e}_{Sn}(k) &= h_{Sn}(k) - \hat{h}_{Sn}(k) \\ &= JL_{Sn}(k) - J\hat{L}_{Sn}(k) \\ &= J\tilde{L}_{Sn}(k) \end{aligned} \tag{33}$$

where $\tilde{L}_{Sn}(k) = L_{Sn}(k) - \hat{L}_{Sn}(k)$

The performance index of the H_2 estimation design of the indoor positioning problem can be expressed by using the mean square error of estimation error $\tilde{e}_{Sn}(k)$ as:

$$\begin{aligned} X_{Sn} &= E\left\{\tilde{e}_{Sn}(k+1)\tilde{e}_{Sn}(k+1)^T\right\} \\ &= E\left\{J\tilde{L}_{Sn}(k+1)\tilde{L}_{Sn}(k+1)^T J^T\right\} \end{aligned} \tag{34}$$

where $\tilde{e}_{Sn}(k+1) = J\tilde{L}_{Sn}(k+1)$

Furthermore, Equation (34) can be presented as:

$$\begin{aligned} X_{Sn} &= E\left\{tr\left(J\widetilde{L}_{Sn}(k+1)\widetilde{L}_{Sn}(k+1)^T J^T\right)\right\} \\ &= tr\left(JE\left\{\widetilde{L}_{Sn}(k+1)\widetilde{L}_{Sn}(k+1)^T\right\}J^T\right) \end{aligned} \tag{35}$$

From Equation (26), the estimation error $\widetilde{L}_{Sn}(k+1)$ at a steady state can be described as:

$$\begin{aligned} \widetilde{L}_{Sn}(k+1) &= L_{Sn}(k+1) - \hat{L}_{Sn}(k+1) \\ &= \Psi_{Sn}L_{Sn}(k) + \Lambda w_{Sn}(k) \\ &\qquad -\left\{\Psi_{Sn}\hat{L}_{Sn}(k) + G_{Sn}\left[Z_{Sn}(k) - \Omega\hat{L}_{Sn}(k)\right]\right\} \\ &= \Psi_{Sn}(L_{Sn}(k) - \hat{L}_{Sn}(k)) + \Lambda w_{Sn}(k) \\ &\qquad -G_{Sn}\left[\Omega L_{Sn}(k) + \Pi v_{Sn}(k) - \Omega\hat{L}_{Sn}(k)\right] \\ &= \Psi_{Sn}\widetilde{L}_{Sn}(k) + \Lambda w_{Sn}(k) - G_{Sn}\left[\Omega\widetilde{L}_{Sn}(k) + \Pi v_{Sn}(k)\right] \\ &= (\Psi_{Sn} - G_{Sn}\Omega)\widetilde{L}_{Sn}(k) + \Lambda w_{Sn}(k) - G_{Sn}\Pi v_{Sn}(k) \end{aligned} \tag{36}$$

After some mathematical derivations, the H_2 estimation design for background noise reduction of the indoor positioning design could be summarized as the following Theorem 1.

Theorem 1. *An H_2 steady state estimator for the indoor positioning problem can be constructed if a positive-definite matrix $E_{Sn} = E_{Sn}^T$ can be found such that the following LMIs hold*

$$\begin{bmatrix} E_{Sn} & E_{Sn}\Lambda & D_{Sn}\Pi & (E_{Sn}\Psi_{Sn} - D_{Sn}\Omega) \\ \Lambda^T E_{Sn} & I & 0 & 0 \\ \Pi^T D_{Sn}^T & 0 & I & 0 \\ (E_{Sn}\Psi_{Sn} - D_{Sn}\Omega)^T & 0 & 0 & E_{Sn} \end{bmatrix} > 0 \tag{37}$$

where $D_{Sn} = E_{Sn}G_{Sn}$, and the steady state covariance of the estimation error is bounded by

$$E\left\{\widetilde{e}_{Sn}(k+1)\widetilde{e}_{Sn}(k+1)^T\right\} < tr\left(JE_{Sn}^{-1}J^T\right) \tag{38}$$

Remark 2. *Proof of Theorem 1 is given in* Appendix A.

From Equation (37), D_{Sn} and E_{Sn} can be computationally calculated. Based on these two parameters, the estimation gain of the H_2 estimation design can be obtained by using $G_{Sn} = E_{Sn}^{-1}D_{Sn}$. Additionally, by substituting G_{Sn} into Equation (32), the H_2 estimator can be constructed.

The process of constructing the proposed H_2 estimation design is summarized in the following steps:

Step 1. Given the R_{Sn}, T_{Sn} as identity matrices, and J as a constant matrix based on the extraction of desired state variables.
Step 2. Solve the LMI form of Equation (37) for obtaining the positive matrix D_{Sn} and E_{2Sn}.
Step 3. Calculate the estimation gain G_{Sn} based on solutions of E_{Sn} and D_{Sn} in Step 2.
Step 4. Substituting the estimation gain G_{Sn} into Equation (32), the H_2 estimation design can be constructed as Equation (39).

$$\begin{aligned} \hat{L}_{Sn}(k+1) &= \Psi_{Sn}\hat{L}_{Sn}(k) + G_{Sn}\left[Z_{Sn}(k) - \Omega\hat{L}_{Sn}(k)\right] \\ \hat{h}_{Sn}(k) &= J\hat{L}_{Sn}(k) \end{aligned} \tag{39}$$

3. Verification of Indoor Positioning Performance

To verify the proposed indoor positioning system, simulation results and practical tests will be discussed and compared. Before discussing the positioning performance of this proposed design, the environmental settings of the hardware and software used will be detailed first. Next, the simulation results, which contain one scenario, will be discussed and analyzed. The practical implementation and testing of this proposed design will be verified after the simulation process. Finally, the comparisons of the simulation results and the practical tests will be discussed.

3.1. Testing Environment Setting

3.1.1. Arrangements of Hardware

Hardware adopted for the indoor positioning verification of this proposed design comprises four speakers: Philips AT10, for broadcasting tagged sound sources, one iPhone6 microphone used as a receiver, and a dual core CPU of an iPhone 6 adopted as the calculator of the proposed indoor positioning algorithm. Specifications of the Philips AT10 and the iPhone 6 are displayed in Figure 5 and are listed in Tables 1 and 2.

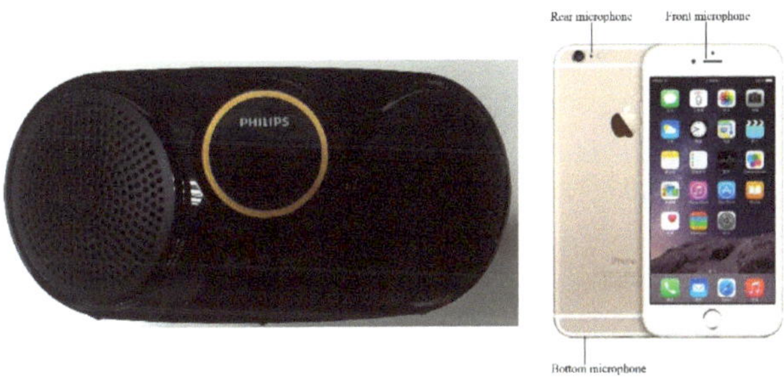

(**a**) (**b**)

Figure 5. The used devices: (**a**) Wireless speaker (Philips AT10). (**b**) iPhone6.

Table 1. Specifications of the Philips AT10.

Total output power	3 W
Frequency response	60–20,000 Hz, ± 3 dB
Signal to noise ratio	>70 dBA
MP3 bit rate	32–320 kbps
Dimensions	239 × 104 × 127 mm
Speaker impedance	6 ohm

Table 2. Specifications of iPhone 6.

Chip	A8 chip with 64 bits
CPU	Dual-core 1.4 GHz Typhoon (ARM v8-based)
Number of microphones	Triple microphones, bottom, front, rear (only one could be arbitrarily used by the developer.)
Dimensions	138.1 × 67 × 6.9 mm

3.1.2. Software Design

After choosing the hardware above, the distinct tag frequencies can be calculated by the FFT method using the measured raw data of the receiver. The analog to digital resolution of iPhone 6 is 16 bits; hence, the measured analog sound pressure of these four tagged sound sources can be presented with 2048 bits.

Distinct tagged frequencies are selected from the following frequency spans:

$$\begin{aligned} 15\text{ kHz} &< f_{S1} < 16\text{ kHz} \\ 16\text{ kHz} &< f_{S2} < 17\text{ kHz} \\ 17\text{ kHz} &< f_{S3} < 18\text{ kHz} \\ 18\text{ kHz} &< f_{S4} < 19\text{ kHz} \end{aligned} \tag{40}$$

The setting parameters for the distinct tagged sound source are shown in Table 3, including tag frequencies, relative distances, and SPL values, respectively. These values are defined as the initial values and are fixed.

Table 3. Setting values of each tagged sound source.

Number n	Sound Source S_n	Frequency f_{Sn}	Relative Distance $r_{Sn}(0)$	SPL Value $l_{Sn}(0)$
1	S1	15.5 kHz	0.1 m	111.5 dB
2	S2	16.5 kHz	0.1 m	117.5 dB
3	S3	17.5 kHz	0.1 m	114.9 dB
4	S4	18.5 kHz	0.1 m	116.1 dB

For analyzing the effects of the inevitable environmental corruptions which infiltrate to the located tagged sound sources. Two background sounds: 1. one recorded frame of a really noisy pub and 2. a song named: Free loops are used as the environmental corruptions for verifying the robust property of this proposed indoor positioning design.

3.2. Test Results

3.2.1. The Configuration for Practical Results

For simulations, a flow chart with three stages each playing specific roles in the indoor positioning process, is illustrated in Figure 6. The corruptions of environments, such as reverberant effects and environmental uncertainties, are set up as random noises, with the magnitude 15 dB for mimicking the practical situations.

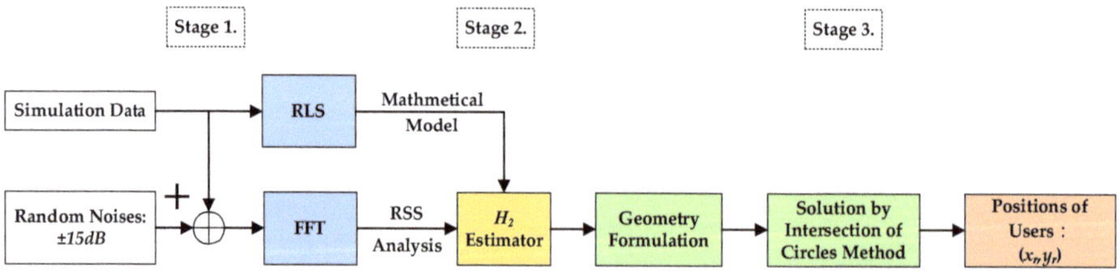

Figure 6. The overall flow chart of these proposed indoor positioning designs.

The parameters for the proposed H_2 estimator in Equation (39) are given in Table 4, and for conserving computational power, the system order m for the SPL model in Equation (23) is selected as 15. The steady state values $\hat{\lambda}_{Sn}(k)$, G_{Sn}, and E_{Sn}, for n=1 to 4, of SPL models with respect to tagged sound sources S1 to S4 are listed in Table 4. For saving space, only $\hat{\lambda}_{S1}(k)$, G_{S1}, and E_{S1} of the tagged sound source S1 are listed in Table 4.

Table 4. Initial values of the proposed H_2 estimation design with respect to the tagged sound source S1.

Variables	Definitions	Values
$\hat{L}_{S1}$	Initialization of estimation states	$0_{m\times 1}$
Q^{+}_{S1}	Error covariance	$I_{m\times m}$
T_{S1}	System disturbance covariance	0.5
R_{S1}	Measurement noise covariance	0.01
J	Constant matrix	$I_{m\times m}$
ε_{S1}	Positive value	20
F_v	Positive definite weighting matrix	80
F_w	Positive definite weighting matrix	80
U	Positive definite weighting matrix	$I_{m\times m}$

The system steady state parameters $\hat{\lambda}_{S1}$ and the weighting matrices of H_2 estimation design under steady-state conditions.

$$\hat{\lambda}_{S1} = \begin{bmatrix} 0.2132 \\ 0.1922 \\ 0.1712 \\ 0.1503 \\ 0.1294 \\ 0.1084 \\ 0.0875 \\ 0.0666 \\ 0.0457 \\ 0.0248 \\ 0.0039 \\ -0.017 \\ -0.0379 \\ -0.0587 \\ -0.0796 \end{bmatrix}, G_{S1} = \begin{bmatrix} 0.2106 \\ 0.9824 \\ 2.274e^{-4} \\ 2.026e^{-4} \\ 1.778e^{-4} \\ 1.531e^{-4} \\ 1.283e^{-4} \\ 1.035e^{-4} \\ 7.879e^{-5} \\ 5.405e^{-5} \\ 2.932e^{-5} \\ 4.609e^{-6} \\ -2.01e^{-5} \\ -4.479e^{-5} \\ -6.948e^{-5} \end{bmatrix}$$

$$E_{S1} = \begin{bmatrix} 0.01682 & 0 & \cdots & \cdots & 0 \\ 0 & 0.01682 & \ddots & \ddots & \vdots \\ \vdots & \ddots & 0.01682 & \ddots & \vdots \\ \vdots & \ddots & \ddots & 0.01682 & 0 \\ 0 & \cdots & \cdots & 0 & 0.01682 \end{bmatrix}_{15\times 15}$$

3.2.2. Testing Scenario

As shown in Figure 7, a prearranged circle path (simulation data) within an indoor area (length: 45 m × width: 40 m), which has 4 tagged sound sources (+) allocated in the four corners of this area, is used. The radius of this prearranged circle path is 10 m, and the initial point of the receiver is at the point (x_r = 25 me, y_r = 20 m). In Figure 7, the mobile phone user (red dot) walks counterclockwise along the circle path (arrow).

Figure 8 shows the indoor positioning result, which only uses measured SPLs and Equations (6) and (20) to calculate the user's position (x_r, y_r), without the help of an estimation design. The root mean square (RMS) error of this simulation is 1.59 m. As for the positioning result of utilizing the H_2 estimation design, it is plotted in Figure 9, and the RMS error is 0.77 m.

From Figures 9 and 10, it is obvious that the indoor positioning accuracy can be effectively improved by the proposed estimation design, and the indoor positioning performance of the proposed method is superior to that without any estimation design.

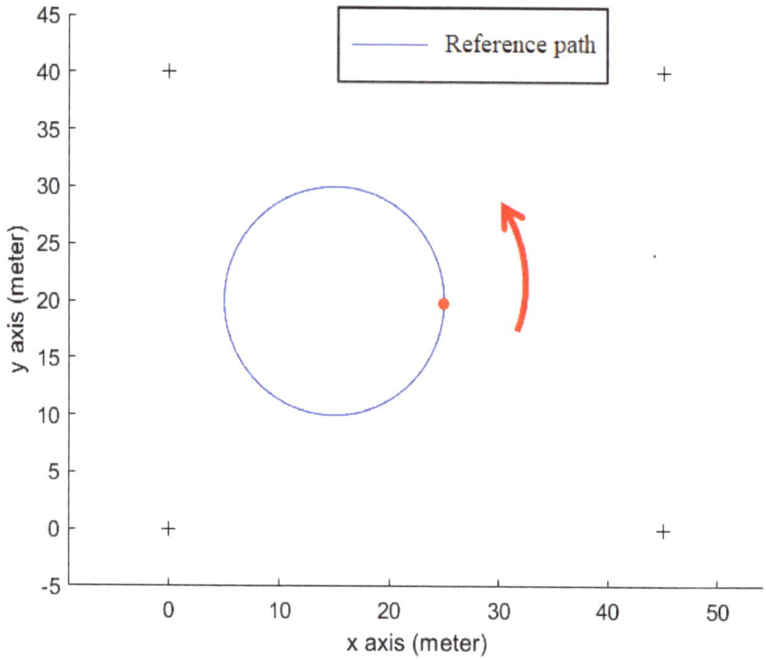

Figure 7. A circle path for the indoor positioning test.

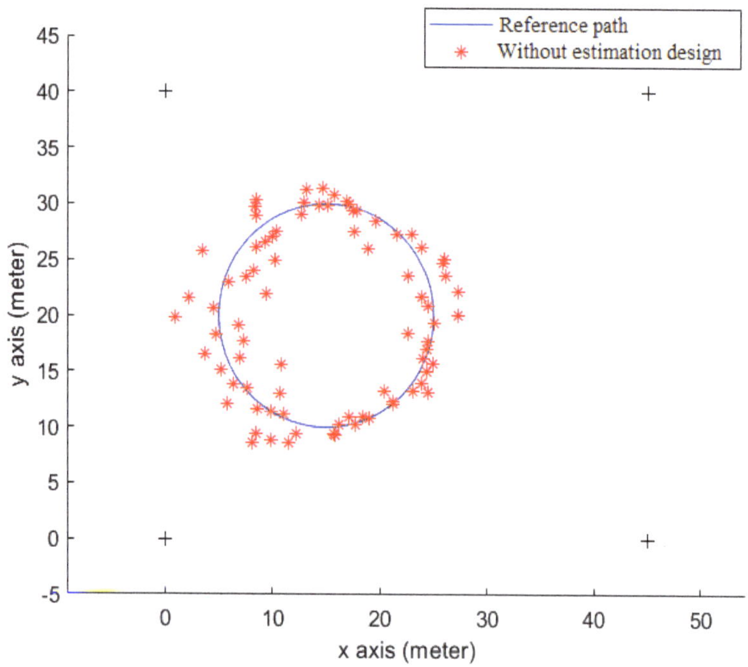

Figure 8. The positioning result without using any estimation design.

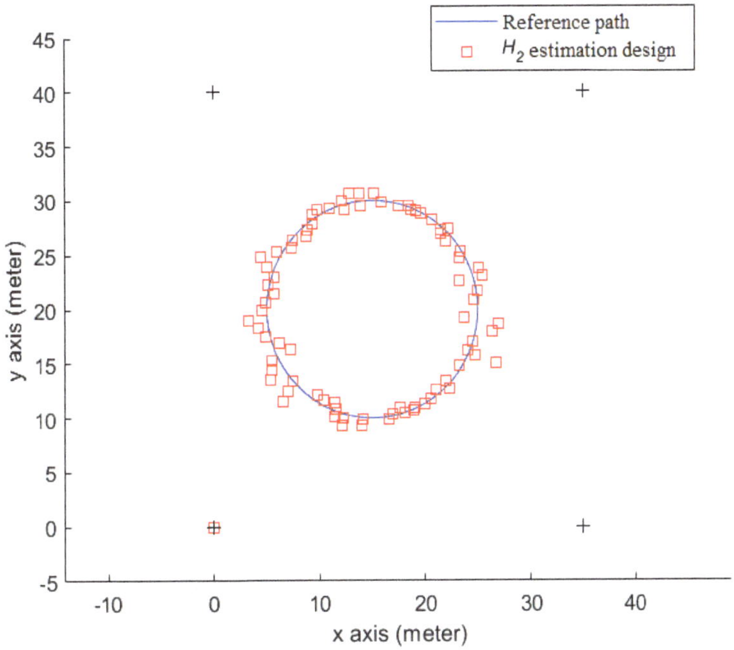

Figure 9. The indoor positioning result of the proposed H_2 estimation design (simulation).

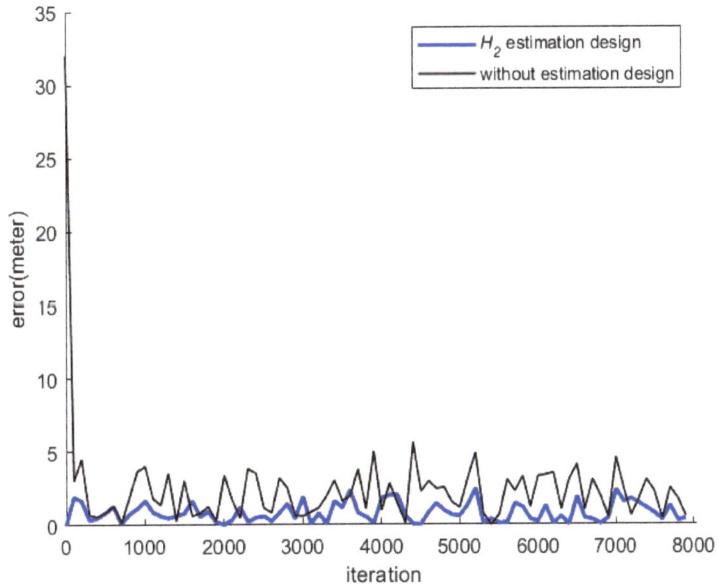

Figure 10. Histories of positioning errors of the proposed estimation design with respect to the positioning scheme without using any estimation design (simulation).

3.3. Practical Experiments

Figure 11 shows the overall testing process of our proposed indoor positioning design. Similarly, this process for practical tests is also divided into three stages. In the first stage, the real time measured SPL data is used to build the mathematical walking model of the mobile phone user, and an RSS analysis using the FFT is adopted for identifying the first two tagged sound sources of the allocated four tagged sound sources. In stage 2, purified SPLs can be delivered, and in the third stage, real time positions of the mobile phone user can be solved by using the intersection of circles method.

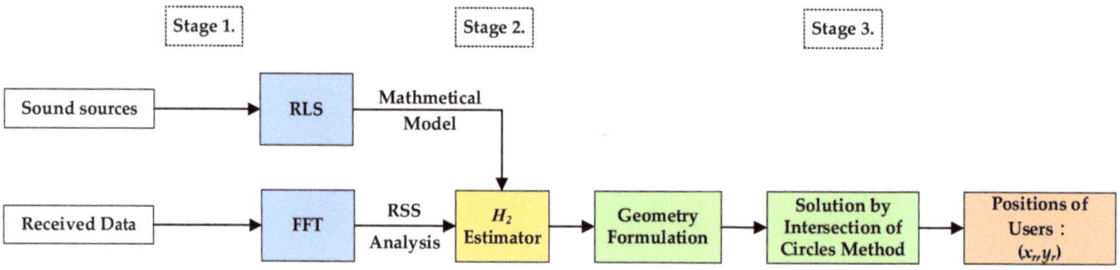

Figure 11. The flow chart of the proposed indoor positioning algorithm for practical tests.

The practical experiment of this proposed design is conducted in the building of the Department of Systems and Naval Mechatronic Engineering, National Cheng Kung University, No.1, University Road, Tainan City, Taiwan. The tested building and layouts of the floors are shown as Figure 12. The iPhone6 is chosen as a receiver, and four tagged sound sources are allocated in appropriated positions, according to the requirements of the experiments. The sampling frequency of the received data is 44,100 Hz. Four distinct frequencies: 15.5 kHz, 16.5 kHz, 17.5 kHz, and 18.5 kHz, are selected for the purpose of identifying the detected sound sources. After the above arrangements in the test environment and apparatus, the same test pattern—a circle path—is assigned beforehand to verify the indoor positioning abilities of this proposed estimation design and the scheme without any estimation design.

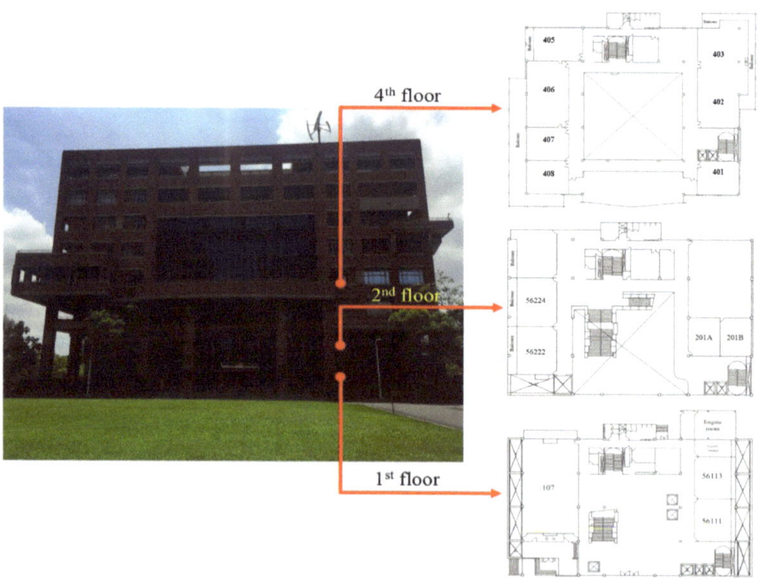

Figure 12. The test building and the layouts of floors for this proposed estimation design.

This real test is conducted in an area with dimensions: (40 m × 45 m) and covered with four tagged sound sources. The relative distance between the tagged sound sources (S1 to S4) is 18 m, as shown in Figure 13, and the cross stars represent the locations of the tagged sound sources. In this scenario, the test path has a circle trajectory, with a radius of 5 m. A total of 7998 SPL data sampled within 40 s are acquired in this test, and each real-time indoor positioning point is drawn per 0.01 Hz. The mobile phone user will walk counterclockwise along this circle path, as illustrated in Figure 14.

Figure 15 shows the total indoor positioning result, without using any estimation design, and the RMS error with respect to the circle reference path is 2.49 m. Figure 16 shows the positioning result of the proposed H_2 estimation design, and the RMS error, which corresponds to the circle reference path, is 0.74 m. Figure 17 shows the indoor positioning error histories of the proposed estimation design with respect to the indoor positioning scheme without any estimation design.

From Figures 16 and 17, it can be seen that the proposed indoor positioning design highly improves the positioning accuracy, even under the effect of the worst background noises, and it outperforms the design that uses only measured SPLs, without any estimation design.

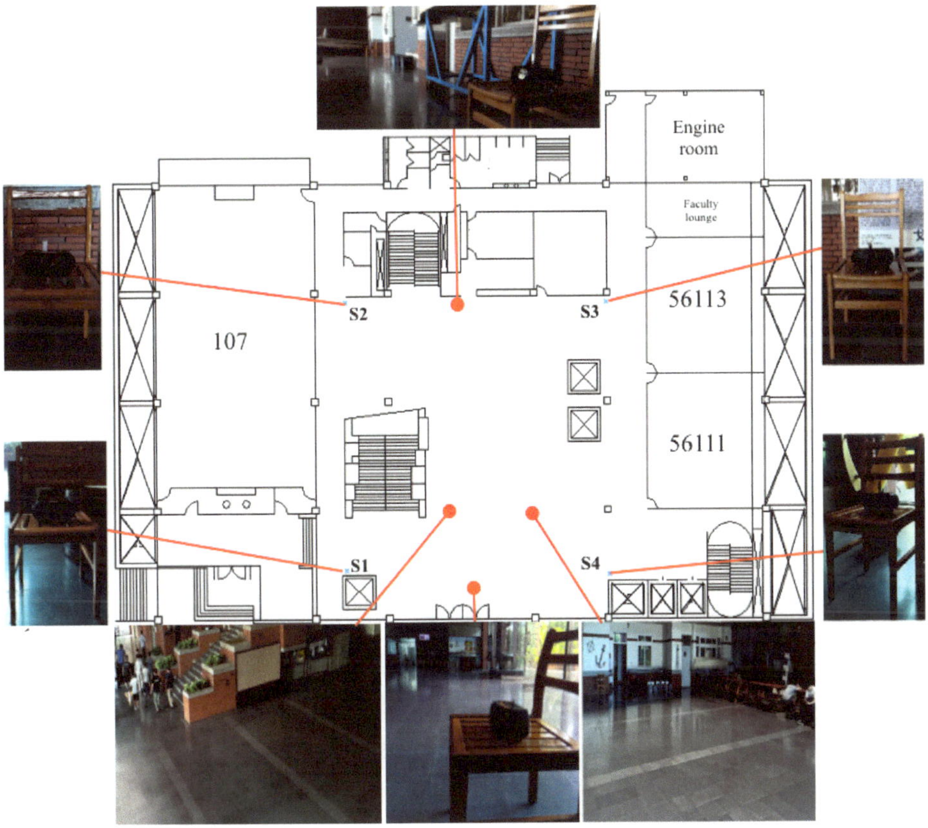

Figure 13. The test environment in the first floor of the building.

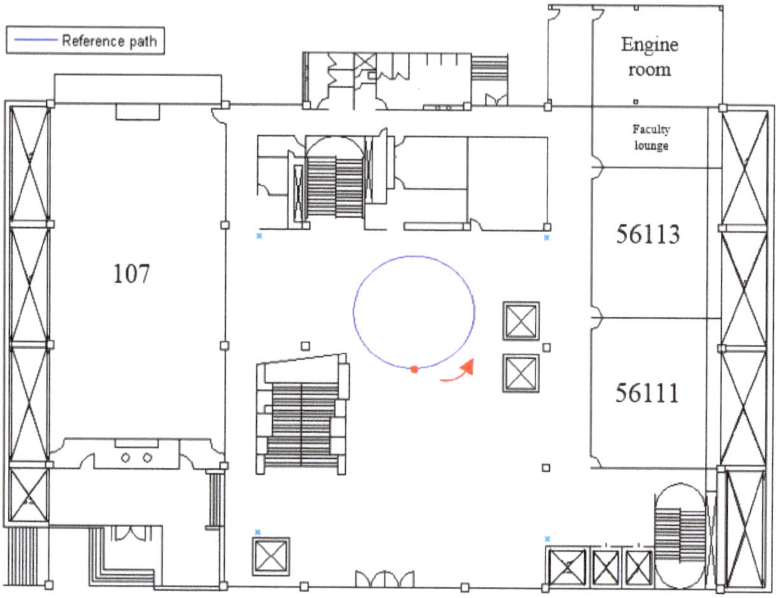

Figure 14. The circle path for the indoor positioning test in an indoor plane.

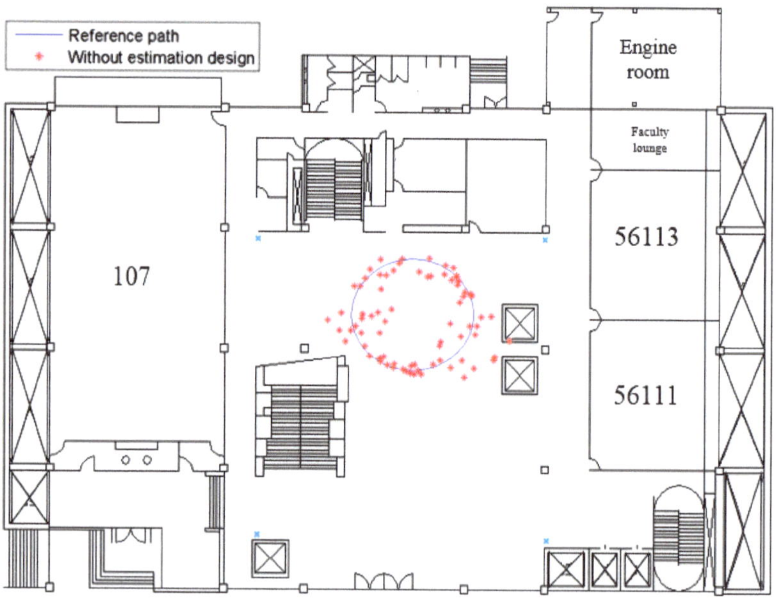

Figure 15. The indoor positioning result without any estimation design.

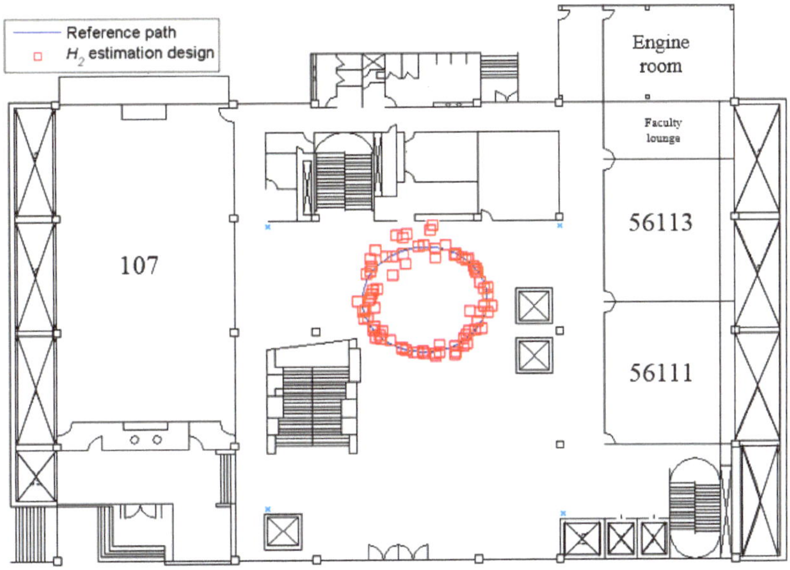

Figure 16. The indoor positioning result of the proposed H_2 estimation design (real test).

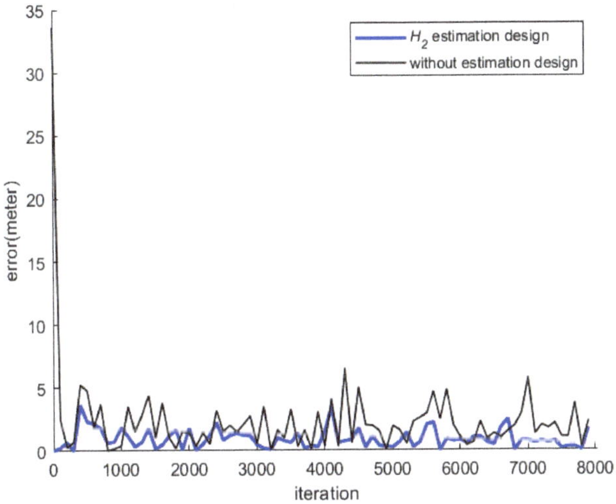

Figure 17. Histories of positioning errors of the proposed estimation design with respect to the positioning scheme without using any estimation design (real test).

3.4. Comparisons of Simulation Results and Practical Results

As shown in Table 5, the RMS error of the indoor positioning result, without the help of estimation designs, is 1.59 m within an area with a length of 45 m and a width of 40 m.

Table 5. The positioning results without estimation designs in the simulation.

Reference Path	Covered Area	RMS Error
Circle	45 m × 40 m	1.59 m

As shown in Table 6, the RMS error of the indoor positioning performance of the H_2 estimation design is 0.77 m, based on the same test environment and condition.

Table 6. The positioning results of the proposed H_2 estimation design in the simulation.

Reference Path	Covered Area	RMS Error
Circle	45 m × 40 m	0.77 m

In real experiments, the covered areas are similar to those of simulations. Table 7 shows the RMS error of the indoor positioning, without using the estimation design, within an area with a length of 45 m and a width of 40 m.

Table 7. The positioning results without any estimation design in the practical experiments.

Reference Path	Covered Area	RMS Error
Circle	45 m × 40 m	2.49 m

As for the H_2 estimation design, the positioning performance is listed in Table 8. Under the same test condition, the proposed H_2 estimation design delivers an indoor positioning performance, with an RMS error of 0.77 m.

Table 8. The indoor positioning of the proposed H_2 estimation design in the practical experiments.

Reference Path	Covered Area	RMS Error
Circle	40 m × 45 m	0.74 m

From comparisons of Tables 5–8, it can be seen that the indoor positioning performances of this proposed method in simulation and practical experiments are similar because the noise levels of the environmental corruptions simulated by Equation (8) have been tuned beforehand, according to off-line measured environmental corruptions for the purpose of approximating the noise level of the true corruption in the real environment. Similar results can be obtained for the indoor positioning scheme without any estimation design. As a whole, the proposed indoor positioning design yields better positioning performance than that without any estimation design, and it is robust in regards to the environmental background noises.

4. Conclusions

An effective and accurate indoor positioning function is necessary for users who carry mobile phones because they spend a lot of time in shielding spaces. For achieving this design target, an indoor positioning scheme with the H_2 estimation design, which can purify the corrupted SPLs of measured sound sources, is successfully developed for mobile phone users in this investigation. Although the test conditions vary randomly, and indoor environments are challenging due to the unknown background noises, simulation results and practical tests obviously show the promising indoor positioning performance of this proposed method: an average positioning RMS error of 0.75 m can be obtained. Two main contributions can be summarized for this investigation: 1. A compact indoor positioning system, with a high indoor positioning accuracy and capable of execution on mobile phones, is developed because the proposed H_2 estimation design possesses the following natural potentials: a low power consumption for computation and an easy-to-implement filter structure. 2. This investigation provides a positioning possibility other than GPS for use in commonly unshielded spaces, as well as indoor environments.

Author Contributions: Conceptualization, Y.-H.C., Y.-Y.C. and P.-Y.C.; methodology, Y.-H.C. and Y.-Y.C.; software, Y.-H.C. and Y.-Y.C.; validation, Y.-H.C., Y.-Y.C. and P.-Y.C.; formal analysis, Y.-H.C.,

Y.-Y.C. and P.-Y.C.; investigation, Y.-H.C., Y.-Y.C. and P.-Y.C.; resources, Y.-H.C. and Y.-Y.C.; data curation, Y.-H.C., Y.-Y.C. and P.-Y.C.; writing—original draft preparation, Y.-H.C., Y.-Y.C. and P.-Y.C.; writing—review and editing, Y.-H.C. and Y.-Y.C.; visualization, Y.-Y.C. and P.-Y.C.; supervision, Y.-H.C. and Y.-Y.C.; funding acquisition, Y.-H.C. All authors have read and agreed to the published version of the manuscript.

Funding: This research was funded by the Ministry of Science and Technology of Taiwan, grant number MOST 111-2221-E-020-023.

Institutional Review Board Statement: Not applicable.

Informed Consent Statement: Not applicable.

Data Availability Statement: Not applicable.

Conflicts of Interest: The authors declare no conflict of interest.

Appendix A

From Equations (35) and (36), the estimation covariance can be further derived as

$$
\begin{aligned}
&E\left\{\widetilde{L}_{Sn}(k+1)\widetilde{L}_{Sn}(k+1)^T\right\} \\
&= E\left\{\begin{array}{l} [(\Psi_{Sn} - G_{Sn}\Omega)L_{Sn}(k) + \Lambda w_{Sn}(k) - G_{Sn}\Pi v_{Sn}(k)] \\ \cdot\left[\widetilde{L}_{Sn}(k)^T(\Psi_{Sn} - G_{Sn}\Omega)^T + w_{Sn}(k)^T\Lambda^T - v_{Sn}(k)^T\Pi^T G_{Sn}^T\right] \end{array}\right\} \\
&= E\left\{\begin{array}{l} (\Psi_{Sn} - G_{Sn}\Omega)\widetilde{L}_{Sn}(k)\widetilde{L}_{Sn}(k)^T(\Psi_{Sn} - G_{Sn}\Omega)^T \\ +\Lambda w_{Sn}(k)w_{Sn}(k)^T\Lambda^T + G_{Sn}\Pi v_{Sn}(k)v_{Sn}(k)^T\Pi^T G_{Sn}^T \end{array}\right\} \\
&= (\Psi_{Sn} - G_{Sn}\Omega)E\left\{\widetilde{L}_{Sn}(k)\widetilde{L}_{Sn}(k)^T\right\}(\Psi_{Sn} - G_{Sn}\Omega)^T \\
&\qquad +\Lambda\Lambda^T + G_{Sn}\Pi\Pi^T G_{Sn}^T
\end{aligned}
\tag{A1}
$$

where $L_{Sn}(k)$, $w_{Sn}(k)$, and $v_{Sn}(k)$ are mutually orthogonal, and the covariance matrix $E\{w_{Sn}(k)w_{Sn}(k)^T\} = I_{m\times m}$ and $E\{v_{Sn}(k)v_{Sn}(k)^T\} = I_{m\times m}$ are assumed as the identity matrix. The other covariance matrix $E\left\{\widetilde{L}_{Sn}(k)\widetilde{L}_{Sn}(k)^T\right\}$, at a steady state in practical design, is constant (i.e., $k \to \infty$) and is represented as $E\left\{\widetilde{L}_{Sn}(k)\widetilde{L}_{Sn}(k)^T\right\} = \Theta_{Sn}$.

Combining Equation (35) and Equation (A1), the mean-square error can be rewritten as:

$$
\begin{aligned}
X_{Sn} &= tr\left(JE\left\{\widetilde{L}_{Sn}(k+1)\widetilde{L}_{Sn}(k+1)^T\right\}J^T\right) \\
&= tr\left(J\left[\begin{array}{c} (\Psi_{Sn} - G_{Sn}\Omega)\Theta_{Sn}(\Psi_{Sn} - G_{Sn}\Omega)^T \\ +\Lambda\Lambda^T + G_{Sn}\Pi\Pi^T G_{Sn}^T \end{array}\right]J^T\right) \\
&= tr\left(J\left[\begin{array}{c} (\Psi_{Sn} - G_{Sn}\Omega)\Theta_{Sn}(\Psi_{Sn} - G_{Sn}\Omega)^T - \Theta_{Sn} \\ +\Lambda\Lambda^T + G_{Sn}\Pi\Pi^T G_{Sn}^T \end{array}\right]J^T\right) \\
&\quad +tr\left(J\Theta_{Sn}J^T\right)
\end{aligned}
\tag{A2}
$$

Equation (A2) shows that the mean-square error X_{Sn} must have an upper bound as below:

$$
X_{Sn} \le tr\left(J\Theta_{Sn}J^T\right) \tag{A3}
$$

under the following inequality holds:

$$
(\Psi_{Sn} - G_{Sn}\Omega)\Theta_{Sn}(\Psi_{Sn} - G_{Sn}\Omega)^T - \Theta_{Sn} + \Lambda\Lambda^T + G_{Sn}\Pi\Pi^T G_{Sn}^T < 0 \tag{A4}
$$

Given that $E_{Sn} = \Theta_{Sn}^{-1}$, multiplying both sides of Equation (A4) by E_{Sn} and selecting $D_{Sn} = E_{Sn}G_{Sn}$, the Equation (A4) can be rewritten as:

$$
\begin{array}{c}
(E_{Sn}\Psi_{Sn} - D_{Sn}\Omega)E_{2Sn}^{-1}(E_{Sn}\Psi_{Sn} - D_{Sn}\Omega)^{-1} \\
-E_{Sn} + E_{Sn}\Lambda\Lambda^T E_{Sn} + D_{Sn}\Pi\Pi^T D_{Sn}^T < 0
\end{array}
\tag{A5}
$$

For acquiring solution E_{Sn} in Equation (A5), the Schur complement is employing to Equation (A5) for finding out a positive solution E_{Sn}, yielding the following equivalent LMI form:

$$\begin{bmatrix} E_{Sn} & E_{Sn}\Lambda & D_{Sn}\Pi & (E_{Sn}\Psi_{Sn} - D_{Sn}\Omega) \\ \Lambda^T E_{Sn} & I & 0 & 0 \\ \Pi^T D_{Sn}^T & 0 & I & 0 \\ (E_{Sn}\Psi_{Sn} - D_{Sn}\Omega)^T & 0 & 0 & E_{Sn} \end{bmatrix} > 0 \tag{A6}$$

This is Equation (37), and the proof is completed.

References

1. Commission, P.C. *FCC Amended Report to Congress on the Deployment of E-911 Phase II Services Tier III Service Providers*, 1st ed.; Wireless Telecommunications; FCC: Washington, DC, USA, 2005.
2. Atia, M.M.; Liu, S.; Nematallah, H.; Karamat, T.B.; Noureldin, A. Integrated Indoor Navigation System for Ground Vehicles With Automatic 3-D Alignment and Position Initialization. *IEEE Tran. on Veh. Technol.* **2015**, *64*, 1279–1292. [CrossRef]
3. Constandache, I.; Agarwal, S.; Tashev, I.; Choudhury, R.R. Daredevil: Indoor location using sound. *SIGMOBILE Mob. Comput. Commun. Rev.* **2014**, *18*, 9–19. [CrossRef]
4. WHERENET Real-Time Locating System. Available online: http://www.wherenet.com/ (accessed on 1 August 2008).
5. Casas, R.; Cuartielles, D.; Marco, A.; Gracia, H.J.; Falco, J.L. Hidden Issues in Deploying an Indoor Location System. *IEEE Perv. Comput.* **2007**, *6*, 62–69. [CrossRef]
6. NDI Measurement You Can Trust. Available online: http://www.ndigital.com/ (accessed on 1 August 2008).
7. States, R.A.; Pappas, E. Precision and repeatability of the Optotrak 3020 motion measurement system. *J. Med. Eng. Technol.* **2006**, *30*, 11–16. [CrossRef] [PubMed]
8. Fernando, X.N.; Krishnan, S.; Sun, H.; Kazemi-Moud, K. Adaptive denoising at infrared wireless receivers. *Prceeding SPIE* **2003**, *5074*, 199–207.
9. Harter, A.; Hopper, A.; Steggles, P.; Ward, A.; Webster, P. The anatomy of a context-aware application. In Proceedings of the 5th Annual ACM/IEEE International Conference on Mobile Computing and Networking, Seattle, WA, USA, 15 August 1999.
10. Chunhan, L.; Yushin, C.; Gunhong, P.; Jaeheon, R.; Seung-Gweon, J.; Seokhyun, P. Indoor positioning system based on incident angles of infrared emitters. Industrial Electronics Society, 2004. ECON 2004. In Proceedings of the 30th Annual Conference of IEEE, Busan, Republic of Korea, 2 November 2004.
11. Kaemarungsi, K.; Krishnamurthy, P. Properties of indoor received signal strength for WLAN location fingerprinting. Mobile and Ubiquitous Systems: Networking and Services, 2004. MOBIQUITOUS 2004. In Proceedings of the First Annual International Conference on, Boston, MA, USA, 26 August 2004.
12. Tsung-Nan, L.; Po-Chiang, L. Performance comparison of indoor positioning techniques based on location fingerprinting in wireless networks. In Proceedings of the 2005 International Conference on Wireless Networks, Communications and Mobile Computing, Maui, HI, USA, 13 June 2005.
13. Chon, H.D.; Jun, S.; Jung, H.; An, S.W. Using RFID for Accurate Positioning. *J. Glob. Post. Sys.* **2004**, *3*, 32–39. [CrossRef]
14. Wi-Fi Design, WLAN Planning and Site Survey Tool, Wi-Fi Spectrum Analysis. Available online: http://www.ekahau.com/ (accessed on 1 August 2008).
15. King, T.; Kopf, S.; Haenselmann, T.; Lubberger, C.; Effelsberg, W. COMPASS: A probabilistic indoor positioning system based on 802.11 and digital compasses. In Proceedings of the 1st International Workshop on Wireless Network Testbeds, Experimental Evaluation & Characterization, Los Angeles, CA, USA, 29 September 2006.
16. Bahl, P.; Padmanabhan, V.N. RADAR: An in-building RF-based user location and tracking system. INFOCOM 2000. In Proceedings of the Nineteenth Annual Joint Conference of the IEEE Computer and Communications Societies, Tel Aviv, Israel, 6 August 2000.
17. Wang, Y.; Jia, X.; Lee, H.K. An indoors wireless positioning system based on wireless local area network infrastructure. In Proceedings of the 6th International Symposium on Satellite Navigation Technology Including Mobile Positioning and Location Services, Melbourne, Australia, 22 July 2003.
18. Kim, Y.; Chon, Y.; Cha, H. Smartphone-Based Collaborative and Autonomous Radio Fingerprinting. *IEEE Trans. Syst. Man Cybern. Part C* **2012**, *42*, 112–122. [CrossRef]
19. Saeed, A.; Kosba, A.E.; Youssef, M. Ichnaea: A Low-Overhead Robust WLAN Device-Free Passive Localization System. *IEEE J. Select. Top. Sig. Proc.* **2014**, *8*, 5–15. [CrossRef]
20. Gu, Y.; Lo, A.; Niemegeers, I. A survey of indoor positioning systems for wireless personal networks. *IEEE Comm. Surv. Tutor.* **2009**, *11*, 13–32. [CrossRef]
21. The Bat Ultrasonic Location System. Available online: http://www.cl.cam.ac.uk/research/dtg/attarchive/bat/ (accessed on 1 August 2008).
22. Priyantha, N.B. *The Cricket Indoor Location System*; Massachusetts Institute of Technology: Cambridge, MA, USA, 2005.

23. Priyantha, N.B.; Chakraborty, A.; Balakrishnan, H. The Cricket location-support system. In Proceedings of the 6th annual international conference on Mobile computing and networking, Boston, MA, USA, 1 August 2000.
24. Hoflinger, F.; Hoppe, J.; Zhang, R.; Ens, A.; Reindl, L.; Wendeberg, J.; Schindelhauer, C. Acoustic indoor-localization system for smart phones. In Proceedings of the 2014 11th International Multi-Conference on Systems, Signals and Devices (SSD14), Barcelona, Spain, 11 February 2014.
25. Liu, K.; Liu, X.; Li, X. Guoguo: Enabling Fine-Grained Smartphone Localization via Acoustic Anchors. *IEEE Trans. Mob. Comput.* **2016**, *15*, 1144–1156. [CrossRef]
26. Polotti, P.; Sampietro, M.; Sarti, A.; Tubaro, S.; Crevoisier, A. Acoustic localization of tactile interactions for the development of novel tangible interfaces. In Proceedings of the 8th Int. Conference on Digital Audio Effects (DAFX-05), Madrid, Spain, 20 September 2005.
27. Crevoisier, A.; Polotti, P. Tangible acoustic interfaces and their applications for the design of new musical instruments. In Proceedings of the 2005 Conference on New Interfaces for Musical Expression, Vancouver, BC, Canada, 26 May 2005.
28. Qi, Y.; Kobayashi, H.; Suda, H. On time-of-arrival positioning in a multipath environment. *IEEE Trans. Veh. Technol.* **2006**, *55*, 1516–1526. [CrossRef]
29. Jin, Y.; O'Donoughue, N.; Moura, J.M. Position location by time reversal in communication networks. In Proceedings of the 2008 IEEE International Conference on Acoustics, Speech and Signal Processing, Las Vegas, NV, USA, 31 March 2008.
30. Carotenuto, R.; Merenda, M.; Iero, D.; Corte, F.G.D. Mobile Synchronization Recovery for Ultrasonic Indoor Positioning. *Sensors* **2020**, *20*, 702. [CrossRef] [PubMed]
31. Gualda, D.; Rubio, M.C.; Urena, J.; Bachiller, S.; Villadangos. J., M.; Hernandez, A.; Garcia, J.J.; Jimenez, A. LOCATE-US:Indoor Positionong for Mobile Device Using Encoded Ultrasonic Signals, Inertial Sensors and Graph-Matching. *Sensors* **2021**, *21*, 1950. [CrossRef] [PubMed]

sensors

MDPI

Article

Research on Smart Tourism Oriented Sensor Network Construction and Information Service Mode

Ruomei Tang [1,2], Chenyue Huang [1], Xinyu Zhao [1] and Yunbing Tang [3,*]

[1] School of Art and Design, Nanjing Forest University, Nanjing 210037, China
[2] Research Center for Digital Innovation Design, Nanjing Forestry University, Nanjing 210037, China
[3] School of Journalism, Fudan University, Shanghai 200433, China
* Correspondence: tangyunbing@fudan.edu.cn

Abstract: Smart tourism is the latest achievement of tourism development at home and abroad. It is also an essential part of the smart city. Promoting the application of computer and sensor technology in smart tourism is conducive to improving the efficiency of public tourism services and guiding the innovation of the tourism public service mode. In this paper, we have proposed a new method of using data collected by sensor networks. We have developed and deployed sensors to collect data, which are transmitted to the modular cloud platform, and combined with cluster technology and an Uncertain Support Vector Classifier (A-USVC) location prediction method to assist in emergency events. Considering the attraction of tourists, the system also incorporated human trajectory analysis and intensity of interaction as consideration factors to validate the spatial dynamics of different interests and enhance the tourists' experience. The system explored the innovative road of computer technology to boost the development of smart tourism, which helps to promote the high-quality development of tourism.

Keywords: sensor data acquisition; smart tourism; data transmission; low energy consumption sensor; monitoring management system

Citation: Tang, R.; Huang, C.; Zhao, X.; Tang, Y. Research on Smart Tourism Oriented Sensor Network Construction and Information Service Mode. *Sensors* **2022**, *22*, 10008. https://doi.org/10.3390/s222410008

Academic Editors: Chien Aun Chan, Ming Yan and Chunguo Li

Received: 17 November 2022
Accepted: 16 December 2022
Published: 19 December 2022

1. Introduction

In 2022, the World Federation of Tourism Cities (WTCF) and the Tourism Research Center of China's Academy of Social Sciences jointly released the World Tourism Economic Trend Report (2022). The report showed that the global tourism revenue in 2021 was USD 3.3 trillion, equivalent to 3.8% of the worldwide GDP. Although this proportion has dropped significantly compared to before the Corona Virus Disease 2019 (COVID-19), it is still an essential part of the worldwide economy. In recent years, with the rapid development of Internet of Things technology, such as relying on big data, mobile devices, and sensors, more ways of data dissemination have been broadened. Physical media such as sensors have been embedded in daily life with unprecedented breadth and depth and extended to more fields. New industrial forms such as the smart city and smart tourism have emerged as the times require. Smart tourism is mainly composed of an intelligent information layer that collects numerical data, an intelligent exchange layer that supports interconnection, and an intelligent processing layer responsible for data analysis, visualization, integration, and smart use [1]. It aims to enhance the tourism experience through the most advanced information technology and big data [2]. In addition, promoting smart tourism development characterized by digitalization, networking and intelligence will help improve the service experience and promote the innovation of the smart tourism public service mode [3].

In this paper, we focus on the use of sensors and the construction of an intelligent system in Jiangsu Horticultural Expo. Sensors are installed in the scenic spots to collect information during the tour to realize the monitoring and management of the scenic spots. Jiangsu Horticultural Expo is rich in ecological environment resources, which is a successful demonstration of the "double-cultivation of cities." It also has practical experience in applying digital-related technologies and has the technical foundation for turning digital research results into innovative tourism application products [1]. Therefore, it is necessary to develop a sensor platform that can monitor and analyze data in real-time, and use cloud computing technology to process and store the generated data to provide a scientific way for managers to solve problems [4].

2. Related Works

As the focus of scientific tourism management, smart tourism has attracted the attention of scholars. Governmental and academic institutions have in recent years attempted to design sustainable, technological, and efficient tourist cities to counter many of these problems. (Gretzel et al.) [5]. At present, a lot of research has been conducted on smart tourism, involving the use of the latest technology. Aguilar et al. developed a new method to perform automatic billing functions in the cafeteria using neural networks [6]. Cacho et al. focused on developing intelligent travel guides for tourists, thus simplifying the travel planning process [7]. Kasnesis et al. have developed wireless acoustic sensors to identify and collect audio signals in cultural sites, and used them for data collection and the protection of cultural heritage [4]. Car et al. found that the adoption of IoT technology has improved the business processes and resource allocation of smart tourism [8]. Online social networks help to collect tourists' preferences. French et al. have connected tourists in social networks, providing them with information, tour guides, and accommodation services [9].

In conclusion, the impact of the IoT and cloud computing on the tourism industry is disruptive. Communication technology has triggered significant changes in the field of the IoT, and the innovative iteration of sensors and cloud-based transmission has become essential features of smart tourism. Data-driven, computing-driven, and scene-driven formats have also become the inherent dynamism of smart tourism industry innovation [10]. At present, the research on improving the service quality of the tourism industry and promoting the transformation of intelligent tourism is still in the initial stage. Therefore, it is necessary to comprehensively improve the service function of the application computing system from the "end" application system and the "cloud" computing system of smart tourism.

Several studies on smart tourism information systems can be found in the literature. For example, by providing intelligent services to tourism departments such as transportation, hotels, and travel attractions and collecting feedback data, we can help build the integrity of the database system and formulate strategies to improve tourism management and services [11]. In addition, smart tourism information systems can be extended, and dynamic systems can offer efficient access to comprehensive sensor nodes and tourism platforms. This system provides a transaction settlement, intelligent guides, innovative marketing, and intelligence management of tourist attractions to support enterprises and tourism business [12,13]. In the dynamic environment of scenic areas, wireless sensor networks may receive a variety of environmental factors such as the electromagnetic field, temperature, humidity, noise, etc., and have a higher node failure rate and data loss rate than traditional networks. Significantly affected by energy consumption and size, the node configuration cannot achieve high sensor accuracy, and information loss is an inevitable problem. The first Reliable Data Aggregation Protocol (RDAT) secure data fusion algorithm based on a trust management mechanism was proposed by Ozdemir et al. [14,15]. By considering the energy efficiency factor, Liu et al. proposed an improved Reliable Trust-Based and Energy-Efficient Secure Data-Aggregation (iRTEDA) algorithm based on the RDAT algorithm [16]. In response to the above problems, an energy-efficient and reliable, secure data fusion algorithm, the Energy-Efficient protocol of Reliable Trust-Based Data Aggregation (ERTDA), is proposed in this paper, which can guarantee the reliability of the data transmission link and effectively extend the network life cycle. (Table 1)

In order to test whether the sensor nodes can realize the functions of wireless sensor network transmission and network convergence, this paper builds the sensor network platform shown in Figure 1. The environmental sensor node's environmental monitoring function can collect information in three environments: temperature, vibration, and sound. The digital temperature sensor DS18B20, the three-week acceleration sensor ADXL345, and the digital MEMS sensor INMP441 are used, respectively. These sensors are small in size and easy for system integration. The output is digital, eliminating the need for noise reduction conversion circuits. Because the cooperation of multiple sensor nodes generally completes the monitoring of the target object, the sensor node sends data to the sink node. After receiving the data, the sink node tells the sensor node that it has received the data and ends the transmission. Because there is a time gap in the data transmission stage, the data is transmitted to the sink node by the multi-hop transmission strategy, which verifies the cooperation between different sensor types.

Table 1. Description of smart tourism application.

Methods	Application Domain	Sensing Accuracy	Energy Efficiency
Data Integration [11]	Management Services	-	-
IoT [13]	Marketing Services	-	-
RDAT Algorithm [14]	Data Transmission	Low	Low
iRTEDA Algorithm [16]	Data Transmission	Low	Low

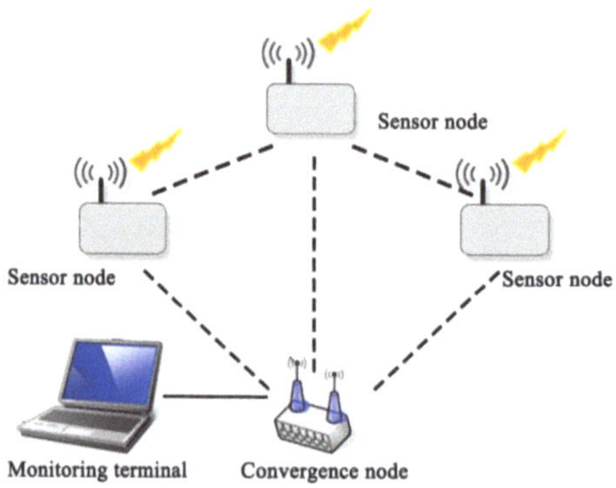

Figure 1. Structure diagram of sensor test platform.

The rest of the paper is organized as follows. Section 2 introduces the research status of smart tourism and the construction of intelligent service systems in scenic spots. Section 3 describes the work method of the super-brain system and proposes the algorithm model of ERTDA and the process to cope with the scenic emergency pairs, the workflow of sensor signal reception, and transmission. Section 4 presents a few suggestions and application models for the challenge of low energy consumption of sensors. Section 5 presents results and discussion. This section introduced the main application results of the super-brain system and discussed the development of smart tourism.

3. Materials and Methods

The core of the intelligent sensor network platform for tourism is as follows:

- High performance data collection and fusion, comprehensive integration to promote tourism services, and information organization depth to monitor the spatial location, critical areas, and meteorological environment data of tourists in the environment in real time, and analyze and process the detection results.
- Enables intelligent sensors to operate with low energy consumption. At present, most wireless sensor nodes are powered by lithium batteries. Once they are used in environmental monitoring, battery replacement will become a big problem. However, the energy consumption of wireless sensor networks is generally consumed by nodes sending, receiving, and fusing data.

Therefore, another critical point in constructing intelligent travel sensor networks is to design low-power wireless transmission of environmental monitoring sensor nodes, which can effectively reduce the energy consumption in the wireless transmission process and significantly improve the running time of nodes.

Wireless sensor technology has triggered a new round of revolution in the field of the IoT. Each data acquisition point in the wireless sensor network is a small embedded system, which is the primary platform of the wireless sensor network, realizing intelligent data acquisition and cloud transmission. When nodes perceive the environment, they must convert non-electrical signals into electrical signals. A/D converters usually provide multiple analog channels, which can realize the conversion of numerous analog quantities. However, only one analog portion can be converted simultaneously, so the multi-way switch is used to select the analog amount in the structure shown in Figure 2. For example, in the clustering structure, the cluster head can fuse the data sent by the members in the cluster. At the same time, the common nodes can do preliminary digital filtering to improve the accuracy of data collection. The result of data processing can be temporarily stored in the node data storage module, sent to other nodes through the communication module, and transmitted to the sink node to realize remote data processing.

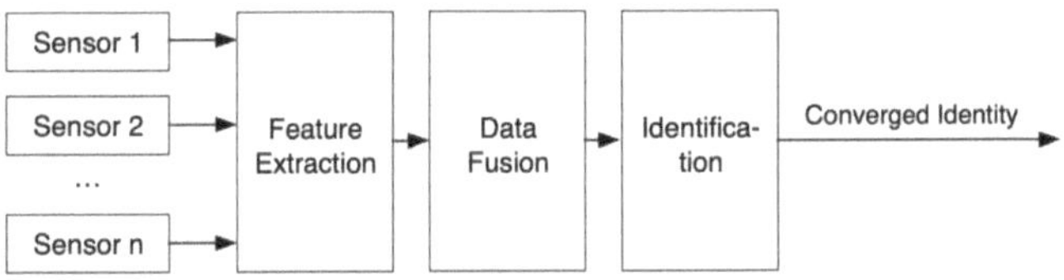

Figure 2. Data-level fusion structure.

3.1. Data Collection

The ERTDA algorithm can effectively organize the captured nodes' behavior of sending malicious messages in the network by calculating the trust value of the obtained target nodes compared with the threshold value, improving the security performance from within transmission network. The advantage of the ERTDA algorithm is that when the wireless sensor is used for environmental monitoring, it can focus on grasping and analyzing the monitoring data of a region rather than the data information of a specific node. It is an efficient, energy-saving, and reliable security data fusion algorithm. To evaluate the performance of the ERTDA algorithm, we run the iRTEDA model and the RDAT model in the same simulation environment and simulate and analyze the energy consumption rate and the life cycle of the network for the three models. In the parameter setting of the simulation, the whole wireless sensor network is set to contain 150 homogeneous sensor nodes randomly deployed in a 500 m $\times$ 500 m area.

We compared the number of dead nodes in adequate time for the three algorithmic models, and it is clear from Figure 3a that the number of dead nodes at the same time is much higher for the RDAT algorithm than for the iRTEDA and ERTDA algorithms. In the iRTEDA algorithm, the number of dead nodes is 50 when the network runs to 2500 s, after which there is a significant and rapid increase in the number of dead nodes. In the ERTDA algorithm, the energy factor of all nodes on the data link is taken into account, so the number of dead nodes does not reach 60 until the network runs until 3000 s, which effectively improves the life cycle of the transmission network.

Figure 3b shows the energy consumption ratio of the three models in the same simulation environment. As can be seen from the figure, the energy consumption ratio of the RDAT algorithm is much higher than that of the other two algorithms from the beginning of the network operation. When the network runs for 2500 s, the energy consumption of the RDAT algorithm network reaches 90%. In comparison, the iRTEDA algorithm and ERTDA algorithm only consume 43.5% and 38.6% of network energy, respectively, which greatly prolongs the life cycle of the network. The ERTDA algorithm considers the energy of other nodes in the data transmission link and the remaining energy of sensor nodes, which significantly slows down the rate of isolated nodes in the transmission network. When the network runs for 3000 s, the energy consumption ratio of the ERTDA algorithm only reaches 54.3%, which is lower than 72.4% of the iRTEDA algorithm.

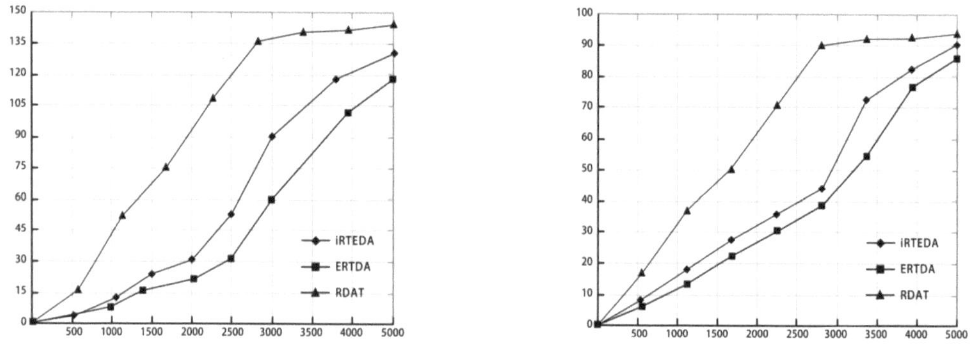

(a) (b)

Figure 3. The consumption comparison of three algorithms. (**a**) The comparison of node death rates of three algorithm models. (**b**) The comparison of energy consumption rates of three algorithm models.

Combining the simulation results and analysis of Figure 3a,b, we can accurately determine that the ERTDA algorithm model can effectively slow down the rate of dead nodes, reduce the proportion of energy consumption, and prolong the life cycle of the transmission network.

The ERTDA algorithm observes and monitors sensor nodes' behavior using the Watchdog mechanism, including data collection, transmission, and fusion. For every other fixed period, the node records the received node data and calculates the trust value of the node by using the Beta distributed management model. Finally, all sensor nodes establish the trust value table through the Watchdog mechanism. The super-brain system needs to deal with four business systems: management, security, internet, and more than 30 kinds of real-time data. To ensure the accuracy of data collection, we introduce the ERTDA algorithm to make a correct or wrong binary evaluation judgment on the behavior of sensor nodes by using Beta distribution and define the Beta probability density function by gamma function Γ:

$$Brta(\varphi|\alpha,\beta)\frac{\Gamma(\alpha+\beta)}{\Gamma(\alpha)\cdot\Gamma(\beta)}\varphi^{\alpha-1}(1-\varphi)^{\beta-1}, \tag{1}$$

where $0 \leq \varphi \leq 1$, $\alpha > 0$, $\beta > 0$, here is the probability of the behavior or event represented by the sending parameter φ. However, when the constraint $\alpha < 1$ is satisfied, $\varphi \neq 0$, and when $\beta < 1$, $\varphi \neq 0$, the expected value of the Beta probability distribution density function is:

$$\mathrm{E}(\varphi) = \frac{\alpha}{\alpha \div \beta}, \tag{2}$$

In the ERTDA algorithm, it is assumed that sensor node i and sensor node j observe each other's state. The parameters α and β are represented by positive correct behavior m and negative wrong behavior n. Once the trust degree of the possibility of an event happening in the entity in the future is obtained, it will be:

$$\alpha = \mathrm{m}+1,\ \beta = \mathrm{n}+1 \tag{3}$$

where $\mathrm{m} \geq 0$ and $\mathrm{n} \geq 0$. m is the number of correct behaviors of the target node j observed by the monitoring node i, n is the number of wrong behaviors of the target node j followed by the monitoring node i. The parameter φ is the reputation value of the target node, therefore it represents the probability that the reputation value of the target node takes different values. It means the expectation of the reputation value of the target, that is, the er point. It indicates the trustworthiness of the node [17].

Under the guidance of wireless sensor network technology, more intelligent and miniaturized sensors bring a wide range of product information into the Internet of Things system. We adopt the improved ERTDA algorithm model to most likely assist sensor devices in efficiently processing user information. It strengthens the timeliness and accuracy of data processing, reduces data delay and broadband pressure, and realizes more accurate service and interaction with participants [10].

3.2. Perception Data Flow on the Intelligent Scene

The super-brain system focuses on the real-time monitoring of scenic area passenger flow and the timely issuance of passenger overload alert in terms of technical application. By analyzing the environmental capacity of scenic spots and setting the overload threshold of passenger flow, the data collected by sensors is converted into a tourist heat map to predict passenger flow. When the actual passenger flow in the scenic spot reaches the upper limit of passenger flow, it will provide a scientific basis for diversifying people in tourist-intensive areas. In addition, the temperature, humidity, wind speed, and other environmental factors in the scenic area are sorted out, and the meteorological changes in a short period of time are predicted to cope with potential meteorological disasters such as rainstorms, floods, and debris flow and ensure the safety of tourists' lives and property.

In crowded scenes, pedestrian counting often cannot get high statistical accuracy because of unreliable detection. To solve this problem, in this paper, based on the use of Convolutional Neural Network (CNN) technology, we carry out pedestrian counting according to head detection. Firstly, we used the cascaded Adaboost detector to get the preliminary head proposals. Then, we used transfer learning technology to retrain CNN and, after that, the head classification model constructed by CNN and Support Vector Machine (SVM) was used to fine-recognize the head to improve the detection accuracy rate. Finally, the track association was used for tracking and counting the head targets. Experimental results show that our proposed method can locate a single pedestrian quickly and accurately, and the process has relatively high statistical accuracy [18].

- Training sample: the positive sample used to train the CNN classifier model is consistent with the positive selection used to prepare the cascaded Adaboost detector;
- Training and detection process: After the test set images pass through the cascaded Adaboost detectors, many target areas are obtained. These target areas are input into the CNN model to get the final human head target;
- After the final head detection target is obtained by the CNN classifier model, the Euclidean distance is limited to the current detection target, the candidate-associated head matching area is obtained, and the associated head track is obtained. When the associated headway crosses the set detection line, a count is performed. During this process, signs of visitor movement are judged by the changes in sign position.

The CNN learning and training algorithm:
Input: training sample set:

$$S\{(x_1, y_1), (x_2, y_2), \ldots, (x_m, y_m)\} \tag{4}$$

while m is the total number of pieces:

$$y_i = \{-1, +1\} \tag{5}$$

Output: the final CNN classifier f(x).

3.3. Data Transfer and Events Processing

With the continuous development of 5G communication technology, the Internet of Things, with intelligent sensors as an essential bridge, has realized the intelligent sensing and cloud transmission of data information. Through the three-screen linkage of the command center, messages, events, and instructions can be efficiently issued and reported. For example, when a tourist is lost, different types of sensors will jointly search and provide helpful decision-making information in terms of a video surveillance query of the lost person, a prediction of its possible location, detection of nearby calling resources, etc., to speed up the handling of the incident of missing people (Figure 4).

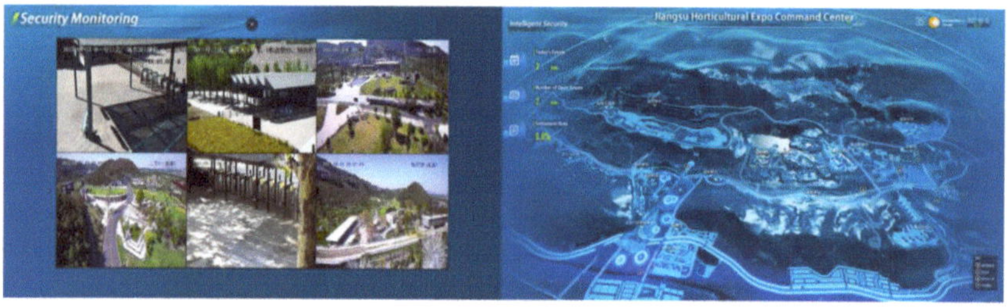

Figure 4. Super-brain system intelligent security system.

- Lost persons can call the emergency number of the park for user location, and the sensor can get the location information of tourists in time. Taking the maximum speed v of position movement in unit time as the radius of a circle as the in-situ sensor within the radiation range randomly samples n possible position points;
- The sensor observation service obtains real-time data of observation attributes or historical observations in a specified period. The trained classification prediction model predicts the next beacon node close to the sampling point per unit of time. It is summarized as the close beacon node. The unknown node is located in the intersection area between the maximum movement speed radius circle of the sampling point and the communication radius circle of the beacon node to which it belongs;
- The mobile sensor obtains the latest position data, or a position data sequence in a particular period of time, and uses the deflection direction of the mobile node to eliminate the impossible coordinate position points;
- Take the mean value of the remaining location points as the result of the location prediction of lost people. The video sensor links the data address and displays the observation data sequence in the form of graph and digital dynamic change.

There are n beacon nodes uniformly distributed in the park sensor network, each node has the same communication radius r and fails to cover the whole network, and each node has a signal-receiving device for receiving and measuring the signal strength from other nodes to that node. Let $P_r(d)$ denote the signal strength received by a receiver (j) at a distance d from source (i), P_r is the transmitted power of the source, G_t, and G_r denote the signal gain when sending and receiving, respectively, and (λ) is the electromagnetic wave wavelength. The relationship between them can be expressed by F_{riis}, the equation as:

$$P_r(d) = \left(\frac{\lambda}{4\pi d}\right)^2 P_t G_t G_r \tag{6}$$

Assuming that the magnitude of the signal strength measured by the signal source beacon node (i) itself is P_i, it will be:

$$\mathrm{A}\left[P_i, P_j\right] = \frac{P_{r(d)}}{P_i} \tag{7}$$

where $A[P_i, P_j]$ is the affiliation degree of an unknown node (j) to beacon node (i). The affiliation degree represents the ratio of the signal strength magnitude received from beacon node (i) at node (j) to the signal strength magnitude of beacon node (i) itself. From this, the node affiliation vector can be established as follows:

$$\underset{B}{\rightarrow} = \{ID,\ (x_i, y_i),\ A[P_i, P_1], A[P_i, P_2], \ldots, A[P_i, P_{n-1}], A[P_i, P_n],\ A \in (0,1)\} \quad (8)$$

where ID denotes the number of the ith beacon node and (x_i, y_i) is the coordinate position of beacon node (i). Since the lost visitor is in motion, each value A in its collected node affiliation vector $\underset{B}{\rightarrow}$ changes at all times, i.e., the closer $A[P_i, P_1]$ is to 1, the closer the two nodes are. The unknown node's motion unknown can be predicted according to the change law of A value, which improves the efficiency and accuracy of the calculation [19].

3.4. Intelligent Scenes and Interactive Experience

According to the scene theory, the scenes refer not only to spatial environmental scenes, but also includes the environmental atmosphere of behavior and psychology created by media information [20]. In the IoT environment, the connotation of the scene has been further extended. The collection, perception, processing, and analysis of scene elements, such as social relations, creates a sense of presence [21]. With the updated iteration of sensors, the amount of information collected by data increases exponentially. Personalized recommendations based on tourists' preferences will become an essential channel for future travel consumption. The scene framework provides an ideal solution for personalized consumption experiences under artificial intelligence and big data technology [10].

The advantage of the super-brain system lies in integrating tourism with science and technology so that tourists can enjoy unmanned technology experiences and intelligent services in all aspects, such as ticket purchase, sightseeing, catering, shopping, transportation, accommodation, etc., and provide tourists with convenient and exciting consumption experiences. The IoT technology group has also improved the accuracy and efficiency of sensors, positioning systems, and big data transmission. For example, through all kinds of sensor devices and positioning systems, consumers' activity information can be instantly collected, consumer services that meet the needs of popular culture can be customized, and personalized intelligent consumption scenes can be built. In addition to innovative services, the super-brain system is committed to improving the public service experience and providing brilliant hardware facilities and intelligent services. For example, in the visitor's small program, you can make reservations and purchase tickets, check the arrival time of the sightseeing bus in real-time, and update the occupancy rate information of smart toilets within the play radius (Figure 5).

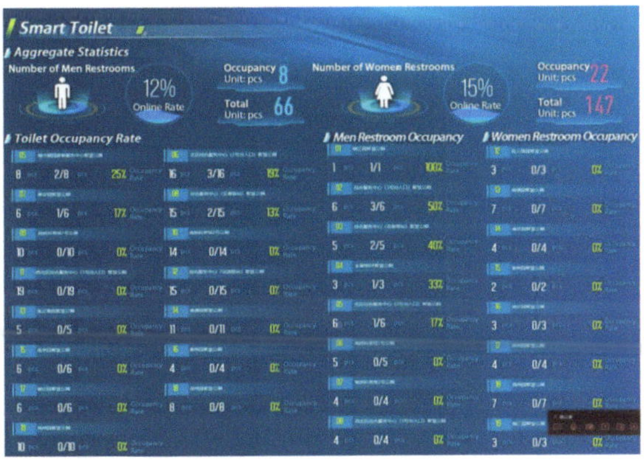

Figure 5. Smart toilet.

4. Discussion

Relying on the super-brain system, several sensor application products have been produced throughout smart tourism. Sensors have been widely used in scenic spot monitoring and management with different functions and scenes, such as panoramic spot running situation monitoring, tourist flow warning, tour route optimization, intelligent service experience, etc., and have achieved specific practical results. In short, integrating and combining sensors and tourist attractions is expected to create more comfortable tourism services and more excellent commercial value. However, it has been found that sensors still face particular challenges in their operation, maintenance, and cooperative processes. Meanwhile, sensors used in tourist attractions cannot make a unique and comprehensive standard according to different stakeholders and organizers, so it is necessary to promote and improve related research and practice continuously.

4.1. Range and Data Transmission Problems of Wireless Sensors

Wireless monitoring sensors solve the problem of collecting various environmental data, and they can realize long-term, timely, and reliable data collection and the wireless transmission of environmental information. However, from the existing practical experience, the power supply efficiency and system reliability of wireless monitoring sensors are limited. Wireless sensors also have obvious shortcomings, such as data fusion technology, power consumption, the dynamic topology of the network, limited node functions, fault tolerance, and so on [22]. The practical application in some scenic spots and the effect of analysis and prediction could be better. For example, the composition and structure of sensors in scenic spots are becoming more and more complex. In some areas, due to the extensive monitoring area and comprehensive sensor coverage, the monitoring devices are limited by wired or battery power supply, which makes it challenging to carry out large-area distributed installation, long-term use, and maintenance. At present, generating an energy model for sensor nodes can accurately predict the energy consumption of nodes, and it is an essential part of protocol development, wireless sensor network design, and Wireless Sensor Network (WSN) performance evaluation [23]. Currently, the practice of using wireless sensors to collect data is abundant, therefore three new schemes for reference are summarized for the research of the low-power sensing systems.

4.1.1. Low-Power Device Selection

To reduce the system's dependence on the power supply, the most direct and effective way is to reduce the power consumption of the system, that is, by designing a low-power energy management circuit. In low-power consumption design, selecting chip devices for hardware circuits is one of the most critical links [24].

The wireless sensor node is the smallest unit of the sensor network, composed of different structures according to other requirements. The technical bottlenecks of current wireless sensor networks are how to reduce the energy consumption of wireless sensor nodes and rationalize the energy of sensors. Due to the wide range of monitoring in the environment, it is necessary to deploy multiple nodes. Therefore, after ensuring that a single node can achieve a more efficient monitoring function, we should try our best to choose low-consumption and low-cost optimization and configuration. The wireless transceiver module exchanges the data sent by the processor module with other sensor nodes through wireless communication to meet information transmission requirements (Figure 6).

To ensure the long-term and stable operation of the monitoring point of the Expo site, it is necessary to select the low-power consumption sensor chip and the main control chip. Through comparison, it is found that using MSP430 as the main control chip can significantly reduce the system's power consumption. At the same time, integrating the DS18B20 temperature sensor and the ADXL345 three-axis acceleration sensor can monitor the temperature and vibration information of the equipment in real-time and provide stable voltage output for the sensing system.

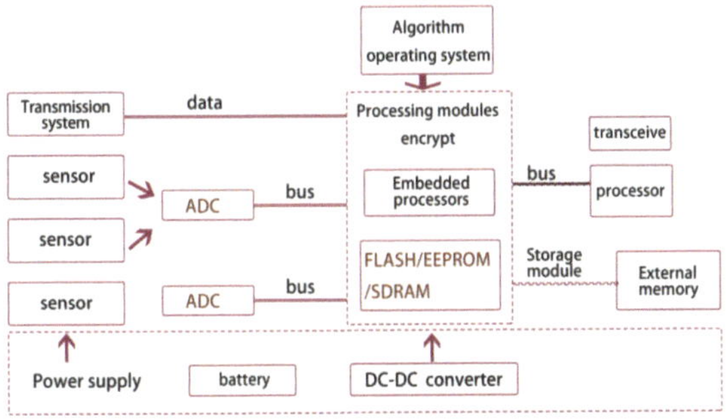

Figure 6. Hardware structure diagram of the wireless sensor node.

4.1.2. Low-Power Circuit Design

Sensor nodes distributed in the Horticultural Expo are the basic units of the wireless sensor network. The design of low-power nodes can prolong the network's life and effectively improve the operational performance of sensors. The sensor nodes contain the main power-consuming components of the circuit. These components only need to be on during operating hours, so in non-operating mode, the power supply to these components needs to be completely turned off to achieve low power consumption in the system [25].

Although the battery power supply of the sensor network is limited, the low-power circuit design has been used in empirical research to maintain the regular operation of the sensor and improve the working efficiency. Energy-aware routing (AODV routing protocol) will strive to keep most nodes running in their maximum lifetime. Each node with a high energy consumption rate and a short remaining life cycle should be shut down for a period of practice. The high energy consumption rate is determined by comparing the energy consumption rate of this node with other nodes. Closing a node will make the energy-aware routing protocol choose to replace the node or change the whole route to the destination node. This repeated process can distribute routing roles among most nodes, thus balancing the network's energy consumption. The steps of the Ad hoc On-Demand Distance Vector Routing (AODV) energy awareness reason protocol are shown in Figure 7. These steps are as follows:

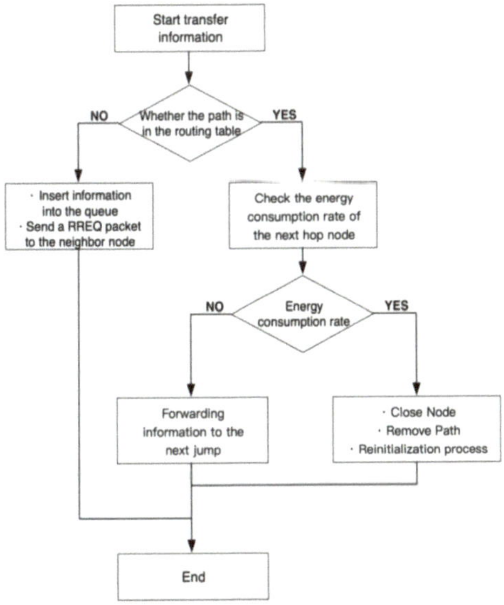

Figure 7. Steps of the AODV energy-aware routing protocol.

- If a sensor node needs to transmit a message, it must check its routing table to find a way to the destination node. Thus, if a route can be found in the routing table, it forwards the message to the next node. Otherwise, the information is kept in the queue, and the source node sends routing request (RREQ) packets to its neighboring nodes to initiate the route discovery process;
- Before forwarding the message to the next hop, the energy consumption rate of the next hop is checked;
- If the energy consumption rate is high, the next hop will be closed for a specified period. The route will be removed from the routing table, which will result in starting the route discovery process again at the source node to find a new route to the destination node;

The synchronization mechanism of the wireless sensor can ensure that all nodes in the network wake up and sleep. On the one hand, it can avoid the waste of network nodes' energy, and at the same time, it can ensure that all nodes in the system have the same energy consumption [26]. The flow chart of the time synchronization mechanism in this paper is shown in Figure 8.

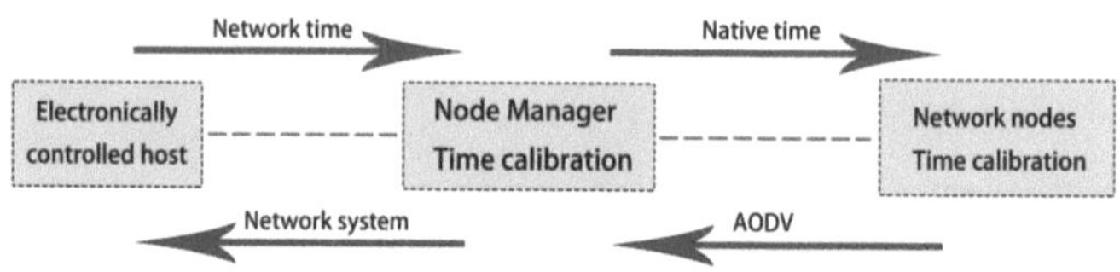

Figure 8. Time synchronization mechanism.

4.1.3. Dynamic Consumption Management Technology

Dynamic power consumption is when the load content is charged and discharged. Dynamic power consumption is the power consumption the digital circuit must calculate when it finishes its work, including flip power consumption and short-circuit power consumption. The short-circuit power consumption is the power consumption of the Complementary Metal Oxide Semiconductor (CMOS) when the Positive Channel Metal Oxide Semiconductor (PMOS) and the N-Metal Oxide Semiconductor (NMOS) transistors are turned on simultaneously. According to the principle of the capacitor charging and discharging point, the power consumption formula for inversion is shown as:

$$P_{switch} = a{\cdot}f{\cdot}c{\cdot}VVD2, \tag{9}$$

where a is the activity factor, f is the signal frequency, c is the load capacitance, and VDD is the supply voltage. The formula shows that switching power consumption is closely related to the load capacitance, activity factor, signal frequency, and supply voltage. Therefore, the supply voltage and frequency can reduce the flip power consumption. The input signal of CMOS is transformed by logic level, and the PMOS and NMOS are connected with the ground, resulting in a short-circuit current, whose consumption formula is shown as:

$$P_{internal} = t_{shart}{\cdot}f{\cdot}VVD{\cdot}l_t, \tag{10}$$

The power consumption can be reduced by adjusting the duration of the simultaneous conduction of PMOS and NMOS. By synthesizing (9) and (10), the total dynamic consumption can be expressed as formula (11).

$$P_{tatal} = a{\cdot}f{\cdot}c{\cdot}VVD^2 + t_{shart}{\cdot}f{\cdot}VVD{\cdot}l_t, \tag{11}$$

For environmental sensors, power consumption is not only a problem of energy consumption but is also affected by transmission and feasibility. Therefore, to reduce the sensor's dynamic consumption during operation, a sensor dynamic power consumption management technology is proposed. According to the energy consumed in different working states, the duration of other active states is set to reduce power consumption. (Table 2)

Dynamic consumption management technology can solve the workload situation in sensor nodes or the situation that idle nodes can not work again because of long-term existence.

Table 2. Sensor working status module table.

Conversion Duration	Control Module	Wireless Module	Functional Mode
10 ms	Rest	Standby	Resting State
20 ms	Rest	Open	Resting State
15 ms	Free	Open	Receiving State
15 ms	Run	Open	Sending State

It can be seen from the table that the conversion time of additional nodes in the intelligent sensor system is different in other active states.

- In the dormant state, the consumption of sensor nodes is minimal. Still, the sensor nodes cannot perceive the surrounding environment, which leads to a lack of environmental monitoring information. At the same time, the transition from the dormant state to other modes takes a long time, which has a particular impact on the transmission of the data monitoring system. It should be designed and optimized in a dormant state, and the duration should be shortened as much as possible to make up for the deficiency of the transition;
- In the standby state, the central board node is in the dormant state, the wireless module is running, and the conversion time is relatively long, therefore it is necessary to sense and receive the information data of the node all the time. In the process of the transition from the receiving state to the standby state, the delay time is long;
- In the receiving state, the wireless sensor is in the state of receiving information data, and its consumption function is second only to the sending state;
- In the sending state, every part of the sensor system usually works, and all modules usually operate, therefore the consumption is also the largest. The running time should be shortened as far as possible based on ensuring the regular operation of the sensor system during optimization;

The number of tourists in the scenic spot will change with the tour's content, so the wireless transmission route will significantly impact the power consumption in the network transmission process. Based on this, we optimized the routing protocol to reduce energy consumption in the network transmission process.

When the data transmission time in the stable transmission phase is longer than that in the cluster establishment phase, the energy consumption of network operation can be saved to a greater extent, and the data acquisition time can be increased. As shown in Figure 9, after measuring the geographical location and residual energy of cluster head nodes, the information collected by cluster head nodes with close distance and more residual energy is directly sent to the sink nodes. In the past, until all the data were transmitted to the sink node after this data transmission was completed, the system would conduct a new round of cluster establishment and stable transmission and re-plan the cluster head election and multi-hop transmission.

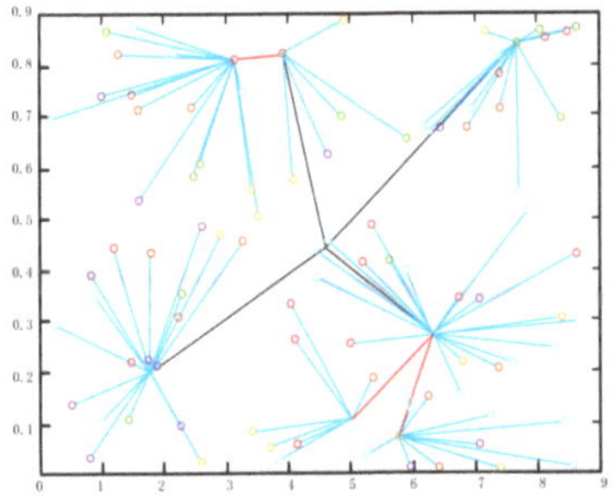

Figure 9. Network node clustering graph.

The working states of the above sensor nodes must be switched by combining software and hardware systems. The system power consumption can be reduced dynamically when the primary receiving, processing, and sending functions are realized.

4.2. Multi-sensor Collaboration Models and Data Fusion Issues

Many kinds of sensors have errors in data fusion, leading to a suboptimal estimation of events, resulting in low efficiency and significant processing problems. Just as the application of sensor data fusion to optimize the decision-making process in empirical research has produced remarkable practical results. According to the research literature, several ways are summarized to combine different data sources, such as decision-making, averaging, guidance, Bayesian statistics, and integration.

In terms of data fusion, data fusion by multiple sensors can accurately analyze and process data information while reducing the amount of data transmission for subsequent decision making and evaluation. In this paper, the data of multiple sensors are fused, which can reflect the mutual support of multiple sensors, thus avoiding the limitation of the measurement performance of a single sensor, thus improving the overall effectiveness and accuracy of the multi-sensor monitoring system. According to the test values of each sensor, the adaptive fusion algorithm finds the optimal weighting factor adaptively corresponding to each sensor. Under the condition of satisfying the minimum total mean square error, the fused result is optimized, thus meeting the needs of the environmental monitoring of the Expo, ensuring that the super-brain system of Horticultural Expo can quickly fuse data, coordinating and helping tourists in the scenic spot to meet various service needs (Figure 10) [27].

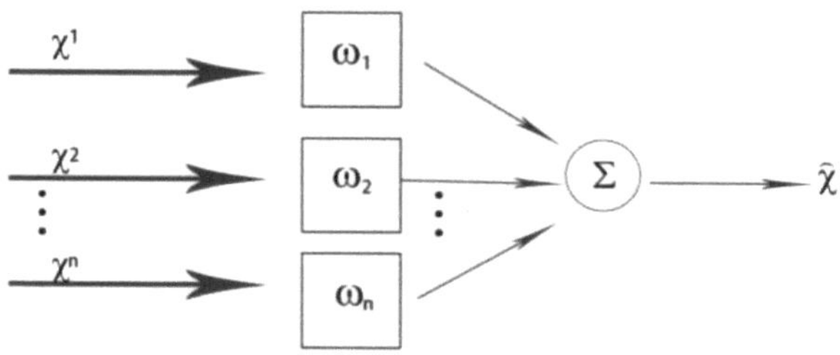

Figure 10. Adaptive weighted fusion algorithm model.

For any security system, the multi-sensor environment acquisition scheme is the key to overcoming the uncertainty caused by a single sensor. The cooperation mode and data fusion of different types of sensors help each group of sensors provide one kind of data for the system. The system fuses these heterogeneous data, thus realizing the mutual complement of sensor types (Figure 11).

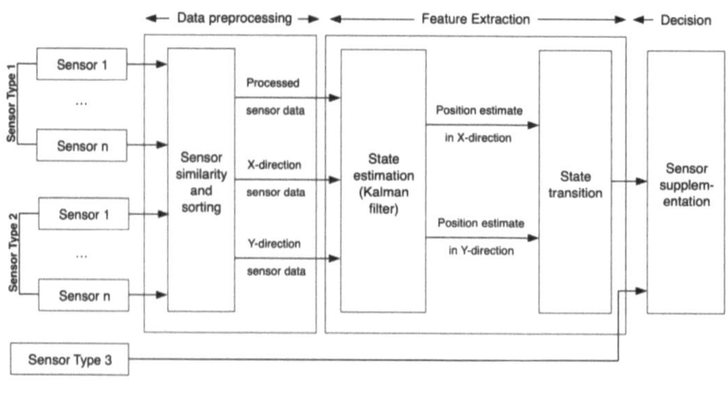

Figure 11. Complementary system block diagram.

The system used different types of data input as a part of the process to make improved decisions. The idea of this stage is to find the mutual complement of different types of input data. These provide an estimate (Kalman Filter) to predict the system's state. Finally, this estimate is provided to the decision-making module, and the final system decision is obtained.

We compare the results with the commonly used data collection technologies, namely Envoy HTTP and Rabbit MQ. The results show that in the request or reply interaction (Figure 12a), when the number of traversed hops increases, the performance of CMOS will be better. The performance of CMOS forwarding realized by proxy is better at the local request rate (Figure 12b).

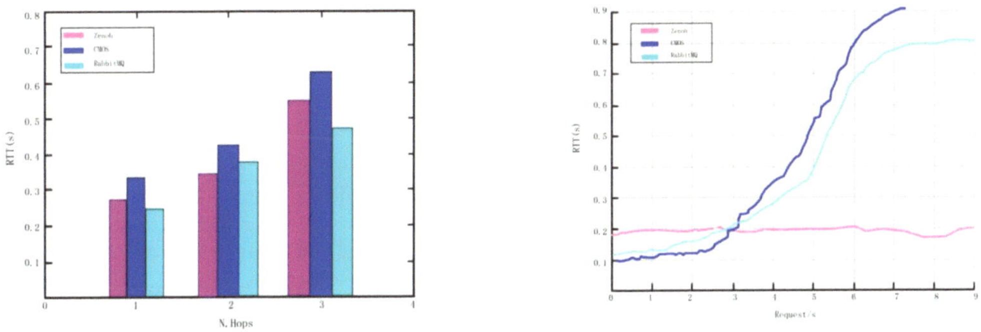

(**a**) (**b**)

Figure 12. Data collection. (**a**) Graph of round-trip times for requests across network hops. (**b**) Request round trip time graph.

In the third test, following the same criteria as above, comparing and verifying that there are different complex infrastructures and network topologies between the two layers, there is routing capacity and delay in distributing increasing information among multiple users. The test results show that the sensor will be delayed even after passing through multiple intermediaries (Figure 13a), but the two solutions are relatively less affected by complexity. The results show that the proposed platform can process and memorize a large amount of data and re-compare, and provide extremely limited total delay (Figure 13b).

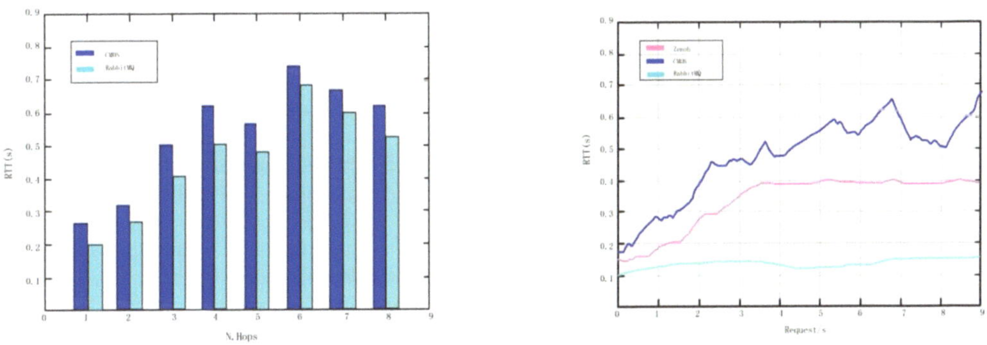

(**a**) (**b**)

Figure 13. Data collection and comparison. (**a**) Sensor registration delay. (**b**) Data integration.

MATLAB 2020a software is used to simulate the experiment, and 2000 round experiments are carried out. The number of surviving nodes and energy consumption is compared to traditional sensors. (Table 3) Set 100 sensor nodes randomly distributed in a 10,000 square area, where the location of the sink node is in the center of the whole network (50m, 50m). The specific parameters are as follows.

Table 3. Simulation parameter table.

Parameter	Value
Convergent Node Position	(50 m, 50 m)
Number of Nodes	100
Network Area	100 m × 100 m
Packet Length	4000
Node Initial Energy E_0	0.5
Power consumption of free channel model signal amplifier E_{FS}	10
Consumption of Signal Amplifier in Multipath Fading Channel Model E_{mp}	0.0013

4.3. Reference Achitecture of Personalized Service and Data Refinement of Sensors

According to the sensor's specialized construction of different stakeholders and the organizer's supervision system, it can reshape the tourism sensor of Horticultural Expo. Based on human-centered interaction and data collection, the personalized intensity experience of different tourists is generated. Sensors gather new service forms based on providing essential services for tourist attractions. They will combine them with vital services according to specific tourist profiles and current needs to create a common-view platform and customize personalized play experiences with different strengths for tourists. At the same time, based on the integration and enhancement of service capabilities, it will add additional value to producers and consumers. The data sensor is responsible for collecting, managing, and analyzing all the information provided by third-party suppliers, realizing the expansion and integration of data, and processing dynamic and non-dynamic real-time information in a timely and rapid manner [28]. For example, according to different tourists' needs and other scene information in the venue, the dynamic route of cultural heritage, the active route of ecological restoration, the active route of sightseeing, the immersive interactive play route, etc., are customized. The application program combines tourists' preferences with urban transportation networks and points of interest generated. Through the development of intelligent sensors in Horticultural Expo, this paper reconsiders and proposes smart tourism examples based on intelligent sensors: the extension and connection of clever technology to tourists' gaze, interactive travel, thoughtful analysis and decision-making, immersion and authenticity, etc.

5. Conclusions and Future Work

In the research of this paper, it is found that data observation and resource fusion based on sensor technology can realize both big data collection for the whole area of Horticultural Expo and provide a reliable data source for the in-depth study of smart tourism. The data show that based on the trust management mechanism, the super-brain system proposed an efficient and energy-saving data fusion algorithm ERTDA, which effectively guaranteed the network security and extended the network running time. The hardware system and sensor technology are effectively combined to provide information technology support for treating emergency events in scenic spots and improve the overall management level of scenic spots. In addition, two referential solutions are summarized, aiming at the low-power consumption sensor system of the intelligent sensor in the Horticultural Expo, as well as the optimization method of multi-sensor and data fusion. In the deep integration of sensors and tourism, the performance of sensors should be optimized again to reduce the energy consumption of sensors.

As a recent example of "smart tourism", the super-brain intelligent sensor system in Jiangsu Horticultural Expo is worth exploring for more possibilities. Although the intelligent system and operation mode of the "super-brain" system is analyzed in detail in this paper, other aspects of intelligent sensors are not discussed further. For example, personalized service customization, applying the super-brain system to the cultural value development of Horticultural Expo, popularizing and installing intelligent sensor systems in different tourism environments, and so on. In future research, consumer demand based on tourists' preferences and personalized service will become an essential channel for smart tourism to upgrade its cultural industry format. Based on collecting and analyzing tourists' data, the sensor group realizes the in-depth mining of cultural tourism consumption behavior. Drawing accurate portraits of tourists and perceiving tourists' consumption preferences builds a scene setting that matches the tourists. Finally, it makes personalized recommendations on their consumption content. With the support of sensor technology, we realize the collection, perception, and processing of scenic scene elements, create multiple scene categories, develop and apply smart technologies to different tourism environments.

Author Contributions: Conceptualization, Y.T., C.H. and X.Z.; methodology, R.T.; software, C.H.; validation, C.H., R.T. and X.Z.; formal analysis, C.H.; investigation, R.T.; resources, R.T.; data curation, X.Z.; writing—original draft preparation, C.H.; writing—review and editing, X.Z.; visualization, R.T.; supervision, C.H.; project administration, R.T. All authors have read and agreed to the published version of the manuscript.

Funding: This research was funded by 2022 National Social Science Foundation Art Project of China, grant number 22BA023, and Research on the Theoretical System of Art Communication in the Context of Digital Media.

Institutional Review Board Statement: Not applicable.

Informed Consent Statement: Not applicable.

Data Availability Statement: Not applicable.

Conflicts of Interest: The authors declare no conflict of interest.

References

1. Beilin, Z. Functional Prospect of Smart Cultural Tourism Application Products in "Digital Humanities". *J. Libr. Inf. Serv.* **2021**, *24*, 35–43.
2. Gretzel, U.; Sigala, M.; Xiang, Z.; Koo, C. Smart Tourism: Foundations and Developments. *J. Electron. Mark.* **2015**, *3*, 179–188. [CrossRef]
3. Qingzhong, M.; Junfen, W. Discussion on Enabling High-quality Development of Tourism Industry by "Blockchain Smart Tourism". *J. Acad. Explor.* **2021**, *9*, 48–54.
4. Kasnesis, P.; Tatlas, N.A.; Mitilineos, S.A.; Patrikakis, C.Z.; Potirakis, S.M. Acoustic Sensor Data Flow for Cultural Heritage Monitoring and Safeguarding. *J. Sens.* **2019**, *19*, 1629. [CrossRef] [PubMed]
5. Gretzel, U.; Reino, S.; Kopera, S.; Koo, C. Smart Tourism Challenges. *J. Tour.* **2015**, *16*, 41–47.
6. Aguilar, E.; Remesciro, B.; Bolanos, M.; Radeva, P. Grab, Pay and Eat: Semantic Food Detection for Smart Restaurnts. *IEEE Trans. Multimed.* **2018**, *20*, 3266–3275. [CrossRef]
7. Cacho, A.; Mendes-Filho, L.; Estaregue, D.; Moura, B.; Cacho, N.; Lopes, F.; Alves, C. Mobile Tourist Guide Supporting: A Smart City Initiative: A Brazilian Case Study. *Int. J. Tour. Cities* **2016**, *2*, 164–183. [CrossRef]
8. Car, T.; Stifanich, L.P.; Šimunić, M. Internet of Things (IoT) in Tourism and Hospitality: Opportunities and Challenges. *Tour. South East Eur.* **2019**, *5*, 16–18.
9. French, A.M.; Robert, L.X.; Bose, R. Toward a Holistic Understanding of Continued Use of Social Networking Tourism: A Mixed-Methods Approach. *J. Inf. Manag.* **2017**, *54*, 802–813. [CrossRef]
10. Yan, M.; Wang, J.; Shen, Y.; Lv, C. A non-photorealistic rendering method based on Chinese ink and wash painting style for 3D mountain models. *Herit. Sci.* **2022**, *10*, 186. [CrossRef]
11. Kumar, B.; Sharma, N. Approaches, Issues and Challenges in Recommender Systems: A Systematic Review. *J. Sci. Technol.* **2016**, *9*, 1–12.
12. Wang, N. Research on Construction of Smart Tourism Perception System and Management Platform. *Appl. Mech. Mater. J.* **2014**, *687*, 1745–1748.
13. Hamid, R.A.; Albahri, A.S.; Alwan, J.K. How Smart Is E-Tourism? A Systematic Review of Smart Tourism Recommendation System Applying Data Management. *J. Comput. Sci. Rev.* **2021**, *39*, 337. [CrossRef]

14. Ozdemir, S. Functional Reputation Based Data Aggregation for Wireless Sensor Networks. In Proceedings of the IEEE International Conference on Wireless & Mobile Computing, Networking & Communication, Washington, DC, USA, 12–14 October 2008; IEEE Press: New York, NY, USA, 2008; pp. 592–597.
15. Ozdemir, S. Functional Reputation Based Reliable Data Aggregation and Transmission for Wireless Sensor Networks. *J. Comput. Commun.* **2008**, *17*, 3941–3953. [CrossRef]
16. Liu, C.; Liu, Y.; Zhang, Z. Improved Reliable Trust-Based and Energy-Efficient Data Aggregation for Wireless Sensor Networks. *Int. J. Distrib. Sens. Netw.* **2013**, *9*, 1–11. [CrossRef]
17. Ganeriwal, S.; Balzano, L.K.; Srivastava, M. Reputation Based Farmwork for High Integrity Sensor Networks. *ACM Trans. Sens. Netw.* **2008**, *3*, 1–37. [CrossRef]
18. Yan, M.; Li, W.; Chan, C.A.; Bian, S.; Chih-Lin, I.; Gygax, A.F. PECS: Towards personalized edge caching for future service-centric networks. *China Commun.* **2019**, *16*, 93–106. [CrossRef]
19. Aoudjit, R.; Belkadi, M.; Dsoui, M. Mobility Prediction Based on Data mining. *Int. J. Database Theory Appl.* **2013**, *2*, 71–78.
20. Meyrowitz, J. *No Sense of Place: The Impact of Electronic Media on Social Behavior*; Oxford University Press: Oxford, UK, 1986.
21. Madden, K.; Elaine, R.; Sharon, L.; Joan, C. Trailgazers: A Scoping Study of Footfall Sensors to Aid Tourist Trail Management in Ireland and Other Atlantic Areas of Europe. *J. Sens.* **2021**, *6*, 2038. [CrossRef]
22. Dong, H.; Longjun, W.; Yufeng, X.; Hongyan, L.; Gombay, N. Fuzzy System for Monitoring Energy Consumption of Wireless Sensor Network Nodes. *J. Intell. Fuzzy Syst.* **2018**, *4*, 4319–4328.
23. Robert, S.; Shel, I. *Age of Context: Mobile, Sensors, Data and the Future of Privacy*; Create Space Independent Publishing Platform: Scotts Valley, CA, USA, 2013.
24. Zhanga, Z.; Hea, J.; Wena, T.; Zhaia, C.; Hana, J.; Mua, J.; Jiab, W.; Zhanga, B.; Zhang, W.; Choua, X. Magnetically Levitated-Triboelectric Nanogenerator as a Self-Powered Vibration Monitoring Sensor. *J. Nano Energy* **2017**, *33*, 88–97. [CrossRef]
25. Liu, M. Modified Monitoring System of Soil Temperature Based on ARM. *J. Environ. Technol. Innov.* **2021**, *21*, 101346. [CrossRef]
26. Chen, J.; Xiao, W.; Li, X.; Zheng, Y.; Huang, X.; Huang, D.; Wang, M. A Routing Optimization Method for Software-Defined Optical Transport Networks Based on Ensembles and Reinforcement Learning. *Sensors* **2022**, *22*, 8139. [CrossRef] [PubMed]
27. Yan, M.; Li, S.; Chan, C.A.; Shen, Y.; Yu, Y. Mobility Prediction Using a Weighted Markov Model Based on Mobile User Classification. *Sensors* **2021**, *21*, 1740. [CrossRef]
28. Andrea, S.; Thomas, V.; Antonio, C. An Architecture for Service Integration to Fully Support Novel Personalized Smart Tourism Offerings. *J. Sens.* **2022**, *22*, 1619.

sensors

Article

A Novel Improved YOLOv3-SC Model for Individual Pig Detection

Wangli Hao, Wenwang Han, Meng Han and Fuzhong Li *

School of Software, Shanxi Agricultural University, Jinzhong 030801, China
* Correspondence: lifuzhong@sxau.edu.cn

Abstract: Pork is the most widely consumed meat product in the world, and achieving accurate detection of individual pigs is of great significance for intelligent pig breeding and health monitoring. Improved pig detection has important implications for improving pork production and quality, as well as economics. However, most of the current approaches are based on manual labor, resulting in unfeasible performance. In order to improve the efficiency and effectiveness of individual pig detection, this paper describes the development of an attention module enhanced YOLOv3-SC model (YOLOv3-SPP-CBAM. SPP denotes the Spatial Pyramid Pooling module and CBAM indicates the Convolutional Block Attention Module). Specifically, leveraging the attention module, the network will extract much richer feature information, leading the improved performance. Furthermore, by integrating the SPP structured network, multi-scale feature fusion can be achieved, which makes the network more robust. On the constructed dataset of 4019 samples, the experimental results showed that the YOLOv3-SC network achieved 99.24% mAP in identifying individual pigs with a detection time of 16 ms. Compared with the other popular four models, including YOLOv1, YOLOv2, Faster-RCNN, and YOLOv3, the mAP of pig identification was improved by 2.31%, 1.44%, 1.28%, and 0.61%, respectively. The YOLOv3-SC proposed in this paper can achieve accurate individual detection of pigs. Consequently, this novel proposed model can be employed for the rapid detection of individual pigs on farms, and provides new ideas for individual pig detection.

Keywords: pig detection; YOLOv3; Convolutional Block Attention Module; Spatial Pyramid Pooling

Citation: Hao, W.; Han, W.; Han, M.; Li, F. A Novel Improved YOLOv3-SC Model for Individual Pig Detection. *Sensors* **2022**, 22, 8792. https://doi.org/10.3390/s22228792

Academic Editors: Chien Aun Chan, Chunguo Li and Ming Yan

Received: 9 October 2022
Accepted: 2 November 2022
Published: 15 November 2022

1. Introduction

Pigs are the most common source of meat products worldwide. With the progress of human society, people pay more and more attention to the quality of pork.

Object detection technology has high value in improving animal welfare. Early in pig production, it can be utilized to monitor pig health to improve pork quality [1]. The dietary behavior of animals is closely related to their health status, and subtle dietary changes are important for animal health observations [2]. When pigs are sick, they usually show reduced feeding, reduced exercise, depression, and lethargy [3]. Leveraging scientific methods to monitor live pigs and, if necessary, human intervention, will help protect animal welfare and prompt pork quality and profitability. Initially, the vast majority of pigs were monitored manually, which led to significant increases in labor intensity. Meanwhile, during the monitoring process, human subjective judgment errors often occur, which was not conducive to the high-quality production of live pigs.

To handle the above problems, in the early days, researchers used RFID systems to monitor pigs' diets. However, the sensitivity of RFID system monitoring was often affected by their surrounding environment as well as their own height, direction, and distance [4]. In pursuit of better accuracy, it is necessary to constantly adjust the position of the antenna [5], and RFID monitoring requires a large number of pig ear tags, which is time-consuming and expensive to maintain. Furthermore, there are also some problems to these approaches that depend on the utilization of the wearable equipment RFID such as being easy to damage, being invasive, and prone to infection [6,7]. Subsequently, the breeder deployed cameras to record the pigs' behavior and manually analyzed the recorded video data, in order to obtain the health status of the pigs. These methods are all based on manual analysis, resulting in a significant increase in the workload of breeders.

Despite some desirable results, the above-mentioned methods suffer from compromised animal welfare and high physical labor intensity. This makes it urgent to leverage more efficient methods, for pig detection. The following are some effective attempts by researchers.

In [8], by leveraging elliptical displacement calculation methods, Kashila et al. has achieved 89.8% accuracy in pig movement detection. Concerning the individual pig classification, Kashila et al. [9] has received 88.7% accuracy in detecting individual pig identification via ellipse fitting technique. Based on traditional computer vision technology [10], Nasirahmadi et al. [11,12] has employed ellipse fitting and the Otus algorithm, to realize the individual pig detection and pig lying position detection. Furthermore, Nasirahmadi [13] has utilized support vector machine (SVM) algorithm to classify pig poses, with 94% classification accuracy achieved. Leveraging the linear discriminant analysis algorithm, Viazzi et al. [14] has achieved 89.0% accuracy in the recognition of aggressive behavior of pigs. Furthermore, a more promising method for individual identification and behavioral recognition of pigs is based on 2D or 3D cameras. For example, Matthews et al. [15] has utilized a depth camera to track the movement of pigs, enabling the effective detection of pigs' standing, eating, and drinking behaviors. Depending on the depth sensor, Kim et al. [16] has realized the pig standing behavior recognition under the complex environment. Meanwhile, the effectiveness of the proposed method in terms of both the cost and the accuracy have been verified. Based on the images captured by the CCD camera, Nasirahmadi [11] et al. have utilized an ellipse fitting approach to locate each pig in the image, while cameras can easily record the pig behavior, factors such as farm environment and lighting conditions can make pig classification challenging.

Currently, deep-learning based approaches [17–19] have achieved promising detection performance, especially in the field of animal phenotype detection. For example, Wu et al. [20] has proposed an effective corbel detection method based on YOLOv3 and relative step size characteristic vector. Specifically, based on the relative step size characteristic vector, the YOLOv3 algorithm was utilized to detect the position of the corbel, and then the LSTM model was employed to identify the normal walking and the lame behavior of the cattle, and an accuracy of 98.57% obtained. Shen et al. [21] has first applied the YOLO model to detect cows, and then an improved AlexNet model has been employed to classify the corresponding detected individual cow. Finally, they obtained 96.65% accuracy of individual cow classification. Tassinari [22] proposed a deep learning-based system for individual cow classification and location analysis. Zhang [23] proposed a lightweight YOLO detection model, using MobileNetV3 to replace the backbone network in the YOLOv3 network, and obtained 96.8% of the cattle key position detection accuracy. Hu et al. [24] employed the YOLO algorithm to extract cow objects, and then a segmentation algorithm was utilized to extract the head, torso, and legs parts of the corresponding cow object. Subsequently, the deep feature fusion was performed on these extracted parts. Finally, the SVM classifier was employed to do the classification, and an accuracy of 98.36% was obtained. Jiang [25] proposed a filter-based YOLOv3 algorithm and achieved 99.18% accuracy in the detecting key parts of cows. Based on an RGB camera and convolutional neural network, Bezen [26] built a computer vision system for measuring cow feeding, and an accuracy of 93.65% was obtained. Achour [27] has built a CNN-based image analysis system for the classification of individual cows, their foraging behavior, and their food. Specifically, their model obtained an accuracy of 97% for individual cow classification and an accuracy of 92% for cow foraging behavior separately. Wu [28] proposed a CNN-LSTM (Fusion of Convolutional Neural Network and Long Short-Term Memory Network) model for cow action recognition. Specifically, the action categories of cows in their experiments included drinking, ruminating, walking, standing, and lying down, and the average classification accuracy of their model has reached 97.6%.

Above all, while the monitoring method using RFID ear tags is simple, it often causes harm to the pig and compromises animal welfare. Although the computer vision technology [29] can improve animal welfare and recognition accuracy, it is not suitable for industrial production requirements due to its slow detection speed. Furthermore, when the pigs are occluded, or the size of the target pigs in the image varies greatly, the detection performance of the model drops significantly.

Attention mechanism is an important means to improve feature robustness [30], among which Convolutional Block Attention Module (CBAM) [31] have shown promising success in a broad range of fields. Based on an intermediate feature map, CBAM captures attention maps in two independent dimensions, including channel and spatial dimensions. Then, the input feature map is multiplied with this attention map for adaptive feature refinement. Since CBAM is a general and lightweight component, it can be seamlessly incorporated into any CNN architecture for end-to-end training with negligible overhead.

Furthermore, the Spatial Pyramid Pooling (SPP) [32] module realizes the feature map-level fusion of local features and global features, enriching the expressiveness of the final feature map.

Consequently, based on YOLOv3, leveraging the advantages of CBAM and SPP, this paper proposes a novel improved pig detection model YOLOv3-SC. The YOLOv3-SC model enables efficient detection of pigs. In addition, the model can achieve effective pig detection in the case of occlusion, and can achieve effective multi-scale pig targets. The main contributions of this paper is summarized as follows:

- We first propose a novel pig detection method YOLOv3-SC based on the CBAM and the SPP modules. The channel attention and the spatial attention units in the CBAM module enable the YOLOv3-SC to focus on the regions of the image that are important for detection, thereby extracting richer, more robust, and more discriminative features. The SPP module endows YOLOv3-SC the capacity of extracting multi-scale features, which enables the model to detect objects of different sizes, thereby improving the model's pig detection performance. Specifically, our model achieves the best performance for pig detection task, with 2.3% improvement of the existing model.
- Numerous ablation experiments have been designed and performed to verify the performance of our model. Specifically, these studies include the comparison of different models, evaluation of the effectiveness of the spp module, evaluation of the effectiveness of the CBAM module, and the evaluation of the superiority of the YOLOv3-SC.

2. Materials and Methods

2.1. Datasets

The individual pig detection dataset utilized in this paper was collected from one pig farm in Jinzhong City, Shanxi Province, China. The breeding method of this farm is captive breeding, surrounded by iron fences to form a closed area, and the ground of the farm is cement concrete. The data collection cameras were installed at a height of 3 m from the ground, at 45° diagonally and directly above the farm. Through this collection strategy, the whole view of the pig and its range can be well captured. The data collection period lasted for two months, from August to October 2020. It should be noted here that videos with poor picture quality are deleted due to factors such as light, and finally a total of about 2 Terabyte video data is obtained. Specifically, the video is sliced into image frames with rgb format at a sampling rate of 25 f/s and some images with no target objects, blurring and poor quality, are deleted. Further, the labelImg tool was employed to label the image frames in the PASCAL VOC format, and the labeled data was saved as an XML file. Finally, we obtained a dataset with a total of 4019 images with 13,996 annotations, some sample images are shown in Figure 1. Figure 1a indicates the camera position above the 45° angle of the farm. Figure 1b shows the camera position above the 45° diagonal of the farm.

Figure 1c indicates that the camera position directly above the farm. In order to evaluate the performance of the proposed model, the dataset is divided as follows, 3255 samples are employed as the training data, the 362 samples are employed for validation data, and the remaining 402 samples are utilized as the test data. The samples in the test data are the unseen data. Further, to increase the diversity of data and allows the model to obtain richer features, this paper adopts the following data augmentation techniques, such as random scaling, random flipping, random cropping, and other operations.

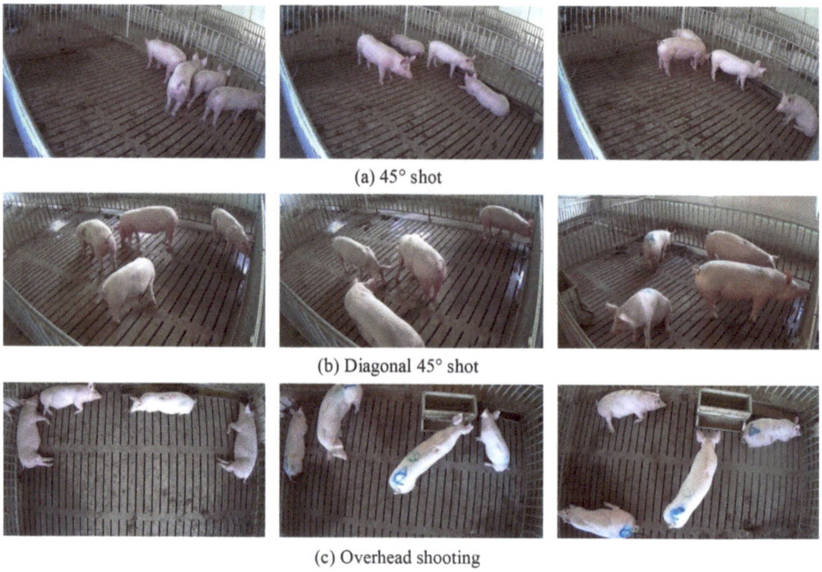

Figure 1. Some examples of the dataset.

2.2. Technical Route

The technical route of the individual pig detection model proposed in this paper is shown in Figure 2. To reduce the noise in the data and enable the model to obtain better detection ability, the samples are first preprocessed and data augmented. Specifically, the image preprocessing operation utilized for data noise reduction refers to deleting samples with poor quality in the data set and and resize the input image to a fixed size 416×416. In order to increase the diversity of data, we have adopted the following data enhancement methods, including the random_distort, random_expand, random_interp, and random_flip, shuffle_gtbox. Subsequently, the processed data is sent to the YOLOv3-SC for model training and evaluation. Finally, an effective individual pig detection model is obtained, which can realize fast and accurate individual pig detection.

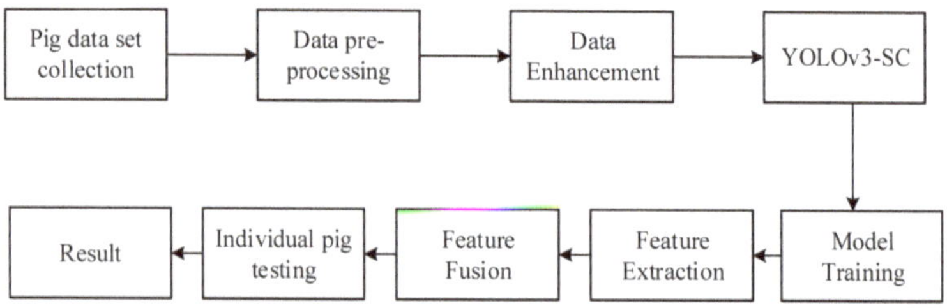

Figure 2. Technical route of the proposed pig detection model.

2.2.1. Feature Extraction

The backbone network DarkNet-53 with the addition of the CBAM of the proposed model, are utilized to extract features with richer spatiotemporal dependencies, which will facilitate the pig detection significantly.

2.2.2. Feature Fusion

The Feature Fusion in Figure 2 refers to two kind of fusions; they are the multiple YOLO-head fusion and the SPP (Spatial Pyramid Pooling) fusion. The detailed feature fusion operations are described in Section 2.5.

2.3. *YOLOv3*

Based on YOLOv1 [33] and YOLOv2 [34], to leverage the anchor mechanism, BN operation, and multi-scale fusion strategies, we proposed YOLOv3 models. The basic principle of YOLOv3 is to divide the input image into $S \times S$ grids, where $S = 7$ and each grid predicts 3 anchors. Each anchor has 5 parameters including (x, y, w, h, c), where x and y are the coordinate positions of the anchor, w and h represent the width and height of the anchor, and c is the confidence level of the predicted object. In addition to the parameters of the anchor, the YOLOv3 algorithm predicts the probability of each category and the confidence level can be achieved by the following Equation (1).

$$Confidence = p\_r(obj) \times IoU \tag{1}$$

where $p\_r(Obj)$ is set as 0 or 1 and the IoU denotes the intersection ratio of the predicted and true frames. The confidence reflects whether the grid contains objects or not, and the accuracy of the prediction frame when the grid contains objects. Finally, the redundant anchor is eliminated by non-maximal suppression (NMS), and the position and size of the corresponding anchors are adjusted to produce the final result. YOLOv3 improves the detection performance via introducing the anchor mechanism based on YOLOv2 and a K-means clustering algorithm. The K-means clustering algorithm can be employed to obtain the suitable prior frame size, which is shown in the following Table 1 [35].

By introducing suitable prior frames, the network no longer needs to randomly generate anchor frames of different sizes to predict objects, thus making the network train faster and converge faster. YOLOv3 leverages a multi-scale strategy for object detection, which can detect more objects and identify smaller objects than those of YOLOv1 and YOLOv2, respectively. Concretely, the YOLOv3 network consists of four parts, including the input unit, backbone network unit, neck unit, and output unit. The backbone network in the YOLOv3 framework is the Darknet-53, and its basic unit is the residual structure [36], which can alleviate the gradient vanishing or explosion problems caused by the deepening of the network layers. The network structure of YOLOv3 is shown in Figure 3.

Table 1. YOLOv3 Anchor Size

Cell Size	Detection Box Size
13 × 13	(10,13) (16,30) (33,23)
26 × 26	(30,61) (62,45) (59,119)
52 × 52	(116,90) (156,198) (373,326)

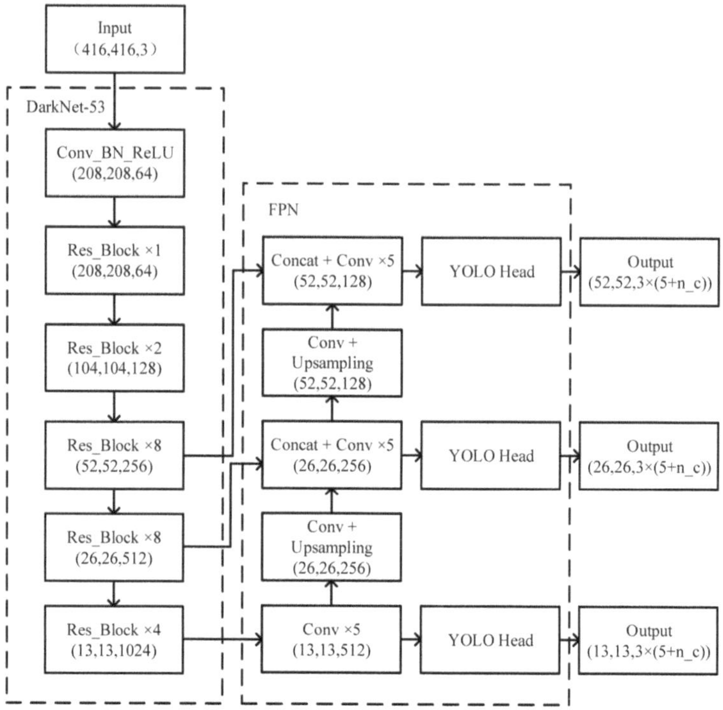

Figure 3. The network structure of YOLOv3.

2.4. Attention Module

Pigs exhibit different actions at different times of the day, and there are often problems with pigs occluding each other. These situations can lead to a lack of distinct behavioral identification features in pig datasets. In order to handle the above-mentioned problems and improve the accuracy of the network by capturing more effective features, it is necessary for the network to learn action features adaptively. Consequently, this paper proposes an attention-enhanced YOLOv3 network, which aims to utilize the attention module to make the neural network pay more attention to the corresponding regions in the image, and these regions play a key role in action discrimination.

The Convolutional Block Attention Module (CBAM) [31] is a lightweight unit that consists of two separate parts, they are the Channel Attention Module (CAM, Channel Attention Module) and the Spatial Attention Module (SAM, Spatial Attention Module). CBAM is a combination of spatial attention and channel attention, which can be utilized to obtain rich semantic information in pig images. CBAM can capture the dependencies between channel feature space features, reduce the weight of unimportant information, and improve the detection performance of individual pigs. The structure diagram of the CBAM module is shown in Figure 4.

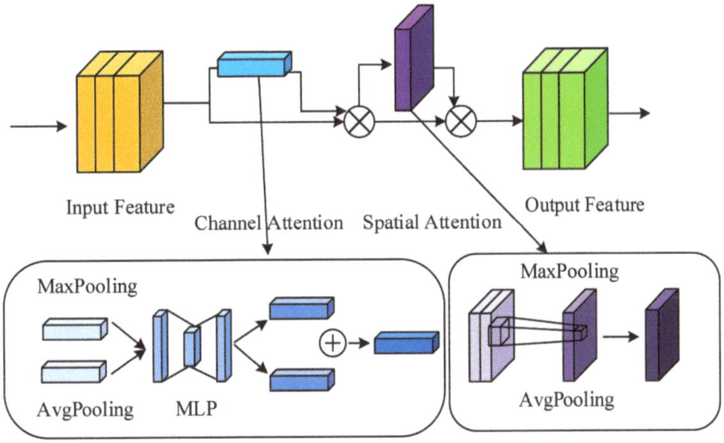

Figure 4. The structure of the CBAM module.

Specifically, CAM first performs a maximum pooling operation and a global average pooling operation on the input feature layer in turn, their output features are sent to a Multi-Layer Perception (MLP) layer. Then the output features of the MLP of two branches are summed together and send to a sigmoid function for fixing the weights between 0 and 1 distribution. The final result is obtained by multiplying the original input feature layer.

SAM first performs a maximum pooling operation and a global average pooling operation on the input feature layer, and then executes a tensor splicing on the corresponding output features. Finally, they are sent to the Sigmoid function to fix the weights between 0 and 1 distribution and then multiplied with the original input feature to achieve the final result.

The CBAM module can be represented by the following equation.

$$F_c = Z_c(F)F \tag{2}$$

$$F_2 = Z_c(F_c)F_c \tag{3}$$

$$Z_c(f) = \sigma\{F_{fc}[AvgPool(F)] + F_{fc}[MaxPool(F)]\} \tag{4}$$

$$Z_s(f) = \sigma\{C_c[AvgPool(F1)] + F_c[MaxPool(F1)]\} \tag{5}$$

where Z_c presents the channel attention module (CAM), Z_s denotes the spatial attention module (SAM), F indicates the feature layer of the input network, and F_c/F_s denotes the feature map after the Channel Attention Module (CAM)/Spatial Attention Module(SAM), respectively. $\times$ represents to perform the pointwise multiplication, and F_{fc} is the fully connected operation. $AvgPool$ denotes the global average pooling operation and the $MaxPool$ indicates the global maximum pooling operation, respectively. C_c presents the tensor splicing Concat operation. $+$ represents the summation operation and the σ denotes the sigmoid activation function.

2.5. *The Proposed Novel YOLOv3-SC Model*

The proposed YOLOv3-SC model is built by leveraging the attention mechanism and Spatial Pyramid Pooling (SPP) module to the YOLOv3 backbone network Darknet-53. Specifically, the CBAM module endows YOLOv3 with powerful feature extraction capabilities. Furthermore, the SPP structure extracts features of different scales in the final stage of the backbone network and fuses them. This design can alleviate network overfitting, increase the robustness of the model, and allow the network to learn richer features. The architecture of the novel proposed YOLOv3-SC is presented in Figure 5.

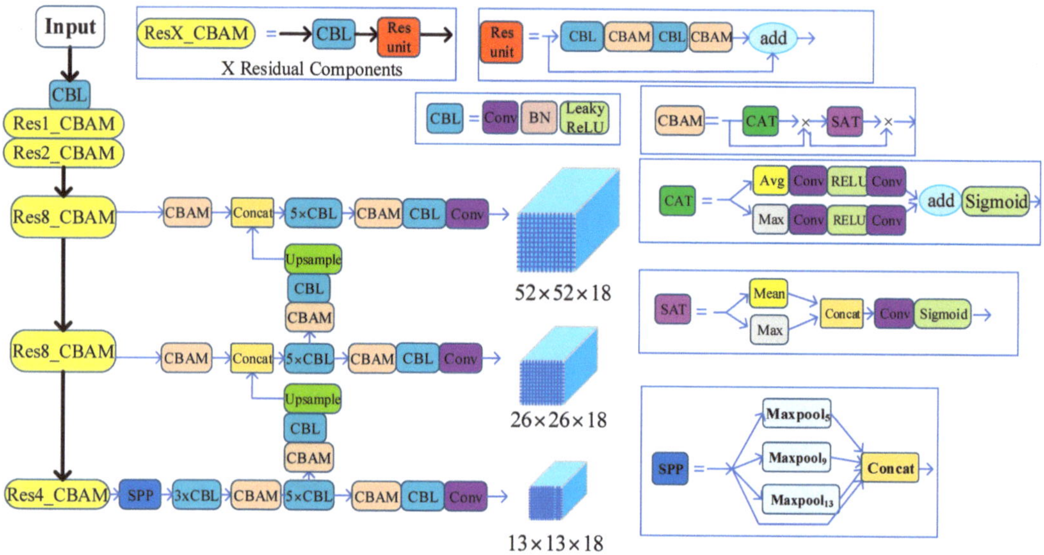

Figure 5. The pipeline of the proposed YOLOv3-SC. Here, we should note that the 5, 9, 13 in the $Maxpool_5$, $Maxpool_9$, $Maxpool_{13}$ indicate the pooling kernel size of the maxpool operation.

Specifically, as in Figure 5 show, the YOLOv3-SC is enhanced by integrating the CBAM module in each *Res_Block* and adding SPP unit in the final stage of the backbone network.

The backbone network with the addition of the CBAM consists of Res1-CBAM, Res2-CBAM, Res4-CBAM, and Res8-CBAM. Res*i*_CBAM $i = 1, 2, 4, 8$ denotes a stack of CBL and *i* residual structures with CBAM modules, where CBL is composed of Conv, Batch Normalization (BN), and activation function (LeakyReLu) unit. The CBAM component allows the model to extract richer spatiotemporal features, which will facilitate the pig detection significantly.

Furthermore, the SPP structure is introduced for feature fusion. Specifically, SPP performs maximum pooling operations at different scales on the input feature maps, and finally all the output feature maps are tensor-spliced with the original feature maps. In this way, the network can perform feature fusion at different scales to prevent overfitting.

The input image is improved with the backbone network and SPP structure to obtain three different scales of feature layers with sizes of 13 × 13, 26 × 26, and 52 × 52. After that, further feature fusion operations are performed. The 13 × 13 feature map is upsampled to obtain the 26 × 26 feature map and the 26 × 26 feature map of the backbone network is concatenated. After that, we perform up-sampling again to obtain 52 × 52 feature map and 52 × 52 feature map in the backbone network for Concat operation to fuse the feature information of different scales. In the feature fusion stage, we first perform CBAM operation on the input of different feature layers, and then perform CBAM operation again after the tensor stitching Concat to obtain the improved feature fusion network.

The training process of the YOLOv3 is presented in Algorithm 1.

Algorithm 1: YOLOv3-SC Model Training

```
Input: Pig image, target box
Output: Predicted box
Initialization(learning rate, epochs);
for i in epoch do
    for train_image, target in train_dataloader do
        output = YOLOv3-SC(train_image)
        loss = Loss function(output, target)
        Optimizer.zero_grad()
        loss.backward()
        Optimizer.step()
    end
    for test_image, target in test_dataloader do
        output = YOLOv3-SC(test_image)
        loss = loss_function(output, target)
    end
    Lr_scheduler() (Adjust the learning rate)
    Save() (Save the weights of the model)
end
```

2.6. The Loss Function

In order to achieve the optimized pig detection model, a reasonable loss function needs to be designed for the network training. Specifically, the loss function consists of three terms: including category loss ($\mathscr{L}_{cls}$), confidence loss ($\mathscr{L}_{conf}$), and locality loss ($\mathscr{L}_{loc}$). Among them, IoU is utilized to calculate the locality loss, and the cross-entropy loss is employed to calculate the confidence loss and class loss separately. The loss function of the proposed model is defined as follows.

$$Loss = \mathscr{L}_{loc} + \mathscr{L}_{cls} + \mathscr{L}_{conf} \tag{6}$$

where $\mathscr{L}_{loc}$ indicates the error between the coordinates and the length and width of the real frame and the coordinates and length and width of the predicted frame, $\mathscr{L}_{conf}$ illustrates the prediction region confidence error, and $\mathscr{L}_{cls}$ denotes the object classification error, respectively.

$$AP = \int_0^1 P(R)dR \tag{7}$$

$$mAP = \frac{\sum_1^n (AP)}{n} \tag{8}$$

$$P = TP/(TP + FP) \tag{9}$$

$$R = TP/(TP + FN) \tag{10}$$

where *TP* represents the number of positive samples predicted to be positive; *FP* indicates the number of negative samples predicted as positive samples; and *FN* illustrates the number of positive samples predicted as negative samples. n represents the number of the detected pig category, and its value is set as 1 here. *AP* is the average precision, denotes the area under the PR curve. mAP indicates the average accuracy over all categories.

2.7. Experiment Setup

In this paper, for fair comparison, all experiments are developed and run based on the PyTorch framework. The stochastic gradient descent optimization algorithm is employed for the model parameter update. Furthermore, the batchsize is 16, the initial learning rate is set as 0.001 and updated based on the cosine descent theory, the momentum is 0.937, and the total number of iterations is 400 epochs.

Additionally, the hardware configuration is as follows: operating system Ubuntu 20.04, CPU Intel(R) Xeon(R) CPU E5-2670 v3, GPU Nvidia Feforce GTX 3060 12G, and memory 16G DDR4.

3. Results

In this section, the experimental results and the discussions will be illustrated in detail. The experiments are organized in the following several parts, including the comparison of different models, the evaluation of the effectiveness of the SPP module, and the evaluation of the effectiveness of the SPP module. The purposes are to verify the effectiveness of the model, the effectiveness of the SPP unit in the YOLOv3-SC and the effectiveness of the CBAM unit in YOLOv3-SC correspondingly.

3.1. Comparison of Different Models

In order to validate the effectiveness of the proposed YOLOv3-SC, several models are utilized for comparison, including YOLOv1, YOLOv2, YOLOv3. Results are shown in Table 2.

Table 2. Comparison of YOLOv3 and YOLOv3-SC.

Model	Mean Average Accuracy (mAP/%)	Precision (P/%)	Recall (R/%)	F1 Score
YOLOv1	97.00	94.00	93.12	0.92
YOLOv2	97.83	94.40	93.55	0.93
Faster-RCNN	98.08	94.92	93.78	0.94
YOLOv3	98.64	95.94	94.12	0.95
YOLOv3-SC	99.24	98.27	94.31	0.97

Table 2 illustrates that the proposed model YOLOv3-SC achieves the best performance on all evaluation criteria. Specifically, YOLOv3-SC achieves 99.24% mAP, which is 2.31%/1.44%/1.18%/0.61% higher than that of YOLOv1/YOLOv2/Faster-RCNN/YOLOv3; the YOLOv3-SPP obtains 98.27% Precision, which is 4.54%/4.10%/3.53%/2.43% better than that of YOLOv1/YOLOv2/Faster-RCNN/YOLOv3; the YOLOv3-SPP achieves 94.31% Recall, which is 1.28%/1.44%/0.81%/0.22% superior than that of YOLOv1/YOLOv2/Faster-RCNN/YOLOv3; and the YOLOv3-SPP obtains 0.96 F1 score, which is 4.35%/3.22%/2.13%/1.05% higher than that of YOLOv1/YOLOv2/Faster-RCNN/YOLOv3. These all results validate the effectiveness of the proposed YOLOv3-SC.

The SPP structure and the CBAM attention component of the YOLOv3-SC allow the model to focus on the discriminant regions of the image in pig detection, fuse multi-scale feature maps, and extract more effective features even in the case of pigs sticking to each other. Consequently, YOLOv3-SC implements fast and efficient pig detection and achieves 99.24% mAP, which is significantly better than other models.

3.2. Evaluation of the Effectiveness of the SPP Module

In order to validate the effectiveness of the SPP module, we compare the YOLOv3 and YOLOv3-SPP. YOLOv3-SPP is built by encompassing the SPP module into the YOLOv3. Comparison results are shown in Table 3.

From Table 3, it can be seen that the YOLOv3-SPP model is superior to the YOLOv3 model on all evaluation criteria. Specifically, the YOLOv3-SPP achieves 99.19% mAP, which is 0.56% higher than that of YOLOv3; the YOLOv3-SPP obtains 97.19% Precision, which is 1.72% better than that of YOLOv3; the YOLOv3-SPP achieves 95.08% Recall, which is 1.02%

superior than that of YOLOv3; and the YOLOv3-SPP obtains 0.96 F1 score, which is 1.05% higher than that of YOLOv3. These results validate the effectiveness of the SPP module.

Table 3. Comparison of YOLOv3 and YOLOv3-SPP.

Model	Mean Average Accuracy (mAP/%)	Precision (P/%)	Recall (R/%)	F1 Score
YOLOv3	98.64	95.94	94.12	0.95
YOLOv3-SPP	99.19	97.19	95.08	0.96

The reason why the YOLOv3-SPP model is superior to YOLOv3 can be attributed to the following reasons. Specifically, the SPP module integrates both local and global features, thereby capturing multi-scale feature information, enhancing the expressiveness of features, and improving the robustness and the performance of the model.

3.3. Evaluation of the Effectiveness of the CBAM Attention Module

To evaluate the effectiveness of the CBAM attention module, some models are utilized for comparison, including YOLOv3 and YOLOv3-CBAM. YOLOv3-CBAM is established by leverage the CBAM module into the backbone of the YOLOv3. The comparison results are shown in Table 4.

Table 4. Comparison of YOLOv3 and YOLOv3-CBAM.

Model	Mean Average Accuracy (mAP/%)	Precision (P/%)	Recall (R/%)	F1 Score
YOLOv3	98.64	95.94	94.12	0.95
YOLOv3-CBAM	99.17	97.46	94.24	0.96

Table 4 illustrates that the YOLOv3-CBAM model is superior to the YOLOv3 model on all evaluation criteria. Specifically, the YOLOv3-CBAM achieves 99.17% mAP, which is 0.54% higher than that of YOLOv3; the YOLOv3-CBAM obtains 97.46% Precision, which is 1.58% better than that of YOLOv3; the YOLOv3-CBAM achieves 94.24% Recall, which is 0.13% superior than that of YOLOv3; and the YOLOv3-CBAM obtains 0.96 F1 score, which is 1.05% higher than that of YOLOv3. These results evaluate the effectiveness of the CBAM module.

3.4. Evaluation of the Superiority of the YOLOv3-SC

To evaluate the superiority of the proposed model YOLOv3-SC, we compare it with the models in the above section, including the YOLOv3, YOLOv3-SPP, and YOLOv3-CBAM. YOLOv3-SC is built by integrating both the CBAM and the SPP modules into the structure of the YOLOv3. The corresponding comparison results are shown in Table 5 and Figure 6.

Table 5. Comparison of YOLOv3-SPP and YOLOv3-CBAM with YOLOv3-SC.

Model	Mean Average Accuracy (mAP/%)	Precision (P/%)	Recall (R/%)	F1 Score
YOLOv3	98.64	95.94	94.12	0.95
YOLOv3-SPP	99.19	97.19	95.08	0.96
YOLOv3-CBAM	99.17	97.46	94.24	0.96
YOLOv3-SC	99.24	98.27	94.31	0.97

Table 5 and Figure 6 illustrate that the YOLOv3-SC model obtains the best results, which verifies the superiority of the integration of SPP and CBAM modules.

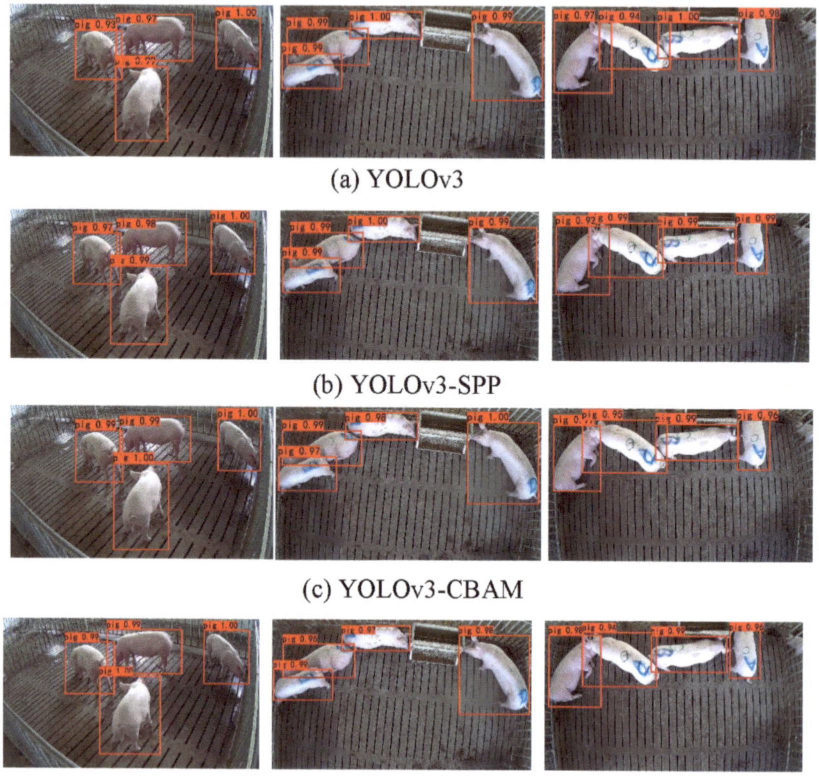

(a) YOLOv3

(b) YOLOv3-SPP

(c) YOLOv3-CBAM

(d) YOLOv3-SC

Figure 6. The detection results of different models.

4. Discussion

In this paper, we propose an improved YOLOv3 model, YOLOv3-SC, to achieve the efficient detection of individual pigs. Specifically, in order to verify the superiority of the proposed model, this paper compares and discusses the performance of the four models YOLOv3, YOLOv3-SPP, YOLOv3-CBAM, and YOLOv3-SC. The experimental results demonstrate that both the YOLOv3-SPP and YOLOv3-CBAM models achieve better performance than those of the YOLOv3 model, which verifies the effectiveness of the SPP module and the CBAM unit. Moreover, the YOLOv3-SC model achieves the best performance, verifying the effectiveness of the proposed model. By leveraging the attention module CBAM, the proposed model can adaptively focus on the important features and reduce the weight information on the non-important features in pig detection. Furthermore, the SPP structure allows the model to combine the multi-scale information, which improves the model's detection ability on small targets and adapts to the changing environment of individual pig detection in pig farms. The utilization of the SPP structure enhances the pig detection effect and performance of the model. Future work will explore more optimized data augmentation methods and more effective attention mechanisms that can be applied to more complex environments.

5. Conclusions

This paper develops a novel effective pig detection model YOLOv3-SC, which encompasses both the CBAM model and the SPP module into the backbone of YOLOv3 framework. The channel attention and the spatial attention units in the CBAM module enable the YOLOv3-SC to focus on the regions of the image that are important for detection, thereby extracting richer, more robust, and more discriminative features. The SPP module endows YOLOv3-SC the capacity of extracting multi-scale features, which enables the model to detect objects of different sizes, thereby improving the model's pig detection performance. Ablation studies validate the superiority of both the CBAM and the SPP modules. Furthermore, experimental results show that the proposed YOLOv3-SC model obtains the promising pig detection performance. Specifically, the YOLOv3-SC achieves 99.24% mAP performance, which is significantly higher than those of the other popular models.

Author Contributions: Conceptualization, W.H. (Wangli Hao); Data curation, W.H. (Wenwang Han); Formal analysis, F.L.; Investigation, W.H. (Wenwang Han); Methodology, W.H. (Wangli Hao) and W.H. (Wenwang Han); Project administration, F.L.; Resources, W.H. (Wangli Hao); Software, M.H.; Validation, M.H.; Writing—original draft, W.H. (Wangli Hao); Writing—review and editing, W.H. (Wangli Hao). All authors have read and agreed to the published version of the manuscript.

Funding: This work was supported by the Shanxi Province Education Science "14th Five-Year Plan" 2021 Annual Project General Planning Project + " Industry-University-Research"-driven Smart Agricultural Talent Training Model in Agriculture and Forestry Colleges (GH-21006); Shanxi Province Higher Education Teaching Reform and Innovation Project (J20220274); Shanxi Agricultural University doctoral research start-up project (2021BQ88); Shanxi Postgraduate Education and Teaching Reform Project Fund (2022YJJG094); and Shanxi Agricultural University 2021 «Neural Network» Course Ideological and Political Project (KCSZ202133).

Data Availability Statement: The datasets generated during and/or analyzed during the current study are available from the corresponding author on reasonable request.

Conflicts of Interest: We declare that this paper has no conflict of interest. Furthermore, we do not have any commercial or associative interest that represents a conflict of interest in connection with the work submitted.

Abbreviations

The following abbreviations are used in this manuscript:

SPP	Spatial Pyramid Pooling
CBAM	Convolutional Block Attention Module
SC	Spatial Pyramid Pooling and Convolutional Block Attention Module
YOLO	You Only Look Once

References

1. Gonzalez, L.A.; Tolkamp, B.J.; Coffey, M.P.; Ferret, A.; Kyriazakis, I. Changes in Feeding Behavior as Possible Indicators for the Automatic Monitoring of Health Disorders in Dairy Cows. *J. Dairy Sci.* **2008**, *91*, 1017–1028. [CrossRef] [PubMed]
2. Hulsen, J.; Scheepens, K. *Pig Signals: Look, Think and Act*; China Agricultural Science and Technology Press: Beijing, China, 2006.
3. Hart, B.L. Biological basis of the behavior of sick animals. *Neurosci. Biobehav. Rev.* **1988**, *12*, 123–137. [CrossRef]
4. Maselyne, J.; Saeys, W.; Ketelaere, B.D.; Mertens, K.; Vangeyte, J.; Hessel, E.F.; Millet, S.; Nuffel, A.V. Validation of a High Frequency Radio Frequency Identification (HF RFID) system for registering feeding patterns of growing-finishing pigs. *Comput. Electron. Agric.* **2014**, *102*, 10–18. [CrossRef]
5. Adrion, F.; Kapun, A.; Eckert, F.; Holland, E.-M.; Staiger, M.; Götz, S.; Gallmann, E. Monitoring trough visits of growing-finishing pigs with UHF-RFID. *Comput. Electron. Agric.* **2017**, *144*, 144–153. [CrossRef]
6. Neethirajan, S. Recent advances in wearable sensors for animal health management.*Sens. Bio-Sens. Res.* **2017**, *12*, 15–29. [CrossRef]
7. Schleppe, J.B.; Lachapelle, G.; Booker, C.W.; Pittman, T. Challenges in the design of a GNSS ear tag for feedlot cattle.*Comput. Electron. Agric.* **2010**, *70*, 84–95 [CrossRef]
8. Mohammad, A.K.; Claudia, B.; Sanne, O.; Christel, P.H.M.; Theo, A.N.; Frank, T.; Daniel, B. Automatic monitoring of pig locomotion using image analysis. *Livest. Sci.* **2014**, *159*, 141–148.

9. Kashiha, M.A.; Bahr, C.; Ott, S.; Moons, C.P.; Niewold, T.A.; Ödberg, F.O.; Berckmans, D. Automatic identification of marked pigs in a pen using image pattern recognition. *Comput. Electron. Agric.* **2013**, *93*, 111–120. [CrossRef]
10. Hernández-Hernández, J.L.; García-Mateos, G.; González-Esquiva, J.M.; Escarabajal-Henarejos, D.; Ruiz-Canales, A.; Molina-Martínez, J.M. Molina-Martínez Optimal color space selection method for plant/soil segmentation in agriculture.*Comput. Electron. Agric.* **2016**, *122*, 124–132. [CrossRef]
11. Nasirahmadi, A. Using machine vision for investigation of changes in pig group lying patterns. *Comput. Electron. Agric.* **2015**, *119*,184–190. [CrossRef]
12. Nasirahmadi, A.; Hensel, O.; Edwards, S.A.; Sturm, B. Automatic detection of mounting behaviours among pigs using image analysis. *Comput. Electron. Agric.* **2016**, *124*, 295–302. [CrossRef]
13. Nasirahmadi, A.; Sturm, B.; Olsson, A.C.; Jeppsson, K.H.; Müller, S.; Edwards, S.; Hensel, O. Automatic scoring of lateral and sternal lying posture in grouped pigs using image processing and Support Vector Machine. *Comput. Electron. Agric.* **2019**, *156*, 475–481. [CrossRef]
14. Viazzi, S.; Ismayilova, G.; Oczak, M.; Sonoda, L.T.; Fels, M.; Guarino, M.; Vranken, E.; Hartung, J.; Bahr, C.; Berckmans, D. Image feature extraction for classification of aggressive interactions among pigs. *Comput. Electron. Agric.* **2014**, *104*, 57–62. [CrossRef]
15. Matthews, S.G.; Miller, A.L.; Pltz, T.; Kyriazakis, I. Automated tracking to measure behavioural changes in pigs for health and welfare monitoring. *Sci. Rep.* **2017**, *7*, 17582. [CrossRef] [PubMed]
16. Kim, J.; Chung, Y.; Choi, Y.; Sa, J.; Kim, H.; Chung, Y.; Park, D.; Kim, H. Depth-based detection of standing-pigs in moving noise environments. *Sensors* **2017**, *17*, 2757. [CrossRef]
17. Zhuo, Y.; Yan, L.; Zheng, W.; Zhang, Y.; Gou, C. A Novel Vehicle Detection Framework Based on Parallel Vision. *Wirel. Commun. Mob. Comput.* **2022**, *2022*, 9667506. [CrossRef]
18. Wang, X.; Wang, W.; Lu, J.; Wang, H. HRST: An Improved HRNet for Detecting Joint Points of Pigs. *Sensors* **2022**, *22*, 7215. [CrossRef]
19. Tu, S.; Yuan, W.; Liang, Y.; Wang, F.; Wan, H. Automatic Detection and Segmentation for Group-Housed Pigs Based on PigMS R-CNN. *Sensors* **2021**, *21*, 3251. [CrossRef]
20. Wu, D.; Wu, Q.; Yin, X.; Jiang, B.; Wang, H.; He, D.; Song, H. Lameness detection of dairy cows based on the YOLOv3 deep learning algorithm and a relative step size characteristic vector. *Biosyst. Eng.* **2020**, *189*, 150–163. [CrossRef]
21. Shen, W.; Hu, H.; Dai, B.; Wei, X.; Sun, J.; Jiang, L.; Sun, Y. Individual identification of dairy cows based on convolutional neural networks. *Multimed. Tools Appl.* **2020**, *79*, 14711–14724. [CrossRef]
22. Tassinari, P.; Bovo, M.; Benni, S.; Franzoni, S.; Poggi, M.; Mammi, L.M.E.; Mattoccia, S.; Di Stefano, L.; Bonora, F.; Barbaresi, A.; et al. A computer vision approach based on deep learning for the detection of dairy cows in free stall barn. *Comput. Electron. Agric.* **2021**, *182*, 106030. [CrossRef]
23. Zhang, X.; Kang, X.; Feng, N.; Gang, L. Automatic recognition of dairy cow mastitis from thermal images by a deep learning detector. *Comput. Electron. Agric.* **2020**, *178*, 105754.
24. Hu, H.; Dai, B.; Shen, W.; Wei, X.; Sun, J.; Li, R.; Zhang, Y. Cow identification based on fusion of deep parts features—ScienceDirect. *Biosyst. Eng.* **2020**, *192*, 245–256. [CrossRef]
25. Jiang, B.; Wu, Q.; Yin, X.; Wu, D.; Song, H.; He, D. FLYOLOv3 deep learning for key parts of dairy cow body detection. *Comput. Electron. Agric.*, **2019**, *166*, 104982. [CrossRef]
26. Ran, B.; Edan, Y.; Halachmi, I. Computer vision system for measuring individual cow feed intake using RGB-D camera and deep learning algorithms. *Comput. Electron. Agric.* **2020**, *172*, 105345.
27. Achour, B.; Belkadi, M.; Filali, I.; Laghrouche, M.; Lahdir, M. Image analysis for individual identification and feeding behaviour monitoring of dairy cows based on Convolutional Neural Networks (CNN). *Biosyst. Eng.* **2020**, *198*, 31–49. [CrossRef]
28. Wu, D.; Wang, Y.; Han, M.; Song, L.; Shang, Y.; Zhang, X.; Song, H. Using a CNN-LSTM for basic behaviors detection of a single dairy cow in a complex environment—ScienceDirect. *Comput. Electron. Agric.* **2021**, *182*, 106016. [CrossRef]
29. Liu, W.; Anguelov, D.; Erhan, D.; Szegedy, C.; Reed, S.; Fu, C.Y.; Berg, A.C. SSD: Single Shot MultiBox Detector. In *European Conference on Computer Vision*; Springer: Amsterdam, The Netherlands, 2016; pp. 21–37.
30. Hao,W.; Han, M.; Li, S.; Li, F. MTAL: A Novel Chinese Herbal Medicine Classification Approach with Mutual Triplet Attention Learning. *Wirel. Commun. Mob. Comput.* **2022**, *2022*, 8034435. [CrossRef]
31. Woo, S.; Park, J.; Lee, J.Y.; Kweon, I.S. CBAM: Convolutional Block Attention Module. In Proceedings of the European Conference on Computer Vision, Munich, Germany, 8–14 September 2018; pp. 3–19.
32. He, K.; Zhang, X.; Ren, S. Sun, J. Spatial pyramid pooling in deep convolutional networks for visual recognition. *IEEE Trans. Pattern Anal. Mach. Intell.* **2015**, *37*, 2015. [CrossRef]
33. Redmon, J.; Divvala, S.; Girshick, R.; Farhadi, A. You Only Look Once: Unified, Real-Time Object Detection. In Proceedings of the IEEE Conference on Computer Vision and Pattern Recognition, Las Vegas, NV, USA, 27–30 June 2016; pp. 779–788.
34. Redmon, J.; Farhadi, A. YOLO9000: Better, Faster, Stronger. In Proceedings of the IEEE Conference on Computer Vision and Pattern Recognition, Honolulu, HI, USA, 21–26 July 2017; pp. 7263–7271.
35. Redmon, J.; Farhadi, A. YOLOv3: An Incremental Improvement. *arXiv* **2018**, arXiv:1804.02767.
36. He, K.; Zhang, X.; Ren, S.; Sun, J. Deep Residual Learning for Image Recognition. In Proceedings of the IEEE Conference on Computer Vision and Pattern Recognition, Las Vegas, NV, USA, 27–30 June 2016; pp. 770–778.

sensors

Article

Learning Moiré Pattern Elimination in Both Frequency and Spatial Domains for Image Demoiréing

Chenming Liu [1,2,3], Yongbin Wang [1,3,*], Nenghuan Zhang [1,3], Ruipeng Gang [2] and Sai Ma [2]

[1] State Key Laboratory of Media Convergence and Communication, Communication University of China, Beijing 100024, China
[2] Academy of Broadcasting Science, National Radio of Television Administration, Beijing 100866, China
[3] Key Laboratory of Convergent Media and Intelligent Technology, Ministry of Education, Communication University of China, Beijing 100024, China
* Correspondence: ybwang@cuc.edu.cn

Abstract: Recently, with the rapid development of mobile sensing technology, capturing scene information by mobile sensing devices in the form of images or videos has become a prevalent recording method. However, the moiré pattern phenomenon may occur when the scene contains digital screens or regular strips, which greatly degrade the visual performance and image quality. In this paper, considering the complexity and diversity of moiré patterns, we propose a novel end-to-end image demoiré method, which can learn moiré pattern elimination in both the frequency and spatial domains. To be specific, in the frequency domain, considering the signal energy of moiré pattern is widely distributed in the frequency, we introduce a wavelet transform to decompose the multi-scale image features, which can help the model identify the moiré features more precisely to suppress them effectively. On the other hand, we also design a spatial domain demoiré block (SDDB). The SDDB module can extract moiré features from the mixed features, then subtract them to obtain clean image features. The combination of the frequency domain and the spatial domain enhances the model's ability in terms of moiré feature recognition and elimination. Finally, extensive experiments demonstrate the superior performance of our proposed method to other state-of-the-art methods. The Grad-CAM results in our ablation study fully indicate the effectiveness of the two proposed blocks in our method.

Keywords: moiré patterns; image demoiré; frequency domain; wavelet transform

Citation: Liu, C.; Wang, Y.; Zhang, N.; Gang, R.; Ma, S. Learning Moiré Pattern Elimination in Both Frequency and Spatial Domains for Image Demoiréing. *Sensors* **2022**, *22*, 8322. https://doi.org/10.3390/s22218322

Academic Editors: Chien Aun Chan, Chunguo Li and Ming Yan

Received: 6 September 2022
Accepted: 26 October 2022
Published: 30 October 2022

1. Introduction

In recent years, mobile phones have greatly changed our lives and are widely used in many scenes [1,2]. They can help us to record important information in time by shooting images or videos. However, when we use mobile phones to record scenes with LED screens or display screens, the images or videos tend to have wavy interference patterns, known as moiré patterns. The reason for this is that when the photosensitive elements in the scanning instrument and digital camera are disturbed by high frequency, two equal amplitude sine waves with similar frequencies are superimposed, and the amplitude of the synthesized signal will change according to the difference of the two frequencies; then, colored and irregular-shaped stripes will appear on the images [3]. One way to avoid moiré patterns is to improve the camera resolution, making it larger than the resolution of the captured screen, but this is very expensive. Another way is to add a low-pass filter in front of the sensing device of the camera to reduce the moiré patterns, but the disadvantage of this is that the addition of the low-pass filter will impair the image details, making the image content blur.

With the rapid development of deep learning, using algorithms based on deep learning for image demoiréing has received considerable attention. For example, Liu et al. [4] considered that the energy distribution of moiré patterns is relatively concentrated in the frequency domain and used discrete cosine transform (DCT) to transfer the image from the spatial domain to the frequency domain. They decomposed the image into texture components and moiré components in the frequency domain so as to better eliminate the high-frequency moiré patterns in the images. However, it is difficult to completely eliminate moiré patterns only through the frequency domain. Yang et al. [5] used layer decomposition on polyphase components (LDPC) to decompose the image into a background layer and a moiré layer. In addition, to better remove the moiré patterns, the method was applied on Y and RGB channels. Sun et al. [6] proposed a multi-resolution network (DMCNN) for the multi-frequency characteristics of moiré patterns. However, this network cannot deal with low frequencies and large color blocks. Some images have poor visual performance. In 2020, He et al. [7] proposed a full high-definition demoiréing network (FHDe^{2}Net) to solve the problem of high-resolution image demoiréing. They used a global-to-local pattern removal strategy for fine detail preservation in high-resolution images and adopted DCT to transform the image into the frequency domain so as to better address the problems of the wider moiré pattern scale range. In 2022, Yu et al. [8] proposed ESDNet, a method for demoiréing in 4K ultra-high definition. In order to eliminate the multi-scale moiré patterns, they built a semantic-aligned scale-aware module. However, most existing methods attempted to eliminate the moiré patterns only in the frequency domain or the spatial domain. Moiré patterns have a certain diversity and complexity. In the frequency domain, moiré patterns span low frequencies and high frequencies. In the spatial domain, moiré patterns mix with the image texture seriously and also cause color distortions. It is difficult to completely remove moiré patterns while keeping the original texture just from one domain.

To address the above-mentioned problems, we propose a novel and effective end-to-end demoiré network, which can eliminate the moiré patterns both in frequency and spatial domains, named FSD-Net. Specifically, in the frequency domain, we introduce the wavelet transform to decompose the multi-scale image features, which can help the network better identify the moiré features so as to suppress the moiré features in the image generation. In the spatial domain, we design a spatial domain demoiré block (SDDB), which can extract the moiré features from the mixed image features. After the extraction, the moiré features will be subtracted from the mixed image feature, which can obtain clean features to generate a clean image without moiré patterns. By demoiréing both in the frequency domain and in spatial domain, our proposed network can obtain superior performance regarding eliminating moiré patterns. Experimental results demonstrate that our proposed method outperforms several state-of-the-art demoiré methods.

The contributions of our research can be summarized as follows:

- We propose a novel method to eliminate moiré patterns both in the frequency domain and spatial domain. Experimental results indicate that our method achieves state-of-the-art performance compared with other methods.
- We introduce wavelet transform to decompose the multi-scale image features, Which may help the network to better identify the moiré features so as to suppress the moiré features during the image generation.
- We design a spatial-domain demoiré block, which can effectively extract moiré features from mixed image features. Then we can subtract the moiré features from the mixed features to obtain clean features, which are used during the image generation.

2. Related Work

2.1. Traditional Methods

Moiré patterns are often caused by the aliasing of two equal-amplitude sine waves with similar frequencies. The direct solution is to change the frequency of one sine wave. In 2000, Nishioka et al. [9] proposed that adding a low-pass filter in front of the digital camera lens could effectively remove the moiré patterns. This method can avoid the moiré patterns during the shooting, though it also loses some details and texture information. Another method is to upgrade the camera with a higher resolution photosensitive element, but such a camera is expensive and cannot fundamentally solve the moiré problem. Sidorov et al. [10] proposed a spectral model, which leverages the magnitude of the Fourier spectrum of the image to identify the moiré patterns. In addition, signal decomposition is also an important method for removing moiré patterns. Yang et al. [11] proposed a novel image demoiréing method by signal decomposition and guided filtering. Firstly, they adopted a low-rank and sparse matrix decomposition model to remove moiré patterns in the green (G) channel. Then, they removed moiré patterns in red (R) and blue (B) channels via guided filtering by the obtained texture layer of the G channel.

2.2. Deep Learning Methods

Unlike with other image restoration tasks, such as image denoising [12–14], image dehazing [15,16], and image demosaicing [17,18], the difficulty of image demoiréing is how to remove moiré patterns with various frequencies and color distortion. With the widespread popularity of deep learning, deep convolutional neural networks have also been applied to image demoiréing. In 2018, Sun et al. [6] first proposed a multi-resolution fully convolutional neural network DMCNN for multi-frequency moiré feature elimination. In 2019, Gao et al. [19] considered that the relationship among multi-scale features is significantly ignored and designed a feature-enhancing branch to fuse high-level features with low-level ones, which can restore the image details during image demoiréing. In 2019, Cheng et al. [20] also dealt with multi-scale features and proposed a dynamic feature-encoding module, which can encode the variations of moiré patterns. He et al. [21] proposed a Moiré pattern removal neural network (MopNet) based on DenseNet. The model integrated the moiré frequency distribution, edge intensities, and appearance categories into design learning modules. However, when irregular and unstable backgrounds such as sand and stone ground are encountered, the edge features of the pattern are difficult to determine and need to be improved. In 2020, Zheng et al. [22] proposed a novel multi-scale bandpass convolutional neural network (MBCNN) , which splits image demoiréing into two steps: moiré texture removal and tone mapping. In 2020, Liu et al. [23] proposed a WDNet network, including a direction perception module. The module can carry out the convolution operation in 8 different directions so as to better capture the spatial distributions of moiré patterns. He et al. [7] proposed a full high-definition demoiréing network (FHDe^{2}Net) to solve the high-resolution image demoiréing by a cascade of two networks focused on global and local level moiré removal, respectively. Moreover, they also proposed a lower high-resolution content separation branch, which can preserve the fine details against the distortions in demoiré processing. In 2021, Park et al. [24] proposed an unsupervised end-to-end moiré pattern removal method based on cyclic moiré learning. Compared with other methods, this method used an unpaired set of clean and moiré images. In 2022, Yu et al. [8] proposed a baseline model, ESDNet, a method for removing moiré patterns in 4K ultra-high-definition images, and constructed a semantic-aligned scale-aware module to solve the moiré elimination effectively. In the same year, Dai et al. [25] proposed the first moiré pattern removal method with implicit feature space alignment and selective feature aggregation for hand-held video.

2.3. Wavelet-Based Methods

The wavelet transform can decompose complex and composite information into elementary simple forms at different positions and scales, which can benefit us in understanding the information. Wavelet-based methods have been explored in several computer vision tasks. For example, Lotfi et al. [26] adopted the Daubechies 4 wavelet transform and first-order color moments to represent the image information, then a neural network was proposed to identify the category of aircraft images. Nayak et al. [27] utilized a two-dimensional discrete wavelet transform for extracting brain magnetic resonance image features. Liu et al. [28] used wavelet-based methods to capture age-related texture details at multiple scales in the frequency domain. Huang et al. [29] used wavelets for face super-resolution, where neural networks were used to predict the wavelet coefficients.

3. Methodology

In this section, we first describe the structure of our proposed FSD-Net for image demoiréing and the overall pipeline. Next, we present the details of the frequency domain demoiré block and the spatial domain demoiré block, which are the crucial components of FSD-Net. After that, we present the loss function.

3.1. Overall Network

The overall structure of our proposed FSD-Net is shown in Figure 1. As can be seen, our network consists of a generator and a discriminator. The generator is a U-shaped network with skip connections between the encoder and the decoder. In the encoder, we adopt four down encoder Blocks and a res block to encode the multi-scale moiré image features. Using the decoder, we gradually remove the moiré patterns from the multi-scale feature maps both in the frequency domain and spatial domain by our proposed demoiré block. The discriminator is designed following the structure of [30], which can help further enhance the quality of the generated images.

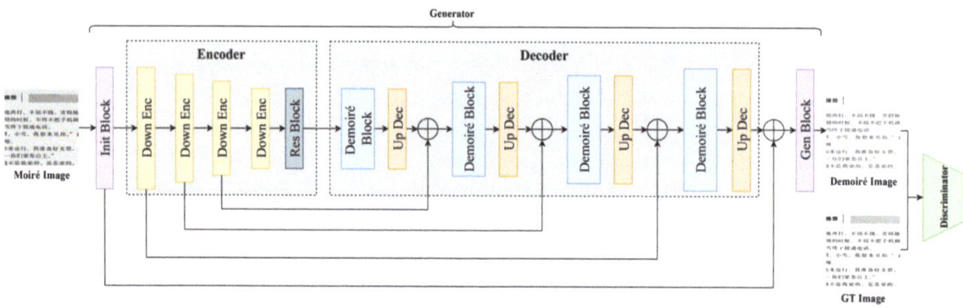

Figure 1. The overall structure of our proposed FSD-Net, which consists of a generator and a discriminator. The generator is designed following the image encode-decode structure.

To be specific, the input of FSD-Net is a moiré image with the size of $I \in \mathbb{R}^{3\times H\times W}$. Firstly, we adopt an init block to obtain the initial image feature maps, denoted as $f_e^0 \in \mathbb{R}^{C\times H\times W}$, where C, H, and W are the channel, height and weight size of the feature maps, respectively. Then, the initial feature maps will be sent into the encoder, which consists of four down encoder blocks and a res block. Each down encoder block consists of a 3×3 convolution with stride 2, an instance normalization, and a ReLU function. The res block is used to further encode the image feature maps; the output of the encoder is $f_e^4 \in \mathbb{R}^{16C\times(H/16)\times(W/16)}$.

Next, following the encoder of the generator, the feature maps f_e^4 are passed through four stages of decoding. Each stage contains a stack of the proposed demoiré block and an up dec block. The motivation of the demoiré block is related to the two major challenges during moiré pattern elimination. One is that the moiré patterns exhibit considerable variation in frequency. The other is that in the spatial domain, the moiré features are seriously mixed with image texture. It is difficult to filter out the moiré patterns under these problems. To address the above-mentioned issues, we built the demoiré block with two core designs: the frequency domain demoiré block (FDDB) and the spatial domain demoiré block (SDDB). In the demoiré block, we first put the decoder feature maps f_d^i into the FDDB, which can recognize the moiré feature maps in the frequency domain by wavelet transform and then weight the moiré feature maps to suppress them during the image generation. Then, the SDDB is added at the end of FDDB to further remove the moiré features, which are mixed with image texture features by subtracting operations, to obtain cleaner features. After the demoiré block, we adopt a transposed convolution operation in the up-decoder block to decode the image features. Finally, we adopt a gen block to restore the image to its original image size.

3.2. Frequency Domain Demoiré Block

The key innovation of our proposed FDDB is to disentangle image decoding feature maps into multiple frequency sub-bands and recognize the moiré features in the frequency domain, then suppress the moiré features by weighting coefficients. We introduce wavelet transform (WT) to transform image features from the spatial domain to the frequency domain. Wavelet transform consists of a low-pass filter and a high-pass filter; it applies the low-pass filter and high-pass filter alternately along feature columns and rows to produce four sub-band frequency features, denoted as LL, LH, HL, and HH, where L indicates the low frequencies and H represents the high frequencies. By using the wavelet transform, the model can better distinguish the moiré features and image features.

The structure of the frequency domain demoiré block is shown in Figure 2. As shown in the figure, we first adopt wavelet transform to decompose features into multiple wavelet sub-bands with different frequency contents. The operation is defined as:

$$f_d^{LL}, f_d^{LH}, f_d^{HL}, f_d^{HH} = WT(f_d), \tag{1}$$

where $f_d \in \mathbb{R}^{C \times H \times W}$ is the input image features, $f_d^{LL}, f_d^{LH}, f_d^{HL}$, and f_d^{HH} represent the four sub-bands, the size of the sub-band feature map is $C \times H/2 \times W/2$, and $WT()$ denotes the wavelet transform function.

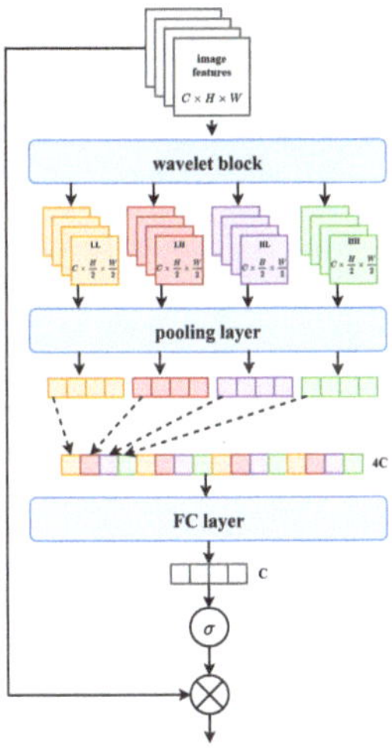

Figure 2. The architecture of the frequency domain demoiré block (FDDB).

Once the four sub-bands are obtained, we adopt a pooling layer to obtain the feature vector. In this step, it is important to align the low frequency and high frequency pooling parameters of the same input feature map. So after the pooling layer, the four sub frequency domain pooling parameters of each feature map is arranged with the corresponding positions.

Finally, the flattened vector will be sent to a FC layer and a sigmoid function to obtain the weighted parameters, which will be used to weight the input image features so as to suppress the moiré features and obtain clean image features.

3.3. Spatial Domain Demoiré Block

The above FDDB based on wavelet transform focuses on the recognition of frequency differences between the feature maps. However, some moiré patterns are mixed with the original image texture, and it is difficult to discriminate one from the other; the FDDB suffers from a limited capability to remove the moiré patterns in this case. To address this limitation, we design a structure that removes moiré patterns directly in the spatial domain. We achieve this by designing a spatial domain demoiré block, which is illustrated in Figure 3. In the SDDB, we leverage the depth-wise convolution operation to extract the moiré features from the mixed features. Then, we use a channel-wise convolution operation to obtain refined moiré features. Finally, we obtain the clean image features by subtracting the moiré features from the original input features. These operations can be defined as follows:

$$f_m = Conv(\hat{f_d}), \tag{2}$$

$$\hat{f_d^*} = \hat{f_d} \ominus f_m, \tag{3}$$

where $Conv()$ denotes both the depth-wise convolution and the channel-wise convolution, and $\ominus$ denotes the feature-wise subtraction operation.

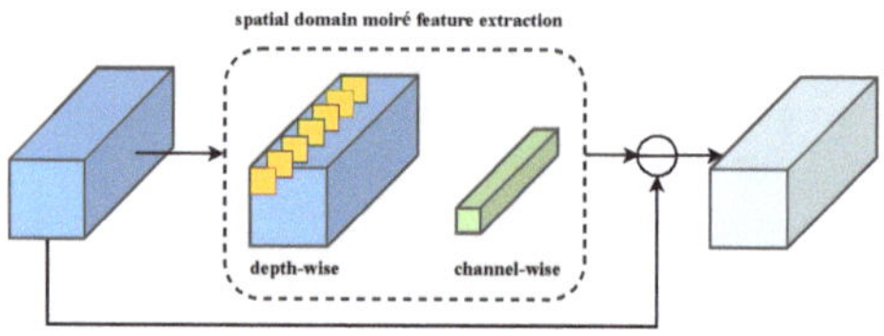

Figure 3. The structure of the spatial domain demoiré block (SDDB).

3.4. Loss Function

In our research, to obtain better demoiréing performance, we adopt a combined loss function, including content loss $\mathcal{L}_{ct}$, perceptual loss $\mathcal{L}_{per}$, and adversarial loss $\mathcal{L}_{adv}$. The overall loss function of the generator can be formulated as:

$$\mathcal{L} = \lambda_{ct} \times \mathcal{L}_{ct} + \lambda_{per} \times \mathcal{L}_{per} + \lambda_{adv} \times \mathcal{L}_{adv}^{G}, \tag{4}$$

where λ_{ct}, λ_{per}, and λ_{adv} are the balancing parameters of content loss, perceptual loss, and adversarial loss, respectively.

Content Loss. We use the $L1$ loss to measure the content loss between the GT image and the demoiréing image generated by our proposed network.

$$\mathcal{L}_{ct} = \|I_{gt} - I_{de}\|_1, \tag{5}$$

where I_{gt} denotes the GT image, and I_{de} denotes the demoiréing image.

Perceptual Loss. To penalize the perceptual and semantic discrepancy, we adopt the pre-trained VGG-19 network to extract the features of the GT image I_{gt} and the features of the demoiréing image I_{de}. Then, we use the $L1$ loss to measure the perceptual loss. The formula is defined as follows:

$$\mathcal{L}_{per} = \sum_i (\|\phi_i(I_{gt}) - \phi_i(I_{de})\|_1), \tag{6}$$

where $\phi()$ denotes the VGG-19 network and i denotes the i-th layer of VGG-19 network. In our experiments, we employ the feature maps of the four layers conv2_2, conv3_4, conv4_4, and conv5_4 to calculate the perceptual loss.

Adversarial Loss. To effectively synthesize a realistic image, we introduce an adversarial loss, which can promote the generator to create a realistic demoiré image. We use the binary cross-entropy criterion with a softmax function to calculate the loss value. The loss function for the generator is defined as:

$$\mathcal{L}_{adv}^{G} = -log(D(I_{de})); \tag{7}$$

and the loss function for training the discriminator is defined as:

$$\mathcal{L}_{adv}^{D} = -log(D(I_{gt})) - log(1 - D(I_{de})); \tag{8}$$

where $D()$ represents the discriminator.

4. Experiments

In this section, we first describe the dataset used in our experiment and the implementation details. Then, we report the subjective and objective evaluation results in comparison with other state-of-the-art methods to demonstrate the effectiveness of our proposed demoiré method. Finally, we conduct some ablation studies to verify the performance benefit brought by each functional component in our method.

4.1. Dataset and Implementation Details

Dataset. Our experiments are conducted on the dataset provided by the Document Image Demoiré Contest, which is a sub-competition of the Baidu NetDisk AI competition. All images in the dataset are collected from real-world scenes. The dataset consists of 1000 training samples and 200 test samples; each sample contains an image with moiré patterns and a ground-truth image without moiré patterns.

Implementation Details. The discriminator in our proposed method has a similar architecture to [30], which has an input size of 256 × 256. To achieve a good discriminative performance and accommodate the input size of 512 × 512 in our paper, we increased the depth of the discriminator network. We set the initial learning rate to 2×10^{-4} and 1×10^{-4} for the generator and the discriminator, respectively. Then, the two learning rates were both decayed by 0.1 in the 20th epoch and 40th epoch, respectively. The total training epoch was set to 60. The batch size was set to 10. We adopted the Adam optimizer [31] with $\beta_1 = 0.5$, $\beta_2 = 0.99$ to optimize our generator and discriminator. All input images were resized into 512 × 512 pixels, and random horizontal flipping was adopted for data enhancement. The balancing parameters $\lambda_{ct}, \lambda_{per}, \lambda_{adv}$ in the loss function were set to 2.0, 1.0, and 1.0, respectively. Our proposed network was implemented with a PyTorch framework and trained with two NVIDIA RTX3090 GPUs.

Evaluation Metrics. Following the previous works, we used common metrics to evaluate the demoiré performance: PSNR (peak signal-to-noise ratio) and SSIM (structural similarity). In addition, we also showed the visual comparison with other state-of-the-art methods to evaluate the effectiveness of our proposed method.

4.2. Comparison to Other Methods

We compared our method with several state-of-the-art methods, including U-Net [32], WDNet [23], MBCNN [22], FHDe^{2}Net [7], and HRDN [33].

U-Net is a very excellent network and is widely used in many image generation and restoration tasks. We chose this model as one of the comparative methods and trained it from scratch.

WDNet is a demoiré method based on wavelet transform. In contrast to our proposed method, WDNet first employs 2D fast wavelet transform to decompose the input RGB image into a sequence of wavelet subbands; these subbands will then be sent to the network to remove moiré patterns. Finally, an inverse wavelet transform is adopted to obtain the final demoiré RGB image. Thus, WDNet works mainly in the wavelet domain.

MBCNN is a demoiré method based on multi-scale features; it removes the moiré patterns in the frequency domain by the discrete cosine transform (DCT).

FHDe^{2}Net is also a method that removes the moiré patterns in the frequency domain by the DCT. This method adopts a cascaded global-to-local moiré pattern removal strategy, which can handle higher-resolution images.

V is a novel high-resolution demoiré network. It also removes the moiré patterns based on frequency domain and multi-scale features. It also fully takes advantage of the relationship among feature maps with different resolutions to exchange information and enhance details.

The qualitative results. In order to verify the effectiveness of our proposed method, we report the qualitative results compared with other methods. Table 1 illustrates the qualitative results. As can be seen, our method outperforms them both in PSNR and SSIM, achieving state-of-the-art performance. The results fully demonstrate the effectiveness of our proposed method.

Table 1. The comparison of qualitative results between our proposed FSD-Net and other methods.

Network	PSNR (dB)	SSIM
U-Net	27.62	0.8289
WDNet	28.66	0.8745
MBCNN	26.83	0.8113
FHDe^{2}Net	26.39	0.8786
HRDN	27.38	0.8626
ours	33.24	0.8970

The visual results. Figures 4 and 5 show visual comparisons with different degrees of moiré patterns. Figure 4 shows the visual results with relatively slight moiré patterns. We note that WDNet and HRDN are unable to totally remove the moiré patterns compared to other methods. U-Net and MBCNN cannot generate clear image content during the demoiréing process. FHDe^{2}Net cannot competently handle the color distortion. In contrast, our method can remove moiré patterns while preserving image details.

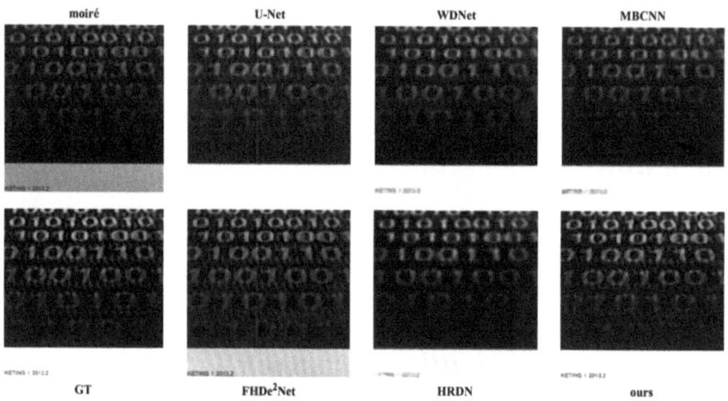

Figure 4. Visual comparisons between our proposed FSD-Net and other methods.

Figure 5 shows the visual results with serious moiré patterns. It can be seen that MBCNN cannot reconstruct text details, resulting in image blur. FHDe^{2}Net and HRDN cannot perfectly remove moiré patterns. U-Net and WDNet also cannot obtain an ideal demoiréing performance. Our proposed FSD-Net greatly outperforms the above-mentioned methods and generates a superior visual performance. The visual comparisons further confirm the effectiveness of our method in terms of preserving high-quality details while eliminating moiré patterns.

4.3. Ablation Study

To validate the effectiveness of the proposed frequency domain demoiré block and spatial domain demoiré block, we conducted an ablation study, which contains the following variants. (1) Baseline: we adopted only the generator without the frequency domain demoiré block and the spatial domain demoiré block. (2) Baseline+FDB: we added the frequency domain demoiré block on the baseline network. (3) Baseline+SDB: we added the spatial domain demoiré block on the baseline network. (4) Baseline+FDB+SDB: we added both the frequency domain demoiré block and the spatial domain demoiré block to the baseline network. We report the quantitative results in Table 2.

Figure 5. Visual comparisons between our proposed FSD-Net and other methods. The example image contains serious moiré patterns.

From Table 2, it can be seen that: (1) compared to the baseline, the Baseline+FDB and Baseline+SDB both clearly improve the PSNR and SSIM scores, demonstrating that the two functional blocks both contribute to a performance gain. (2) The combination of the two functional blocks remarkably improves the metrics, further proving the demoiréing solution in both the frequency domain and spatial domain is correct and feasible.

Table 2. Quantitative results of different variants of FSD-Net.

Network	PSNR (dB)	SSIM
Baseline	32.01	0.8717
Baseline+FDB	32.62	0.8841
Baseline+SDB	32.88	0.8896
Baseline+FDB+SDB	33.24	0.8970

In addition, we further provide the visualization results to better illustrate the effectiveness of our proposed frequency domain demoiré block and spatial domain demoiré block.

Figure 6 shows the Grad-CAM results generated by the 3×3 convolutional operation within the spatial domain demoiré block. It can be seen that the active areas are the salient moiré regions, not the image content areas. It indicates that our proposed spatial domain demoiré block can focus on capturing the moiré features and suppress the moiré features during the image generation to obtain a clean image.

To verify the impact of our proposed frequency domain demoiré block, we show the comparison results between the original feature maps and the weighted feature maps by the frequency domain demoiré block. The visualization results are shown in Figure 7. We can see that: (1) in the first row, the left side of (a), (b), and (c) mainly activate the image content, while the corresponding weighted suppression effects in the right side of (a), (b), and (c) are not obvious; (2) In the second row, the features of (d) and (e) mainly represent the moiré regions, so that the weighted suppression is relatively significant. The feature of (f) contains the image content and the moiré content. Our proposed frequency domain demoiré block can effectively identify and suppress the moiré feature. Such visual results prove that our proposed frequency domain demoiré block can assist the model in recognizing and suppressing moiré features.

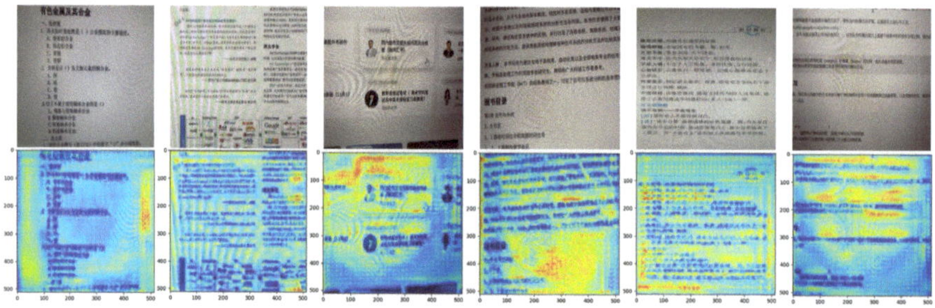

Figure 6. The total visualization results of the four spatial domain demoiré blocks. The first row is the original moiré images. The second row is the Grad-CAM results.

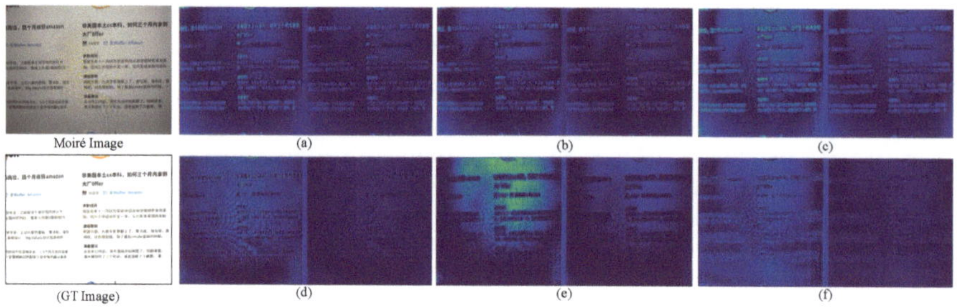

Figure 7. The visualization results of the last frequency domain demoiré block. In (**a**–**f**), the left side is the original feature maps, and the right side is the weighted feature maps by the last frequency domain demoiré block.

5. Discussion and Conclusions

In this paper, we study how to eliminate moiré patterns more effectively. We summarize and find that previous methods mainly remove moiré patterns in one domain. However, moiré patterns cover a wide range in frequency and will appear in any area with different colors and shapes, making it difficult to eliminate moiré patterns from one domain. To this end, we explore to eliminate moiré patterns both in frequency and spatial domains. In the frequency domain, we introduce wavelet transform to help identify moiré pattern features so as to suppress them. In the spatial domain, we design an SDDB module to remove the moiré features that are mixed with image features. The comparable experiments show that our proposed method is better than other state-of-the-art methods in moiré pattern elimination.

However, there are still several limitations of our demoiré method. First, the reconstruction quality of the demoiré image still has a gap with the ground-truth image, especially in the regions where moiré patterns and image content are seriously mixed. Therefore, the research of content enhancement is our future work, including image deblurring and super-resolution. In addition, we note that our method cannot well preserve the background color if it is similar with some moiré colors. Thus, how to further improve the ability of our method to distinguish moiré features from image features still needs exploration.

Author Contributions: Conceptualization, C.L., Y.W. and N.Z.; methodology, C.L., N.Z. and R.G.; software, R.G. and S.M.; validation, C.L., N.Z. and R.G.; formal analysis, Y.W. and S.M.; investigation, C.L. and N.Z.; resources, C.L. and R.G.; data curation, R.G. and S.M.; writing—original draft preparation, C.L., N.Z. and R.G.; writing—review and editing, C.L. and S.M.; visualization, R.G. and S.M.; supervision, Y.W.; project administration, Y.W.; funding acquisition, C.L. All authors have read and agreed to the published version of the manuscript.

Funding: This research was funded by two projects of the Academy of Broadcasting Science, National Radio and Television Administration of China under projects: "Video Super-Resolution Algorithm Design and Software Development for Face Blur Problem" (JBKY20220210) and "Research on 8K UHD + Multi-Camera + Interactive Viewing Cloud Broadcasting System Technologies and Solutions for Performing Arts Scenes" (JBKY20220250).

Institutional Review Board Statement: Not applicable.

Informed Consent Statement: Not applicable.

Data Availability Statement: Not applicable.

Conflicts of Interest: The authors declare no conflict of interest.

References

1. Yan, M.; Li, S.; Chan, C.A.; Shen, Y.; Yu, Y. Mobility Prediction Using a Weighted Markov Model Based on Mobile User Classification. *Sensors* **2021**, *21*, 1740. [CrossRef] [PubMed]
2. Jiang, Y.; Song, L.; Zhang, J.; Song, Y.; Yan, M. Multi-Category Gesture Recognition Modeling Based on sEMG and IMU Signals. *Sensors* **2022**, *22*, 5855. [CrossRef]
3. Oster, G.; Wasserman, M.; Zwerling, C. Theoretical interpretation of moiré patterns. *Josa* **1964**, *54*, 169–175. [CrossRef]
4. Liu, F.; Yang, J.; Yue, H. Moiré pattern removal from texture images via low-rank and sparse matrix decomposition. In Proceedings of the 2015 Visual Communications and Image Processing (VCIP), Singapore, 13–16 December 2015; pp. 1–4.
5. Yang, J.; Zhang, X.; Cai, C.; Li, K. Demoiréing for screen-shot images with multi-channel layer decomposition. In Proceedings of the 2017 IEEE Visual Communications and Image Processing (VCIP), St. Petersburg, FL, USA, 10–13 December 2017; pp. 1–4.
6. Sun, Y.; Yu, Y.; Wang, W. Moiré photo restoration using multiresolution convolutional neural networks. *IEEE Trans. Image Process.* **2018**, *27*, 4160–4172. [CrossRef]
7. He, B.; Wang, C.; Shi, B.; Duan, L.Y. FHDe2Net: Full high definition demoireing network. In Proceedings of the European Conference on Computer Vision, Glasgow, UK, 23–28 August 2020; Springer: Cham, Switzerland, 2020; pp. 713–729.
8. Yu, X.; Dai, P.; Li, W.; Ma, L.; Shen, J.; Li, J.; Qi, X. Towards Efficient and Scale-Robust Ultra-High-Definition Image Demoireing. *arXiv* **2022**, arXiv:2207.09935.
9. Nishioka, K.; Hasegawa, N.; Ono, K.; Tatsuno, Y. Endoscope System Provided with Low-Pass Filter for Moire Removal. U.S. Patent 19970917429, 26 August 1997.
10. Sidorov, D.N.; Kokaram, A.C. Suppression of moiré patterns via spectral analysis. In Proceedings of the Visual Communications and Image Processing 2002, San Jose, CA, USA, 19 January 2002; SPIE: Bellingham, WA, USA, 2002; Volume 4671, pp. 895–906.
11. Yang, J.; Liu, F.; Yue, H.; Fu, X.; Hou, C.; Wu, F. Textured image demoiréing via signal decomposition and guided filtering. *IEEE Trans. Image Process.* **2017**, *26*, 3528–3541. [CrossRef] [PubMed]
12. Elad, M.; Aharon, M. Image denoising via sparse and redundant representations over learned dictionaries. *IEEE Trans. Image Process.* **2006**, *15*, 3736–3745. [CrossRef] [PubMed]
13. Zhang, K.; Zuo, W.; Zhang, L. FFDNet: Toward a fast and flexible solution for CNN-based image denoising. *IEEE Trans. Image Process.* **2018**, *27*, 4608–4622. [CrossRef] [PubMed]
14. Tian, C.; Xu, Y.; Li, Z.; Zuo, W.; Fei, L.; Liu, H. Attention-guided CNN for image denoising. *Neural Netw.* **2020**, *124*, 117–129. [CrossRef] [PubMed]
15. Ren, W.; Liu, S.; Zhang, H.; Pan, J.; Cao, X.; Yang, M.H. Single image dehazing via multi-scale convolutional neural networks. In Proceedings of the European Conference on Computer Vision, Amsterdam, The Netherlands, 11–14 October 2016; Springer: Cham, Switzerland, 2016; pp. 154–169.
16. Shao, Y.; Li, L.; Ren, W.; Gao, C.; Sang, N. Domain adaptation for image dehazing. In Proceedings of the IEEE/CVF Conference on Computer Vision and Pattern Recognition, Seattle, WA, USA, 14–19 June 2020; pp. 2808–2817.
17. Mei, K.; Li, J.; Zhang, J.; Wu, H.; Li, J.; Huang, R. Higher-resolution network for image demosaicing and enhancing. In Proceedings of the 2019 IEEE/CVF International Conference on Computer Vision Workshop (ICCVW), Seoul, Korea, 27–28 October 2019; pp. 3441–3448.
18. Shao, L.; Rehman, A.U. Image demosaicing using content and colour-correlation analysis. *Signal Process.* **2014**, *103*, 84–91. [CrossRef]

19. Gao, T.; Guo, Y.; Zheng, X.; Wang, Q.; Luo, X. Moiré pattern removal with multi-scale feature enhancing network. In Proceedings of the 2019 IEEE International Conference on Multimedia & Expo Workshops (ICMEW), Shanghai, China, 8–12 July 2019; pp. 240–245.
20. Cheng, X.; Fu, Z.; Yang, J. Multi-scale dynamic feature encoding network for image demoiréing. In Proceedings of the 2019 IEEE/CVF International Conference on Computer Vision Workshop (ICCVW), Seoul, Korea, 27–28 October 2019; pp. 3486–3493.
21. He, B.; Wang, C.; Shi, B.; Duan, L.Y. Mop moire patterns using mopnet. In Proceedings of the IEEE/CVF International Conference on Computer Vision, Seoul, Korea, 27 October–2 November 2019; pp. 2424–2432.
22. Zheng, B.; Yuan, S.; Slabaugh, G.; Leonardis, A. Image demoireing with learnable bandpass filters. In Proceedings of the IEEE/CVF Conference on Computer Vision and Pattern Recognition, Seattle, WA, USA, 14–19 June 2020; pp. 3636–3645.
23. Liu, L.; Liu, J.; Yuan, S.; Slabaugh, G.; Leonardis, A.; Zhou, W.; Tian, Q. Wavelet-based dual-branch network for image demoiréing. In Proceedings of the European Conference on Computer Vision, Glasgow, UK, 23–28 August 2020; Springer: Cham, Switzerland, 2020; pp. 86–102.
24. Park, H.; Vien, A.G.; Koh, Y.J.; Lee, C. Unpaired image demoiréing based on cyclic moiré learning. In Proceedings of the 2021 Asia-Pacific Signal and Information Processing Association Annual Summit and Conference (APSIPA ASC), Tokyo, Japan, 14–17 December 2021; pp. 146–150.
25. Dai, P.; Yu, X.; Ma, L.; Zhang, B.; Li, J.; Li, W.; Shen, J.; Qi, X. Video Demoireing With Relation-Based Temporal Consistency. In Proceedings of the IEEE/CVF Conference on Computer Vision and Pattern Recognition (CVPR), New Orleans, LA, USA, 19–24 June 2022; pp. 17622–17631.
26. Lotfi, M.; Solimani, A.; Dargazany, A.; Afzal, H.; Bandarabadi, M. Combining wavelet transforms and neural networks for image classification. In Proceedings of the 2009 41st Southeastern Symposium on System Theory, Washington, DC, USA, 15–17 March 2009; pp. 44–48.
27. Nayak, D.R.; Dash, R.; Majhi, B. Brain MR image classification using two-dimensional discrete wavelet transform and AdaBoost with random forests. *Neurocomputing* **2016**, *177*, 188–197. [CrossRef]
28. Liu, Y.; Li, Q.; Sun, Z. Attribute-aware face aging with wavelet-based generative adversarial networks. In Proceedings of the IEEE/CVF Conference on Computer Vision and Pattern Recognition, Tullahoma, TN, USA, 15–17 March 2019; pp. 11877–11886.
29. Huang, H.; He, R.; Sun, Z.; Tan, T. Wavelet-srnet: A wavelet-based cnn for multi-scale face super resolution. In Proceedings of the IEEE International Conference on Computer Vision, Venice, Italy, 22–29 October 2017; pp. 1689–1697.
30. Wang, X.; Xie, L.; Dong, C.; Shan, Y. Real-esrgan: Training real-world blind super-resolution with pure synthetic data. In Proceedings of the IEEE/CVF International Conference on Computer Vision, Montreal, QC, Canada, 10–17 October 2021; pp. 1905–1914.
31. Kingma, D.P.; Ba, J. Adam: A method for stochastic optimization. *arXiv* **2014**, arXiv:1412.6980.
32. Ronneberger, O.; Fischer, P.; Brox, T. U-net: Convolutional networks for biomedical image segmentation. In Proceedings of the International Conference on Medical Image Computing and Computer-Assisted Intervention, Munich, Germany, 5–9 October 2015; Springer: Cham, Switzerland, 2015; pp. 234–241.
33. Yang, S.; Lei, Y.; Xiong, S.; Wang, W. High resolution demoire network. In Proceedings of the 2020 IEEE International Conference on Image Processing (ICIP), Abu Dhabi, United Arab Emirates, 25–28 October 2020; pp. 888–892.

sensors

MDPI

Article

A Routing Optimization Method for Software-Defined Optical Transport Networks Based on Ensembles and Reinforcement Learning

Junyan Chen [1,2], Wei Xiao [1], Xinmei Li [1], Yang Zheng [3,*], Xuefeng Huang [1], Danli Huang [1] and Min Wang [1]

1 School of Computer Science and Information Security, Guilin University of Electronic Technology, Guilin 541004, China
2 School of Computer Science and Engineering, Northeastern University, Shenyang 110819, China
3 Institute of Automation, Chinese Academy of Sciences, Beijing 100190, China
* Correspondence: yang.zheng@ia.ac.cn

Abstract: Optical transport networks (OTNs) are widely used in backbone- and metro-area transmission networks to increase network transmission capacity. In the OTN, it is particularly crucial to rationally allocate routes and maximize network capacities. By employing deep reinforcement learning (DRL)- and software-defined networking (SDN)-based solutions, the capacity of optical networks can be effectively increased. However, because most DRL-based routing optimization methods have low sample usage and difficulty in coping with sudden network connectivity changes, converging in software-defined OTN scenarios is challenging. Additionally, the generalization ability of these methods is weak. This paper proposes an ensembles- and message-passing neural-network-based Deep Q-Network (EMDQN) method for optical network routing optimization to address this problem. To effectively explore the environment and improve agent performance, the multiple EMDQN agents select actions based on the highest upper-confidence bounds. Furthermore, the EMDQN agent captures the network's spatial feature information using a message passing neural network (MPNN)-based DRL policy network, which enables the DRL agent to have generalization capability. The experimental results show that the EMDQN algorithm proposed in this paper performs better in terms of convergence. EMDQN effectively improves the throughput rate and link utilization of optical networks and has better generalization capabilities.

Keywords: optical transport network; software-defined networking; deep Q-network; message-passing neural network; ensemble learning

Citation: Chen, J.; Xiao, W.; Li, X.; Zheng, Y.; Huang, X.; Huang, D.; Wang, M. A Routing Optimization Method for Software-Defined Optical Transport Networks Based on Ensembles and Reinforcement Learning. *Sensors* **2022**, *22*, 9139. https://doi.org/10.3390/s22218139

Academic Editors: Chien Aun Chan, Ming Yan and Chunguo Li

Received: 11 October 2022
Accepted: 20 October 2022
Published: 24 October 2022

1. Introduction

The optical transport network (OTN) is a transport network that enables the transmission, multiplexing, route selection, and monitoring of service signals in an optical domain, ensuring its performance index and survivability. The OTN can support the transparent transmission of customer signals, high-bandwidth multiplexing, and configuration. It also provides end-to-end connectivity and networking capabilities. With the rapid development of network communication technology, the demand for OTN networks has increased significantly in terms of the scale of information volume, demand complexity, and dynamic spatio-temporal distribution. Unlike traditional networks, the OTN can meet more network requirements due to its suitable transmission medium, which has a high transmission speed, more data transmission, and a long transmission distance.

Traditional routing design schemes manually model network demand characteristics and design routing policies in a focused way. The traditional routing protocol is designed for wired networks, with a fixed bandwidth allocation pattern and low bandwidth utilization. It cannot provide differentiated services based on the level of assistance, nor can it cope with the rapid changes in topology and link quality standards in optical network environments.

Additionally, because OTN demand has complex spatio-temporal distribution fluctuations, the optimization problem of its routing is an NP-hard problem [1]. In this case, traditional network routing design schemes do not apply to the OTN.

With the development of new network architectures, such as the software-defined networking (SDN) and the maturation of deep reinforcement learning (DRL) techniques in recent years, software-defined optical transport networks (SD-OTNs) based on the SDN are gaining popularity in the industry. Recent studies have used the DRL to address SDN-related problems, such as QoS-aware secure routing for the SDN-IoT [2], SDN routing optimization problems [3], and the SDN demand control [4]. However, due to DRL agents' lack of generalization capabilities, they do not achieve good results in new network topologies. Thus, DRL agents cannot make correct routing decisions when presented with unexplored network scenarios during the training phase. The main reason behind this phenomenon is that graphs essentially represent computer networks. In addition, traditional DRL algorithms use typical neural network (NN) architectures (e.g., fully connected convolutional neural networks), which are unsuitable for modeling information about graph structures. Due to the computational effort and high time complexity of the routing optimization problem, traditional DRL algorithms are challenging for the DRL agent to converge quickly when addressing the network routing optimization problem. Additionally, OTN network problems are incredibly complex and have high trial-and-error costs, making it difficult to implement DRL algorithms in real optical networks.

This paper proposes an ensembles- and message-passing neural-network-based deep Q-network (EMDQN) method to solve the SD-OTN routing decision problem. The message-passing neural network (MPNN) is a deep learning (DL) method based on a graph structure [5]. The MPNN contributes to learning the relationship between graph elements and their rules. In this paper, the MPNN is used to capture information about the relationship between the demand on links and network topology, which can improve the model's generalization ability. Despite computationally complex network problems, ensemble learning has a unique advantage that can increase sample utilization. We reweigh the sample transitions based on the uncertainty estimates of ensemble learning. This method can improve the signal-to-noise ratio during Q-network updates, and stabilize the learning process of the EMDQN agent, which helps the deep Q-network (DQN) [6] operate stably in OTN networks.

The main contributions of this paper are as follows:

1. We propose an SD-OTN routing optimization algorithm based on the reinforcement learning model of the EMDQN. To effectively improve the extrapolation capability of DRL decision-makers, we design a more refined state representation and a limited set of actions.
2. We use the MPNN algorithm instead of the traditional DQN's policy networks, which can capture the relationship between links and network topology demand and improve the DRL decision-maker performance and generalization capability. Additionally, we exploit the advantages of efficient exploration through ensemble learning to explore the environment in parallel and improve convergence performance.
3. We design practical comparison experiments to verify the superior performance of the EMDQN model.

The rest of this paper is structured as follows. In Section 2, this paper discusses research related to the proposed solution for the network problem. Section 3 describes the software-defined network system architecture and the OTN optimization scenarios and tasks. In Section 4, this paper describes the design of DRL-based routing optimization decisions. In Section 5, this paper presents an extensive evaluation of DRL-based solutions in some realistic OTN scenarios. Finally, in Section 6, we present our conclusion and directions for future work.

2. Related Research

Traditional routing optimization schemes are usually based on the OSPF (open shortest path first) [7] or ECMP (equal-cost multipath routing) [8]. The OSPF protocol routes all flow requests individually to the shortest path. The ECMP protocol increases transmission bandwidth using multiple links simultaneously. However, these approaches, based on fixed forwarding rules, are prone to link congestion and cannot meet the demand of exponential traffic growth. Recently, most heuristic algorithm-based approaches have been built under the architecture of the SDN. The authors in [9] proposed a heuristic ant-colony-based dynamic layout algorithm for SDNs with multiple controllers, which can effectively reduce controller-to-switch and controller-to-controller communication delays caused by link failures. The authors in [10] applied a random-based heuristic method called the alienated ant algorithm, which forces ants to spread out across all available paths while searching for food rather than converging on a single path. The authors in [11] analytically extract historical user data through a semi-supervised clustering algorithm for efficient data classification, analysis, and feature extraction. Subsequently, they used a supervised classification algorithm to predict the flow of service demand. The authors in [12] proposed a heuristic algorithm-based solution for DWDM-based OTN network planning. The authors in [13] proposed a least-cost tree heuristic algorithm to solve the OTN path-sharing and load-balancing problem. However, because of a lack of historical experience in data learning, heuristic algorithms can only build models for specific problems. When the network changes, it is difficult to determine the network parameters and there is limited scalability to guarantee service quality. Furthermore, because of the tremendous computational effort and high computational complexity of these methods, heuristic algorithms do not perform well on OTN networks.

With SDN's maturity and large-scale commercialization, the SD-OTN based on the SDN is becoming increasingly popular in the industry. SD-OTN adapts the reconfigurable optical add-drop multiplexer (ROADM) nodes through the southbound interface protocol and establishes a unified resource and service model. The SD-OTN controller can realize topology and network status data collection, routing policy distribution, and network monitoring. Therefore, many researchers deploy artificial intelligence algorithms in the controller. Deep learning, with its powerful learning algorithms and excellent performance advantages, has gradually been applied to the SDN. To solve the SDN load-balancing problem, Chen et al. [14] used the long short-term memory (LSTM) to predict the network traffic in the SDN application plane. The authors in [15] proposed a weighted Markov prediction model based on mobile user classification to optimize network resources and reduce network congestion. The authors in [16] proposed an intrusion detection system based on SDN and deep learning, reducing the burden of security configuration files on network devices. However, deep learning requires many datasets for training and has poor generalization abilities due to its inability to interact with the environment. These factors make it difficult to optimize the performance of dynamic networks. Compared with deep learning, reinforcement learning uses online learning for model training, changing agent behaviors through continuous exploration, learning, and experimentation to obtain the best return. Therefore, reinforcement learning does not require the model to be trained in advance. It can change its action according to the environment and reward feedback. The authors in [17] designed a Q-learning-based localization-free routing for underwater sensor networks. The authors in [18] proposed a deep Q-routing algorithm to compute the path of any source-destination pair request using a deep Q-network with prioritized experience replay. The authors in [19] proposed traction control ideas to solve the routing problem. The authors in [20] proposed a routing optimization algorithm based on the proximal policy optimization (PPO) model in reinforcement learning. The authors in [21] discussed a solution for automatic routing in the OTN using DRL. Although the studies described above have been successful for the SDN demand-routing optimization problem, they do not perform as well in new topologies because they do not consider the model's generalization capability.

The traditional DRL algorithms use a typical neural network (NN) as the policy network. The NN can extract and filter the features of the input information and data layer by layer to finally obtain the results of tasks, such as classification and prediction. However, as research advances, conventional neural networks are unable to solve all network routing problems and will struggle to handle non-Euclidean-structured graph data. Therefore, we need to optimize the traditional reinforcement learning algorithm to improve its ability to extract the information features of the sample. Off-policy reinforcement learning (Off-policy RL) algorithms significantly improve sample utilization by reusing past experiences. The authors in [22] propose an off-policy actor–critic RL algorithm based on a maximum entropy reinforcement learning framework. The participants' goal in this framework is to maximize the expected reward while maximizing the entropy. They achieved state-of-the-art sample efficiency results by combining a maximum entropy framework. However, in practice, the commonly used off-policy approximate dynamic programming methods based on the Q-learning and actor–critic methods are susceptible to data distribution. They can only make limited progress without collecting additional on-policy data. To address this problem, the authors in [23] proposed bootstrap error accumulation reduction to reduce off-policy algorithm instability caused by accumulating backup operators via the Bellman algorithm. The authors in [24] developed a new estimator called offline dual reinforcement learning, which is based on the cross-folding estimation of Q-functions and marginalized density ratios. The authors in [25] used a framework combining imitation learning and deep reinforcement learning, effectively reducing the RL algorithm's instability. The authors in [26] used the DQN replay datasets to study off-policy RL, effectively reducing the off-policy algorithm's instability. The authors in [27] proposed an intelligent routing algorithm combining the graph neural network (GNN) and deep deterministic policy gradient (DDPG) in the SDN environment, which can be effectively extended to different network topologies, improving load-balancing capabilities and generalizability. The authors in [28] combined GNN with the DQN algorithm to address the lack of generalization abilities in untrained OTN topologies. OTN topology graphs are non-Euclidean data, and the nodes in their topology graphs typically contain useful feature information that most neural networks are unable to comprehend. They use MPNN to extract feature information between OTN topological nodes, which improves the generalization performance of the DRL algorithm.

However, it is a challenge for a single DRL agent to balance exploration and development, resulting in limited convergence performance. Ensemble learning solves a single prediction problem by building several models. It works by generating several classifiers or models, each of which learns and predicts independently. These predictions are finally combined into a combined prediction, which outperforms any single classification for making predictions [29]. There are two types of integrated base learning machines. One type involves using various learning algorithms on the same dataset to obtain a base learning machine, which is usually referred to as heterogeneous [30–32]. The other type applies the same learning algorithm on a different training set (which can be obtained by random sampling based on the original training dataset, etc.), and the base learning machine obtained using this method is said to be a homogeneous type. However, because of the high implementation difficulty and low scalability of heterogeneous types of base learning machines, expansion to high-dimensional state and action spaces is difficult, making it unsuitable for solving OTN routing optimization problems. Table 1 summarizes the description of the papers reviewed, whether SDN and RL are considered, and the evaluation indicators. The EMDQN algorithm we propose applies the same reinforcement learning algorithm to different training sets to generate the base learning machine. We combine multiple EMDQN agents to construct an ensemble learning machine and generate diverse samples to effectively generate learning machines with high generalization abilities and significant differences.

Table 1. Related work.

Paper	Description	RL	DL	OTN	Evaluating Indicator
[7]	Performance analysis of OSPF				Network convergence, traffic dropped
[8]	Embarks upon a systematic algorithmic study of traffic engineering with ECMP				Throughput
[9]	Allocation of computational resources based on heuristic ant colony algorithm				Latency, load balancing, task completion time.
[10]	A load-balancing algorithm based on the alienated ant algorithm				Throughput, delay, packet loss rate
[11]	SDN routing solution about flow feature extraction, requirement prediction and route selection				Routing efficiency
[12]	An OTN network planning solution over DWDM based on heuristic algorithms			√	Network resource consumption
[13]	A heuristic algorithm of minimum cost tree for path sharing and load balancing			√	Tree cost, run time, degree of load balancing
[14]	A network traffic prediction model based on LSTM		√		Throughput, load-balancing degree
[16]	Deep learning classifier for detection of anomalies		√		Precision, recall, accuracy of classification
[17]	A Q-learning-based localization-free anypath routing	√			Delay, network lifetime, packet delivery ratio
[19]	Combines the control theory and DRL technology to achieve an efficient network control scheme	√	√		Transmission delay
[20]	An RL routing algorithm to solve a traffic engineering	√	√		Throughput and delay, transmission time
[21]	Designing state and action to simplify the DRL algorithm	√	√	√	Link utilization
[27]	A set of extensions to the MQTT protocol that meet application-defined real-time requirements	√	√		Latency
[28]	A DRL algorithm combined with GNN	√	√	√	Network capacity
[30]	A new method of data missing estimation with tensor heterogeneous ensemble learning		√		Data missing rates
[32]	A method to automatically learn long-term associations between traffic samples		√		Calculates precision, recall and F1-score

3. SD-OTN Architecture

In this paper, the designed SD-OTN architecture consists of the application, control, and data planes, as shown in Figure 1. The description of each part of the network architecture is as follows:

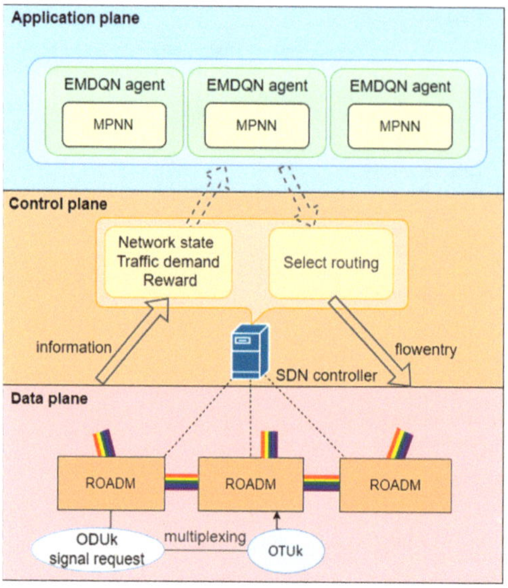

Figure 1. The SD-OTN architecture. The architecture consists of the application plane, control plane, and data plane.

1. Data plane. The data plane consists of the ROADM nodes and the predefined optical paths connecting them. In the data plane, the capacity of the links and the connection status of the ROADM nodes are predefined. The data plane must collect the current optical data unit (ODU) signal requests and network status information, which it must then send to the control plane via the southbound interface. The data plane implements the new routing forwarding policy after receiving it from the control plane. It communicates the new network state and traffic demand to the control plane, from which decision-makers in the application plane learn.
2. Control plane. The control plane consists of the SDN controller. The control plane obtains the ODU signal request and network status information via the southbound interface and calculates the reward using the reward function. Through the northbound interface, the control plane sends the network state, traffic demand, and reward to the application plane via the northbound interface. When receiving optimized routing action from the application plane, the control plane sends a routing forwarding policy to the data plane based on the routing action.
3. Application plane. The application plane manages the EMDQN agents. The agents obtain network state information from the control plane, encode it, and feed it into the agents' policy network, which generates optimized routing actions. Subsequently, the routing actions are sent down to the control plane.

4. EMDQN-Based Decision Design for Routing Optimization

In this section, we describe in detail the EMDQN algorithm proposed in this paper.

4.1. DRL-Based Routing Optimization in SD-OTN

Based on the system architecture described above, the DRL agent's role is to assign routes to incoming traffic demands for a specific sequence of optical paths (i.e., end-to-end paths) to maximize network utility. Because the DRL agent operates in the electrical domain, traffic demands are treated as requests for ODU signals. These signals, which may originate from different clients, are multiplexed into an optical transform unit (OTU), as shown in Figure 1. The final OTU frames are transmitted through the optical channels in the OTN [33].

We use G to refer to an optical transmission network, as shown in Equation (1):

$$G = (V, E) \tag{1}$$

where V and E represent the set of n ROADM nodes and m optical links in the network topology, respectively, as shown in Equations (2) and (3).

$$V = [v_1, v_2, \ldots, v_n]. \tag{2}$$

$$E = [e_1, e_2, \ldots, e_m]. \tag{3}$$

We use C to denote the set of link bandwidth capacity, as shown in Equation (4), where $|C| = |E| = m$:

$$C = [c_1, c_2, \ldots, c_m]. \tag{4}$$

The path k from node v_i to node v_j is defined as a sequence of links, as shown in Equation (5), where $e_{k(i)} \in E$:

$$p_k = \left\{e_{k(0)}, e_{k(1)}, \ldots, e_{k(n)}\right\}. \tag{5}$$

We use d_k to denote the traffic demand of the path k, and define D as the set of all traffic demands, as shown in Equation (6):

$$D = [d_1, d_2, \ldots, d_{n*n}]. \tag{6}$$

The traffic routing problem in OTN is a classical resource allocation problem [26]. If the bandwidth capacity of the distributed routing path is greater than the size of the bandwidth requirement, the allocation is successful. After successfully allocating bandwidth capacity for a node pair's traffic demand, the routing path will not be able to release the bandwidth occupied by that demand until the end of this episode. We use rb_i to describe the remaining bandwidth of the link e_i, which is the link bandwidth capacity c_i minus the traffic demands of all paths passing through link e_i, as shown in Equation (7). RB is the set of the remaining bandwidth of all links, as shown in Equation (8).

$$rb_i = c_i - \sum d_k. \tag{7}$$

$$RB = [rb_1, rb_2, \ldots, rb_m]. \tag{8}$$

We use q_k to denote the allocating traffic demand of the path k, as shown in Equation (9). Q is the set of all allocating traffic demands, as shown in Equation (10).

$$q_k = \begin{cases} d_k, & if\ \forall e \in p_k\ and\ r_e > d_k \\ 0, & else \end{cases}. \tag{9}$$

$$Q = [q_1, q_2, \ldots, q_{n*n}]. \tag{10}$$

The optimization objective in this paper is to successfully allocate as much of the traffic demand as possible, as shown in Equation (11):

$$\max\left(\sum_{q_i \in Q} q_i\right). \tag{11}$$

In view of the above optimization objective, the routing optimization can be modeled as a Markov decision process, defined by the tuple $\{S, A, P, R\}$, where S is the state space, A is the action space, P is the set of transfer probabilities, and R is the set of rewards. The specific design is as follows:

1. Action space: The action space is designed as k shortest hop-based paths of source-destination nodes. The action selects one of the k paths to transmit the traffic demand

of source–destination nodes. The parameter k is customizable and varies according to the topology's complexity. The action space is invariant to the arrangement of nodes and edges, which is discretely distributed, allowing the DRL agent to understand the actions on arbitrary network states easily.

2. State space: The state space is designed as the remaining bandwidth RB, the traffic demand D, and the link betweenness. The link betweenness is a centrality metric, which indicates how many paths are likely to cross the link. For each node pair in the topology, k candidate shortest routes are calculated, with the link betweenness value being the number of shortest routes passing through the link divided by the total number of paths, as shown in Equation (12), where bn_i represents the betweenness of the link e_i, N represents he total number of paths, p_i^k represents the number of shortest routes passing through the link e_i in k candidate shortest routes:

$$bn_i = p_i^k / N. \tag{12}$$

3. Reward function: The reward function returns a positive reward if the selected link has sufficient capacity to support the traffic demand in an episode; otherwise, it returns no reward and terminates the episode. According to the optimization objective in Equation (11), the final reward for the episode is the sum of the rewards of all successfully assigned traffic demand tuples $\{src,\ dst,\ demand\}$, as shown in Equation (13), where N is the number of traffic demand tuples, r_t represents the reward after the action at time t, q_i represents the i-th traffic demand successfully assigned, and q_{max} represents the maximum traffic demand successfully assigned. The higher the reward, the more bandwidth demands are successfully allocated in that time step, and the better the network load-balancing capability.

.

$$r_t = \sum_{i=1}^{N} q_i / q_{max}. \tag{13}$$

4.2. DQN Algorithm

Based on the above DRL-based optimization solution, this paper selects the DQN algorithm to implement a reinforcement learning agent. The DQN is a classical DRL algorithm based on value functions. It combines a convolutional neural network (CNN) with the Q-learning algorithm, using the CNN model to output the Q-value corresponding to each action to ascertain which to perform [6].

The DQN algorithm uses two network models containing CNNs for learning: the prediction network $Q(s,a,\theta)$ and the target network $\hat{Q}(s,a,\overline{\theta})$, where θ and $\overline{\theta}$ are the network parameters of the prediction and target networks, respectively. The prediction network outputs the predicted Q-value corresponding to the action, whereas the target network calculates the target value and updates the parameters of the prediction network based on a loss function. The DQN copies the parameters of the prediction network model to the target network after each C-round iteration.

The prediction network approximates the action value function through the CNN model $Q_\pi(s,a)$, as shown in Equation (14):

$$Q(s,a,\theta) \approx Q_\pi(s,a). \tag{14}$$

The DQN agent selects and executes an action based on an ϵ-greedy policy. The policy generates a random number in [0, 1] interval through a uniform distribution. If the number is less than $1 - \epsilon$, it selects an action that maximizes the Q-value; otherwise, it selects an action randomly, as shown in Equation (15):

$$a_t = \begin{cases} \underset{a}{\text{argmax}}\ Q(s_t, a, \theta), & \text{with probability } 1{-}\varepsilon \\ \text{random action}, & \text{otherwise} \end{cases}. \tag{15}$$

The target network calculates the target value y by obtaining a random mini-batch storage sample from the replay buffer, as shown in Equation (16), where r is the reward value and γ is the discount factor:

$$y = r + \gamma \underset{a'}{\max} \hat{Q}\left(s', a', \overline{\theta}\right). \tag{16}$$

The DQN defines the loss function of the network using the mean-square error, as shown in Equation (17). The parameter θ is updated by the mini-batch semi-gradient descent, as shown in Equations (18) and (19):

$$L(\theta) = \mathbf{E}\left[(y - Q(s, a, \theta))^2\right], \tag{17}$$

$$\nabla_\theta L(\theta) \approx \frac{1}{N} \sum_i^N (y - Q(s, a, \theta)) \nabla Q(s, a, \theta), \tag{18}$$

$$\theta \leftarrow \theta - \alpha \nabla_\theta L(\theta), \tag{19}$$

where N represents the number of samples and α represents the update parameter.

The target network is used by the DQN to keep the target Q-value constant over time, which reduces the correlation between the predicted and target Q-values to a certain extent. This operation reduces the possibility of loss value oscillation and divergence during training and improves the algorithm's stability.

4.3. Message-Passing Neural Network

The CNN model has better results in extracting features from Euclidean spatial data (e.g., picture data), characterized by a stable structure and dimensionality. However, graph-structured or topologically structured data are infinitely dimensional and irregular, and the network surrounding each node may be unique. Such structured data renders traditional CNNs ineffective and unable to extract data features effectively. To address this problem, we use the MPNN rather than the CNN as a network model for the DQN. The MPNN is a type of GNN that is suitable for extracting spatial features of topological graph data [5].

Through repeated iterations of the process of passing data about the link's hidden state, the MPNN algorithm abstract information about the characteristics of the network. The characteristic values of the hidden state h_i include the remaining bandwidth rb_i, the link betweenness bn_i, and the traffic demand feature df_i. The traffic demand feature df_i represents the quantitative characteristics of the traffic demand d_i. Because the traffic demand of the OTN environment is discrete and finite, the traffic demand feature is denoted by an n-element one-hot encoding, and link characteristics that are not included in the k routes have a zero value. Additionally, the size of the hidden state is usually larger than the size of the feature values in the hidden state; thus, we use zero values to populate the feature value vector, as shown in Equation (20):

$$h_i = [rb_i, bn_i, df_i, 0, \ldots, 0]. \tag{20}$$

The MPNN workflow is shown in Figure 2. We perform a message-passing process between all links which will be executed T times. First, the MPNN receives link hidden features as the input. Second, each link iterates over all of its adjacent links to obtain the link features. In the message-passing process, for each link k, we generate messages by entering the hidden state h_k of the link and the hidden state h_i of all neighboring links into the message function $m(\cdot)$. The message function $m(\cdot)$ is a fully connected CNN. After iterating over all links, the link k receives messages from all neighboring links (denoted by $N(k)$). It generates a new feature vector M_k using message aggregation, as shown in Equation (21):

$$M_k^{t+1} = \sum_{\mathrm{i} \in \mathrm{N(k)}} m\left(h_k^t, h_i^t\right), \tag{21}$$

where $N(k)$ represents all neighboring links of the link k.

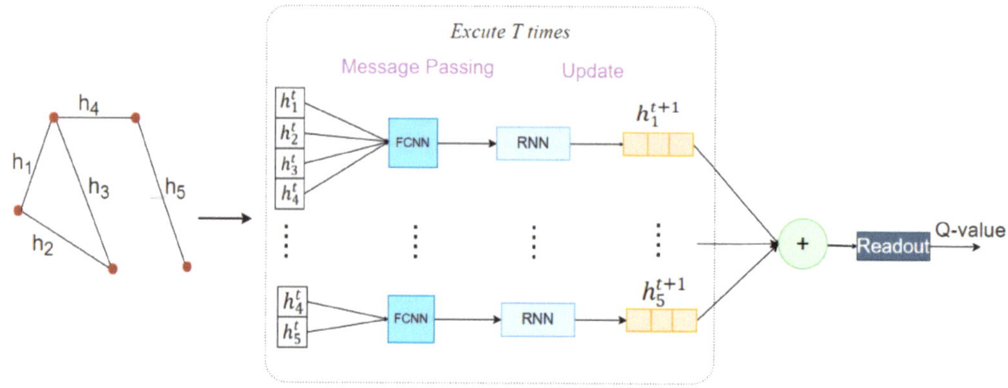

Figure 2. The MPNN workflow.

Second, we update the hidden state of the link by aggregating the feature vector M_k^{t+1} with the link-hidden state h_k^t through the update function $u(\cdot)$, as shown in Equation (22). The update function $u(\cdot)$ is the Gate Recurrent Unit (GRU), which has the characteristics of high training efficiency.

$$h_k^{t+1} = u\left(h_k^t, M_k^{t+1}\right). \tag{22}$$

Finally, after the T-step message transmission, we use the readout function $R(\cdot)$ to aggregate the hidden state of all links and obtain the Q-value, as shown in Equation (23):

$$Q(s, a, \theta) = R\left(\sum_{k \in E} h_{\mathrm{k}}\right), \tag{23}$$

where E represents the set of all links in the topology.

4.4. Ensemble Learning

In the DQN algorithm, it is challenging for a single agent to balance exploration and development, resulting in limited convergence performance. Furthermore, errors in the DQN target values can increase the overall error in the Q-function, leading to an unstable convergence. In this paper, we use ensemble learning to solve the above problems. Ensemble learning has the advantage of efficient exploration and can reduce uncertainty in new samples.

As shown in Figure 3, ensemble learning is realized by a set of multiple EMDQN agents $\{Q(s, a, \theta_i)\}_{i=1}^N$, where θ_i represents the parameter of the i-th agent. To diversify the training of the EMDQN agents, we randomly initialize the policy network of all EMDQN agents. In the training phase, we employ the ϵ-greedy-based upper-confidence bound (UCB) exploration strategy [34], as shown in Equation (24):

$$a_t = \begin{cases} \max\limits_a \{Q_{\mathrm{mean}}(s_t, a, \theta) + \lambda Q_{\mathrm{std}}(s_t, a, \theta)\}, & \text{with probability } 1-\varepsilon \\ \text{random action,} & \text{otherwise} \end{cases}, \tag{24}$$

where $Q_{\text{mean}}(s_t, a, \theta)$ and $Q_{\text{std}}(s_t, a, \theta)$ are the mean and standard deviation of the Q-values output by all MPNN policy networks $\{Q(s, a, \theta_i)\}_{i=1}^{N}$. The exploration reward $\lambda > 0$ is a hyper-parameter. When λ increases, the EMDQN agents become more active in accessing unknown state–action pairs.

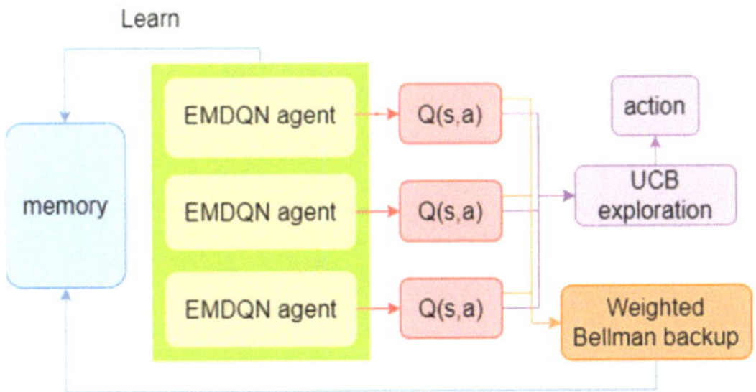

Figure 3. EMDQN workflow.

The traditional DQN loss function (Equation (6)) may be affected by error propagation, that is, it propagates the target Q-network $\hat{Q}(s', a', \overline{\theta})$ error to the current state of the Q-network $Q(s, a, \theta)$. This error propagation can lead to an unstable convergence. To alleviate this problem, for each EMDQN agent i, this paper uses Bellman weighted backups, as shown in Equation (25):

$$L_{WQ}^{\text{EMDQN}}(\theta_i) = w(s)\left(r + \gamma \max_{a'} \hat{Q}(s', a', \overline{\theta_i}) - Q(s, a, \theta_i)\right)^2, \tag{25}$$

where $w(s)$ represents the confidence weight of the set of target Q-networks in the interval [0.5, 1.0]. $w(s)$ is calculated from Equation (26), where the weight parameter W is a hyper-parameter, σ is a sigmoid function, $\hat{Q}_{std}(s)$ is the empirical standard deviation of all target Q-networks $\left\{\max_a \hat{Q}(s, a, \overline{\theta})\right\}_{i=1}^{N}$. $L_{WQ}^{EMDQN}(\cdot)$ reduces the weights of sample transitions with high variance between target Q-networks, resulting in better signal-to-noise ratios for network updates.

$$w(s) = \sigma(-\hat{Q}_{std}(s) * W) + 0.5. \tag{26}$$

4.5. The Working Process of the EMDQN Agent

The working process of the EMDQN agent at each iteration is described in Algorithm 1. We first reset the environment and obtain the environment link capacity and traffic demand tuple $\{src,\ dst,\ demand\}$ (line 1). Subsequently, we execute a loop to continuously assign traffic demands. In the process, we compute k shortest links (Line 3) and allocate the traffic demand for each shortest link through k cycles (Lines 4–8). Based on this, we can compute the Q-value for each action. We select actions using an ϵ-greedy-based UCB exploration strategy (Line 9); subsequently, we apply the chosen route to the environment (Line 10). We store the rewards and state transfer during the interaction with the environment in the experience replay buffer (Line 11) while applying the transferred state (Line 12). The cycle stops when any link is unable to carry the traffic demand. Next, we execute the agent learning phase. For the sampled batch (Line 15), we plot the mask using the Bernoulli distribution (Line 16) and calculate the batch weight using all EMDQN agents. Following that, we multiply the sample by the weight and mask to minimize L_{WQ}^{EMDQN} (Line 18). Finally, we evaluate the set of EMDQN agents in the environment (Line 21) and collect the rewards, as well as the status of the environment, in the evaluation process to analyze the training situation of the EMDQN agent.

Algorithm 1: working process of the EMDQN agent

```
s, demand, src, dst ← env.reset()
while (Done != False) do
    k_path ← compute_k_path(k, src, dst)
    for i ← 1 to k do
        path ← get_path(i, k_path)
        s′ ← allocate(s, path, src, dst, demand)
        k_Q[i] ← compute_Q(s′, path)
    end for
    a ← act(k_Q, ε, k_path, s)
    s′, r, done, demand′, src′, dst′ ← env.step(s, a)
    agent.remember(s, a, r, s′, done)
    s, demand, src, dst ← s′, demand′, src′, dst′
end while
for i ← 1 to STEP do
    batch ← sample()
    m ← bernoulli()
    for each agent i do
        Update agent by minimizing L^{WQ}_{EMDQN}(θ_i)
    end for
end for
agent.evaluate()
```

5. Experiments and Analysis

In this section, we simulate the SD-OTN routing scenario using the OpenAI gym framework to train and evaluate the EMDQN algorithm. Furthermore, we conduct experiments and analyses by adjusting the hyper-parameters and evaluating the algorithm load-balancing ability and generalization ability.

5.1. Experimental Environment

The computer used for the experiments has an AMD R5 5600G processor with a base frequency of 2900 MHz, a 2 TB solid-state drive, and 32 GB of RAM. The experiment uses the Tensorflow deep learning framework to implement the EMDQN algorithm. We select NSFNET, GEANT2, and GBN for the optical transmission network topology, with the lightpath bandwidth being 200 ODU0 bandwidth units, as shown in Figure 4. Among these, the NSFNET network contains 14 ROADM nodes and 21 lightpaths, the GEANT2 network contains 24 ROADM nodes and 36 lightpaths, and the GBN network contains 17 ROADM nodes and 27 lightpaths.

In this paper, the lightpath bandwidth requirements are expressed as multiples of ODU0 signals, i.e., 8, 32, and 64 ODU0 bandwidth units. In each episode, the environment generates a traffic demand tuple {src, dst, demand} at random. Additionally, the EMDQN agent should assign the appropriate route for each tuple received. If the assignment is successful, it will receive a reward as defined in Equation (1). Otherwise, it will not be rewarded. Since the new traffic demand is randomly generated, the routing policy designed by the EMDQN agent does not rely on traffic demand distribution information, reducing the EMDQN agent's overfitting to the particular network scenario used for training.

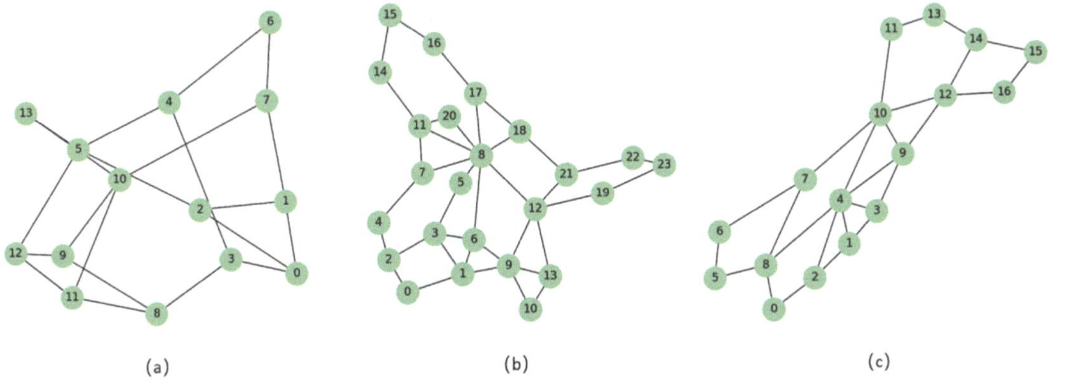

Figure 4. The optical transmission network topologies: (**a**) the NSFNET topology; (**b**) the GEANT2 topology; (**c**) the GBN topology.

5.2. Hyper-Parameters Settings

We experimentally select suitable hyper-parameters for the EMDQN agent, as shown in Figure 5. In the experiments, we chose the NSFNET as the experimental network topology. The size of the link-hidden state is related to the amount of coding information. We set the size of the link-hidden state to twenty and the number of feature values to five, and filled the rest with zero. To facilitate observation, we smoothed the data when drawing the graph.

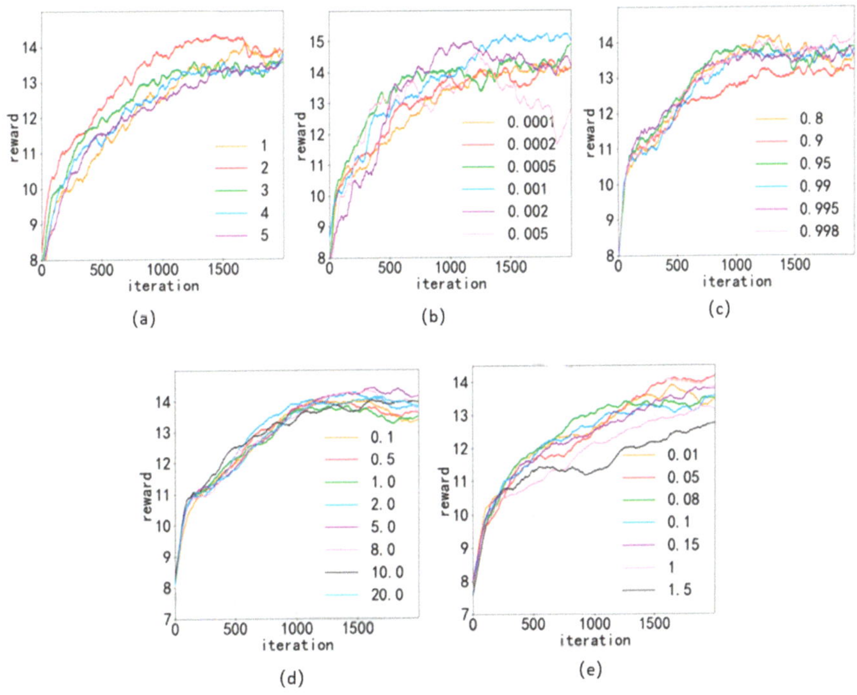

Figure 5. Comparison of the effect of some super-references: (**a**) ensemble number; (**b**) learning rate; (**c**) ϵ-decay; (**d**) UCB exploration reward λ; (**e**) weight parameter W.

Figure 5a shows the training results for the different numbers of EMDQN agents. When the number of agents is high, the training slows down and aggravates the overfitting of DRL agents in the application scenario, resulting in poorer results. The performance is optimal when the number of EMDQN agents is two. Figure 5b shows the training results of the stochastic gradient descent algorithm with different learning rates. When the learning rate was 0.001, the algorithm reward achieved the highest value. Figure 5c shows the training results for different decay rates of ϵ. In the initial stage of training, ϵ is close to 1. We executed 70 iterations and started to reduce ϵ exponentially using ϵ-decay until it decreased to 0.05. During the process of ϵ reduction, the training curve tends to flatten out, finally reaching convergence. The training results show that the reward value curve is most stable after convergence when ϵ-decay is 0.995. Figure 5d shows the training results for different λ values in Equation (11). λ denotes the exploration reward of the EMDQN agent. From the results in Figure 5d, it is clear that the algorithm reward value is highest when λ value is 5. Figure 5e depicts the training results for different weight parameters W in Equation (14). In this paper, we set the size of samples to 32. As W increases, the sample weights converge and become less than one, which affects the sample efficiency of the EMDQN. The reward of this algorithm reaches its highest value when the value of W is 0.05.

Table 2 shows some relevant parameters of the EMDQN and values taken after tuning the parameters.

Table 2. Some relevant parameters of the EMDQN.

Parameter	Value
Batch size	32
Learning rate	0.001
Soft wights copy α	0.08
Dropout rate	0.01
State hidden	20
Ensemble number	2
UCB exploration reward λ	5
Weight parameter W	0.05
ϵ-decay	0.995
Discount factor γ	0.95

5.3. Load-Balancing Performance Evaluation

In this section, we experimentally evaluate the EMDQN in the three network topologies, as described in Section 5.1. The DRL agent runs 2000 iterations. In each iteration, the agent trains 50 episodes and evaluates 40 episodes. Furthermore, the DRL agent updates the network during the training period. During the evaluation, the DRL agent does not update the network; rather, it applies the action to the environment intending to maximize the Q-function, and subsequently records network state data, such as rewards, link utilization, and throughput for each episode.

We implement other SDN solutions for performance comparison with EMDQN algorithms, such as OSPF [7], ECMP [8], DQN [17], PPO [19], and DQN+GNN [26]. The DQN+GNN is an ablation experiment among the compared algorithms, i.e., a performance comparison of the EMDQN model with ensemble learning removed. The DQN and PPO are classic DRL algorithms that use a fully connected feedforward NN as a policy network. The OSPF is an open shortest path algorithm that performs an action selection by calculating the shortest number of hops of the link traversed between the source and destination nodes. The ECMP algorithm is an equal-value multipath routing protocol that allows the use of multiple links simultaneously in the network. The ECMP algorithm distributes the bandwidth demand equally over k lightpaths in this experiment. Furthermore, OUD0 signals are not divisible, but we can verify the performance in other network scenarios in this way.

Figure 6 shows the average reward of all algorithms for the three evaluation scenarios, where the confidence interval is 95%. In this paper, we design the reward based on whether the bandwidth demand can be successfully allocated. The greater the reward, the more bandwidth demand is successfully allocated, and the better the network load-balancing capability. In all three evaluation scenarios, the EMDQN algorithm proposed in this paper performs better than other algorithms after convergence. The EMDQN algorithm outperforms the DQN+GNN algorithm with ensemble learning removed after convergence by more than 7%, demonstrating that the multi-agent ensemble learning approach can effectively improve the convergence performance of the DQN. Additionally, the EMDQN and DQN+GNN outperform the classical reinforcement learning algorithms (DQN and PPO) by more than 25% in all three evaluated scenarios. This indicates that the MPNN can effectively improve the decision performance of the reinforcement learning model by capturing information about the relationship between the demand on links and network topology. The DQN and PPO algorithms perform about as well as the ECMP algorithm after convergence. The OSPF algorithm, on the other hand, routes all flow requests singularly to the shortest path. Since this method is based on fixed forwarding rules, it can easily lead to link congestion. Therefore, the OSPF algorithm is only close to ECMP in the GBN scenario and the lowest in other scenarios.

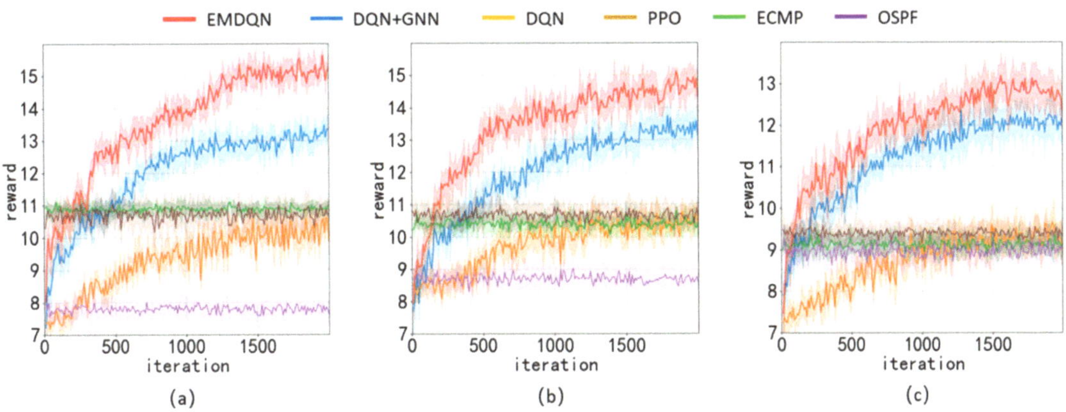

Figure 6. Comparison of the rewards of each algorithm in different scenarios: (**a**) NSFNET scenario evaluation; (**b**) GEANT2 scenario evaluation; (**c**) GBN scenario evaluation.

Table 3 shows the average throughput of each algorithm in ODU0 bandwidth units for the three network topologies. Table 4 displays the average link utilization of each algorithm across the three network topologies. The average throughput and link utilization of the EMDQN are higher than those of other algorithms under various network topologies, indicating that the EMDQN algorithm has a better load-balancing capability for the network after convergence. The performance of the EMDQN algorithm is higher than that of the DQN+GNN algorithm, which is a good indication that ensemble learning can improve the convergence performance of the model. The results show that the EMDQN has excellent decision-making abilities.

Table 3. A comparison of the average throughput of each algorithm in different scenarios.

	EMDQN	DQN+GNN	DQN	PPO	ECMP	OSPF
NSFNET	**1028.17 ± 27.45**	899.28 ± 24.21	709.47 ± 35.47	737.48 ± 26.92	747.68 ± 13.50	548.86 ± 20.97
GEANT2	**995.56 ± 35.26**	903.48 ± 35.36	721.97 ± 31.07	726.89 ± 24.09	717.35 ± 19.59	605.27 ± 21.53
GBN	**864.77 ± 30.28**	826.06 ± 30.19	646.56 ± 29.71	652.05 ± 15.51	636.10 ± 18.69	627.70 ± 23.02

Table 4. A comparison of the average link utilization of each algorithm in different scenarios.

	EMDQN	DQN+GNN	DQN	PPO	ECMP	OSPF
NSFNET	**56 ± 0.90%**	50 ± 1.28%	40 ± 1.69%	43 ± 0.65%	41 ± 1.04%	15 ± 1.16%
GEANT2	**40 ± 0.92%**	36 ± 1.46%	29 ± 1.49%	30 ± 0.60%	21 ± 0.90%	11 ± 0.83%
GBN	**45 ± 0.92%**	42 ± 1.54%	34 ± 1.48%	35 ± 0.76%	25 ± 1.22%	15 ± 1.07%

5.4. Generalization Performance Evaluation

In a real OTN scenario, there is the possibility that the network topology changes due to a broken lightpath. In this case, the DRL model usually needs to be retrained, resulting in a network state that is low-load-balanced for an extended period, which is intolerable for real network situations. To verify the generalization performance of the EMDQN in this paper, we simulated the light path breakage case in the training environment. We randomly break 0–10 lightpaths in each network scenario and evaluate 100 iterations using the converged EMDQN model while ensuring that the network topology remains connected.

In the generalization experiments, we compared and analyzed the EMDQN, DQN, and OSPF. The classical DQN algorithm is implemented using a fully connected network and will fail if the network topology changes. To avoid retraining the DQN model, we removed some network parameters and applied the same evaluation method after adjusting the state inputs. Figure 7 shows the experimental results of the model's evaluation of randomly malfunctioning lightpaths in different network scenarios. When a lightpath malfunctions, a new route needs to be found to avoid the failed lightpath. As the number of faulty lightpaths increases, fewer routes become available, resulting in a reduction in network transmission traffic and a decrease in the load capacity of the network. The OSPF and the classical DQN algorithms have progressively lower rewards as the number of faulty lightpaths increases and have worse performance than the EMDQN model. In contrast, the EMDQN algorithm can still understand the state of the network and obtain a higher reward. The results demonstrate that the MPNN can still improve the model's generalization ability in the case of network failure, which allows the EMDQN agent to maintain a good performance.

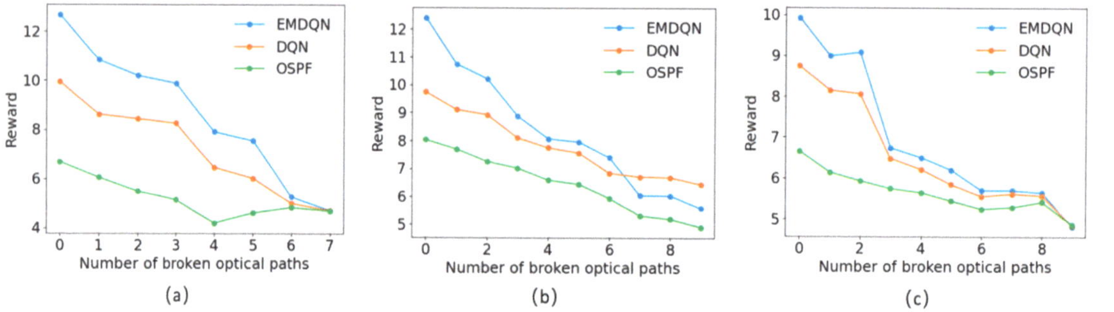

Figure 7. Evaluation of the model in different network scenarios with randomly broken lightpaths: (**a**) broken lightpath at NSFNET; (**b**) broken lightpath at GEANT2; (**c**) broken lightpath at GBN.

To further verify the generalization performance of EMDQN agents, we use the EMDQN model trained to converge in NSFNET to transfer to GEANT2 and GBN network topologies for evaluation. Because of the generalization capabilities of the MPNN, the converged EMDQN agents can directly transfer to operate in different network topologies. The experimental results are shown in Figure 8. The classical DQN algorithm does not perform effectively as the traditional routing algorithm OSPF when the network topology is changed. However, the EMDQN model in this paper still works stably, and the reward section, average value, and stability are significantly better than the DQN and OSPF algorithms. This confirms that the EMDQN agent can still maintain excellent decision-making abilities in the case of network connectivity changes.

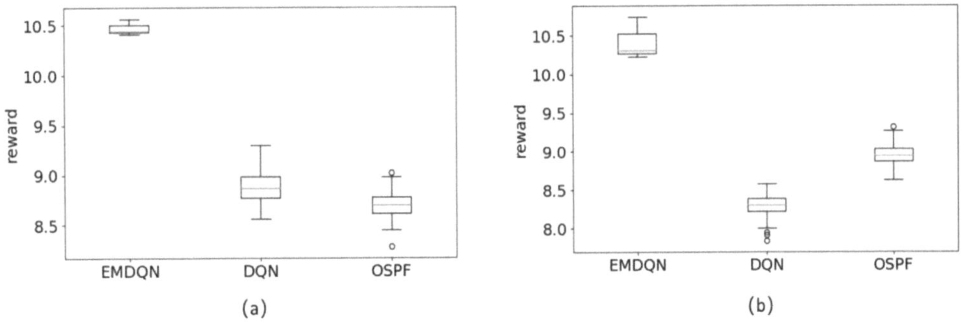

Figure 8. Performance of the algorithm after changing the network topology: (**a**) train in the NSFNET and evaluate in the GEANT2; (**b**) train in the NSFNET and evaluate in the GBN.

6. Conclusions and Future Work

In this paper, we proposed the EMDQN algorithm, which uses the MPNN as a policy network for the DRL to improve the DRL agent's decision-making and generalization abilities, allowing the EMDQN agent to efficiently generalize the unknown topology. We verify the convergence and generalization performance of the EMDQN algorithm by SD-OTN simulating experiments, analyzing and comparing traditional routing protocols with some SDN solutions based on other DRL algorithms. The experimental results show that the EMDQN model can generalize unknown network topologies and outperform other SDN solutions. Furthermore, integrating multiple agents using integrated learning and combining weighted Bellman backup as well as UCB exploration strategies improves the convergence performance while alleviating the DRL agents' unstable operation when converging.

However, the difficulty of adjusting parameters is one challenge faced by the DRL. As seen in Section 5.3, the EMDQN has more hyper-parameters and requires several experiments to complete the adjustment parameters. Therefore, in our future work, we will continue to improve the DRL algorithm and reduce the parameter sensitivity of the DRL method to reduce the reality gap in the DRL method's landing.

Author Contributions: Conceptualization, J.C. and Y.Z.; methodology, J.C.; software, W.X. and X.L.; validation, J.C., W.X., and X.L.; formal analysis, X.H.; investigation, M.W.; resources, D.H.; data curation, J.C.; writing—original draft preparation, J.C., W.X., and X.L.; writing—review and editing, J.C. and Y.Z.; visualization, X.H.; supervision, Y.Z.; project administration, Y.Z.; funding acquisition, J.C. All authors have read and agreed to the published version of the manuscript.

Funding: This work is supported by the major program of Guangxi Natural Science Foundation (No.2020GXNSFDA238001) and the Middle-aged and Young Teachers' Basic Ability Promotion Project of Guangxi (No.2020KY05033).

Institutional Review Board Statement: Not applicable.

Informed Consent Statement: Not applicable.

Data Availability Statement: The authors confirm that the data supporting the findings of this study are available within the article.

Conflicts of Interest: The authors declare that they have no conflict of interest.

References

1. Karakus, M.; Durresi, A. Quality of service (QoS) in software defined networking (SDN). *J. Netw. Comput. Appl.* **2017**, *80*, 200–218. [CrossRef]
2. Guo, X.; Lin, H.; Li, Z.; Peng, M. Deep-Reinforcement-Learning-Based QoS-Aware Secure Routing for SDN-IoT. *IEEE Internet Things J.* **2020**, *7*, 6242–6251. [CrossRef]
3. Sun, P.; Lan, J.; Guo, Z.; Xu, Y.; Hu, Y. Improving the Scalability of Deep Reinforcement Learning-Based Routing with Control on Partial Nodes. In Proceedings of the 2020 IEEE International Conference on Acoustics, Speech and Signal Processing (ICASSP 2020), Barcelona, Spain, 4–8 May 2020; pp. 3557–3561.
4. Nguyen, T.G.; Phan, T.V.; Hoang, D.T.; Nguyen, T.N.; So-In, C. Federated Deep Reinforcement Learning for Traffic Monitoring in SDN-Based IoT Networks. *IEEE Trans. Cogn. Commun. Netw.* **2021**, *7*, 1048–1065. [CrossRef]
5. Gilmer, J.; Schoenholz, S.S.; Riley, P.F.; Vinyals, O.; Dahl, G.E. Neural message passing for quantum chemistry. In Proceedings of the 34th International Conference on Machine Learning (ICML 2017), Sydney, Australia, 4–11 August 2017; Volume 70, pp. 1263–1272.
6. Mnih, V.; Kavukcuoglu, K.; Silver, D.; Graves, A.; Antonoglou, I.; Wierstra, D.; Riedmiller, M. Playing Atari with deep reinforcement learning. *arXiv* **2013**, arXiv:1312.5602.
7. Ali Khan, A.; Zafrullah, M.; Hussain, M.; Ahmad, A. Performance analysis of OSPF and hybrid networks. In Proceedings of the International Symposium on Wireless Systems and Networks (ISWSN 2017), Lahore, Pakistan, 19–22 November 2017; pp. 1–4.
8. Chiesa, M.; Kindler, G.; Schapira, M. Traffic engineering with Equal-Cost-Multipath: An algorithmic perspective. *IEEE/ACM Trans. Netw.* **2017**, *25*, 779–792. [CrossRef]
9. Li, C.; Jiang, K.; Luo, Y. Dynamic placement of multiple controllers based on SDN and allocation of computational resources based on heuristic ant colony algorithm. *Knowl. Based Syst.* **2022**, *241*, 108330. [CrossRef]
10. Di Stefano, A.; Cammarata, G.; Morana, G.; Zito, D. A4SDN—Adaptive Alienated Ant Algorithm for Software-Defined Networking. In Proceedings of the 10th International Conference on P2P, Parallel, Grid, Cloud and Internet Computing (3PGCIC 2015), Krakow, Poland, 4–6 November 2015; pp. 344–350.
11. Chen, F.; Zheng, X. Machine-learning based routing pre-plan for sdn. In *International Workshop on Multi-Disciplinary Trends in Artificial Intelligence*; Springer: Cham, Switzerland, 2015; pp. 149–159.
12. Xavier, A.; Silva, J.; Martins-Filho, J.; Bastos-Filho, C.; Chaves, D.; Almeida, R.; Araujo, D.; Martins, J. Heuristic planning algorithm for sharing restoration interfaces in OTN over DWDM networks. *Opt. Fiber Technol.* **2021**, *61*, 102426. [CrossRef]
13. Fang, C.; Feng, C.; Chen, X. A heuristic algorithm for minimum cost multicast routing in OTN network. In Proceedings of the 19th Annual Wireless and Optical Communications Conference (WOCC 2010), Shanghai, China, 14–15 May 2010; pp. 1–5.
14. Chen, J.; Wang, Y.; Huang, X.; Xie, X.; Zhang, H.; Lu, X. ALBLP: Adaptive Load-Balancing Architecture Based on Link-State Prediction in Software-Defined Networking. *Wirel. Commun. Mob. Comput.* **2022**, *2022*, 8354150. [CrossRef]
15. Yan, M.; Li, S.; Chan, C.A.; Shen, Y.; Yu, Y. Mobility Prediction Using a Weighted Markov Model Based on Mobile User Classification. *Sensors* **2021**, *21*, 1740. [CrossRef]
16. Wani, A.; Revathi, S.; Khaliq, R. SDN-based intrusion detection system for IoT using deep learning classifier (IDSIoT-SDL). *CAAI Trans. Intell. Technol.* **2021**, *6*, 281–290. [CrossRef]
17. Zhou, Y.; Cao, T.; Xiang, W. Anypath Routing Protocol Design via Q-Learning for Underwater Sensor Networks. *IEEE Internet Things J.* **2021**, *8*, 8173–8190. [CrossRef]
18. Jalil, S.Q.; Rehmani, M.; Chalup, S. DQR: Deep Q-Routing in Software Defined Networks. In Proceedings of the 2020 International Joint Conference on Neural Networks (IJCNN), Glasgow, UK, 19–24 July 2020; pp. 1–8.
19. Sun, P.; Guo, Z.; Lan, J.; Li, J.; Hu, Y.; Baker, T. ScaleDRL: A scalable deep reinforcement learning approach for traffic engineering in SDN with pinning control. *Comput. Netw.* **2021**, *190*, 107891. [CrossRef]
20. Che, X.; Kang, W.; Ouyang, Y.; Yang, K.; Li, J. SDN Routing Optimization Algorithm Based on Reinforcement Learning. *Comput. Eng. Appl.* **2021**, *57*, 93–98.
21. Suárez-Varela, J.; Mestres, A.; Yu, J.; Kuang, L.; Feng, H.; Barlet-Ros, P.; Cabellos-Aparicio, A. Routing based on deep reinforcement learning in optical transport networks. In Proceedings of the 2019 Optical Fiber Communications Conference and Exhibition (OFC), San Diego, CA, USA, 3–7 March 2019; pp. 1–3.
22. Haarnoja, T.; Zhou, A.; Abbeel, P.; Levine, S. Soft actor-critic: Off-policy maximum entropy deep reinforcement learning with a stochastic actor. In Proceedings of the 35th International Conference on Machine Learning (ICML 2018), Stockholm, Sweden, 10–15 July 2018; Volume 80, pp. 1861–1870.
23. Kumar, A.; Fu, J.; Soh, M.; Tucker, G.; Levine, S. Stabilizing off-policy Q-learning via bootstrapping error reduction. In Proceedings of the 33rd International Conference on Neural Information Processing Systems, Vancouver, BC, Canada, 8–14 December 2019; pp. 11784–11794.
24. Kallus, N.; Uehara, M. Double reinforcement learning for efficient off-policy evaluation in Markov decision processes. *J. Mach. Learn. Res.* **2002**, *21*, 6742–6804.
25. Qiang, F.; Xin, X.; Xitong, W.; Yujun, Z. Target-driven visual navigation in indoor scenes using reinforcement learning and imitation learning. *CAAI Trans. Intell. Technol.* **2022**, *7*, 167–176.
26. Agarwal, R.; Schuurmans, D.; Norouzi, M. An optimistic perspective on offline reinforcement learning. In Proceedings of the 37th International Conference on Machine Learning (ICML 2020), Virtual Event, 13–18 July 2020; pp. 104–114.

27. Shahri, E.; Pedreiras, P.; Almeida, L. Extending MQTT with Real-Time Communication Services Based on SDN. *Sensors* **2022**, *22*, 3162. [CrossRef] [PubMed]
28. Almasan, P.; Suárez-Varela, J.; Badia-Sampera, A.; Rusek, K.; Barlet-Ros, P.; Cabellos-Aparicio, A. Deep Reinforcement Learning meets Graph Neural Networks: Exploring a routing optimization use case. *arXiv* **2020**, arXiv:1910.07421v2. [CrossRef]
29. Dong, X.; Yu, Z.; Cao, W.; Shi, Y.; Ma, Q. A survey on ensemble learning. *Front. Comput. Sci.* **2020**, *14*, 241–258. [CrossRef]
30. Zhang, T.; Zhang, D.; Yan, H.; Qiu, J.; Gao, J. A new method of data missing estimation with FNN-based tensor heterogeneous ensemble learning for internet of vehicle. *Neurocomputing* **2021**, *420*, 98–110. [CrossRef]
31. Fang, Z.; Wang, Y.; Peng, L.; Hong, H. A comparative study of heterogeneous ensemble-learning techniques for landslide susceptibility mapping. *Int. J. Geogr. Inf. Sci.* **2021**, *35*, 321–347. [CrossRef]
32. Lei, L.; Kou, L.; Zhan, X.; Zhang, J.; Ren, Y. An Anomaly Detection Algorithm Based on Ensemble Learning for 5G Environment. *Sensors* **2022**, *22*, 7436. [CrossRef] [PubMed]
33. Strand, J.; Chiu, A.; Tkach, R. Issues for routing in the optical layer. *IEEE Commun. Mag.* **2001**, *39*, 81–87. [CrossRef]
34. Chen, R.; Sidor, S.; Abbeel, P.; Schulman, J. UCB exploration via Q-ensembles. *arXiv* **2017**, arXiv:1706.01502.

sensors

MDPI

Article

Multi-Category Gesture Recognition Modeling Based on sEMG and IMU Signals

Yujian Jiang [1,2,3,4,*], Lin Song [1,2,3,4], Junming Zhang [1,2,3,4], Yang Song [1,2,3,4] and Ming Yan [1,2,3,4]

1 State Key Laboratory of Media Convergence and Communication, Communication University of China, Beijing 100024, China
2 Key Laboratory of Acoustic Visual Technology and Intelligent Control System, Ministry of Culture and Tourism, Communication University of China, Beijing 100024, China
3 Beijing Key Laboratory of Modern Entertainment Technology, Communication University of China, Beijing 100024, China
4 School of Information and Communication Engineering, Communication University of China, Beijing 100024, China
* Correspondence: yjjiang@cuc.edu.cn

Abstract: Gesture recognition based on wearable devices is one of the vital components of human–computer interaction systems. Compared with skeleton-based recognition in computer vision, gesture recognition using wearable sensors has attracted wide attention for its robustness and convenience. Recently, many studies have proposed deep learning methods based on surface electromyography (sEMG) signals for gesture classification; however, most of the existing datasets are built for surface EMG signals, and there is a lack of datasets for multi-category gestures. Due to model limitations and inadequate classification data, the recognition accuracy of these methods cannot satisfy multi-gesture interaction scenarios. In this paper, a multi-category dataset containing 20 gestures is recorded with the help of a wearable device that can acquire surface electromyographic and inertial (IMU) signals. Various two-stream deep learning models are established and improved further. The basic convolutional neural network (CNN), recurrent neural network (RNN), and Transformer models are experimented on with our dataset as the classifier. The CNN and the RNN models' test accuracy is over 95%; however, the Transformer model has a lower test accuracy of 71.68%. After further improvements, the CNN model is introduced into the residual network and augmented to the CNN-Res model, achieving 98.24% accuracy; moreover, it has the shortest training and testing time. Then, after combining the RNN model and the CNN-Res model, the long short term memory (LSTM)-Res model and gate recurrent unit (GRU)-Res model achieve the highest classification accuracy of 99.67% and 99.49%, respectively. Finally, the fusion of the Transformer model and the CNN model enables the Transformer-CNN model to be constructed. Such improvement dramatically boosts the performance of the Transformer module, increasing the recognition accuracy from 71.86% to 98.96%.

Keywords: sEMG; IMU; hand gesture recognition; convolutional neural network; recurrent neural network; transformer; residual networks

Citation: Jiang, Y.; Song, L.; Zhang, J.; Song, Y.; Yan, M. Multi-Category Gesture Recognition Modeling Based on sEMG and IMU Signals. *Sensors* **2022**, *22*, 5855. https://doi.org/10.3390/s22155855

Academic Editor: Jan Cornelis

Received: 26 June 2022
Accepted: 2 August 2022
Published: 5 August 2022

1. Introduction

Human–computer interaction (HCI) is the study of information exchange and the mutual influence of technology between humans and computers. The Gesture Recognition System is one of the crucial components of the HCI system. Many wearable devices with human–computer interaction functions have been released in recent years. For example, Thalmic Lab's Myo armband is a wearable device that can collect surface electromyography (sEMG) and Inertial Measurement Unit (IMU) and wirelessly transmits the two kinds of data to a server via Bluetooth signals. Since gesture recognition data are generated through sensors attached to the skin, the recognition results are not susceptible to light changes and object occlusions in the environment when using the Myo armband. Compared to gesture recognition in computer vision, the Myo-based gesture recognition method has become a hot topic of research in the field of HCI.

Many studies have proposed machine learning or deep learning algorithms to implement Myo-based gesture recognition tasks. Among them, support vector machine (SVM) [1–4], k-nearest neighbor (KNN) [5–9], decision tree (DT) [10], convolutional neural network (CNN) [11–17], recurrent neural network (RNN) [18–24], and artificial neural networks (ANN) [25–30] are the most popular algorithms with good recognition accuracy; however, there are still some challenges in this field of research. First, most studies build their datasets for specific application scenarios; these datasets involve mostly less than 10 gesture actions, and there is a lack of publicly available datasets for more classification tasks. Second, most existing studies only identify the sEMG signal during hand motion, and very little related research applies both sEMG and IMU to gesture action recognition. In addition, none of these studies evaluate the application of the Transformer model [31] to gesture recognition. Finally, for multi-gesture recognition tasks, there is still potential for research on building deep learning models with high accuracy and low time consumption for classification. The multi-gesture recognition should consider both dynamic and static arm and finger gestures. Hence, to meet the demands of the multi-gesture interaction scene, multiple deep learning models or their combinations are adopted to establish models for multi-gesture recognition with high accuracy and less time-consuming.

In this paper, the CNN-based, RNN-based, and Transformer-based two-stream gesture recognition models are proposed, respectively. Our self-built 20-category gesture dataset, including sEMG and IMU signals, far exceeds the normal 5-category and 6-category gesture datasets. According to the difference in characteristics between sEMG and IMU data for sampling frequency and data length, two-stream architecture is adopted to build basic CNN, RNN, and Transformer models. All of them aim to classify these 20 gestures at the same time, including both static and dynamic gestures. Afterward, the CNN-Res model, RNN-Res model, and Transformer-CNN model are established based on those three basic models. All of the improved models yielded exciting experimental results. By comparing the built models according to the experimental results, this paper selects the most suitable models to cope with different application scenarios for the best gesture recognition performance.

2. Related Work

The task of gesture recognition is to build a robust gesture recognition model with the acquired gestures data to obtain the ideal recognition results; it aims to contribute to the application in various human–computer interaction scenarios. After manufacturing a customizable wearable 3D-printed bionic arm that can be applied to amputees, Said S [1] successfully used the SVM to control the artificial bionic hand. In ref. [21], Nasri N took the Conv-GRU model as the classifier and created an sEMG-Controlled 3D game for rehabilitation therapies; moreover, the random forest (RF) model was also utilized by Mendes [32] to recognize Brazilian Sign Language in Sign Language recognition systems.

As mentioned in the introduction, most deep learning methods are based on the sEMG signals. The mainstream methods can be classified into three categories: CNN models [11–17], RNN models [18–24], and ANN models [25–28]. The single-layer CNN proposed by Zia ur Rehman M accomplished the classification task of 7 gestures [12], which pioneered the application of CNN models to gesture recognition. Ulysse Côté-Allard further optimized the basic CNN model and proposed a CNN model based on transfer learning in [13], resulting in the enhancement of the classifier's performance in the recognition task of 7 gestures with higher accuracy. Similarly, the five-layer CNN model in [13] achieved good classification results based on sEMG signals.

Since the sEMG is a temporal signal, RNN classification models can also play a significant role in the classification based on sEMG. The RNN models applied to gesture recognition include the long short term memory (LSTM) and the gate recurrent unit (GRU). Nadia Nasri first introduced the GRU model to the 6-classification task of sEMG signals in [18], showing the feasibility of using recurrent neural networks to classify the data collected by the Myo armband. Zhen built a 21-classification sEMG dataset [20] and made it publicly available, which is a rare dataset for gesture recognition with more than 20 classifications in existing research. In [20], Zhen proposed a two-layer GRU model connected with fully connected layers. Although the average classification accuracy of the model is only 89.6%, Zhen strongly promotes the progress of GRU model application in multi-classification tasks. Additionally, some researchers have applied the recurrent-convolution neural network (RCNN) model proposed by Lai S for text classification [33] to classify sEMG signals. For example, Nadia Nasri [21] improved the former GRU model by adding a convolutional layer and improved the accuracy in a 7-classification task of sEMG signals, which implies that the combination of CNN models with RNN models can further extract the features of sEMG signals and strengthen the training efficiency of deep learning models. Currently, the Transformer model [31] has not been used in any research on gesture recognition.

With increasing categories in gesture recognition, the rising similarity between each gesture makes the classification more difficult. Capturing sEMG signals alone to recognize gestures seems not enough. Ulysse Côté-Allard, for instance, applied the CNN model to the task of a dataset with 11 classifications [16] but obtained test accuracies that were clearly lower than his previous results in 7 categories [13]; thus, naturally, some researchers are attracted to another signal captured by the Myo armband, the IMU signal. The IMU signal contains motion information and reflects the position variation characteristics during the execution. For illustration, Chiu [34] provided a thresholding method to determine the active signal segment of the motion. In his work, the IMU data were fed into a Long Short-Term Memory (LSTM) network for classification, demonstrating that there are possibilities existing in research for gesture recognition training through IMU signals.

Some scholars have begun to pay attention to combining sEMG and IMU signals for gesture recognition by machine learning algorithms [2,4,34,35] or deep learning algorithms [23,24]. Xiaoliang [24] imposed the LSTM model to solve the gesture recognition problem based on the combination of sEMG and IMU signals. Although Xiaoliang achieved to complete the classification task of 10 gestures, the accuracy remains to be uplifted. Williams [24] also proposed the RCNN model to settle the 5-class task with these two signals, reaching an accuracy of 99%. Both Xiaoliang and Williams took several wearable devices to obtain more adequate data, such as Smart Glove or several Myo armbands. The usage of more devices in the data acquisition was reasonable but inconvenient and limited the number of practical applications.

In conclusion, most studies on gesture recognition based on bioelectrical signals only built datasets of sEMG signals. When facing multiple gesture classification tasks, existing deep learning approaches with high gesture recognition accuracy are still not powerful enough to meet the practical needs of complex interaction situations. How to build a dataset with more variety of gestures and how to build a deep learning model that achieves multi-gesture classification with great precision is the target of future work in most related studies.

Inspired by the relevant work, a 20-class gesture dataset is constructed to meet the demands of multi-classification gesture recognition. For high accuracy and efficiency, the CNN-Res, RNN-Res, and Transformer-CNN models are proposed.

3. Methods

This Section describes how our gesture dataset and deep learning models are constructed. First of all, Section 3.1 describes the construction progress of the dataset. Then, Section 3.2 describes the establishment of the basic models. Finally, Section 3.3 describes how to optimize and combine the basic models to build the improved models.

3.1. Construction of Dataset

3.1.1. Acquisition Tool

The data resource was a wearable acquisition tool, the Myo armband. The data collected were recorded by a workstation, and a large light-emitting diode (LED) screen was used to display the gesture instruction video.

The Myo armband has eight electrode sensors that can capture sEMG signals from the skin surface. There is also an inertial measurement unit with a three-axis gyroscope, three-axis accelerometer, and three-axis magnetometer. Data captured is sent wirelessly to the workstation via built-in Bluetooth.

The position of sEMG's electrode sensors and IMU is shown in Figure 1.

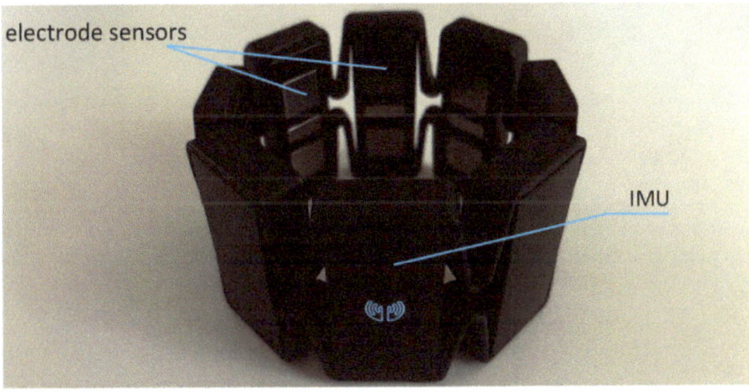

Figure 1. Outlook of Myo armband.

3.1.2. Acquisition Process

We collected sEMG and IMU signals generated during gesture execution and designed 20 gestures, including translation in 6 directions, rotation in 3 directions, making a fist, and numbers 0~9. As shown in Figure 2, the former nine movements are dynamic, and the last 11 movements are static. Dynamic gestures contain an activation gesture. During the execution of dynamic gestures, the activation gesture is maintained all the time.

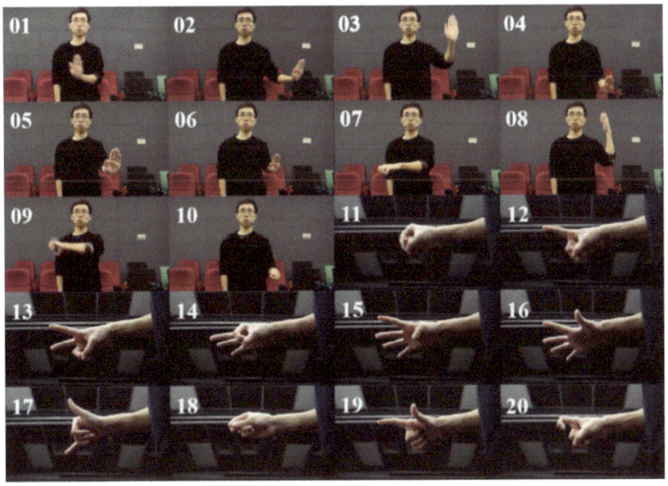

Figure 2. 01~09 are dynamic movements. Their names are 'Push left', 'Push right', 'Push up', 'Push down', 'Push forward', 'Push back', 'Turn left and right', 'Turn back and forth', 'Turn up and down'. 10~20 are static movements. Their names are 'Making a fist', 'Number 0', 'Number 1', 'Number 2', 'Number 3', 'Number 4', 'Number 5', 'Number 6', 'Number 7', 'Number 8', 'Number 9' in order.

Fifty volunteers were recruited to participate in the experiment, all healthy undergraduate or graduate students. Among them, 14 were male, and 36 were female. All volunteers wore the Myo armband in the same position presented in Figure 3. The IMU was located on the medial side of the arm.

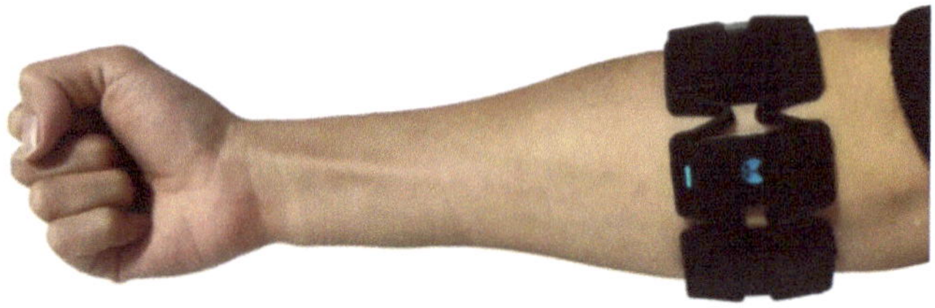

Figure 3. Position of Myo armband.

The volunteers were asked to make gestures following the recorded video of gesture instructions. Each gesture was performed once and took 5 s, and the interval between each gesture was 5 s. Following instructions on the video, the volunteers were prepared to carry out the gestures at the appropriate rhythm. The experiment started after ensuring that the volunteers were familiar with the gestures. To avoid muscle fatigue caused by the long-time performing the movements, the experiment was divided into two groups. The first group collected ten gestures and the second group collected another ten gestures, with a 5–10 min rest break in both groups. During the experiments, we checked the quality of completed gestures and the recorded data. The volunteers would be asked to re-capture the data after the experiment when there were unqualified movements and abnormal data.

The actual time to complete the translational and rotational movements for dynamic gestures was within 2 s and 4 s, respectively. For static movements, all movements were kept within 5 s. Therefore, we only kept the valid length of each type of gesture to guarantee that all the gestures could be correctly classified. At last, we got 400 actions per volunteer, for a total of 20,000 actions. After eliminating the data of 271 incorrectly executed actions, the data of 19,729 gestures were stored in our dataset.

The sEMG data acquired is 8-channel with a sampling frequency of 200 Hz, while the IMU data obtained is 6-channel with a sampling frequency of 50 Hz. All data are sent by Bluetooth to the server and saved in the Myo Data Capture tool.

3.1.3. Data Labeling

We take the Plotly Express extension library in Python to visualize the waveforms of the collected data and manually label the data. Before labeling, we need to get a clear waveform to distinguish a gesture's start and end points to define the signal's active segment.

First, we do full-wave rectification on the sEMG signal so that the data takes absolute values. That makes the negative half axis data roll over to the positive half axis. Then, we sum the data of all the channels (n = 8) and find the arithmetic square root of the sum as the new channel, as shown in function (1); these two operations are intended to display the 8-channel sEMG signal in one waveform plot and reduce the dispersion of the amplitude of it; this new waveform is then smoothed with a moving average algorithm. Each data point is replaced by the average of $M/2$ points before and after (including the point itself), as shown in function (2). Smoothing is applied to obtain a more intuitive waveform diagram, which is convenient for our subsequent labeling.

$$x_{sum} = \sqrt{\sum \left| x^{(k)} \right|},\ k = 1, 2, 3, \ldots, n \tag{1}$$

$$x_{new}(i) = \frac{\sum_{i-\frac{M}{2}}^{i+\frac{M}{2}} x_{sum}(i)}{M+1}, \ i = 1,2,3, \ldots, m \quad (2)$$

where n is the number of channels, m is the number of data points in the merged channel, $x^{(k)}$ is the data of each original channel, x_{sum} is the data after merging channels. $x_{sum}(i)$ is each data point of the merged channel, $M+1$ is the number of data points used in each smoothing process, and $x_{new}(i)$ is the value after smoothing.

After our verification, the waveform is observed when M is taken as 50. As shown in Figure 4a, the blue line represents the value of the signal, and the orange box is the active segment of a signal which is also the data segment we need to label. The corresponding amount of data is taken to label the segments according to the sEMG data length for the different gestures. Labels of the gesture category are 1–20.

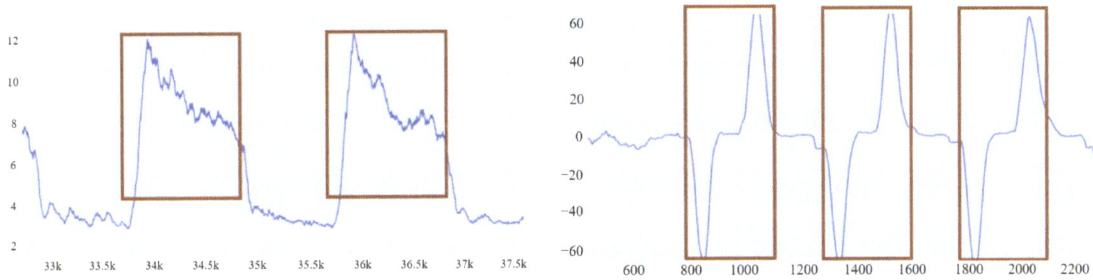

(a) (b)

Figure 4. Active segments of sEMG and IMU. (**a**) sEMG, (**b**) IMU.

For the signals measured by IMU, their six channels are processed by the moving average algorithm in function (2), respectively. Furthermore, it turns out that the data of the y-channel of the gyroscope is relatively easy to identify, as shown in Figure 4b. Therefore, there is no need to merge channels, and the data of the y-channel can be directly regarded as the labeling basis of the signal activity segment. Depending on the length of IMU data for different gestures, the corresponding amount of data is taken. Labels of the gesture category are 1–20, the same as for the sEMG data.

3.1.4. Data Segmentation

The gesture movements are continuous. To meet the real-time requirements of gesture interaction, the active segment data should be segmented with a sliding window.

According to Mendes [32], the sliding window length is selected as 1 s (containing 200 points for sEMG and 50 points for IMU) with a sliding step of 100 ms to segment the sEMG and IMU signals. Eventually, 601,489 sEMG samples of size 200 $\times$ 8 and 601,489 IMU samples of size 50 $\times$ 6 are generated.

Before inputting data into the model, we conducted the normalization operation. Since the sEMG signal and IMU signal differ in values and the IMU signals contain two types of data: accelerometer and gyroscope, we normalized the sEMG and IMU between $(-1, 1)$ with function (3), respectively.

$$x'(j) = 2 \times \frac{x(j) - x_{min}}{x_{max} - x_{min}} - 1, j = 1,2,3, \ldots N \quad (3)$$

where N refers to the number of data points in the sEMG or IMU signal. x_{min} refers to the minimum value of sEMG or IMU signal. x_{max} refers to the maximum value of the sEMG or IMU signal. $x(j)$ refers to the value of each data point. $x'(j)$ refers to the value of each data point after normalization.

3.2. Basic Models

CNN and RNN are the most common models in gesture recognition research; however, the Transformer has not been introduced in any associated research so far. In this Section, a two-stream CNN model, a two-stream RNN model, and a two-stream Transformer model are built to classify the dataset to test the feasibility of these basic models for the sEMG and IMU-based gesture recognition missions.

3.2.1. Two-Stream CNN Model

The basic CNN consists of a convolutional layer, a pooling layer, and a fully connected layer.

To transform the data into a form suitable for a 2D convolutional kernel, dimensional change should be performed on each data. First, the channel dimension is retained, and only the first 196 data points of the sEMG and the first 49 data points of the IMU are reserved. Then, we reshape these points into data of size 14×14 and 7×7, respectively. After that, the size of sEMG data is changed to $14 \times 14 \times 8$, and the size of IMU data is transformed to $7 \times 7 \times 6$.

As shown in Figure 5, the sEMG and IMU data are divided into two streams. In turn, there is a convolutional layer, a maximum pooling layer, and a convolutional layer in the first stream. A Relu layer follows each convolutional layer. The first and second convolutional layer contains 24 and 48 filters of size 3×3, respectively. The maximum pooling layer has a window size of 2×2, which can compress the data size while preserving the key features. The second stream has a convolutional layer and a maximum pooling layer. The convolutional layer also consists of 24 filters of size 3×3, connected to a Relu layer. The pooling layer also has a window of size 2×2. The outputs of two streams are expanded into a single column and are concatenated together after passing through the maximum pooling layer. The output of the last fully connected layer is the classification result corresponding to the input data.

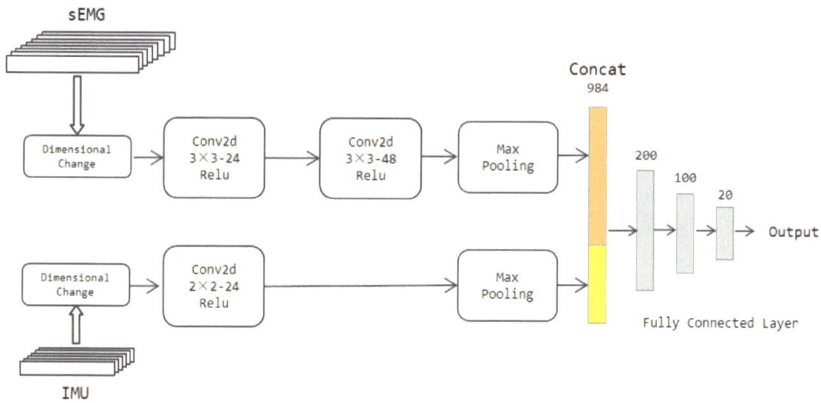

Figure 5. The architecture of the two-stream CNN model.

3.2.2. Two-Stream RNN Model

The LSTM [36] and the GRU [37] models are the most frequent RNN models. For the LSTM model, there are three ways of transferring information between neural units: the long-term information of the previous moment, the output information of the previous moment, and the input information of the current moment; these three information paths are controlled by the forgetting gate, the input gate, and the output gate. The GRU model [37] is a variation of the LSTM model. GRU optimizes the three gates of the LSTM into an update gate and a reset gate, and these two gates control the hidden state of the current moment and the post-selected hidden state, respectively. The GRU model has fewer parameters and a faster training time than the LSTM model; moreover, in many kinds of application cases, GRU can achieve comparable results to LSTM.

We take the two-stream LSTM model as an example to introduce the RNN architecture. As shown in Figure 6, the data are directly input to the two-stream LSTM model in two streams. Each stream has two layers of LSTM network with 50 hidden units. The hidden layer of the LSTM network can capture the state information of the data in the time dimension. We take the information output from the last time node of the LSTM as a vital feature for the classification. The features output from the two-stream network are also expanded into a column and then concatenated together. The fully connected layer has 20 neurons and can be utilized to output classification results depending on the training dataset.

The two-stream GRU model is formed by replacing the LSTM layer in the two-stream LSTM model with the GRU layer. The structure and parameter settings of the model are the same as those of the two-stream LSTM model.

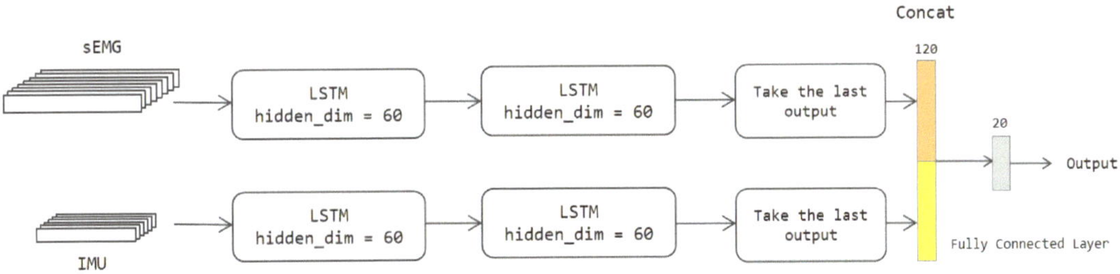

Figure 6. The architecture of the two-stream LSTM model.

3.2.3. Two-Stream Transformer Model

Transformer is a very innovative network proposed by Google Brain in [31], which abandons the circular structure and employs a more interpretable self-attention mechanism to extract the relationships between data. The self-attention mechanism involves three quantities: Query, Key, and Value; they aim to compute the relationship between input data X.

Query represents the query vector, Key represents the vector of the relevance of the queried information to other information, and Value represents the queried information vector. The specific calculation is shown in function (4).

$$\begin{cases} Q = XW^Q \\ K = XW^K \\ V = XW^V \end{cases} \tag{4}$$

where Q refers to Query. K refers to Key. V refers to Value. X refers to the input data. W^Q, W^K, and W^V are the weight matrices, updated with the training, corresponding to these three quantities.

After the values of the three vectors Query, Key and Value are calculated, the self-attention mechanism computes the eigenvalues of the output of the attention layer with function (5).

$$Attention\ (Q,\ K,\ V) = softmax\left(\frac{QK^T}{\sqrt{d_k}}\right)V \tag{5}$$

where d_k refers to the dimension of the vector K.

For more connections among the input data, the Transformer extracts features of the input data through a multi-headed attention mechanism. The multi-head attention mechanism consists of multiple self-attention layers, and each self-attention layer is computed in parallel. Then, the output of each head is spliced and multiplied by the weight matrix. The specific calculation is shown in function (6) and function (7).

$$MultiHead(Q,\ K,\ V) = Concat(head_1, \ldots,\ head_h)W^O \tag{6}$$

$$\text{where } head_i = Attention\left(QW_i^Q, KW_i^K, VW_i^V\right) \quad (7)$$

The Encoder module in the Transformer model is applied in our model to extract the features of the sEMG and IMU signals. The input is first positional encoded so that it has the sequential position information; it subsequently enters a multi-headed attention layer to compute the correlation features and an Add&Norm layer consisting of a residual layer such as ResNet [38] and a normalization layer. Then it proceeds to the Feed Forward layer with two linear units. The Feed Forward layer strengthens the expressiveness of the model, and the output has the exact dimensions as the input. The next is the same Add&Norm layer as the previous one. In the Encoder Layer, the number of heads in the multi-attention mechanism and the number of hidden neural units (*d_model*) of the FF layer are parameters that can be defined. The complete model is shown in Figure 7.

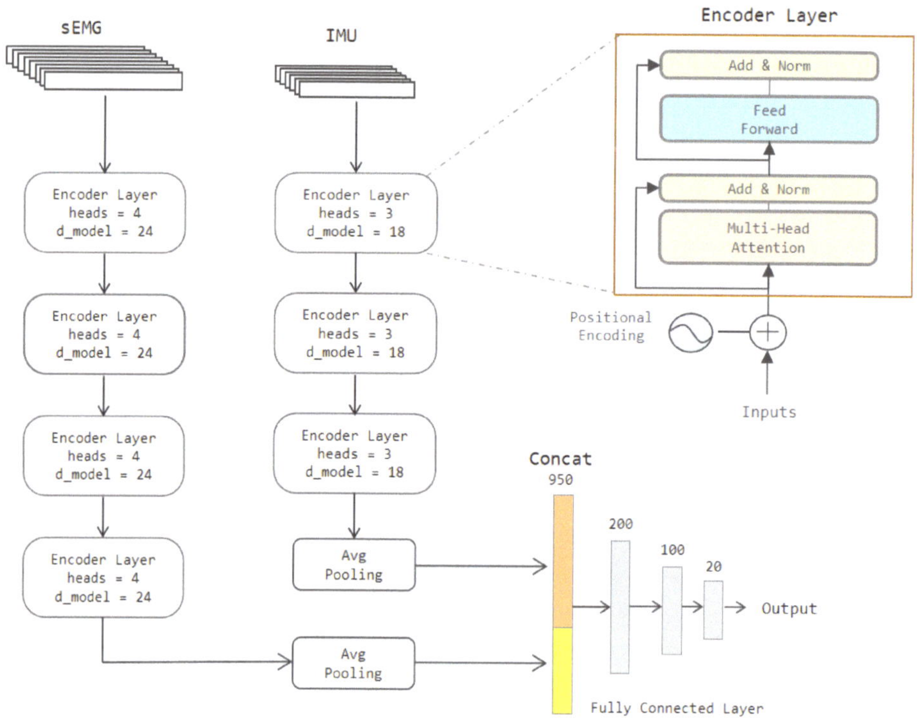

Figure 7. The architecture of the two-stream Transformer model.

As presented in Figure 7, the sEMG data pass through 4 layers of Encoder Layers to retrieve features, followed by the average pooling layer. The average pooling layer is a 1D pooling layer with a window length of 2. That means the features are compressed to half of their original size. The IMU data also pass through 3 Encoder Layers. The features are also squeezed to half the initial size by the same pooling layer. The outputs of these two pooling layers are stitched together and fed to the three fully connected layers to output the classification results.

3.3. Improved Models

At first, the two-stream CNN is regarded as the base model, and a residual network is applied to build the CNN-Res model. Next, the two-stream RNN model is combined with the CNN-Res model to construct the two-stream LSTM-Res model and the two-stream GRU-Res model. Finally, the two-stream Transformer model is fused with the CNN model, forming the two-stream Transformer-CNN model.

3.3.1. Two-Stream CNN-Res Model

In the basic CNN model, only one or two convolutional layers are extracting features of sEMG and IMU data; it is prone to insufficient extraction of features when the number of convolutional layers is small. More layers are a necessity for retrieving more features of significance.

Providing the network is deeper, there is an exposure of gradient disappearance or gradient explosion. At the same time, the deeper network may also trigger the problem of network degradation. Therefore, the residual network [38] is introduced into our network to prevent degradation. A standard residual unit is shown in Figure 8.

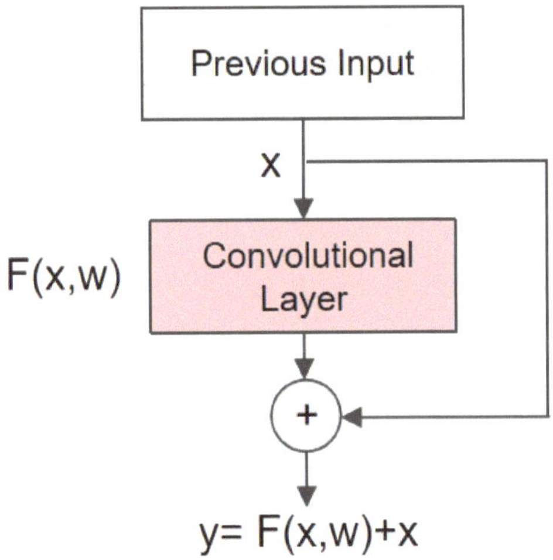

Figure 8. The structure of a Residual unit.

The layer-hopping connection in the residual network allows the input signal to propagate directly from any lower layer to a higher layer; it enables the network to converge faster during training. Deepening the network layers often comes at the cost of increased training time. Thus, the introduction of residual networks compensates for the drawback of a longer training time for deeper networks to some extent.

Motivated by the above analysis, a two-stream CNN-Res model is proposed, and its specific architecture is shown in Figure 9. The data are also input to the network after the dimensional change mentioned in Section 3.2. The first stream network is fed with the sEMG signal and the second stream with the IMU signal.

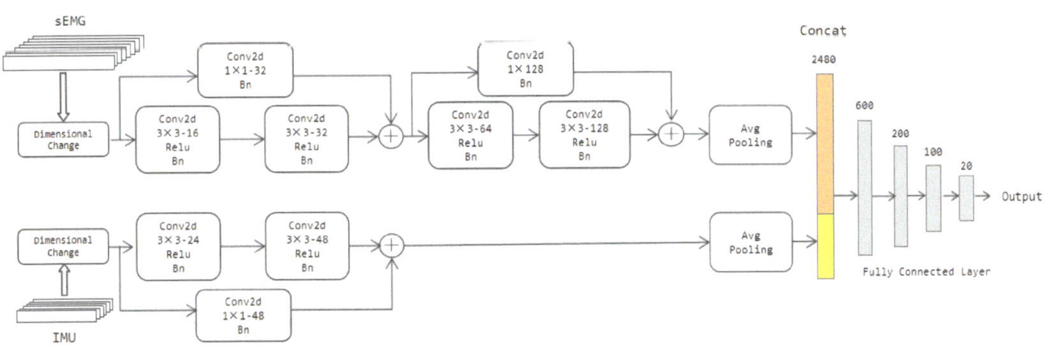

Figure 9. The architecture of the two-stream CNN-Res model.

The parameter settings in the architecture were obtained through tweaking, where we wanted to find a model with high recognition accuracy and relatively low complexity. For the number of network layers, we chose a range from two to six and found that when it was four for the first stream and two for the second stream, the complexity and accuracy of the model reached a balanced state. Two strategies were tried out for the number of channels, namely keeping it constant in all layers or doubling it when going from one layer to the next deeper one; it turned out that the recognition accuracy of the latter was better than the former; thus, after having fixed this strategy, the only choice we had to make was determining the number of filters in the first convolutional layer. Considering the differences in input channel numbers of the two streams, they had better be selected in different ranges. Therefore, we chose them from 4 to 24 and 8 to 48, respectively. Within these ranges, taking 16 for the first and 24 for the second was the optimal choice. For convolution kernel size, 2×2 and 3×3 were tried, and 3×3 was chosen.

The mainstay of the first stream network is composed of 4 convolutional layers and an adaptive averaging pooling layer. Each convolutional layer contains a different number of filters of 3×3, but all of them have a Relu layer and a batch normalization (bn) layer. The filters of these four convolutional layers are 16D, 32D, 64D, and 128D, respectively. The second stream backbone encompasses two convolutional layers with the same content as the first stream network. The number of filters in these two convolutional layers is 24 and 48, respectively.

After every two convolutional layers, a residual unit is added. To ensure the number of channels between the data at the residual cells matches well, we equip a convolutional layer with a 1×1 convolution kernel to change the number of channels of the previous input. Padding operation is also applied in every layer of convolution. The objective is to prevent the data size from changing with the convolution operation so that the data size has consistency. At the output of the two feature streams, they are squeezed by the adaptive average pooling layer to size $128 \times 4 \times 4$ and $48 \times 3 \times 3$, respectively. The maximum pooling layer is not applied here. The rationality is that the average pooling layer can retain more information compared to the maximum pooling layer. The squeezed features are stretched into a column and then stitched together into four fully connected layers to get the classification results.

3.3.2. Two-Stream RNN-Res Model

We built the two-stream RNN-Res model by combining the two-stream RNN model with the two-stream CNN-Res model.

For sEMG and IMU signal data, the LSTM and the GRU are good at extracting long-term dependence features that reflect the global significance of each data point. On the other hand, CNN is a network adept at extracting local features of the data [12]; thus, the CNN network is placed behind the RNN, allowing for further extracting the local features of these long-term dependence features. The combination of long-term dependence features and local features improves the model's characterization ability. Features output by the LSTM and GRU models have already reflected the data's features in the temporal dimension; thus, the 2D convolutional layer in the CNN model does not lose the location information in the original data.

The 2D convolutional layer cannot directly operate on the RNN model output form. Therefore, there is a dimensional transformation operation in front of the CNN model. A dimension of size one is inserted before the temporal dimension of the output features; this action not only preserves original features' content and order but also makes it extracted in the correct form by the CNN-Res network. Furthermore, to increase the depth of the model while reducing the time required for convergence during training, we equally add a residual network to the CNN. The specific network structure is shown in Figure 10.

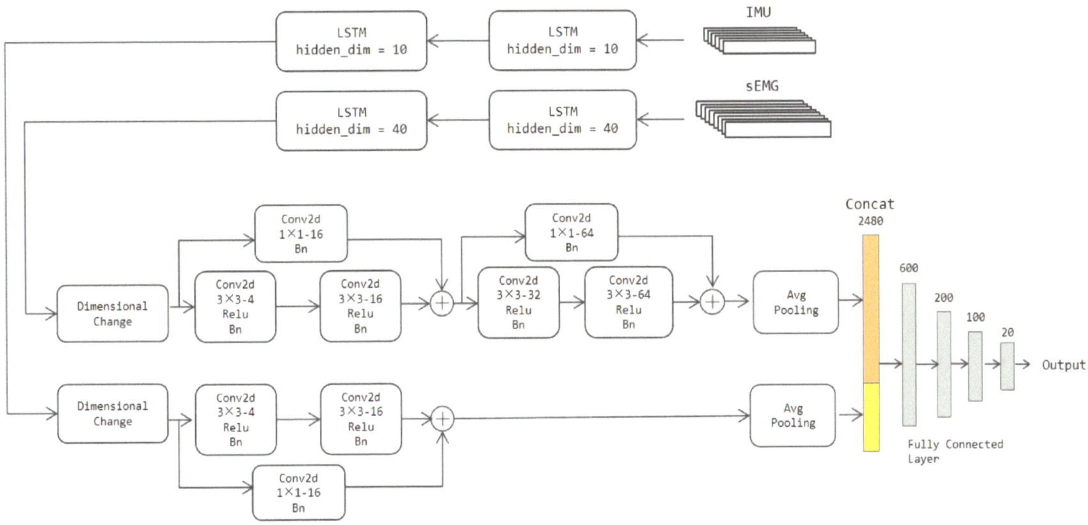

Figure 10. The architecture of the two-stream LSTM-Res model.

The LSTM module has the same structure as Section 3.3.1; however, considering that the LSTM-Res model is complicated, we make the following improvements to reduce the complexity of the network:

(a) Hidden units in the first stream reduce from 60 to 40 and in the second stream from 60 to 10.

(b) Inter-layer maximum pooling layers to the CNN model for down-sampling are added. Their window sizes of the first stream are set to 10×2 and 2×2 after the second and fourth convolutional layers, respectively. Similarly, a maximum pooling layer of size 5×1 is placed after the second convolutional layer in the second stream; these pooling layers squeeze the size of data transmitted.

The rest of the structure in the CNN block is almost consistent with the CNN-Res model. The parameter tweaking was also conducted according to the parameter tweaking process of the CNN-Res model; however, the data's channel numbers of the two streams are both one after the dimensional change, so the number of initial convolution kernels at their starts are chosen to be the same. After tweaking, we choose the combination of "4D, 16D, 32D, 64D" for filters in the first stream and "4D, 16D" in the second stream. After the same adaptive pooling layer, the two streams are concatenated together, and then the classification results are output by the fully connected layers.

As for the GRU-Res model, its architecture is the same as the LSTM-Res model, except for the different RNN network types.

3.3.3. Two-Stream Transformer-CNN Model

In the Transformer, matrix computation eliminates the need for step-by-step computation to obtain features for long-distance data; however, this feature extraction method ignores local details, making the Transformer less capable of capturing local features. In contrast to text information, the sEMG and IMU signals are continuous data generated during action execution. Therefore, ignoring local details can lead to inadequate extraction of features in the temporal dimension of the data.

To enhance the model's ability to extract local features, we improve the Transformer model as follows:

(a) A 1D convolutional layer is added between each encoder layer; these convolutional layers not only extract local features but also increase the number of channels of features. Each convolutional layer doubles the number of channels.

(b) The values of heads and *d_model* in each encoder layer are changed with the deepening of the network. The increase of parameters allows the model to extract deeper global information as the number of network layers deepens.

The above modifications strengthen the model's ability to extract local features as well as increase the information contained in the global features. Given that the convolutional layer's outbound features must be operated by Encoder Layers, temporal messages of the data cannot be destroyed. That is why we take the 1D convolution to obtain local features instead of the 2D convolution. The convolutional layer also contains a padding operation to keep the data length from changing with convolution. To avoid corrupting the location information of the features, we also use an average pooling layer instead of a maximum pooling layer. The amount of information on the features is kept to the maximum extent. The sEMG and IMU data are also input as two streams, and the specific network structure is shown in Figure 11.

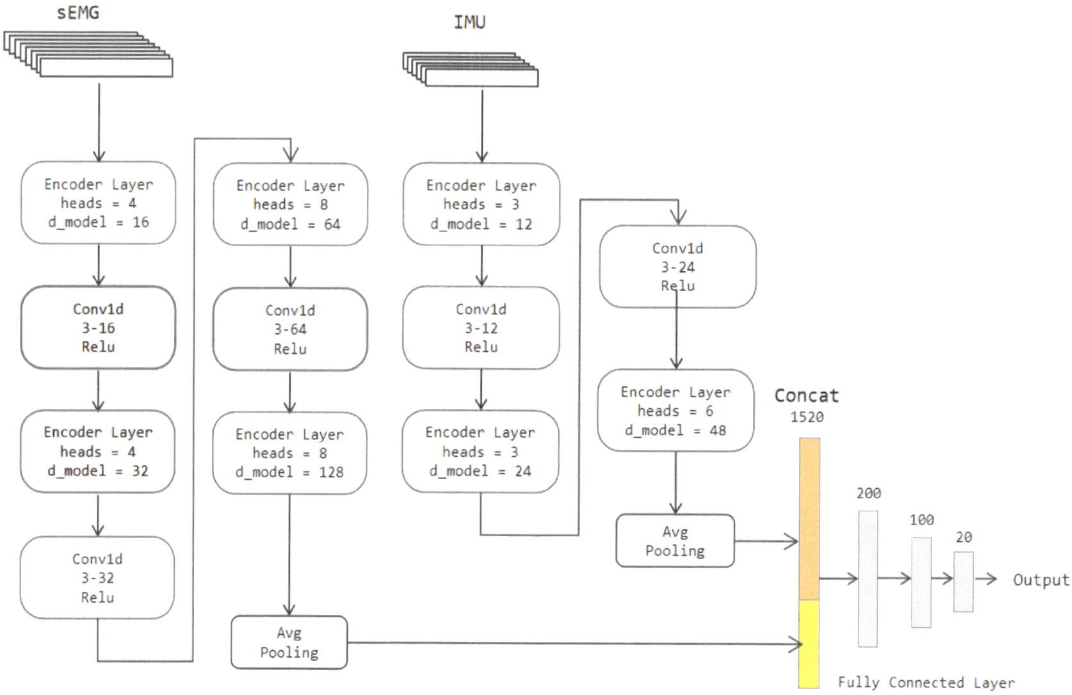

Figure 11. The architecture of the two-stream Transformer-CNN model.

The parameter tweaking of the Transformer-CNN model mainly included: the value of heads and d_model in Encoder Layers and the numbers of filters in 1D convolutional layers. The value of heads must be divisible by the input channel numbers and should not be too large. The choices contained 2, 4, and 8 for the first stream and 3 and 6 for the second stream. After our experiments, when the values of the heads were set to "4, 4, 8, 8" and "3, 3, 6", the accuracy was the highest, and the complexity of the model was relatively low. The parameter settings of 1D convolutional layers were similar to the model defined in Section 3.3.1. We also decided to double the number of channels when signals passed through convolutional layers. The number of filters for the first convolutional layer was selected from 4 to 24 for the first stream and 8 to 48 for the second stream. According to our test results, the best choice is to take 16 for the first and 12 for the second; the tweaking of d_model was identical to them.

The average pooling layer of the first stream has a sliding window length of 10, and the average pooling layer of the second stream has a sliding window length of 5. After the pooling operation, the length of the sEMG data is compressed to 20. The length of the IMU data is shortened to 10. The features of these data are concatenated and input to a four-layer fully connected network to get the classification outcome.

4. Experiments and Results

This Section introduces our experiments and the results of our models. Not only was the experiment of simultaneous classification of 20-category gestures carried out, but also the sEMG signal and IMU signal were applied to identify arm movements and finger movements, respectively. To begin, Section 4.1 states our experimental conditions. Then, Section 4.2 states 20-category experimental results of our basic models, and Section 4.3 states 20-category experimental results of our improved models. After that, Section 4.4 states comparisons and analyses of results of basic models and improved models. Furthermore, Section 4.5 states the results of 9 types of arm movements and 11 types of finger movements, which were recognized separately by two signals.

4.1. Experimental Conditions

Classification experiments on our dataset are performed with the basic and improved models defined in Section 3. As samples generated from the dataset after data segmentation exceeded 600,000, the dataset was divided into training and test sets in the ratio of 49:1. 589,459 samples are in the training set, and 12,030 samples are in the test set.

All the models were trained with the training set and tested on the test set after training. The training loss function was the cross-entropy loss function, and the stochastic gradient descent method was the approach to update the model weights in all the experiments. The batch_size was set to 32. The learning rate was set to 0.005 for the two-stream RNN model and 0.001 for other models (including improved models). Additionally, all of the training processes for the experiments were performed by a GeForce RTX 3090 GPU for 100 epochs; it should be noted that the test set was not involved in updating the model parameters and was only available for testing the model recognition accuracy.

4.2. Results of Basic Models

The experiments of the base models were carried out first, and the results are shown in Table 1. Table 1 contains the training time, testing time, and testing accuracy of all the base models in the experiments of this paper. As can be seen from the table, the two-stream LSTM model and the two-stream GRU model have the highest test accuracy of 97.10% and 95.91%, respectively. The test accuracy of the two-stream CNN model reaches 95.43%. The test accuracy of the two-stream Transformer is the lowest, at 71.68%.

Table 1. Performance results of basic models.

Model	Training Time (h)	Test Time (s)	Test Accuracy (%)
CNN	2.0	0.25	95.43
LSTM	13.3	2.12	97.10
GRU	13.8	2.45	95.91
Transformer	12.5	1.47	71.68

It is evident that the two-stream CNN and RNN models complete the 20-classification task on our dataset with high accuracy, but there is still room for further improvement. Nevertheless, the basic Transformer model is far from satisfying our requirements for recognition accuracy.

4.3. Results of Improved Models

The experiments with improved models were then conducted, and the new results were included in Table 2; it shows an overview of all models' training time, testing time, and accuracy. The ES refers to the experiment incorporating the Early Stopping mechanism into the training process. Single test time means how long it requires to classify one single sample.

As presented in Table 2, it is found that the test accuracy of the CNN-Res model achieves 98.24%; the test accuracy of the LSTM-Res and GRU-Res models reach 99.67% and 99.49%, respectively. Surprisingly, the test accuracy of the Transformer-CNN model attains 98.69%. That is considered sufficiently accurate for gesture recognition. Compared with the basic models, the improved models are significantly more excellent than the basic models. The recognition precision has been enhanced to different degrees. Among the results, the Transformer model has the most striking accuracy gain, about 27.28%.

Because of the improved models' grown complexity, the training and test times naturally become longer; however, the introduction of residual units allows the convergence speed of model training to be significantly boosted, which means the Early Stopping can be implemented to shorten the training time. Because of the Early Stopping, the actual training time of CNN-Res, LSTM-Res, and GRU-Res is even shorter than the training time of the basic models, which is 1.43 h, 8.6 h, 11.7 h, respectively; it turns out that the CNN-Res model has the smallest training and test time among improved models. Therefore, it has the highest training and testing efficiency among all the models. As for the single test time, the Transformer-CNN has the shortest single test time, which means it holds the quickest recognition response.

Table 2. Performance results of improved models.

Model	Training Time (h)	Test Time (s)	Single Test Time (s)	Test Accuracy (%)
CNN	2.0	0.25	0.0303	95.43
CNN-Res	4.33/1.43 (ES)	0.64	0.0343	98.24
LSTM	13.3	2.12	0.0115	97.10
LSTM-Res	18.7/8.6 (ES)	3.22	0.0312	99.67
GRU	13.8	2.45	0.0120	95.91
GRU-Res	18.9/11.7 (ES)	2.82	0.0318	99.49
Transformer	12.5	1.47	0.0126	71.68
Transformer-CNN	14.1	2.45	0.0149	98.96

4.4. Comparison and Analysis

The experimental performance of the improved and basic models will be compared by the variation of the models' loss and test accuracy during the training process. Their test confusion matrixes are also drawn to analyze the details of each gesture recognition result.

4.4.1. CNN vs. CNN-Res

As shown in Figures 12 and 13, the CNN model maintains the test accuracy of around 94% after stabilization, and the loss value stays within 0.004. In contrast, the accuracy of the CNN-Res model is significantly raised and can be maintained above 98.13% after stabilization; moreover, the Loss value also decreases significantly after stabilization and always stays within 0.001. Finally, at the 97th epoch, its test accuracy reaches the highest value of 98.25%.

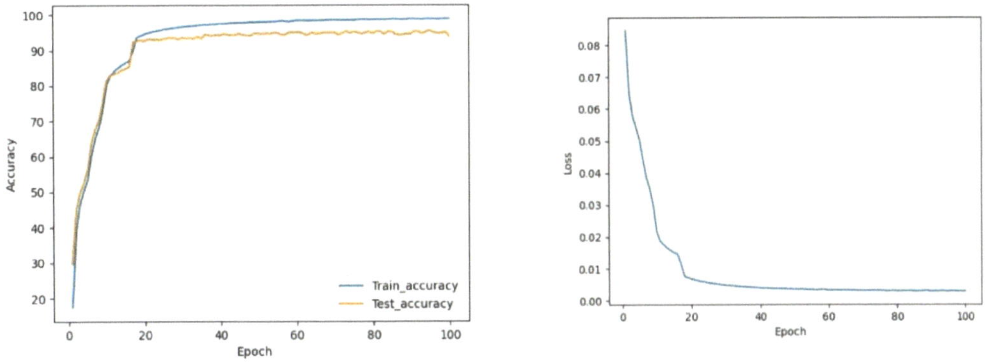

Figure 12. The curve of accuracy and loss during training of CNN.

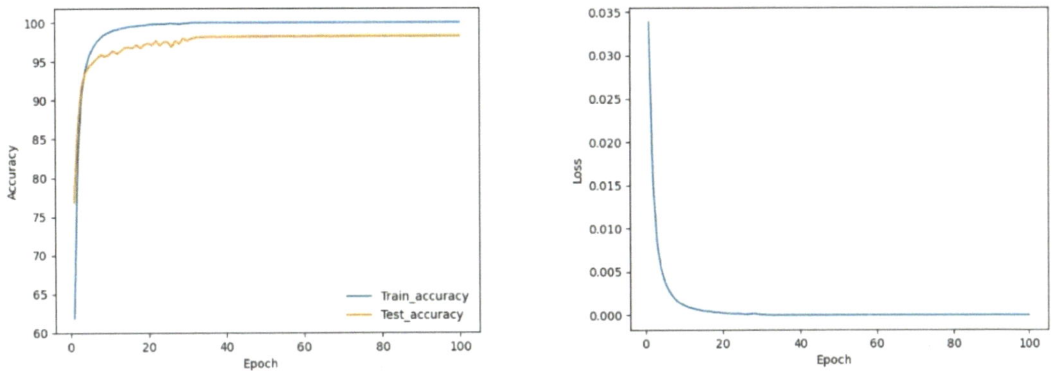

Figure 13. The curve of accuracy and loss during training of CNN-Res.

The test accuracy graph indicates that the CNN-Res model's training process converges at the 33rd epoch, which takes only about 1/3 of the entire training time. Thus, it can be seen that one primary advantage of the introduction of the residual network is that the CNN-Res model significantly improves the convergence speed of training when the network becomes deeper; its training efficiency is strengthened a lot.

According to Figure 14, the recognition accuracy of various gestures is not uniform to the CNN model. The identification accuracy of 'Number 0′', 'Number 2′', and 'Number 4′' is 87%, 89%, and 78%, respectively; this deficiency could be attributed to the similarity of these three static actions, which brings about more difficulties in recognition. As for the recognition results of the CNN-Res model as shown in Figure 15, the recognition accuracy of 'Number 0′', 'Number 2′', and 'Number 4′' is enhanced to 99%, 98%, and 96%, respectively. Furthermore, the recognition accuracies of the other 17 actions are also promoted.

After expanding the layers and introducing residuals, the CNN-Res model achieves better feature extraction with a higher convergence speed.

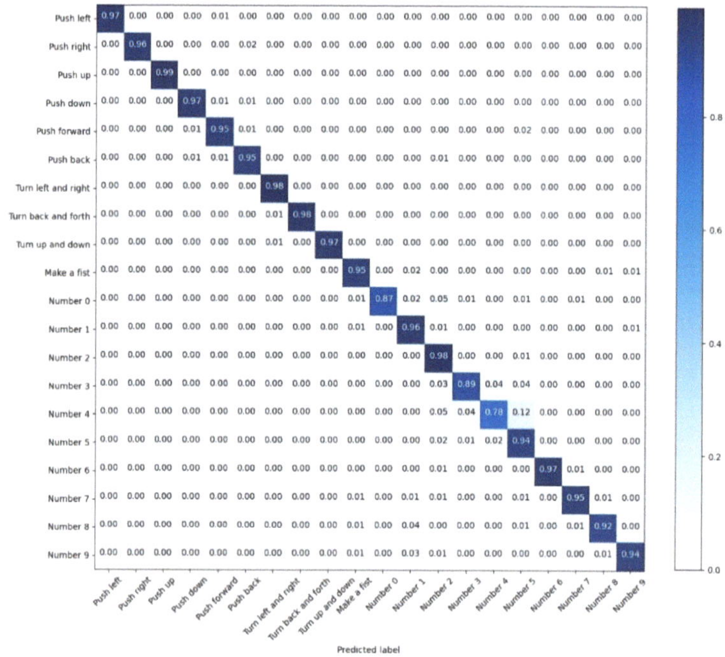

Figure 14. The confusion matrix for the test results of CNN.

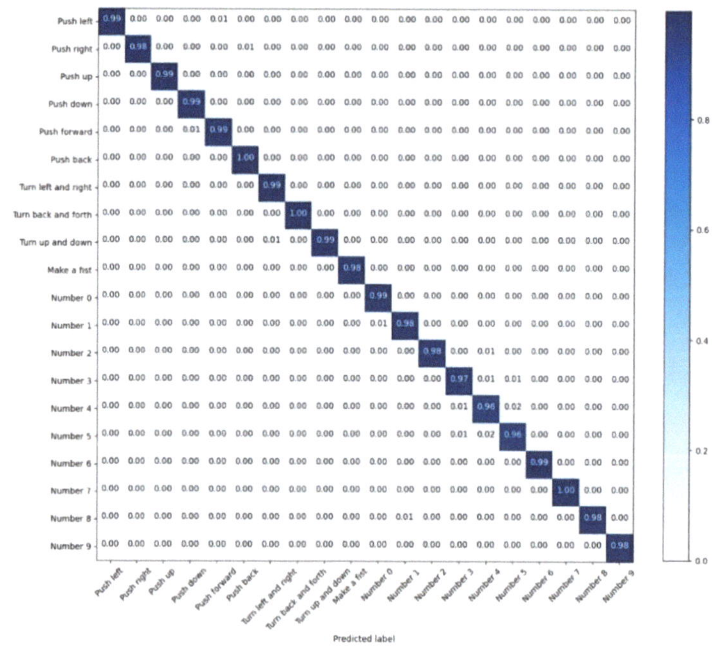

Figure 15. The confusion matrix for the test results of CNN-Res.

4.4.2. RNN vs. RNN-Res

We take the comparison of the LSTM model and LSTM-Res model as an example. As illustrated in Figure 16, the LSTM model keeps the test accuracy above 96% after stabilization, with the highest classification accuracy of 97.10%. Also, the Loss value is held within 0.003. The result suggests that the LSTM model already possesses a good classification effect. To our excitement in Figure 17, however, the test accuracy of the LSTM-Res model reaches up to 99.67% after being stabilized; moreover, the Loss value is almost 0 at a steady state.

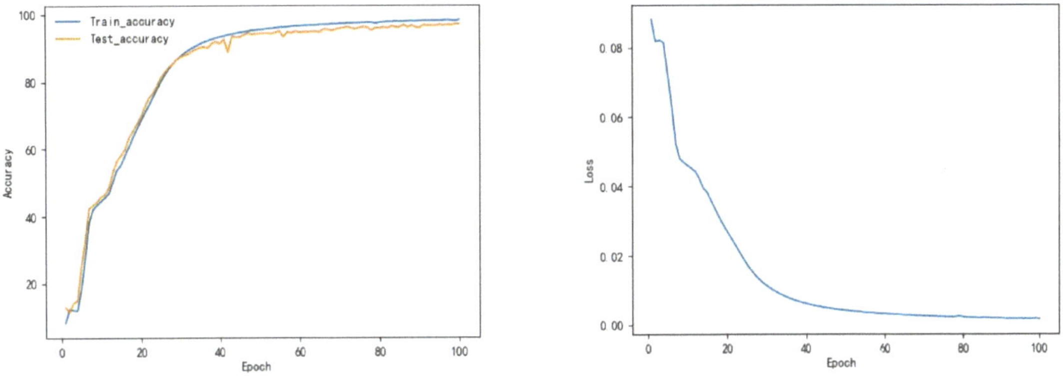

Figure 16. The curve of accuracy and loss during training of LSTM.

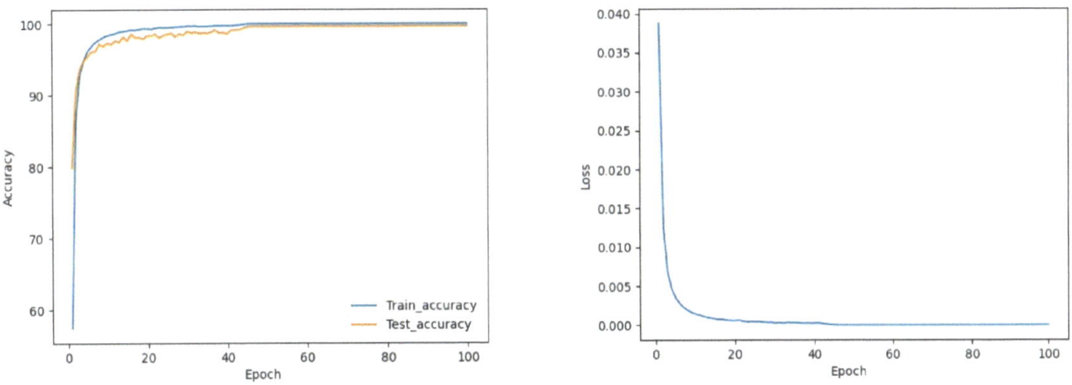

Figure 17. The curve of accuracy and loss during training of LSTM-Res.

What is more, the training of the LSTM-Res model is fully converged at the 46th epoch, accounting for only about 1/2 of the whole training time; it shows that after combining the CNN-Res model, the LSTM-Res model not only benefits the precision but also doubles the training convergence speed based on the LSTM model.

Due to the high test accuracy of LSTM-Res, we keep the values of the test confusion matrix of LSTM-Res with three decimal places.

Among the recognition results of the LSTM model in Figure 18, there are 19 gestures whose recognition accuracy has attained more than 95%; however, the recognition accuracy of the 'Number 4' is relatively low, which is 90%; it reveals that the LSTM model has achieved high accuracy in gesture recognition. Still, the accuracy of individual gestures is not high enough. Nevertheless, as exhibited in Figure 19, apart from "Push up", the other 19 actions are classified with an accuracy of over 99%.

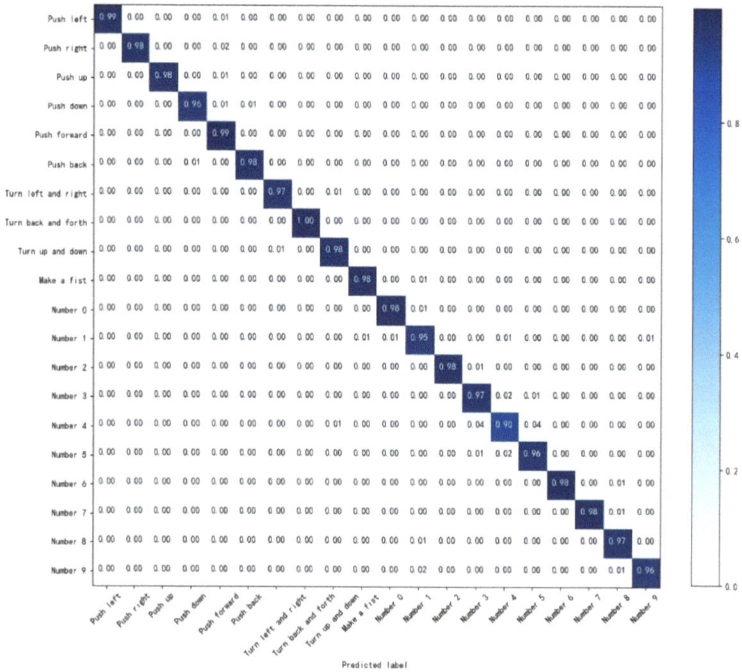

Figure 18. The confusion matrix for the test results of LSTM.

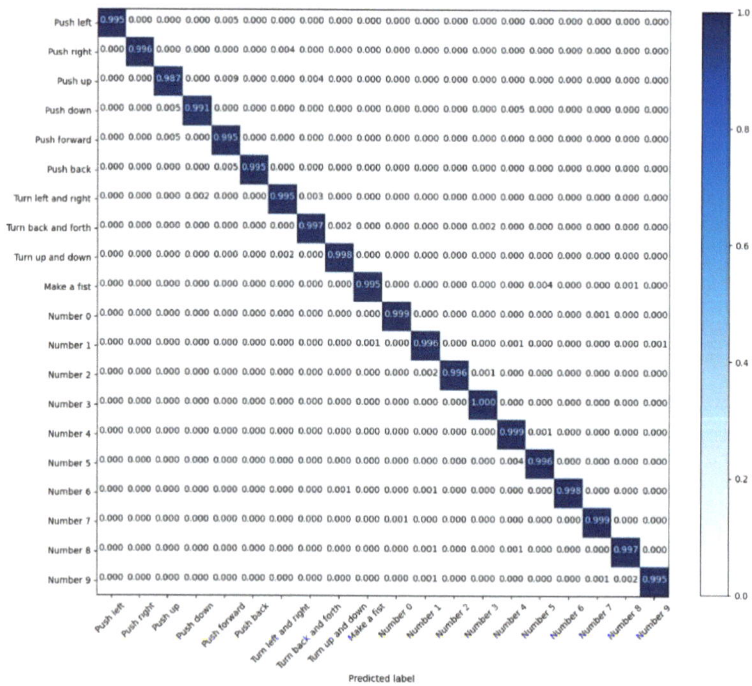

Figure 19. The confusion matrix for the test results of LSTM-Res.

The performance improvement of the LSTM-Res model is credited to the CNN-Res model. The LSTM-Res model can accomplish the 20-classification task with outstanding accuracy and less training time by further extracting the local features with the CNN.

4.4.3. Transformer vs. Transformer-CNN

Figure 20 manifests that the Transformer model does not perform well in classification. Even after undergoing 100 epochs of training, the test accuracy is still only 71.68%; it indicates that the basic Transformer model for gesture recognition is far from sufficient in gesture recognition; however, with the CNN model's fusion, the Transformer-CNN model's accuracy is extensively promoted. As shown in Figure 21, the test accuracy of the Transformer-CNN model stabilized as high as 98.96%, but the convergence time did not change obviously.

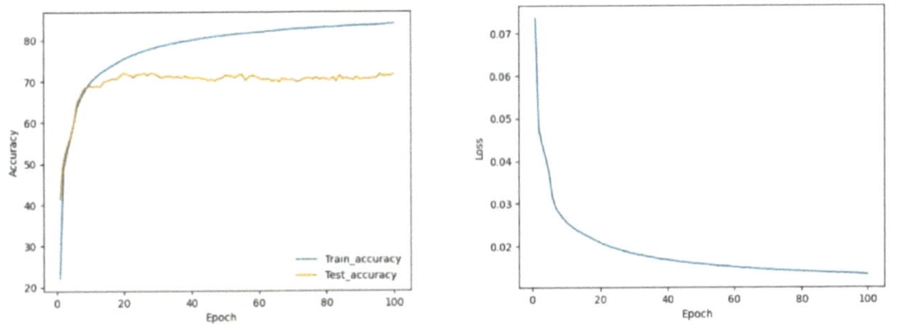

Figure 20. The curve of accuracy and loss during training of Transformer.

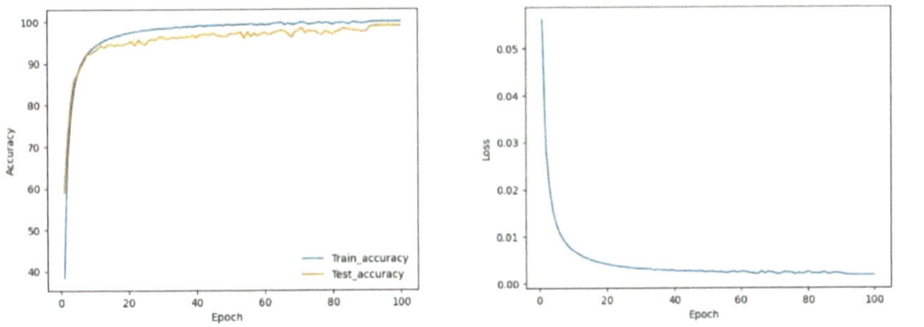

Figure 21. The curve of accuracy and loss during training of Transformer-CNN.

As shown in Figure 22, almost half of the gesture recognition accuracies are under 75%., not to mention that the recognition accuracy of 'Number 1', 'Number 4' and 'Number 5' is only about 50%, which means half of them are incorrectly recognized; however, the Transformer-CNN model's confusion matrix is much higher quality. Figure 23 shows that the classification accuracy of almost all gestures is around 99%. Remarkably, three gestures are correctly identified with 100% accuracy. The lowest accuracy comes from 'Number 4', which is still 97%.

The above comparisons substantiate that the incorporation of the 1D convolutional module ameliorates the performance of the Transformer. As a result, it is apparent that the Transformer-CNN model behaves better than the Transformer model in the task of 20-gesture classification.

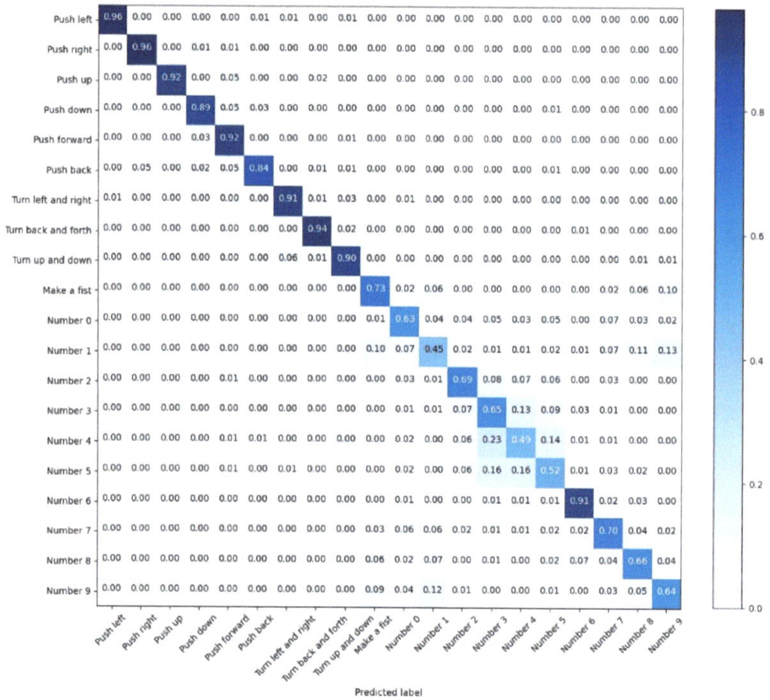

Figure 22. The confusion matrix for the test results of Transformer.

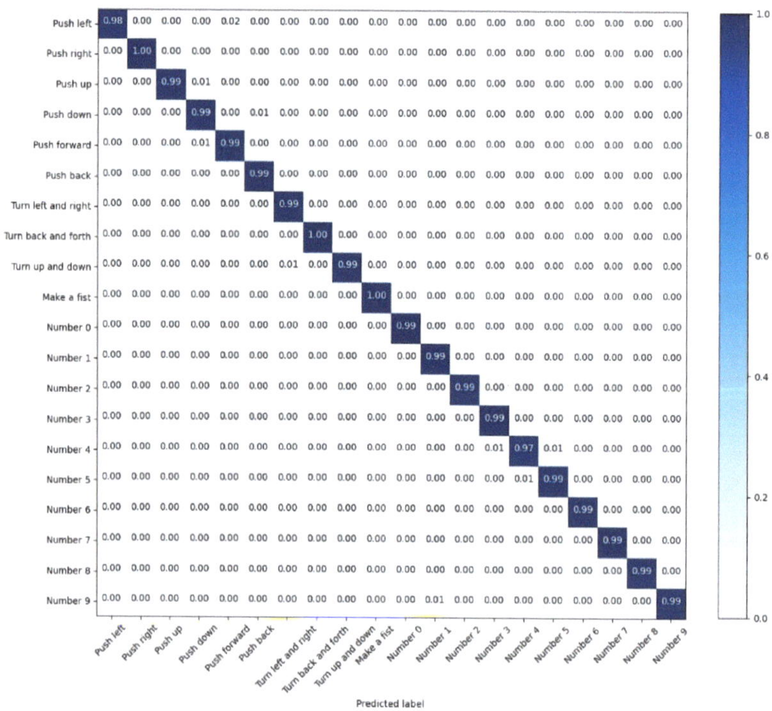

Figure 23. The confusion matrix for the test results of Transformer-CNN.

In summary, the LSTM-Res/GRU-Res model is preferable if the system's objective is to achieve the highest recognition accuracy because they have the best precision; however, the CNN-Res model is the best choice if the system requires a high training efficiency and less testing time, not with a highly demanding requirement for accuracy; it has the least time for training and testing, and its recognition accuracy is up to 98.24%. On top of that, if the system needs high accuracy and the most rapid real-time response to a single gesture, the Transformer-CNN model is the ideal option because of the highest recognition speed and test accuracy of 98.96%.

4.5. Separate Recognition

Our self-built dataset of 20 categories contains nine arm movements and eleven gesture movements; they can be divided into a 9-category sub-dataset and an 11-category sub-dataset.

To further explore the effect of sEMG and IMU signals on the recognition of arm and finger movements, we also split sEMG and IMU signals to identify arm and finger movements, respectively. The improved CNN-Res model, RNN-Res model, and Transformer-CNN model in Section 4.4 have been implemented in three groups of experiments. Each group includes using the sEMG signal to recognize arm and finger movements and the IMU signal to recognize arm and finger movements. The separated experimental results are shown in Table 3, where 'Together' represents the result of simultaneous recognition in Section 4.4.

Table 3. Accuracy results from the separate recognition.

Models / Gestures	CNN-Res		RNN-Res		Transformer-CNN	
	sEMG	IMU	sEMG	IMU	sEMG	IMU
Arm	90.16%	95.52%	98.13%	96.55%	98.13%	96.39%
Finger	95.48%	41.80%	99.12%	42.89%	98.33%	18.21%
Together	98.24%		99.67%		98.96%	

Table 3 reveals that the three models can recognize arm movements with more than 90% accuracy when using sEMG or IMU signals independently; however, for finger movements, although the recognition accuracy of the three models can reach over 95% with sEMG signals, the accuracy with IMU signals is too low. Only using IMU signals can't successfully recognize finger movements. In addition, our proposed models' recognition accuracy of arm and finger movements with one signal alone is lower than that of identifying all 20 categories of movements with the two signals together.

Thus, if the target is to recognize only arm movement, using the sEMG signal or IMU signal alone can achieve good recognition results. If the target is to recognize only finger movements, applying sEMG signals alone is also reachable to high accuracy, but applying IMU signals alone to recognition is not feasible. If the gesture recognition system aims to recognize 20-category gestures simultaneously, the two signals are recommended to be combined. The combination of sEMG and IMU signals enables the system to recognize more gestures and accomplishes better precision.

5. Conclusions

This work conducts a gesture recognition modeling study based on the sEMG and IMU signals.

The conclusions drawn in the paper are as follows:

(1) A dataset containing sEMG signals and IMU signals is built through the Myo armband. The dataset includes 20 different hand gestures with a total of nearly 20,000 actions; these actions involve dynamic movements dominated by arms and static movements dominated by fingers.

(2) Based on the baseline gesture recognition models, including the two-stream CNN model, RNN model, and Transformer model, the two-stream CNN-Res model, RNN-Res model, and Transformer-CNN model are proposed, respectively. The CNN-Res model introduces the residual units and has more profound network layers; it achieves a test accuracy of 98.24% and the shortest training and test time. The RNN-Res model combines the RNN model and the CNN-Res model to enhance the degree of extracting local features, accomplishing the highest recognition accuracy. The LSTM-Res model and the GRU-Res model test accuracy are 99.67% and 99.46%, respectively. The Transformer model is incorporated with the CNN model to enhance its ability to capture local information. The modified Transformer-CNN model improves its accuracy from 71.86% to 98.96%; moreover, its shortest recognition response time of 0.0149 s for a single sample makes it highly applicable in real-time recognition and interaction systems.

(3) Through the separate recognition of arm and finger movements, the effectiveness of the combination of sEMG signals and IMU signals in the multi-category mission of this paper is proved; it turns out that simultaneously adopting two signals allows us to recognize 20 gestures and achieves the highest recognition accuracy.

Future work needs to concentrate on optimizing the parameter settings of the model. Although our proposed models achieve a high recognition precision, their training time is at the level of hours. Therefore, more research is required to establish more efficient models. In addition, deploying models in embedded systems such as real-time interfaces will also be the focus of our future research. After sorting out our dataset, we will make it publicly available on GitHub soon.

Author Contributions: Conceptualization, Y.J., M.Y. and L.S.; methodology, Y.J. and L.S.; software, L.S. and J.Z.; validation, L.S., J.Z. and Y.S.; formal analysis, Y.J. and L.S.; investigation, Y.J. and J.Z.; resources, Y.J. and L.S.; data curation, L.S. and J.Z; writing—original draft preparation, Y.J. and L.S.; writing—review and editing, Y.J., M.Y. and Y.S.; visualization, Y.J. and L.S.; supervision, Y.J.; project administration, Y.J.; funding acquisition, Y.J. All authors have read and agreed to the published version of the manuscript.

Funding: This research was funded by Funds for the National Key R&D Program of China (2021YFF0307603).

Institutional Review Board Statement: Not applicable.

Informed Consent Statement: Informed consent was obtained from all subjects involved in the study.

Conflicts of Interest: The authors declare no conflict of interest.

References

1. Said, S.; Boulkaibet, I.; Sheikh, M.; Karar, A.S.; Kork, S.; Nait-Ali, A. Machine-learning-based muscle control of a 3D-printed bionic arm. *Sensors* **2020**, *20*, 3144. [CrossRef] [PubMed]
2. Colli Alfaro, J.G.; Trejos, A.L. User-Independent Hand Gesture Recognition Classification Models Using Sensor Fusion. *Sensors* **2022**, *22*, 1321. [CrossRef] [PubMed]
3. Zhang, Z.; Tang, Y.; Zhao, S.; Zhang, X. Real-Time Surface EMG Pattern Recognition for Hand Gestures Based on Support Vector Machine. In Proceedings of the 2019 IEEE International Conference on Robotics and Biomimetics (ROBIO), Dali, China, 6–8 December 2019; pp. 1258–1262.
4. López, L.B.; Caraguay, V.; Vimos, V.; Zea, J.; Vásconez, J.; Álvarez, M.; Benalcázar, M. An energy-based method for orientation correction of EMG bracelet sensors in hand gesture recognition systems. *Sensors* **2020**, *20*, 6327. [CrossRef]
5. Sattar, N.Y.; Kausar, Z.; Usama, S.A.; Farooq, U.; Khan, U.S. EMG based control of transhumeral prosthesis using machine learning algorithms. *Int. J. Control. Autom. Syst.* **2021**, *19*, 3522–3532. [CrossRef]
6. Bisi, S.; De Luca, L.; Shrestha, B.; Yang, Z.; Gandhi, V. Development of an EMG-controlled mobile robot. *Robotics* **2018**, *7*, 36. [CrossRef]
7. Wahid, M.F.; Tafreshi, R.; Al-Sowaidi, M.; Langari, R. Subject-independent hand gesture recognition using normalization and machine learning algorithms. *J. Comput. Sci.* **2018**, *27*, 69–76. [CrossRef]
8. Totty, M.S.; Wade, E. Muscle activation and inertial motion data for noninvasive classification of activities of daily living. *IEEE Trans. Biomed. Eng.* **2017**, *65*, 1069–1076.

9. Su, H.; Ovur, S.E.; Zhou, X.; Qi, W.; Ferrigno, G.; De Momi, E. Depth vision guided hand gesture recognition using electromyographic signals. *Adv. Robot.* **2020**, *34*, 985–997. [CrossRef]
10. Amrani, M.Z.; Borst, C.W.; Achour, N. Multi-sensory assessment for hand pattern recognition. *Biomed. Signal Processing Control.* **2022**, *72*, 103368. [CrossRef]
11. Yan, M.; Lou, X.; Wang, Y. Channel noise optimization of polar codes decoding based on a convolutional neural network. *Wirel. Commun. Mob. Comput.* **2021**, *2021*, 10. [CrossRef]
12. Zia ur Rehman, M.; Waris, A.; Gilani, S.O.; Jochumsen, M.; Niazi, I.K.; Jamil, M.; Farina, D.; Kamavuako, E.N. Multiday EMG-based classification of hand motions with deep learning techniques. *Sensors* **2018**, *18*, 2497. [CrossRef] [PubMed]
13. Côté-Allard, U.; Fall, C.L.; Drouin, A.; Campeau-Lecours, A.; Gosselin, C.; Glette, K.; Laviolette, F.; Gosselin, B. Deep learning for electromyographic hand gesture signal classification using transfer learning. *IEEE Trans. Neural Syst. Rehabil. Eng.* **2019**, *27*, 760–771. [CrossRef] [PubMed]
14. Pinzón-Arenas, J.O.; Jiménez-Moreno, R.; Rubiano, A. Percentage estimation of muscular activity of the forearm by means of EMG signals based on the gesture recognized using CNN. *Sens. Bio-Sens. Res.* **2020**, *29*, 100353. [CrossRef]
15. Lu, L.; Mao, J.; Wang, W.; Ding, G.; Zhang, Z. A study of personal recognition method based on EMG signal. *IEEE Trans. Biomed. Circuits Syst.* **2020**, *14*, 681–691. [CrossRef]
16. Côté-Allard, U.; Gagnon-Turcotte, G.; Laviolette, F.; Gosselin, B. A low-cost, wireless, 3-D-printed custom armband for sEMG hand gesture recognition. *Sensors* **2019**, *19*, 2811. [CrossRef]
17. Chen, K.; Yao, L.; Zhang, D.; Wang, X.; Chang, X.; Nie, F. A semisupervised recurrent convolutional attention model for human activity recognition. *IEEE Trans. Neural Netw. Learn. Syst.* **2019**, *31*, 1747–1756. [CrossRef]
18. Nasri, N.; Orts-Escolano, S.; Gomez-Donoso, F.; Cazorla, M. Inferring static hand poses from a low-cost non-intrusive sEMG sensor. *Sensors* **2019**, *19*, 371. [CrossRef]
19. Guo, H.; Sung, Y. Movement estimation using soft sensors based on Bi-LSTM and two-layer LSTM for human motion capture. *Sensors* **2020**, *20*, 1801. [CrossRef] [PubMed]
20. Zhang, Z.; He, C.; Yang, K. A novel surface electromyographic signal-based hand gesture prediction using a recurrent neural network. *Sensors* **2020**, *20*, 3994. [CrossRef]
21. Nasri, N.; Orts-Escolano, S.; Cazorla, M. A semg-controlled 3D game for rehabilitation therapies: Real-time time hand gesture recognition using deep learning techniques. *Sensors* **2020**, *20*, 6451. [CrossRef]
22. Nasri, N.; Gomez-Donoso, F.; Orts-Escolano, S.; Cazorla, M. Using Inferred Gestures from sEMG Signal to Teleoperate a Domestic Robot for the Disabled. In *International Work-Conference on Artificial Neural Networks*; Springer: Cham, Switzerland, 2019; pp. 198–207.
23. Zhang, X.; Yang, Z.; Chen, T.; Chen, D.; Huang, M.-C. Cooperative sensing and wearable computing for sequential hand gesture recognition. *IEEE Sens. J.* **2019**, *19*, 5775–5783. [CrossRef]
24. Williams, H.E.; Shehata, A.W.; Dawson, M.R.; Scheme, E.; Hebert, J.; Pilarski, P. Recurrent Convolutional Neural Networks as an Approach to Position-Aware Myoelectric Prosthesis Control. *IEEE Trans. Biomed. Eng.* **2022**, *69*, 2243–2255. [CrossRef] [PubMed]
25. Li, C.; Ren, J.; Huang, H.; Wang, B.; Zhu, Y.; Hu, H. PCA and deep learning based myoelectric grasping control of a prosthetic hand. *Biomed. Eng. Online* **2018**, *17*, 107. [CrossRef] [PubMed]
26. Sun, L.; An, H.; Ma, H.; Gao, J. Real-time human intention recognition of multi-joints based on MYO. *IEEE Access* **2019**, *8*, 4235–4243. [CrossRef]
27. Cascarano, G.D.; Loconsole, C.; Brunetti, A.; Lattarulo, A.; Buongiorno, D.; Losavio, G.; di Sciascio, E.; Bevilacqua, V. Biometric handwriting analysis to support Parkinson's Disease assessment and grading. *BMC Med. Inform. Decis. Mak.* **2019**, *19*, 252. [CrossRef]
28. Motoche, C.; Benalcázar, M.E. Real-Time Hand Gesture Recognition Based on Electromyographic Signals and Artificial Neural Networks. In *International Conference on Artificial Neural Networks*; Springer: Cham, Switzerland, 2018; pp. 352–361.
29. Huang, D.; Yang, C.; Ju, Z.; Dai, S.-L. Disturbance observer enhanced variable gain controller for robot teleoperation with motion capture using wearable armbands. *Auton. Robot.* **2020**, *44*, 1217–1231. [CrossRef]
30. Tepe, C.; Erdim, M. Classification of surface electromyography and gyroscopic signals of finger gestures acquired by Myo armband using machine learning methods. *Biomed. Signal Processing Control* **2022**, *75*, 103588. [CrossRef]
31. Vaswani, A.; Shazeer, N.; Parmar, N.; Uszkoreit, J.; Jones, L.; Gomez, A.N.; Kaiser, Ł.; Polosukhin, I. Attention is All You Need. In *Advances in Neural Information Processing Systems*; The MIT Press: Cambridge, MA, USA, 2017; p. 30.
32. Mendes Junior, J.J.A.; Freitas, M.L.B.; Campos, D.P.; Farinelli, F.A.; Stevan, S.L., Jr.; Pichorim, S.F. Analysis of influence of segmentation, features, and classification in sEMG processing: A case study of recognition of brazilian sign language alphabet. *Sensors* **2020**, *20*, 4359. [CrossRef]
33. Lai, S.; Xu, L.; Liu, K.; Zhao, J. Recurrent Convolutional Neural Networks for Text Classification. In Proceedings of the Twenty-Ninth AAAI Conference on Artificial Intelligence, Austin, TX, USA, 25–30 January 2015.
34. Chiu, C.; Shih, T.K.; Lin, C.; Hui, L.; Utaminingrum, F.; Yang, T. Application of Hand Recognition System Based on Electromyography and Gyroscope Using Deep Learning. In Proceedings of the 2019 Twelfth International Conference on Ubi-Media Computing (Ubi-Media), Bali, Indonesia, 5–8 August 2019; pp. 96–101.

35. Romero, R.; Cruz, P.J.; Vásconez, J.P.; Benalcázar, M.; Álvarez, R.; Barona, L.; Valdivieso, L. Hand Gesture and Arm Movement Recognition for Multimodal Control of a 3-DOF Helicopter. In *International Conference on Robot Intelligence Technology and Applications*; Springer: Cham, Switzerland, 2022; pp. 363–377.
36. Hochreiter, S.; Schmidhuber, J. Long short-term memory. *Neural Comput.* **1997**, *9*, 1735–1780. [CrossRef] [PubMed]
37. Cho, K.; van Merriënboer, B.; Gulcehre, C.; Bahdanau, D.; Bougares, F.; Schwenk, H.; Bengio, Y. Learning phrase representations using RNN encoder-decoder for statistical machine translation. *arXiv* **2014**, arXiv:1406.1078.
38. He, K.; Zhang, X.; Ren, S.; Sun, J. Deep Residual Learning for Image Recognition. In Proceedings of the IEEE Conference on Computer Vision and Pattern Recognition, Las Vegas, NV, USA, 27–30 June 2016; pp. 770–778.

sensors

MDPI

Article

Image Segmentation Using Active Contours with Hessian-Based Gradient Vector Flow External Force

Qianqian Qian [1], Ke Cheng [1,*], Wei Qian [2], Qingchang Deng [1] and Yuanquan Wang [3,*]

[1] School of Computer Science, Jiangsu University of Science and Technology, Zhenjiang 212003, China; 202070044@stu.just.edu.cn (Q.Q.); 199070046@stu.just.edu.cn (Q.D.)
[2] School of Electronics and Information, Jiangsu University of Science and Technology, Zhenjiang 212003, China; 211110303118@stu.just.edu.cn
[3] School of Artificial Intelligence, Hebei University of Technology, Tianjin 300401, China
* Correspondence: chengke1972@just.edu.cn (K.C.); wangyuanquan@scse.hebut.edu.cn (Y.W.); Tel.: +86-139-5294-5091 (K.C.); +86-139-2061-3363 (Y.W.)

Abstract: The gradient vector flow (GVF) model has been widely used in the field of computer image segmentation. In order to achieve better results in image processing, there are many research papers based on the GVF model. However, few models include image structure. In this paper, the smoothness constraint formula of the GVF model is re-expressed in matrix form, and the image knot represented by the Hessian matrix is included in the GVF model. Through the processing of this process, the relevant diffusion partial differential equation has anisotropy. The GVF model based on the Hessian matrix (HBGVF) has many advantages over other relevant GVF methods, such as accurate convergence to various concave surfaces, excellent weak edge retention ability, and so on. The following will prove the advantages of our proposed model through theoretical analysis and various comparative experiments.

Keywords: gradient vector flow; Hessian matrix; image structure; anisotropy

Citation: Qian, Q.; Cheng, K.; Qian, W.; Deng, Q.; Wang, Y. Image Segmentation Using Active Contours with Hessian-Based Gradient Vector Flow External Force. *Sensors* **2022**, *22*, 4956. https://doi.org/10.3390/s22134956

Academic Editors: Chien Aun Chan, Chunguo Li and Ming Yan

Received: 7 May 2022
Accepted: 20 June 2022
Published: 30 June 2022

1. Introduction

Image segmentation is a key step from image processing to image analysis. Traditional segmentation methods include threshold [1], clustering [2], active contour model [3], region growth [4], etc. Since someone proposed the snake or active contour model in 1988, the snake or active contour model has become one of the mainstream models of image segmentation [3]. Generally, an active contour performs image segmentation by minimizing the combination of internal and external energy and deforming the curve on the image plane; the internal energy keeps the curve continuous and smooth, while the external energy attracts the curve to the boundary of the object to be segmented on the image. Therefore, the problem of finding the boundary of the segmented object can be transformed into the problem of minimizing the internal and external energy. According to the representation of the curve, the active contour is divided into a parametric contour and geometric contour. The parametric model uses explicit parameter representation [3,5–9], and uses image edge mapping to stop the evolution of contour. Parametric models rely heavily on high gradient amplitudes to extract object boundaries, and are effective only when the contrast between background and foreground is clear enough. The geometric model [10–24] is based on the theory of level set technology and usually adopts specific regional homogeneity criteria to guide the evolution of contour.

External force plays a leading role in the evolution of parametric snake contour, so people have invested a lot of energy in the research of external force to improve the robustness of active contour. At present, the proposed gradient vector flow (GVF) [25] is still one of the most successful methods. It spreads the gradient vector from the object boundary to the rest of the image, which not only expands the capture range, but also weakens the influence of noise to a certain extent. Due to its effectiveness, a large number of fast algorithms for the GVF model have been proposed, including vector field convolution (VFC) [26], BVF [27], GVF based on augmented Lagrange [28], the multi-grid method of GVF [29], and efficient numerical format of GVF [30]. Some other efforts focus on improving the initial edge map, for example, a guided filter is employed to enhance the initial edge map [31,32] and a directional edge map is coined for the GVF model [33]; in the literature, the GVF is modified by using the initial contour position and introducing additional boundary conditions of Dirichlet type [34]. Many efforts pay attention to reformulating the energy functional of the GVF model, among others, examples include the harmonic gradient vector flow (HGVF) [35], harmonic surface [32,36], 4DGVF external force field [37], NGVF [38], EPGVF [39], MGVF [40], and CN-GGVF [41]. Recently, the GVF model also has some interesting applications, as well as some interesting work on GVF snake initialization for ultrasonic image segmentation, such as walking particles [42,43]. Very recently, Jaouen proposed an image enhancement vector field based on the partial differential equation (PDE) [44], and pointed out the similarity between the vector field and gradient vector flow, which allows a natural connection between impulse filtering and a large number of work on GVF like fields. It is important to note that the deep learning method plays a very important role for image-based applications presently, such as image segmentation [45–49], detection [50,51], and classification [52,53], and it needs big data for training and the active contour is still of importance for image segmentation.

We can see that although the above contents provide various methods to improve the GVF model, they do not consider the characteristics of image structure. Ref. [54] pointed out that the "Hessian method is a method to extract the direction of image features through high-order differentiation". Inspired by this principle, we express the smooth constraint formula in the GVF model in matrix form, then incorporate the Hessian matrix into the energy functional of the GVF model, and finally get the GVF based on the Hessian matrix, that is HBGVF. Compared with other methods, we experimentally prove that HBGVF has many advantages, such as accurately converging to various concave surfaces while maintaining weak edges. There is more information related to this work in the literature [55,56].

The rest of this paper is arranged as follows: in the next section, we briefly review the snake model and four famous GVF-based external forces, including GVF [25], GGVF [57], VEF [58], NGVF [38], and CN-GVF [41], and compare them with these GVF-based methods through experiments. Section 3 details the HBGVF model proposed in this paper. In Section 4, we prove the advantages of the proposed model through a large number of experiments, and finally draw a conclusion in Section 5.

2. Backgrounds

2.1. Traditional Model: Active Contours

When the early active contour was proposed, it was defined as the elastic curve $\mathbf{c}(s) = [\mathrm{x}(s), \mathrm{y}(s)], s \in [0, 1]$ and the following is its energy function formula:

$$E_{\text{snake}} = \int \frac{1}{2}\left(\alpha|\mathbf{c}'|^2 + \beta|\mathbf{c}''|^2\right) + E_{ext}(\mathbf{c}(s))ds \tag{1}$$

in Formula (1), $\mathbf{c}'(s)$, $\mathbf{c}''(s)$ are the first and second derivatives of $\mathbf{c}(s)$, which are, respectively, positively weighted by α and β. $E_{ext}(\mathbf{c}(s))$ is the image potential, which may be caused by various things, such as edges. The Euler equation for minimizing E_{snake} can be obtained by deformation calculus as follows:

$$\alpha\mathbf{c}''(s) - \beta\mathbf{c}''''(s) - \nabla E_{ext} = 0 \tag{2}$$

Formula (2) is a force balance equation in reference [8],

$$\mathbf{F}_{\text{int}} + \mathbf{F}_{ext} = 0 \tag{3}$$

in Formula (3), $\mathbf{F}_{\text{int}} = \alpha\mathbf{c}''(s) - \beta\mathbf{c}''''(s)$ and $\mathbf{F}_{ext} = -\nabla E_{ext}$. The internal force $\mathbf{F}_{\text{int}}$ keeps the snake contour smooth, while the external force $\mathbf{F}_{ext}$ shrinks the snake contour to the desired image object.

In image **I**, $\mathbf{F}_{ext}$ is often used as the gradient vector of image edge mapping, as shown in the following formula: **I**, as follows,

$$\mathbf{F}_{ext} = -\nabla E_{ext} = \nabla|\nabla G_{\sigma} \otimes \mathbf{I}|^2 \tag{4}$$

In fact, this gradient vector is local, unable to take into account the overall situation, and is not regular enough, so the snake can not evolve effectively under its guidance.

2.2. Gradient Vector Flow (GVF)

Due to the obvious disadvantage of external force in Formula (4), $\mathbf{F}_{ext}$ is replaced by a new vector field $\mathrm{v} = [\mathrm{u}(\mathrm{x},\mathrm{y})\mathrm{V}(\mathrm{x},\mathrm{y})]$ in the GVF model, which can be derived by minimizing the following function,

$$E_{GVF} = \iint \mu\left(u_x^2 + u_y^2 + v_x^2 + v_y^2\right) + |\nabla f|^2|\mathbf{v} - \nabla f|^2 dxdy \tag{5}$$

In (5), μ is a positive weight, f is the edge map of the image **I**, and ∇ is the gradient operator. The newly obtained vector field is a gradient vector flow (GVF) field. The GVF field can be obtained by solving the following equation iteratively,

$$\begin{cases} u_t = \mu\Delta u - |\nabla f|^2(u - f_x) \\ v_t = \mu\Delta v - |\nabla f|^2(v - f_y) \end{cases} \tag{6}$$

where Δ is the Laplacian operator. The diffusion equation is isotropic.

The generalized GVF (GGVF) is an extension of the GVF by replacing μ and $f_x^2 + f_y^2$ in (6) with two spatially varying functions $g(|\nabla f|) = \exp(-|\nabla f|^2/k^2)$ and $h(|\nabla f|) = 1 - g(|\nabla f|)$, respectively [57], k acts as a threshold and controls the smoothing effect. The introduction of such terms makes the GGVF snake behave better than the GVF snake on thin concavity convergence.

2.3. Virtual Electric Field (VEF)

Reference [58] proposed a virtual electric field model (VEF). In this method, each pixel in the image is regarded as an electron, the charge is the size of the image edge, and the virtual electric field at (x_0, y_0) is derived from the sum of all other electrons in the surrounding area D, which is expressed by the following formula,

$$E_{VEF}(x_0, y_0) = \sum_{(x,y)\in D}\left(\frac{(x_0 - x)}{\left(\sqrt{(x_0 - x)^2 + (y_0 - y)^2}\right)^3}, \frac{(y_0 - y)}{\left(\sqrt{(x_0 - x)^2 + (y_0 - y)^2}\right)^3}\right) \cdot f(x,y) \tag{7}$$

in Formula (7), $D = \{(x,y) \mid -t \le x_0 - x \le t, -t \le y_0 - y \le t\}$, f is the size of the image edge image. Fast Fourier transform (FFT) is applied to the VEF model, so Formula (7) is usually written in convolution form, as follows,

$$E_{VEF}(x,y) = \left(-\frac{x}{\left(\sqrt{x^2 + y^2}\right)^3}, -\frac{y}{\left(\sqrt{x^2 + y^2}\right)^3}\right) \otimes f(x,y) \tag{8}$$

in Formula (8), $\otimes$ represents the convolution operation.

Thanks to the use of FFT, the VEF model can be realized in real time. In addition, the VEF model also has some characteristics better than the GVF model, such as a large capture range and more sensitive concave convergence.

2.4. Gradient Vector Flow in Normal Direction (NGVF)

It was pointed out in [59] that the Laplace operator can be decomposed into two terms, as shown below,

$$\Delta u = u_{TT} + u_{NN} \tag{9}$$

Taking $u(x,y)$ as an example, in Formula (9), u_{TT} and u_{NN} are the second derivatives of $u(x,y)$ in the tangential and normal directions of the isophotes, respectively. It was pointed out in [60] that, as an interpolation operator, u_{NN} has the best performance, Δu second and u_{TT} third. The diffusion process in (6) is regarded as the interpolation process, and the NGVF is proposed using the optimal interpolator, as shown in the following formula,

$$\begin{cases} u_t = \mu u_{NN} - (u - f_x)|\nabla f|^2 \\ v_t = \mu v_{NN} - (v - f_y)|\nabla f|^2 \end{cases} \tag{10}$$

where μ is also a positive weight as in (6).

2.5. Component-Normalized Generalized Gradient Vector Flow (CN-GGVF)

In the CN-GGVF model, the diffusion equations are modified in the following form,

$$\begin{cases} u_t = g(|\nabla f|) \cdot (g(|\nabla f|)u_{NN} + h(|\nabla f|)u_{TT}) - h(|\nabla f|) \cdot (u - f_x) \\ v_t = g(|\nabla f|) \cdot (g(|\nabla f|)v_{NN} + h(|\nabla f|)v_{TT}) - h(|\nabla f|) \cdot (v - f_y) \end{cases} \tag{11}$$

where the $g(|\nabla f|)$ and $h(|\nabla f|)$ are identical to those in the GGVF model, and u_{TT} and u_{NN} are identical to those in the NGVF model. Based on deep analysis of the behavior of the GGVF model, Qin et al. proposed to normalized the GVF vector in a component-wise manner, such that the CN-GGVF model can converge to a deep and thin notch, the component-normalized (CN) GGVF field reads,

$$u_{CN-GVF} = \text{sign}(u) = \begin{cases} 1, & u > 0 \\ 0, & u = 0 \\ -1, & u < 0 \end{cases} \tag{12}$$

$$v_{CN-GVF} = \text{sign}(v) = \begin{cases} 1, & v > 0 \\ 0, & v = 0 \\ -1, & v < 0 \end{cases} \tag{13}$$

3. The HBGVF Model

3.1. Gradient Vector Flow Expressed in Matrix Form

By observing Equation, we first reformulate the smoothness constraint in the GVF model into matrix form as follows, $u_x^2 + u_y^2 = \begin{pmatrix} u_x & u_y \end{pmatrix}\begin{pmatrix} u_x \\ u_y \end{pmatrix} = \begin{pmatrix} u_x & u_y \end{pmatrix}\begin{pmatrix} 1 & 0 \\ 0 & 1 \end{pmatrix}\begin{pmatrix} u_x \\ u_y \end{pmatrix}$, we first reformulate the smoothness constraint in the GVF model into matrix form as follows,

$$E_{GVF} = \iint \mu\left[(\nabla u)^T \cdot \mathbf{W} \cdot \nabla u + (\nabla v)^T \cdot \mathbf{W} \cdot \nabla v\right] + |\nabla f|^2 |\mathbf{v} - \nabla f|^2 dxdy \tag{14}$$

in Equation (14), $\mathbf{W}$ is the identity matrix. It can be seen from the above formula that due to the existence of this identity matrix, it induces the scalar L2 norm, so that the GVF model fails to take into account the image characteristic of image structure. We completely replace all **W** with matrix **D** related to the image structure, so we use Hessian matrix to construct, as shown below,

$$E = \iint \mu\left[(\nabla u)^T \cdot \mathbf{D} \cdot \nabla u + (\nabla v)^T \cdot \mathbf{D} \cdot \nabla v\right] + |\nabla f|^2 |\mathbf{v} - \nabla f|^2 dxdy \tag{15}$$

where $\mathbf{D} = \begin{pmatrix} a & b \\ b & c \end{pmatrix}$ is a symmetric and positive semi-definite matrix. The reconstructed model is called the Hessian-based GVF (HBGVF for short). Using the variational method, the HBGVF field can be obtained by solving the following equation, as shown below,

$$\begin{cases} u_t = \mu div(\mathbf{D}\nabla u) - |\nabla f|^2 (u - f_x) = 0 \\ v_t = \mu div(\mathbf{D}\nabla v) - |\nabla f|^2 (v - f_y) = 0 \end{cases} \tag{16}$$

in Equation (16), div is the divergence operator.

3.2. Using the Hessian Matrix to Construct Diffusion Matrix

Through the observation of Formula (16), we can know that its equation is exactly the tensor based diffusion in [61]. The "Hessian method proposed in reference [54] regards the direction of the maximum second-order directional derivative as the direction passing through the image feature, and its vertical direction is regarded as the direction along the image feature." Inspired by this principle, we use Hessian matrix to reconstruct the diffusion matrix **D** in Formula (16). Taking image **I** as an example, its Hessian matrix is represented by the following formula,

$$\mathbf{H} = \begin{pmatrix} I_{xx} & I_{xy} \\ I_{xy} & I_{yy} \end{pmatrix} \tag{17}$$

using the derivative in [61], the two eigenvalues of **H** can be solved by the following formula, expressed by λ_1 and λ_2,

$$\begin{cases} \lambda_1 = \frac{1}{2}\left[(I_{xx} + I_{yy}) + \sqrt{(I_{xx} - I_{yy})^2 + 4I_{xy}^2}\right] \\ \lambda_2 = \frac{1}{2}\left[(I_{xx} + I_{yy}) - \sqrt{(I_{xx} - I_{yy})^2 + 4I_{xy}^2}\right] \end{cases} \tag{18}$$

the eigenvectors corresponding to λ_1 and λ_2 are $\mathbf{e}_1$ and $\mathbf{e}_2$, which are obtained by the following formula:

$$\mathbf{e}_1 = \begin{pmatrix} 2I_{xy} \\ I_{yy} - I_{xx} + \sqrt{(I_{xx} - I_{yy})^2 + 4I_{xy}^2} \end{pmatrix} \tag{19}$$

$$\mathbf{e}_2 = \begin{pmatrix} 2I_{xy} \\ I_{yy} - I_{xx} - \sqrt{(I_{xx} - I_{yy})^2 + 4I_{xy}^2} \end{pmatrix} \tag{20}$$

Obviously, through the observation of Formula (19), we can see that $\lambda_1 \geq \lambda_2$. In reference [54], it is pointed out that because $\lambda_1 \geq \lambda_2$, the feature vector $\mathbf{e}_1$ has the largest second-order directional derivative direction in all directions, which is considered as the direction passing through the image feature, and $\mathbf{e}_2$ is considered as the direction along the image feature. Using the eigenvalues and eigenvectors of the derived Hessian matrix, we construct the diffusion matrix **D** in formula (16). The eigenvector of **D** is used as the eigenvector of **H**. We use η_1,η_2 to represent the two eigenvalues of **D**, as shown in the following formula:

$$\begin{cases} \eta_1 = \frac{1}{1+(|\nabla I|/K)^2} \\ \eta_2 = 1 \end{cases} \tag{21}$$

where K serves as a threshold, and finally, the $\mathbf{D}$ takes the following form,

$$\mathbf{D} = (\mathbf{e}_1\mathbf{e}_2)\begin{pmatrix} \eta_1 & 0 \\ 0 & \eta_2 \end{pmatrix}(\mathbf{e}_1\mathbf{e}_2)^{\mathrm{T}} \quad (22)$$

From Formula (22) we can get some information: (I) when $|\nabla I| \to \infty, \quad \eta_1 \to 0$, the HBGVF snake will give up continuing to spread along the image gradient direction on the boundary and spread on the boundary. Therefore, the noise on the image edge can be eliminated while the image edge is preserved; (II) when $|\nabla I| \to 0, \quad \eta_1 \to 1 = \eta_2$, that is, in the homogeneous region, the diffusion is isotropic, which is beneficial to the elimination of noise.

Through the above methods, the HBGVF model will have anisotropy, so it can accurately converge all kinds of concave surfaces and retain the weak edge of the image. The methods in reference [61] are used for reference to solve the model proposed in this paper, and the source code in Matlab is available to the public upon request. We note that, since the Hessian matrix and the diffusion matrix should be calculated, the computation time of the proposed HBGVF model is longer than the original GVF model.

4. Corresponding Comparative Experiments

In the experimental part, we show the important characteristics of the HBGVF model by comparing the HBGVF model with GVF [25], GGVF [57], VEF [58], NGVF [38], and CN-GGVF [41]. We normalized the image intensity to the [0,1] range, set and α, β to 0.1, and set the time step for all snakes with the size of $\tau = 0.5$. For an image of size $M \cdot N$, the iteration for the calculation of all GVF-like models is $\sqrt{M \cdot N}$, and the time step is 1(less than $1/(4\mu)$). In order to get a large capture range, μ is 0.2 for GVF, NGVF, and HBGVF, k is 0.5 for the GGVF and CN-GGVF, the region D for the VEF model is of size $M \cdot N$, k for the HBGVF is 0.1, unless otherwise stated.

4.1. Common Concerns for the GVF-Like Snakes

The GVF model was originally proposed to overcome the shortcomings of traditional gradient-based external force, such as narrow capture range and poor convergence on concave surfaces. Through the following experiments, we will prove some excellent characteristics of HBGVF snake compared with the GVF snake, such as large capture range, accurate convergence to concave, and insensitive to image initialization. Figure 1 shows the convergence results of HBGVF snake on a oom image, U-shaped image, and main body contour respectively. The gray dotted line is the initial contour, and the red solid line is the convergence result. It can be observed from the figure that the HBGVF snake converges to the U-shaped concave surface and is automatically connected to the subject contour. It can be seen from the initialization results in the figure that the HBGVF snake has the advantages of being insensitive to the initial contour and large capture range.

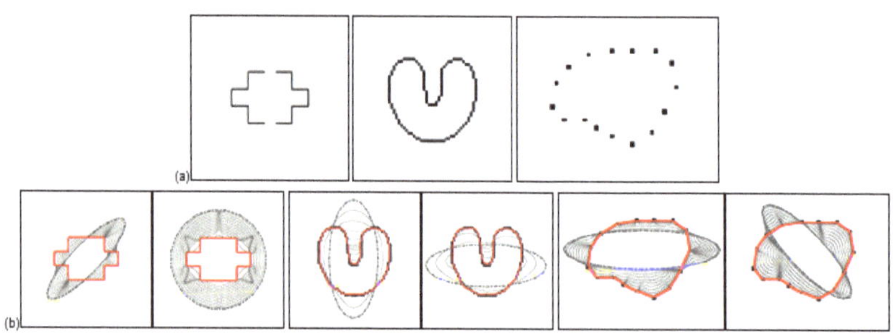

Figure 1. (**a**) Test images: room image, U image, and subject contour. (**b**) Convergence results with different initializations and evolutions of the HBGVF snakes.

4.2. Convergence to Concavities

It can be seen from Figure 1 that the HBGVF snake performs well on converging to the U-shape image. Next, in order to better test the advantages of the HBGVF snake, we use the other three images with different concave surfaces to compare with other methods similar to GVF. Figure 2 presents the convergence results of the corresponding approaches. One can see that just the HBGVF and GGVF snakes can converge on the three images, the reason behind this observation is that the HBGVF model takes into account the image structure that was characterized by the Hessian matrix, and the GGVF model emphasizes the image structure by paying more attention to the edges by using two varying weighting functions. However, the CN-GGVF model also adopts the two varying weighting functions that are identical to those in the GGVF model, the CN-GGVF snake cannot converge to the various concavities at all, the reason is that the component normalization operation changes the direction of the vector field. Taking the heart image as an example, Figure 2i presents the associated GGVF vector field around the entrance of the concavity, one can see that the vector field in the blue circle is approximately horizontal, since the vector left to the blue circle is downward, it drives the snake contour into the concavity. Figure 2h presents the associated CN-GGVF vector field, where the vectors in black and red are these before and after component normalization, respectively, it is clear that the CN-GGVF field in the yellow circle before component normalization (in black) is similar to the GGVF vector, however, due to the component normalization, the CN-GGVF vector (in red) is upward, and pushes the snake contour out of the concavity. As a result, the CN-GGVF snake stops at the upper half of the concavity. This example tells us that component normalization is not always beneficial to the evolution of the snake contour. The concavities in the man and cat images are semi-close, and the CN-GGVF snake is also not good at converging to these concavities. Therefore, the improper use of the weighting function may cause the opposite effect. Of course, the appropriate use can greatly improve the accuracy of the model, such as the application in [62]. The GVF and VEF snakes just failed in one case, and we will see later that the shortcoming of the VEF snake is that it does not perform well when preserving weak edges. The NGVF snake just works well on the man concavities, and since the limited capture range, the initial contour for the cat image is very close to the cat at the left-bottom corner.

4.3. Weak Edge Preserving

Figure 3a is an example of testing the ability of the HBGVF model to retain the weak edge of the image. The outer ring of the image is seriously blurred in the upper right corner. Refer to the edge diagram in Figure 3b. It can be seen that the contour of the snake is easily attracted to the inner ring of the strong edge. Since it is a pair of contradictions to enlarge capture range and to preserve weak edge simultaneously, the regularization parameters are tuned to μ is 0.1 for GVF, NGVF, and HBGVF, k is 0.01 for the GGVF and CN-GGVF, the size of region D for VEF model is just one twenty-fifth of that of the image, k for HBGVF is 0.01. One can see that the HBGVF snakes can preserve the weak edge well although the diffusion parameter μ is identical to those of the GVF and NGVF; the reason behind this observation is that the HBGVF model takes into account the image structure. Although the kernel size for the VEF is very small and the initial contour for the VEF snake is close to the object, the snake contour yet collapsed at the weak edge. The CN-GGVF and GGVF snakes also stop at the weak edge and the convergence results are almost identical due to the similar diffusion mechanism, when compared with that of the HBGVF snake, the result of the HBGVF snake is smoother, this observation implies the HBGVF field is more regular.

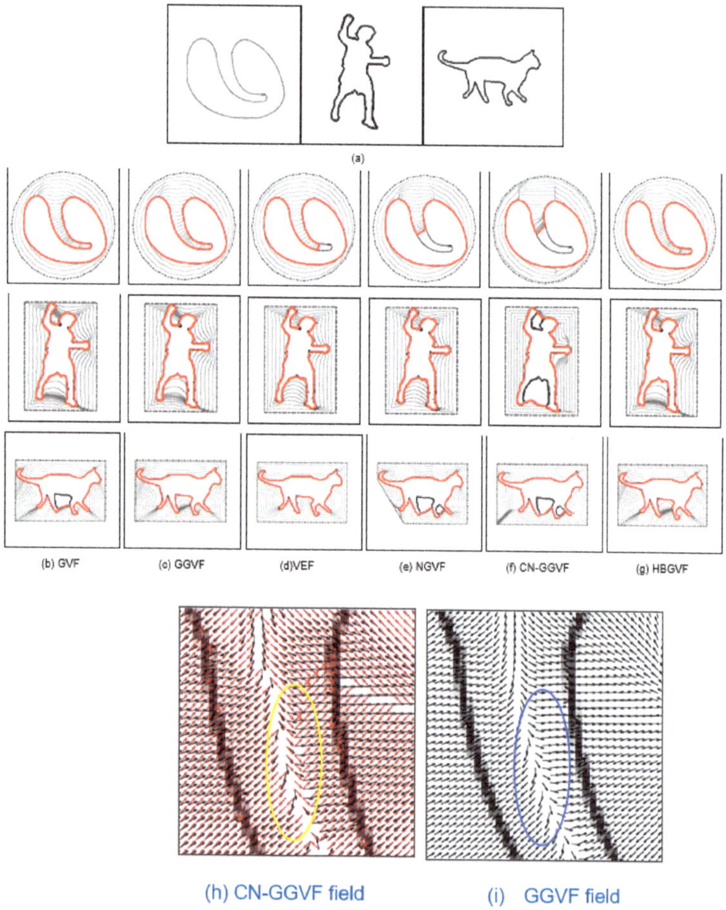

Figure 2. Convergence to concavities. (**a**) Test images: heart image, man image, and cat image. Evolution and convergence results of the (**b**) GVF snake, (**c**) GGVF snake, (**d**) VEF snake, (**e**) NGVF snake, (**f**) CN-GGVF snake, and (**g**) HBGVF snake. (**h**) The CN-GGVF field, the vectors in black and red are these before and after component-normalization, respectively, (**i**) the GGVF field.

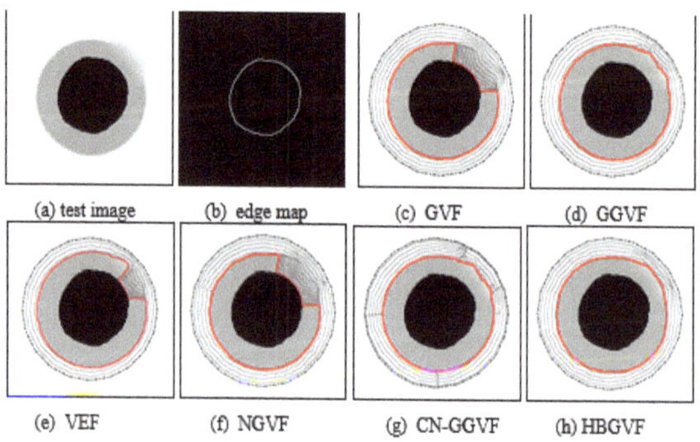

Figure 3. (**a**) Test image, (**b**) edge map. Convergence results of each model: (**c**) the GVF snake, (**d**) the GGVF snake, (**e**) the VEF snake, (**f**) the NGVF snake, (**g**) the CN-GGVF snake, and (**h**) the HBGVF snake.

4.4. Test Results of HBGVF Model on Real Images

In order to further highlight the comprehensive performance of the HBGVF snake, we used several real images for comparison. Figure 4 presents a gear image, where there are more than ten semi-close concavities with order number.

The parameter k is 0.2 for the GGVF and 0.3 for the CN-GGVF in order to get a balance between entering the concavities and preserving a weak edge, the parameters for other models are identical to those in Figures 1 and 2. One can see that the GVF snake converges to the concavities from #0 to #9, although it collapses at the two teeth around concavity 5. The GGVF snake converges to the concavities from #0 to #8, and it seems that the GGVF snake is good at preserving a weak edge, in fact, one can see that there is contour entanglement from the right part in Figure 4d, which is a zoomed-in version of the blue rectangle in the left part. The VEF snake suffers from weak edge leakage, and collapses at most of the teeth, see Figure 4e. Figure 4f is the result of the NGVF snake treatment, which manifests that the NGVF snake is not good at concave convergence, and this observation agrees with that in Figure 2. The CN-GGVF snake performs similarly to the NGVF snake, see Figure 4g. Since the HBGVF takes into account the image structure, the HBGVF snake converges to the concavities from #0 to #12 except the 11th one. However, from the zoomed-in part of the blue rectangle in the left part, HBGVF snakes also performed poorly, as shown in the right part of Figure 4h; in fact, the performance in this example can be enhanced by decreasing the parameter k in HBGVF.

Figure 5 presents a second real image, a flying eagle, and the feathers on the wings are difficult for the active contour to extract. In order to get a balance between extracting the feathers on the wings and enlarging the capture range, the regularization parameter μ is 0.05 for GVF, NGVF, and HBGVF, k is 0.05 for the GGVF and CN-GGVF, the size of region D for the VEF model is just one sixty-forth of that of the image, k for HBGVF is 0.01. The white dash-point lines are the initial contour and the red solid lines are the convergent results. As can be seen from Figure 5a, the GVF snake works well except for the feathers on the right wing. Figure 5b shows that the GGVF snake yields good results in extracting the feathers on both wings, however, it is trapped in local minimum behind the tail. The result of the VEF snake is reported in Figure 5c, and it is obvious from the results that the snake contour is trapped in a local minimum and also fails on extracting the feathers. The NGVF and CN-GGVF snakes are also trapped in a local minimum, see Figure 5d,e, respectively, and the CN-GGVF snake cannot enter the concavities formed by the feathers. On the contrary, Figure 5f shows that the HBGVF snake works well on extracting the feathers and is not trapped in a local minimum, which manifests that the HBGVF field is regular.

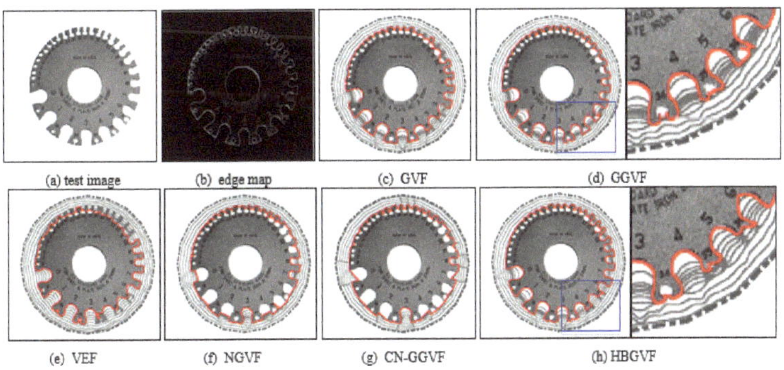

Figure 4. (**a**) Original test metal gauge image; (**b**) edge map; the convergence results of each model: (**c**) the GVF snake, (**d**) the GGVF snake, (**e**) the VEF snake, (**f**) the NGVF snake, (**g**) the CN-GGVF snake, and (**h**) the HBGVF snake.

Figure 6 presents a medical image, and for the weak edge shown in the white box in Figure 6a, the snake contour is prone to leakage here, the intensity inhomogeneity is also a difficulty. In order to achieve a balance between maintaining the weak edge and overcoming inhomogeneity, the regularization parameter μ is 0.02 for GVF, 0.03 for NGVF and HBGVF, k is 0.03 for the GGVF, and 0.07 for the CN-GGVF, the size of region D for VEF model is just 1/144 of that of the image, k for HBGVF is 0.01. One can see that there is weak-edge leakage and local minimum trap simultaneously for the GVF, VEF, and NGVF snakes. The GGVF, CN-GGVF, and HBGVF snakes yield similar results, where there is no weak-edge leakage or local minimum trap. It is clear that the μ for HBGVF and NGVF are identical, and even larger than that for the GVF, however, the HBGVF snake preserves a weak edge well, the reason behind this observation is that the HBGVF model takes into account the image structure. Figure 7 shows more results of the HBGVF snake on real images, the initial contours are dash-point lines and the convergence results are the solid red lines. The first row presents flowers and leaves and the HBGVF snake extracts the objects accurately, the second row shows three eagles and the difficulty for the HBGVF snake is similar to that in Figure 5, the HBGVF snake also yields satisfactory results. There are three medical images in the third row, and in each panel, the image on the left is the original image with initial contour, from which one can see the blurred and weak boundaries of the objects. The display result on the right shows that the HBGVF snake can satisfactorily delineate the object boundaries.

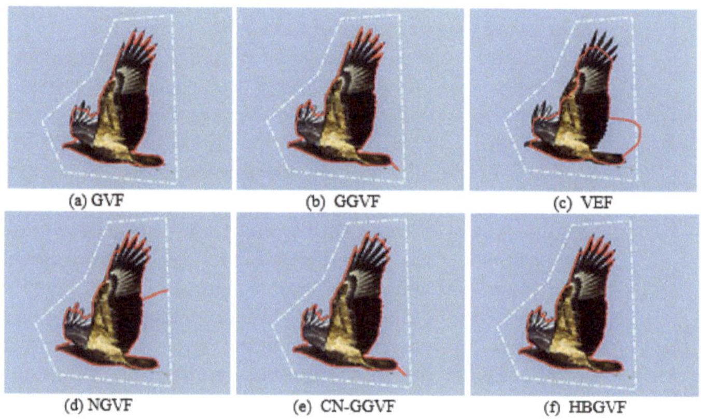

Figure 5. The convergence results of each model: (**a**) the GVF snake, (**b**) the GGVF snake, (**c**) the VEF snake, (**d**) the NGVF snake, (**e**) the CN-GGVF snake, and (**f**) the HBGVF snake. In order to get a balance between preserving the feathers on the wings and enlarging the capture range, the regularization parameter μ is 0.05 for GVF, NGVF, and HBGVF, k is 0.05 for the GGVF and CN-GGVF, the size of region D for VEF model is just one sixty-forth of that of the image, k for HBGVF is 0.01.

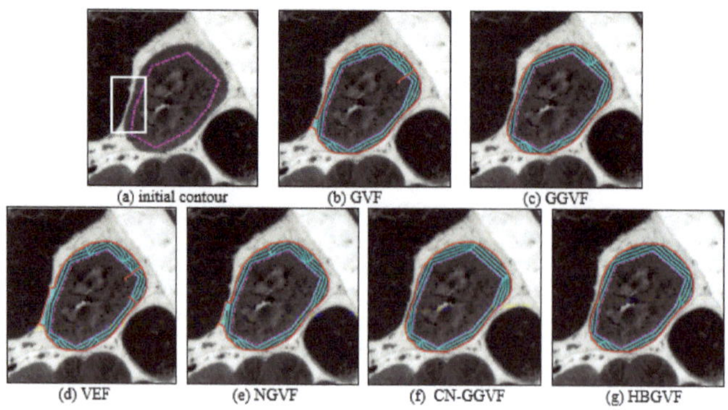

Figure 6. (**a**) Test medical image; the convergence results of each model: (**b**) the GVF snake, (**c**) the GGVF snake, (**d**) the VEF snake, (**e**) the NGVF snake, (**f**) the CN-GGVF snake, and (**g**) the HBGVF snake.

Figure 7. More examples of the convergence results of the HBGVF snake.

5. Conclusions

To sum up, the smoothness constraint formula is expressed in the form of a matrix, and the image structure represented by the Hessian matrix is introduced into the GVF model. This GVF model based on the Hessian matrix is abbreviated as HBGVF. Through the above theoretical analysis and experimental comparison, it can be proved that compared with other GVF-based models, the HBGVF snake has many advantages, such as excellent convergence on various concave surfaces, retaining weak edges, and so on. The above experiments include synthetic images and real images in real life. These experiments have proved the excellent characteristics of the HBGVF model. The proposed HBGVF model can also be employed for other applications such as those in [63–72], and this is our next goal.

Author Contributions: Conceptualization, K.C. and Y.W.; methodology, Y.W., K.C. and Q.Q.; software, Y.W., K.C., Q.Q., W.Q. and Q.D.; validation, Q.Q., W.Q. and Q.D.; formal analysis, K.C., Q.Q., W.Q. and Q.D.; investigation, Q.Q., W.Q. and Q.D.; resources, Q.Q. and W.Q.; data curation, Q.Q. and Q.D.; writing—original draft preparation, Q.Q. and Y.W.; writing—review and editing, Q.Q. and Y.W.; visualization, Q.Q. and Q.D.; supervision, K.C. and Y.W.; project administration, K.C., Y.W. and Q.Q. All authors have read and agreed to the published version of the manuscript.

Funding: This work was supported in part by the National Science Foundation Program of China (NSFC) (grant number: 61976241), and the International Science and technology cooperation plan project of Zhenjiang (grant number: GJ2021008).

Data Availability Statement: Dateset is available at request.

Conflicts of Interest: We declare that we have no financial and personal relationships with other people or organizations that can inappropriately influence our work, there is no professional or other personal interest of any nature or kind in any product, service and/or company that could be construed as influencing the position presented in, or the review of, the manuscript entitled.We hereby declare that the collection, analysis and interpretation of the data in this article and the writing of the report were done by all authors of this article.

References

1. Sahoo, P.K.; Soltani, S.; Wong, A.K.C.; Chen, Y.C. A Survey of Thresholding Techniques. *Comput. Vis. Graph. Image Process.* **1998**, *41*, 142–149. [CrossRef]
2. Cai, W.L.; Chen, S.C.; Zhang, D.Q. Fast and robust fuzzy c-means clustering algorithms incorporating local information for image segmentation. *Pattern Recognit.* **2007**, *40*, 825–838. [CrossRef]
3. Kass, M.; Witkin, A.; Terzopoulos, D. Snakes: Active contour models. *Int. J. Comput. Vis.* **1988**, *1*, 321–331. [CrossRef]
4. Shih, F.Y.; Cheng, S.X. Automatic seeded region growing for color image segmentation. *Image Vis. Comput.* **2005**, *23*, 877–886. [CrossRef]

5. Yu, S.; Lu, Y.; Molloy, D. A Dynamic-Shape-Prior Guided Snake Model With Application in Visually Tracking Dense Cell Populations. *IEEE Trans. Image Process.* **2019**, *8*, 1513–1527. [CrossRef]
6. Zhou, S.; Li, B.; Wang, Y.; Wen, T.; Li, N. The Line- and Block-like Structures Extraction via Ingenious Snake. *Pattern Recognit. Lett.* **2018**, *112*, 324–331. [CrossRef]
7. Nakhmani, A.; Tannenbaum, A. Self-Crossing Detection and Location for Parametric Active Contours. *IEEE Trans. Image Process.* **2012**, *21*, 3150–3156. [CrossRef]
8. Zhao, S.; Li, G.; Zhang, W.; Gu, J. Automatical Intima-media Border Segmentation on Ultrasound Image Sequences using a Kalman filter snake. *IEEE Access* **2018**, *6*, 40804–40810. [CrossRef]
9. Manno-Kovacs, A. Direction Selective Contour Detection for Salient Objects. *IEEE Trans. Circuits Syst. Video Technol.* **2018**, *29*, 375–389. [CrossRef]
10. Paragios, N.; Deriche, R. Geodesic Active Contours and Level Sets for the Detection and Tracking of Moving Objects. *IEEE Trans. Pattern Anal. Mach. Intell.* **2000**, *22*, 266–280. [CrossRef]
11. Zhu, S.C.; Yuille, A. Region Competition: Unifying Snakes, Region Growing, and Bayes/MDL for Multi-band Image Segmentation. *IEEE Trans. Pattern Anal. Mach. Intell.* **1996**, *18*, 884–900.
12. Chan, T.F.; Vese, L.A. Active contours without edges. *IEEE Trans. Image Process.* **2001**, *10*, 266–277. [CrossRef]
13. Brox, T.; Cremers, D. On Local Region Models and a Statistical Interpretation of the Piecewise Smooth Mumford-Shah Functional. *Int. J. Comput. Vis.* **2009**, *84*, 184–193. [CrossRef]
14. Adam, A.; Kimmel, R.; Rivlin, E. On Scene Segmentation and Histograms-Based Curve Evolution. *IEEE Trans. Pattern Anal. Mach. Intell.* **2009**, *31*, 1708–1714. [CrossRef]
15. Ni, K.; Bresson, X.; Chan, T.; Esedoglu, S. Local Histogram Based Segmentation Using the Wasserstein Distance. *Int. J. Comput. Vis.* **2009**, *84*, 97–111. [CrossRef]
16. Zhao, W.; Xu, X.; Zhu, Y.; Xu, F. Active contour model based on local and global Gaussian fitting energy for medical image segmentation. *Optik* **2018**, *158*, 1160–1169. [CrossRef]
17. Ge, Q.; Li, C.; Shao, W.; Li, H. A hybrid active contour model with structured feature for image segmentation. *Signal Process.* **2015**, *108*, 147–158. [CrossRef]
18. Wang, H.; Huang, T.; Du, Y. An adaptive weighting parameter selection for improved integrated active contour model. *Optik* **2015**, *126*, 5331–5335. [CrossRef]
19. Li, C.; Kao, C.Y.; Gore, J.C.; Ding, Z. Minimization of Region-Scalable Fitting Energy for Image Segmentation. *IEEE Trans. Image Process.* **2008**, *17*, 1940–1949.
20. Darolti, C.; Mertins, A.; Bodensteiner, C.; Hofmann, U.G. Local region descriptors for active contours evolution. *IEEE Trans. Image Process.* **2008**, *17*, 2275–2288. [CrossRef]
21. Zhang, K.; Song, H.; Zhang, L. Active contours driven by local image fitting energy. *Pattern Recognit.* **2010**, *43*, 1199–1206. [CrossRef]
22. Estellers, V.; Zosso, D.; Bresson, X.; Thiran, J.P. Harmonic active contours. *IEEE Trans. Image Process.* **2014**, *23*, 69–82. [CrossRef]
23. Gao, Y.; Bouix, S.; Shenton, M.; Tannenbaum, A. Sparse Texture Active Contour. *IEEE Trans. Image Process.* **2013**, *22*, 3866–3878. [CrossRef]
24. Caselles, V.; Kimmel, R.; Sapiro, G. Geodesic Active Contours. *Int. J. Comput. Vis.* **1997**, *22*, 61–79. [CrossRef]
25. Xu, C.; Prince, J. Snakes, shapes, and gradient vector flow. *IEEE Trans. Image Process* **1998**, *7*, 359–369.
26. Li, B.; Acton, S. Active contour external force using vector field convolution for image segmentation. *IEEE Trans. Image Process.* **2007**, *16*, 2096–2106. [CrossRef]
27. Sum, K.W.; Cheung, P.Y.S. Boundary vector field for parametric active contours. *Pattern Recognit.* **2007**, *40*, 1635–1645. [CrossRef]
28. Ren, D.; Zuo, W.; Zhao, X.; Lin, Z.; Zhang, D. Fast gradient vector flow computation based on augmented Lagrangian method. *Pattern Recognit. Lett.* **2013**, *34*, 219–225. [CrossRef]
29. Han, X.; Xu, C.; Prince, J. Fast numerical scheme for gradient vector flow computation using a multigrid method. *IET Image Process.* **2007**, *1*, 48–55. [CrossRef]
30. Boukerroui, D. Efficient numerical schemes for gradient vector flow. *Pattern Recognit.* **2012**, *45*, 626–636. [CrossRef]
31. Zhao, F.; Zhao, J.; Zhao, W.; Qu, F. Guide filter-based gradient vector flow module for infrared image segmentation. *Appl. Opt.* **2015**, *54*, 9809–9817. [CrossRef] [PubMed]
32. Zhu, S.; Bu, X.; Zhou, Q. A Novel Edge Preserving Active Contour Model Using Guided Filter and Harmonic Surface Function for Infrared Image Segmentation. *IEEE Access* **2018**, *6*, 5493–5510. [CrossRef]
33. Cheng, J.; Foo, S.W. Dynamic directional gradient vector flow for snakes. *IEEE Trans. Image Process.* **2006**, *15*, 1563–1571. [CrossRef] [PubMed]
34. Ray, N.; Acton, S.T.; Ley, K. Tracking leukocytes in vivo with shape and size constrained active contours. *IEEE Trans. Med. Imaging* **2002**, *21*, 1222–1235. [CrossRef] [PubMed]
35. Wang, Y.; Jia, Y.; Liu, L. Harmonic gradient vector flow external force for snake model. *Electron. Lett.* **2008**, *44*, 105–106. [CrossRef]
36. Wu, Y.; Wang, Y.; Jia, Y. Adaptive diffusion flow active contours for image segmentation. *Comput. Vis. Image Underst.* **2013**, *117*, 1421–1435. [CrossRef]
37. Jaouen, V.; González, P.; Stute, S. Variational Segmentation of Vector-Valued Images With Gradient Vector Flow. *IEEE Trans. Image Process.* **2014**, *3*, 4773–4785. [CrossRef]

38. Ning, J.; Wu, C.; Liu, S.; Yang, S. NGVF: An improved external force field for active contour model. *Pattern Recognit. Lett.* **2007**, *28*, 58–63.
39. Li, C.; Liu, J.; Fox, M.D. Segmentation of external force field for automatic initialization and splitting of snakes. *Pattern Recognit.* **2005**, *38*, 1947–1960. [CrossRef]
40. Ray, N.; Acton, S.T. Motion gradient vector flow: An external force for tracking rolling leukocytes with shape and size constrained active contours. *IEEE Trans. Med. Imaging* **2004**, *23*, 1466–1478. [CrossRef]
41. Qin, L.; Zhu, C.; Zhao, Y.; Bai, H.; Tian, H. Generalized Gradient Vector Flow for Snakes: New Observations, Analysis, and Improvement. *IEEE Trans. Circuits Syst. Video Technol.* **2013**, *23*, 883–897. [CrossRef]
42. Kirimasthong, K.; Rodtook, A.; Lohitvisate, W.; Makhanov, S.S. Automatic initialization of active contours in ultrasound images of breast cancer. *Pattern Anal. Appl.* **2018**, *21*, 491–500. [CrossRef]
43. Rodtook, A.; Kirimasthong, K.; Lohitvisate, W. Automatic Initialization of Active Contours and Level Set Method in Ultrasound Images of Breast Abnormalities. *Pattern Recognit.* **2018**, *79*, 172–182. [CrossRef]
44. Jaouen, V.; Bert, J.; Boussion, N.; Fayad, H.; Hatt, M.; Visvikis, D. Image enhancement with PDEs and nonconservative advection flow fields. *IEEE Trans. Image Process.* **2019**, *28*, 3075–3088. [CrossRef]
45. Minaee, S.; Boykov, Y.; Porikli, F.; Plaza, A.; Kehtarnavaz, N.; Terzopoulos, D. Image Segmentation Using Deep Learning: A Survey. *IEEE Trans. Pattern Anal. Mach. Intell.* **2022**, *44*, 3523–3542. [CrossRef]
46. Wang, W.; Wu, Y.W.Y.; Li, S.; Chen, B. Quantification of Full Left Ventricular Metrics via Deep Regression Learning with Contour-Guidance. *IEEE Access* **2019**, *7*, 47918–47928. [CrossRef]
47. Zhang, T.; Zhang, X. A Full-Level Context Squeeze-and-Excitation ROI Extractor for SAR Ship Instance Segmentation. *IEEE Geosci. Remote Sens. Lett.* **2022**, *19*, 1–5. [CrossRef]
48. Shen, W.; Xu, W.; Sun, Z.; Ma, J.; Ma, X.; Zhou, S.; Guo, S.; Wang, Y. Automatic Segmentation of the Femur and Tibia Bones from X-ray Images Based on Pure Dilated Residual U-Net. *Inverse Probl. Imaging* **2021**, *15*, 1333–1346. [CrossRef]
49. Zhang, H.; Zhang, W.; Shen, W.; Li, N.; Chen, Y.; Li, S.; Chen, B.; Guo, S.; Wang, Y. Automatic segmentation of the left ventricle from MR images based on nested U-Net with dense block. *Biomed. Signal Process. Control.* **2021**, *68*, 102684. [CrossRef]
50. Zhang, T.; Zhang, X. ShipDeNet-20: An Only 20 Convolution Layers and <1-MB Lightweight SAR Ship Detector. *IEEE Geosci. Remote Sens. Lett.* **2021**, *18*, 1234–1238.
51. Zhang, T.; Zhang, X.; Shi, J.; Wei, S.; Wang, J.; Li, J.; Su, H.; Zhou, Y. Balance Scene Learning Mechanism for Offshore and Inshore Ship Detection in SAR Images. *IEEE Geosci. Remote Sens. Lett.* **2020**, *19*, 1–5. [CrossRef]
52. Zhang, T.; Zhang, X. Squeeze-and-Excitation Laplacian Pyramid Network With Dual-Polarization Feature Fusion for Ship Classification in SAR Images. *IEEE Geosci. Remote Sens. Lett.* **2021**, *19*, 1–5. [CrossRef]
53. Zhang, T.; Zhang, X.; Ke, X.; Liu, C.; Xu, X.; Zhan, X.; Wang, C.; Ahmad, I.; Zhou, Y.; Pan, D.; et al. HOG-ShipCLSNet: A Novel Deep Learning Network With HOG Feature Fusion for SAR Ship Classification. *IEEE Trans. Geosci. Remote Sens.* **2021**, *60*, 1–22. [CrossRef]
54. Carmona, R.; Zhong, S. Adaptive Smoothing Respecting Feature Directions. *IEEE Trans. Image Process.* **1998**, *7*, 353–358. [CrossRef]
55. Wang, Y.; Chen, W.; Yu, T.; Zhang, Y. Hessian based image structure adaptive gradient vector flow for parametric active contours. In Proceedings of the 2010 IEEE International Conference on Image Processing, Hong Kong, China, 26–29 September 2010; pp. 649–652.
56. Cheng, K.; Xiao, T.; Chen, Q.; Wang, Y. Image segmentation using active contours with modified convolutional virtual electric field external force with an edge-stopping function. *PLoS ONE* **2020**, *15*, e0230581. [CrossRef]
57. Xu, C.; Prince, J.L. Generalized gradient vector flow external forces for active contours. *Signal Process.* **1998**, *71*, 131–139. [CrossRef]
58. Park, H.K.; Chung, M.J. External force of snake: Virtual electric field. *Electron. Lett.* **2002**, *38*, 1500–1502. [CrossRef]
59. You, Y.; Xu, W.; Tannenbaum, A.; Kaveh, M. Behavioral analysis of anisotropic diffusion in image processing. *IEEE Trans. Image Process.* **1996**, *5*, 1539–1552.
60. Caselles, V.; Morel, J.; Sbert, C. An axiomatic approach to image interpolation. *IEEE Trans. Image Process.* **1998**, *7*, 376–386. [CrossRef]
61. Weickert, J. Coherence-enhancing diffusion filtering. *Int. J. Comput. Vis.* **1999**, *31*, 111–127. [CrossRef]
62. Yan, M.; Li, S.; Chan, C.A.; Shen, Y.; Yu, Y. Mobility prediction using a weighted Markov model based on mobile user classification. *Sensors* **2021**, *21*, 1740. [CrossRef]
63. Yu, H.; Chua, C. GVF-based anisotropic diffusion models. *IEEE Trans. Image Process.* **2006**, *15*, 1517–1524.
64. Hassouna, M.; Farag, A. Variational curve skeletons using gradient vector flow. *IEEE Trans. Pattern Anal. Mach. Intell.* **2009**, *31*, 2257–2274. [CrossRef]
65. Prasad, V.; Yegnanarayana, B. Finding axes of symmetry from potential fields. *IEEE Trans. Image Process.* **2004**, *13*, 1559–1566. [CrossRef]
66. Battiato, S.; Farinella, G.M.; Puglisi, G. Saliency-Based Selection of Gradient Vector Flow Paths for Content Aware Image Resizing. *IEEE Trans. Image Process.* **2014**, *23*, 2081–2095. [CrossRef]
67. Shivakumara, P.; Phan, T.; Lu, S.; Tan, C.L. Gradient vector flow and grouping based method for arbitrarily-oriented scene text detection in video images. *IEEE Trans. Circuits Syst. Video Technol.* **2013**, *23*, 1729–1739. [CrossRef]
68. Wang, Y.; Jia, Y.; Wu, Y. Segmentation of the left ventricle in cardiac cine MRI using a shape constrained snake model. *Comput. Vis. Image Underst.* **2013**, *117*, 990–1003.

69. Zhu, S.; Gao, J.; Li, Z. Video object tracking based on improved gradient vector flow snake and intra-frame centroids tracking method. *Comput. Electr. Eng.* **2014**, *40*, 174–185. [CrossRef]
70. Li, Q.; Deng, T.; Xie, W. Active contours driven by divergence of gradient vector flow. *Signal Process.* **2016**, *120*, 185–199. [CrossRef]
71. Abdullah, M.; Dlay, S.; Woo, W.; Chambers, J. Robust Iris Segmentation Method Based on a New Active Contour Force with a Noncircular Normalization. *IEEE Trans. Syst. Man Cybern. Syst.* **2017**, *47*, 3128–3142. [CrossRef]
72. Miri, M.S.; Robles, V.A.; Abràmoff, M.D.; Kwon, Y. H.; Garvin, M.K. Incorporation of gradient vector flow field in a multimodal graph-theoretic approach for segmenting the internal limiting membrane from glaucomatous optic nerve head-centered SD-OCT volumes. *Comput. Med. Imaging Graph.* **2017**, *55*, 87–94. [CrossRef]

sensors

MDPI

Review

A Systematic Review of Mobile Phone Data in Crime Applications: A Coherent Taxonomy Based on Data Types and Analysis Perspectives, Challenges, and Future Research Directions

Mohammed Okmi [1,2], Lip Yee Por [1,*], Tan Fong Ang [1], Ward Al-Hussein [1] and Chin Soon Ku [3,*]

1 Faculty of Computer Science and Information Technology, Universiti Malaya, Kuala Lumpur 50603, Malaysia; wva180034@siswa.um.edu.my (W.A.-H.)
2 Department of Information Technology and Security, Jazan University, Jazan 45142, Saudi Arabia
3 Department of Computer Science, Universiti Tunku Abdul Rahman, Kampar 31900, Malaysia
* Correspondence: porlip@um.edu.my (L.Y.P.); kucs@utar.edu.my (C.S.K.)

Abstract: Digital technologies have recently become more advanced, allowing for the development of social networking sites and applications. Despite these advancements, phone calls and text messages still make up the largest proportion of mobile data usage. It is possible to study human communication behaviors and mobility patterns using the useful information that mobile phone data provide. Specifically, the digital traces left by the large number of mobile devices provide important information that facilitates a deeper understanding of human behavior and mobility configurations for researchers in various fields, such as criminology, urban sensing, transportation planning, and healthcare. Mobile phone data record significant spatiotemporal (i.e., geospatial and time-related data) and communication (i.e., call) information. These can be used to achieve different research objectives and form the basis of various practical applications, including human mobility models based on spatiotemporal interactions, real-time identification of criminal activities, inference of friendship interactions, and density distribution estimation. The present research primarily reviews studies that have employed mobile phone data to investigate, assess, and predict human communication and mobility patterns in the context of crime prevention. These investigations have sought, for example, to detect suspicious activities, identify criminal networks, and predict crime, as well as understand human communication and mobility patterns in urban sensing applications. To achieve this, a systematic literature review was conducted on crime research studies that were published between 2014 and 2022 and listed in eight electronic databases. In this review, we evaluated the most advanced methods and techniques used in recent criminology applications based on mobile phone data and the benefits of using this information to predict crime and detect suspected criminals. The results of this literature review contribute to improving the existing understanding of where and how populations live and socialize and how to classify individuals based on their mobility patterns. The results show extraordinary growth in studies that utilized mobile phone data to study human mobility and movement patterns compared to studies that used the data to infer communication behaviors. This observation can be attributed to privacy concerns related to acquiring call detail records (CDRs). Additionally, most of the studies used census and survey data for data validation. The results show that social network analysis tools and techniques have been widely employed to detect criminal networks and urban communities. In addition, correlation analysis has been used to investigate spatial–temporal patterns of crime, and ambient population measures have a significant impact on crime rates.

Keywords: mobile phone data; call detail records (CDRs); urban human mobility patterns; human communication behavior; urban dynamics; criminal networks; social networks; urban crime prediction; urban sensing; systematic literature review

Citation: Okmi, M.; Por, L.Y.; Ang, T.F.; Al-Hussein, W.; Ku, C.S. A Systematic Review of Mobile Phone Data in Crime Applications: A Coherent Taxonomy Based on Data Types and Analysis Perspectives, Challenges, and Future Research Directions. *Sensors* **2023**, *23*, 4350. https://doi.org/10.3390/s23094350

Academic Editors: Chien Aun Chan, Ming Yan and Chunguo Li

Received: 7 March 2023
Revised: 23 April 2023
Accepted: 24 April 2023
Published: 28 April 2023

1. Introduction

Even though digital technologies have become more advanced in recent times and allowed for the development of social networking sites, computer software applications, and emails, research findings have shown that phone calls and text messages still represent the greatest proportions of mobile data usage. Statista, a German company specializing in market and consumer data, estimated that the global number of mobile subscriptions would exceed 8 billion as of 2020. Whenever mobile phone users initiate an activity (e.g., calling, texting, and connecting to the Internet), this action is recorded by the mobile network operator. The information saved includes such details as each call's duration, timestamp, and location at which the interaction started. When considering the aforementioned points along with the mobile subscription figure, it could be concluded that tremendous amounts of mobile phone data are generated every day. Maintaining such rich data, which comprise the details of individuals' behaviors and activities, is advantageous in the sense that human communication behavior and mobility patterns can be studied at a low cost. The accessibility of this data has been reflected in various studies and disciplines over the years in terms of the ubiquitous use of mobile phone data [1]. For instance, publications have focused on urban sensing, safety, health, emergencies, transportation planning, and criminology.

Mobile phone data are log files collected from the users by mobile network operators during the service provision process. They contain all of the interactions that the user has initiated with the network, whether actively (e.g., when making a phone call, sending a text message, or accessing the internet) or passively (e.g., when switching the phone on or off, receiving a signal from the mobile network, or changing the type of connection). They also contain the details of each of these interactions, such as the phone numbers of the caller and receiver, the timestamp, and the duration and location of the interaction (i.e., the cell tower ID). Every telecommunications service provider (TSP) records users' interactions with the cellular network whenever they engage in an activity on their mobile devices; here, the data are recorded in the service provider's database.

Mobile phone data have proven to be the most prominent form of data, helping us understand the microscopic details of social networks, human mobility, and human behavioral patterns [1]. For instance, they have enabled us to understand how members of a target population (i.e., users) change their communication (e.g., calling) behavior and mobility patterns following an emergency event, such as a terrorist attack [2]. Unsurprisingly, mobile phone data have become a topic of discussion in various studies and the centerpiece of many real-world applications [3,4]. In such contexts, they are used to infer social ties and interactions among individuals [5], estimate daily population dynamics [6], map tourist travel behaviors [7,8], identify suspects [9], detect criminal networks based on communication behaviors [10,11], detect criminals based on mobility patterns [12], predict crimes and criminal behaviors [13,14], understand human mobility patterns in urban environments [15], and estimate human mobility and behavior under emergency events, such as natural disasters [16], migration streams [17], reprisals of organized criminals or militia, and the spread of infectious diseases.

When considering the applications described above, it becomes apparent that mobile phone data are among the most reliable sources of information that could help sense and record human activities. Moreover, they have a great potential for being used to reveal many aspects of human mobility patterns and communication behaviors. Therefore, using these data can help us to accurately and effectively predict and understand individual friendship relationships, criminal relationships, social ties, and interactions based on calling behaviors, as well as humans' way of living, which has always been inextricably linked [18] with movement patterns.

In the last decade, mobile phone data have been used as sensors for detecting human mobility and communication behaviors. Due to the wide use of smartphones and the fast growth of telecommunication networks, a large quantity of data on how people move and behave across space and time has been recorded. The digital traces left by smartphones provide valuable, real-time information about various human activities.

For example, mobile phone data have been used to indicate the presence or absence of humans during certain times and at specific locations [19]. Thus, in criminal investigations, location-based mobile phone data could be used to indicate the presence of suspects in an area at a certain time where a crime has taken place, to monitor the spatial and temporal fluctuations of a population's activities in a given area, to estimate the mobility flow of visitors, and to infer land use (i.e., commercial, industrial, residential, and educational) based on the total call volume or number of calls managed by a given cell tower over a given period of time [4,20]. Because these measures and assessments are extracted from mobile data, the data have become a topic of academic discussion and the centerpiece of many real-world applications.

For example, Refs. [21–23] depicted human mobility patterns from mobile phone data by extracting spatiotemporal features in the form of timestamps and cell tower IDs. These features were used to estimate or count the number of times a mobile phone device communicated with a given cell tower. These parameters and measurements aided in the investigation of the relationship between human mobility patterns and crime patterns.

Similarly, spatiotemporal features, such as cell tower IDs, timestamps, and call logs, have been extracted to depict other aspects of human activities, such as identifying residential and working activity to evaluate adherence to NPI policies, such as stay-at-home regulations or recommendations [24–26]; to estimate migration flow [18]; and to calculate the number of trips made between an origin (e.g., home) and destination (workplace) [27]. These measurements were calculated based on the definition of home and work locations, where home is indicated as the most frequently used or contacted cell tower during night-time hours (7 p.m. to 7 a.m.) and work is indicated by the most frequently used cell tower during the day.

Another example of a practical application of human activity characteristics that can be extracted from mobile phone data is the detection of criminal social interactions. For example, social networks can be created among individuals making or receiving calls or messages who are classified as actors (nodes) within the network; each link between actors is represented by the type of communication (call or message). Some studies [28,29] have diagrammed criminal networks by analyzing criminal communication (calling) behaviors, such as call frequency, maximum and minimum numbers of incoming or outgoing calls and messages, and temporal changes in mobile phone call patterns. This process, wherein specific social groups are identified along with their internal structures and communities, is referred to as social network analysis (SNA). SNA can be harnessed to determine the relationships and interactions between criminals by reconstructing the communication relationships that are obtained from mobile phone data as a network, where a node represents a criminal and an edge represents a communication (i.e., a phone call or a message). This method of analysis has been widely adopted in mobile phone data studies since it can help criminal investigators determine who belongs to a criminal organization, who heads it, and the relationships that exist within it. Using this approach in the study of criminal networks allows criminal investigators and experts to understand a network's hierarchy, its key leader, and subordinate leaders, and label the various levels of the criminal organization.

Here, we review existing studies that utilize mobile phone data with a particular focus on detecting and predicting criminal behaviors from people- and place-centric perspectives [13]. This includes studies that employ data on the prediction of crimes and criminal activities, the identification and detection of suspects and criminals, and other studies related to criminological research, such as exploring the relationship between human mobility patterns and crime patterns. We also shed light on the methods that employ mobile phone data to understand the dynamics of human behavior and mobility in urban sensing.

Although studies [1,3] made impressive contributions by exploring the applications of mobile phone data in social networking and urban sensing, knowledge about the use of mobile phone data in criminology research is lacking. Thus, a systematic review is needed to fill this gap by investigating the current state of mobile phone data use and applications in criminology research. Such an investigation would help to offer an overview of current approaches used to fight crime, prevent criminal activities, and detect criminal organizations. In addition, investigating these approaches can generate significant information about the tools and methods previously used to analyze mobile phone data, as well as provide a broader understanding of people's actions and activities in the areas in which they live and socialize and categorize individuals according to their mobility patterns so that the authorities can determine population flows in these zones before and during crimes. Thus, this study was motivated by a desire to enable researchers to create effective methods for extracting useful information from mobile phone data. These well-designed methodologies will enhance the process of identifying suspects and predicting crimes and provide a more complete picture of the dynamics of criminal behaviors from a people- and place-centric perspective.

Thus, this review aims to examine and explore the applications derived from human behavioral patterns extracted from mobile phone data in criminology research and evaluate the characteristics of multiple analysis perspectives (mobility patterns, communication behaviors, and social interactions) that have been derived from mobile phone data to depict aspects of human behavior and activity. The review also focuses on analyzing and explaining the choice of human features and characteristics, such as spatiotemporal and call features, that have been extracted to model human mobility and communication patterns in the context of criminology.

1.1. Survey Analysis

Limited types of reviews and survey articles related to mobile phone data have summarized applications in the mobile phone data domain. Notably, Refs. [1,3,4] presented comprehensive surveys about different applications of mobile phone data. Blondel et al. [1] reviewed social network applications that can be derived from mobile phone data in various disciplines and domains, such as social relationships, urban sensing, epidemics, public transportation, data protection, and criminology, with a major focus on studies that construct social networks according to communications behavior (calling information). Calabrese et al. [3] presented a comprehensive review of mobile phone data applications in the urban sensing domain by discussing the different types of mobile phone datasets and processing techniques that have been created in this domain. Okmi et al. [4] presented surveys about different methods, characteristics, and features used for assessing and predicting human behaviors in various domains such as urban sensing, criminology, transportation, and health. Bhattacharya and Kaski [30] reviewed the human social network application, one of the social network applications in the mobile phone data domain. Ghahramani et al. [31] reviewed another survey paper about mobile phone data in the urban sensing domain. The authors presented a survey of the techniques and methods that have been used with mobile phone data in urban sensing applications, such as urban planning and public safety, by discussing the strengths and weaknesses of various approaches and comparing their advantages and disadvantages with those of other mobility datasets that capture people's mobility patterns, including GPS, handover records, and location data.

Nevertheless, multiple analytical perspectives on mobile phone data that consider human communication behavior, social networks, and mobility patterns at various levels of mobile phone data (i.e., individual, aggregated, and cell tower data) have yet to be fully investigated. Studies of urban sensing domains are typically based on the use of mobile phone data that have been aggregated at the cell tower level, which provides only spatiotemporal information. Therefore, studies focusing on urban sensing domains have mostly discussed the analytical perspective of human mobility analysis patterns. Although [1] sheds light on various mobile phone data applications derived from different types of mobile phone data (i.e., individual, aggregated, and cell tower data), crime applications in mobile phone data have not been fully reviewed or discussed. Furthermore, new crime applications in mobile phone data have not been explored or investigated since that review.

This study differs from the previous survey framework in the sense that it examines and investigates various processing techniques and analytical perspectives that have been built based on mobile phone data at various levels to capture many aspects of human behaviors. These analytical perspectives, such as human mobility patterns, communication behaviors, social interactions, and mobile phone usage activities, have been built on multiple spatiotemporal and call characteristics extracted from mobile phone data. Even though Blondel et al. [1] aimed to review social network applications built on analyzing human social interactions that can be derived from mobile phone data and Calabrese et al. [3] presented a survey of urban sensing applications that are built on analyzing mobility patterns, knowledge about the use of mobile phone data in crime applications is still lacking. Thus, this study is the first to review the research focused on human mobility patterns, social interactions, and communication behaviors in crime applications and urban sensing applications. Figure 1 illustrates the multiple analytical perspectives and applications that this study has investigated and evaluated.

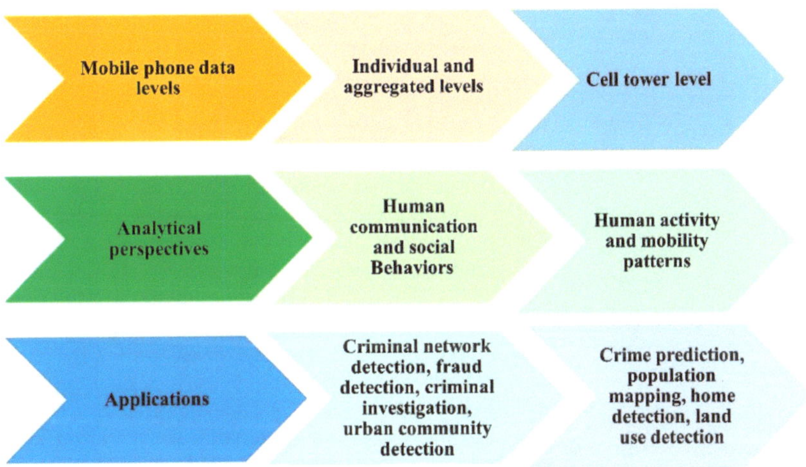

Figure 1. Multiple analytical perspectives and applications based on mobile phone data at various levels capture aspects of human behaviors.

The primary contribution of this systematic literature review (SLR) is to provide a comprehensive overview of the applications of mobile phone data in crime-control research. As a comprehensive SLR on this topic is lacking, the present study represents the first attempt to carry out a critical analysis of this topic. To achieve this goal, a thorough search of eight top scientific databases (i.e., the Association for Computing Machinery Digital Library, Institute of Electrical and Electronics Engineers Xplore, Multidisciplinary Digital Publishing Institute, Sage, Science Direct, Scopus, SpringerLink, and Web of Science) was performed, and 107 primary studies that met the study's scope and criteria were retrieved. This study involved four steps. The first was to extensively and systematically review the current state of mobile phone data use in crime applications, especially in those involving the identification and detection of criminals and the prediction of crimes. The second step was to investigate empirical research using mobile phone data to predict human behavior and mobility patterns in urban sensing applications. The third step involved providing a taxonomy for the final dataset of articles based on the scientific approach used and the research questions answered. The last step was to point out the potential challenges faced by this body of literature's state-of-the-art techniques and to provide potential directions for future research.

1.2. Mobile Phone Data Types (Levels)

Generally, the mobile phone data record the users' interactions on a mobile network and include details such as the IDs of the caller and the callee, the duration and timestamp of the interaction, and the location of the parties involved in the interaction (as determined by the cell tower ID). However, the data can be further divided into two types: a type that records the details of an interaction between a mobile device and the network, known as event-driven mobile phone data, and another type based on the cell tower location updates of mobile phones, known as network-driven mobile phone data (see Okmi et al. [4], Section 3, for more details about mobile phone data types).

The structure of the paper is organized as follows: In Section 2, we present the research methodology. In Section 3, we present the results of the SLR. In Section 4, the study taxonomy is presented. Section 5 addresses the research questions and discusses recent advances in detection methods. Section 6 discusses privacy concerns, investment behavior, and challenges. Section 7 defines the current problem and proposes a system model. Section 8 offers recommendations for future research and concludes the review.

2. Methodology

This section outlines the research methodology used for the study. A systematic literature review was conducted by adopting Kitchenham's guidelines [32] for search processes, inclusion and exclusion criteria, and data extraction. This study follows the reporting guidelines of "PRISMA" ("*Preferred Reporting Items for Systematic Reviews*"), which consist of a 27-item checklist and a 4-phase flow diagram for the selection of papers. The PRISMA statement by Liberati et al. [33] was used for the study selection process. This study also performed a bibliometric analysis along with the SLR to provide more thorough insights into the topic. Figure 2 shows the roadmap of the SLR, which clarified the planning of the review regarding the following points: the formulation of research questions, the study selection process, eligibility criteria (inclusion and exclusion criteria), bibliometrics and data extraction and synthesis strategies, study taxonomies, research questions, and future work. The systematic literature review road map begins with defining the main contributions and objectives of the review to allow the formulation of the research questions needed to achieve the study objectives. To answer these questions, a systematic review and bibliometric analysis were conducted to provide a thorough analysis of the topic. In the next stage, the studies were summarized, and a taxonomy based on the scientific approach was produced. This taxonomy helped to answer the research questions while establishing the current state of the research trends and applications of mobile phone data. In the final step of the road map, study limitations and future work were discussed.

2.1. Research Questions, Explanations, and Motivations

The comprehensive, systematic literature review presented in this research will focus on studies that have explored the use of mobile phone data to detect suspicious movements, crimes, and suspects; to predict human behaviors; and to understand human communication behavior. Two primary research questions have been developed for the present work, which are designed to determine the current status of mobile phone data and to investigate the different characteristics of studies that have employed mobile phone data in a variety of fields. As far as we are aware, there are no previous studies that have reviewed advancements in the field of mobile phone data using the specific inclusion criteria of the present research. This is a significant reason for which we wish to carry out this review. Below, we formulate two research questions with their explanations, as presented in Table 1 of this study.

1. RQ1: What are the current state-of-the-art methods and techniques regarding the use of mobile phone data in crime applications, especially in identifying suspects and predicting crimes?
2. RQ2: How can identifying empirical mobile phone data studies to predict human behavior and mobility patterns contribute to a clearer understanding of the dynamics of criminal behavior contexts from a people- and place-centric perspective?

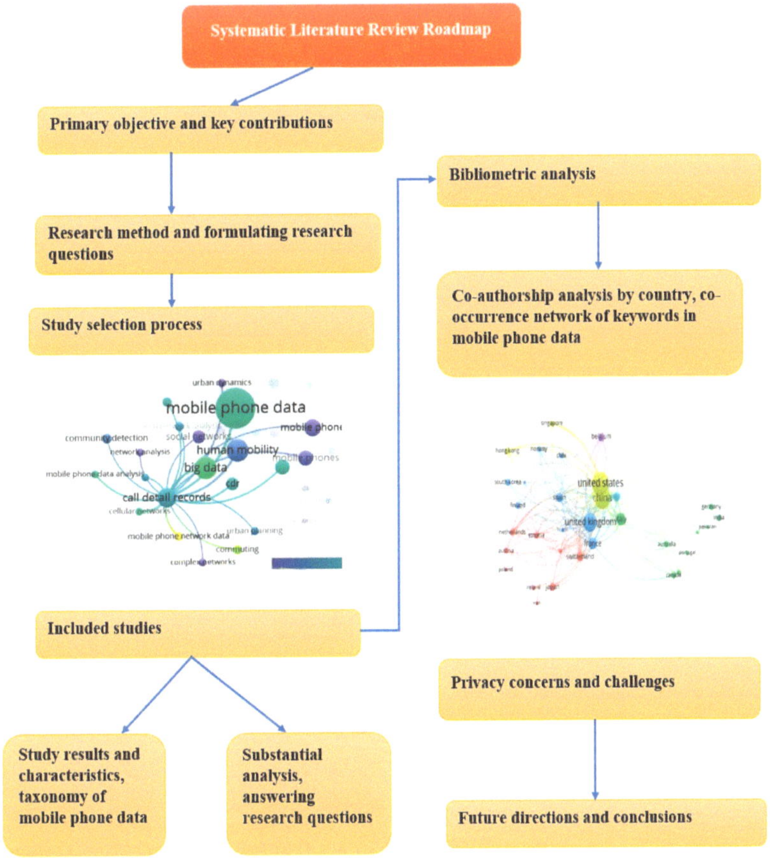

Figure 2. A systematic literature review roadmap that details the phases and stages investigated in this review will provide a more complete picture of the current state of research methods and techniques using mobile phone data.

Table 1. Research questions, explanations, and motivations.

Research Questions	Explanation
RQ1	Several studies have employed mobile phone data to predict crimes and criminal behaviors and to identify criminals and suspects. The present review offers a new and deeper insight into the advanced methods used nowadays in crime applications based on mobile phone data and the benefits of using such data to predict crimes and identify suspects.
RQ2	Mobile phone data have been used in a variety of studies to understand human behaviors and mobility patterns. More precisely, the spatiotemporal information provided by mobile phone data can provide clearer insights into human movements in various applications and academic fields. For instance, mobile phone data has been used to explore human mobility patterns and detect certain types of behaviors in cities and urban zones where criminal activities are much more likely to occur. Mobile phone data have thus been used in different crime and urban sensing applications to serve different purposes, such as defining the actual populations at risk, investigating the relationship between human dynamics and crimes, and inferring land use types based on human dynamics and interactions. For all the above reasons, the defined research question stimulated this investigation of mobile phone data usage in urban sensing, and the results should enable researchers to create more effective methods for extracting useful information from mobile phone data.

2.2. Study Selection

The flow diagram for selecting candidate articles consists of the following phases: identification, screening, eligibility, and inclusion. Figure 3 shows the PRISMA flow diagram for the study selection processes.

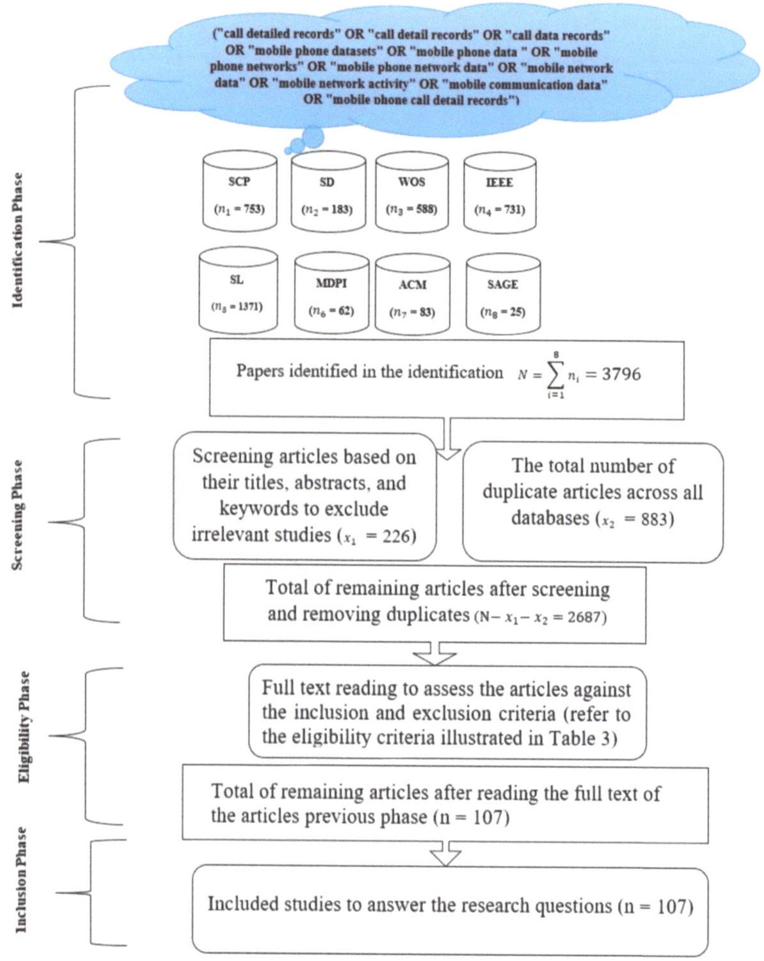

Figure 3. The four-phase flow diagram for the selection of papers.

The first phase comprised the process of identifying the most relevant research articles from reliable database sources by running the following search query ("call detailed records" OR "call detail records" OR "call data records" OR "mobile phone datasets" OR "mobile phone data" OR "mobile phone networks" OR "mobile phone network data" OR "mobile network data" OR "mobile network activity" OR "mobile communication data" OR "mobile phone call detail records") under the "Search Within Title", "Abstract", and/or "Keywords" filters. The search query parameters were adjusted appropriately to account for the default configurations of the databases. A notable case is the AND operator, which is implemented by default between the Search Within terms. This makes it difficult to combine two search terms by using the OR operator against the Title, Abstract, and Keywords filters. For instance, the searches within the SAGE and MDPI databases were run against the Abstract filter because this yielded more results (i.e., publications) as compared to the Title and Keywords filters. Accordingly, the results obtained from the Abstract filter were ensured to be inclusive of the results produced by both the Title and Keywords filters. To ensure a comprehensive search of articles, the search query was run on the following eight database sources: ACM Digital Library, IEEE Xplore, MDPI, SAGE, Science Direct, Scopus, SpringerLink, and Web of Science. These databases offer highly advanced search options that allow the researchers to fine-tune their queries, in addition to their ability to produce accurate citation data, remove duplicated results, and exclude certain materials such as patents and gray literature. In contrast to the aforementioned databases, Google Scholar (GS) was excluded from the present study due to its limited search functionality, inaccurate reporting of metadata, and inability to remove duplicated results.

The performed search yielded a total of 3796 publications based on the given criteria. The obtained data were imported into a Microsoft Excel spreadsheet and EndNote and later ordered by relevance in preparation for the subsequent phases. Figure 4 illustrates a flow chart for retrieving relevant studies through the search of databases.

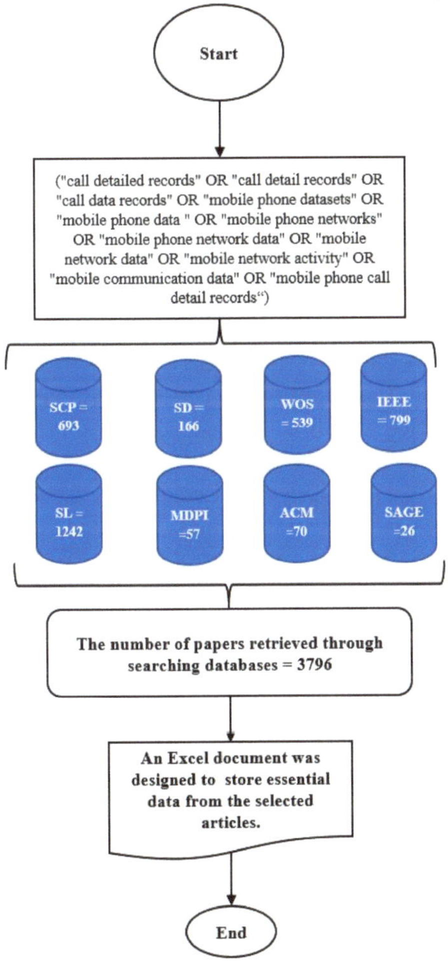

Figure 4. Flow chart for retrieving relevant articles through the search of databases.

The parameters that were used to search the database sources in the identification phase are presented in Table 2. These parameters facilitated the subsequent phases by setting the content language to English and the publication time to the period from 2014 through 2022. This ensured that articles written in languages other than English or before 2014 were omitted from the search results. It is noteworthy that, for the ScienceDirect database, the search query was iterated twice since the database only allows up to 8 OR operators to be included at once. Accordingly, the search was first performed with 8 keywords, then again with the remaining keywords. Subsequently, the results from each search iteration were merged together into one list.

Table 2. Set of parameters that were applied at the identification phase to filter each database.

Phases in the Selection of Papers	Database	Number of Returned Articles	Timespan	Content Type	Search Within
Identification Phase	Scopus (SCP)	n = 753	2014–2022	Article, Review Article	Title, Abstract, and Keywords
	Elsevier ScienceDirect (SD)	n = 183	2014–2022	Article, Review Article	Title, Abstract, and Keywords
	Web of Science (WoS)	n = 588	2014–2022	Article, Review article	Topic (Title, Abstract, and Keywords)
	IEEE Xplore (IEEE)	n = 731	2014–2022	Article, Conference paper	Metadata (Title, Abstract, and keywords)
	SpringerLink (SL)	n = 1371	2014–2022	Article, Conference paper	Abstract
	Multidisciplinary Digital Publishing Institute (MDPI)	n = 62	2014–2022	Article, Review Article	Abstract
	ACM Digital Library (ACM)	n = 83	2014–2022	Article	Title, Abstract
	Science And Geography Education (SAGE)	n = 25	2014–2022	Article	Abstract
Total	Papers identified in the identification phase (n = 3796)				

The second phase was the screening phase, which incorporated the process of removing duplicates from the obtained list of publications across all databases, followed by a manual screening to exclude irrelevant articles based on their titles, abstracts, and keywords. This step is crucial because many articles may fall under the given search query but are published in irrelevant fields. This phase yielded a total of 2687 publications after removing duplication and irrelevant articles.

The third phase was the eligibility phase, which involved reading the full text of the articles selected from the previous phase. This phase assessed the articles against the inclusion criteria to determine their eligibility (refer to Table 3). Therefore, the total number of articles remaining after reading the full text is N = 107.

Finally, in the last phase, the articles chosen from the third phase were used to answer the research questions of the present study. To avoid bias, all of the phases were reviewed and performed by one author, and then a test–retest analysis was conducted by the second author to assess reliability.

Table 3. Eligibility criteria.

Inclusion Criteria (IC)	Exclusion Criteria (EC)
Studies that present novel scientific contributions regarding the use of mobile phone data in detecting and identifying suspects and criminals.	Articles using mobile phone data in the context of smart marketing; the transportation sector (such as transportation planning, transport mode detection, and traffic prediction); economic forecasting; and health sciences research.
Studies that incorporate mobile phone data to predict crimes, perform spatial–temporal crime analysis, or have any other bearing on criminological research.	Studies using mobile phone data to measure human mobility in relation to the epidemiology of infectious diseases.
Studies that investigate the use of mobile phone data in home and work location detection; mapping human population density; classifying land use types; **detecting social interaction networks**; and others.	Publications that are not written in the English language.

2.3. Data Extraction and Synthesis Strategy

This section is essential for any SLR to aid in designing the data extraction form for the study results, and it is needed to help answer the research question. For this purpose, an Excel spreadsheet was created to store essential data from the selected articles. The data extraction form includes four parameters. The following data points were manually extracted:

DE1) The title of the article, the authors, the publishing journal, and other publication details.
DE2) Information related to mobile phone data types and their characteristics.
DE3) Information related to the mobile phone data domain and its applications; study area.
DE4) Information related to methods and techniques used in the mobile phone data domain.

The data synthesis was performed to accumulate and summarize the results of the included primary studies as well as to extract quantitative and qualitative data from the latter in forms that can be represented by tables, pie charts, bar and clustered bar charts, and scatter charts. VOSviewer software was used to obtain a visual representation of the data.

3. Results

This section presents a summary of the results obtained from the study selection process and includes details about the search results, the distribution of mobile phone data types, and a visualization of the co-occurrence of keywords. The distribution of publication type, publisher's locations, most cited publications, distribution of analysis perspectives, and publication years are also provided.

3.1. Search Results

A total of 3796 studies were initially identified from the eight databases in the identification phase. In total, 2687 studies were subsequently excluded through the screening phase based on the filtering of the titles, abstracts, and keywords, which resulted in greatly decreasing the number of papers and removing duplicate papers obtained across all databases. Then, the results were further refined according to the eligibility criteria, and 2584 were removed based on the exclusion criteria. Eventually, 107 studies were included as the final set of selected articles in this review. Figure 3 depicts the four-phase flow diagram for the selection of papers.

3.2. Publications Years

Figure 5 shows the publication years of selected studies as being between 2014 and 2022. It can be seen clearly that the research on mobile phone data has a steady indication of publications.

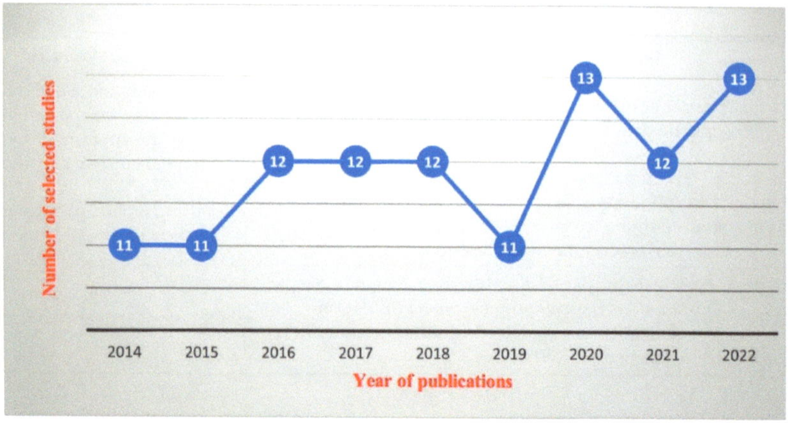

Figure 5. The distribution of selected articles by year of publication.

3.3. *Publication Type*

Figure 6 illustrates the distribution of publication types in mobile phone data. Out of the 107 primary studies selected, we observed that 86 (80%) appeared in articles and 21 (20%) were published in conferences. These statistics demonstrate that articles are the most active publication in the mobile phone data domain.

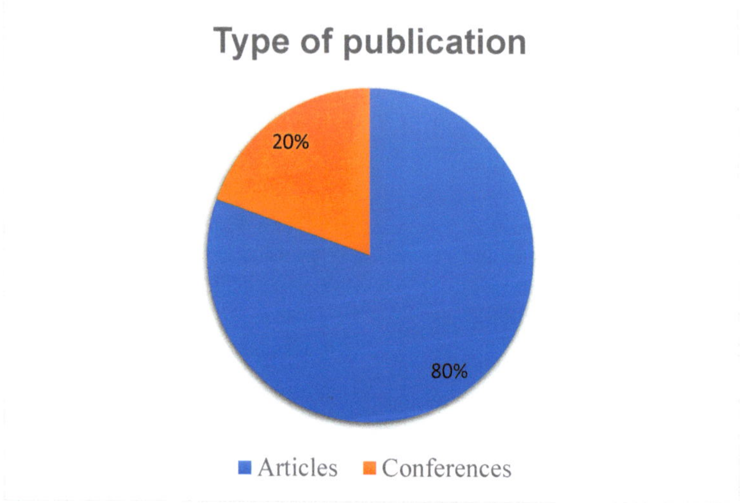

Figure 6. The distribution of publication types.

3.4. *Mobile Phone Data Levels*

Figure 7 illustrates the distribution of all mobile phone data types: mobile phone data aggregated at the cell tower level; mobile phone data at the individual level, known as call detailed records (CDRs) data; and mobile phone data at the aggregate level, known as aggregated CDRs data.

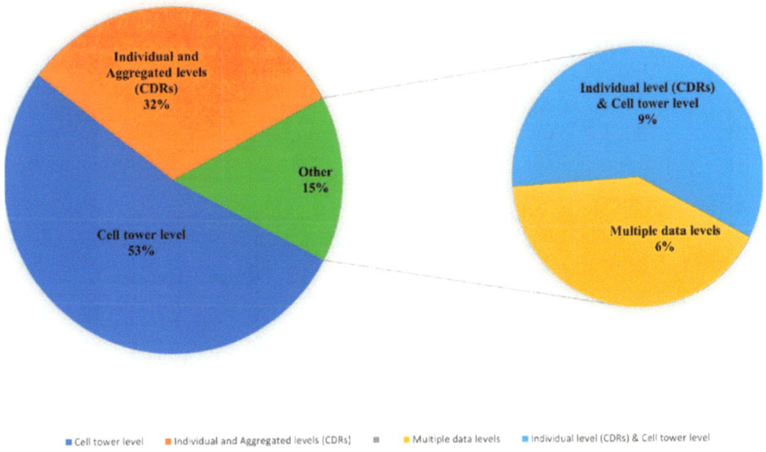

Figure 7. Distribution of mobile phone data types.

The charts illustrate that mobile phone data aggregated at the cell tower level exceed mobile phone data at the individual and aggregated levels. A total of 58 studies (53% of the 107 primary studies) investigated mobile phone data aggregated at the cell tower level, while 35 studies (32% of the primary studies) examined mobile phone data at the individual and aggregated levels, usually called CDRs and aggregated CDRs, respectively. Almost 15% of the studies combined multiple data types; 7 (6% of the 107 primary studies) were survey papers in which the authors studied and reviewed all mobile phone data levels, while 10 (9% of the primary studies) utilized mobile phone data at the individual and cell tower levels. These studies used mobile phone data at these levels to investigate individuals' social networks based on the calling information and to examine human mobility patterns based on the spatial and temporal characteristics. For the sake of simplicity and clarity, during the process of collecting information about the specific types of mobile phone data explored in the literature, we found that there was misunderstanding, confusion, and misuse of the correct terms for each mobile phone data type. For example, most studies refer to mobile phone data that are aggregated at the cell tower level as CDRs data, while CDRs data actually refer to mobile phone data at the individual level. For that reason, we devoted time to clarifying which terms were used for what types of data, finding that researchers and academics have referred to the vast majority of mobile phone data types as CDRs data. To solve this issue, we relied on three things. The first was the attributes that were utilized or investigated by a given study, where each mobile phone data type has different attributes. For example, mobile phone data aggregated at cell tower level have the following attributes: timestamp, user ID, and cell tower ID with the corresponding latitude and longitude coordinates. On the other hand, mobile phone data at the individual level (CDRs data) have the following attributes: caller and callee IDs; caller's connected cell tower ID; callee's connected cell tower ID; duration; and timestamp. Second, they showed whether the authors illustrated how the data were collected and generated, and third, what application was investigated by a given study; for example, mobile phone data aggregated at the cell tower level can capture users' spatiotemporal change patterns based on the spatiotemporal information provided by this data type, which is thus mostly related to mobile phone data applications concerning mobility patterns. This procedure was tedious and time-consuming, but our efforts should help future researchers differentiate different types of mobile phone data and consider these points in the future.

3.5. Citation Count

Table 4 shows the top 13 most cited articles from the primary studies. However, the citation count is a time-variant variable, meaning that it will likely change over time. For the data collection process, the Science Citation Index, Social Science Citation Index, Web of Science, and Scopus are well-known tools for performing bibliometric analyses. However, the data used in this study were collected from Google Scholar, which provided higher citation counts than the aforementioned tools due to its ranking algorithm being more inclusive by including non-peer reviewed papers, working papers, and preprint papers. The total citation count of the 10 most cited papers in this domain is 3588.

Table 4. Top 10 most cited articles and reviews.

Reference	Domain/application	Citation Count	Year
[6]	Mapping human population density	786	2014
[1]	Constructing social networks from mobile phone data	574	2015
[34]	Detecting cities' hotspots	397	2014
[20]	Classifying urban land uses	347	2014
[13]	Predicting crime	325	2014
[3]	Developing urban sensing applications based on mobile phone data	306	2014
[35]	Inferring home and work locations	292	2014
[36]	Mapping society-wide interaction networks of two European countries	268	2014
[10]	Detecting criminal networks	181	2014
[21]	Investigating correlations between human mobility patterns and crime rates (i.e., crime statistics)	112	2016

Studies such as [1,3,6,20] have been the most influential due to, for example, Deville et al. [6] and Pei et al. [20] being the first to present the ideas described in their research. Pei et al. [20] solved the problem of inferring urban land use from mobile phone data by improving the existing classification of different urban land uses, while Deville et al. [6] was one of the first to use mobile phone data to map human population distributions instead of employing traditional datasets, such as censuses and surveys. A criminology study by [13] additionally employed mobile phone data to predict crime hotspots. Notably, Refs. [1,3] provided thorough overviews (research surveys) of how mobile phone data are used in different applications and domains. This table can be helpful for researchers and scholars as an index or reference for not only the most highly cited papers, but also for pinpointing papers on mobile phone data that they can use as starting points for further research in this domain.

3.6. Place of Publication

Figure 8 shows the number of selected studies grouped by place of publisher (journal publishing companies). It can be seen that the selected primary studies are chosen from a variety of different academic publishers. However, as the bar graph demonstrates, Elsevier, Springer, and IEEE have the highest share among 19 academic publishers with 56% (60 out of 107), which is not surprising. As a matter of fact, Scopus, which belongs to the same publisher as Elsevier, IEEE, and Springer, returned the highest share during the identification phase (the initial search result) with 75.2% (2855 papers out of 3796). We also observed that famous world-class publishers such as PNAS and the Royal Society, which are among the most prestigious and highly cited multidisciplinary research journals, are among the publishers in the mobile phone data domain.

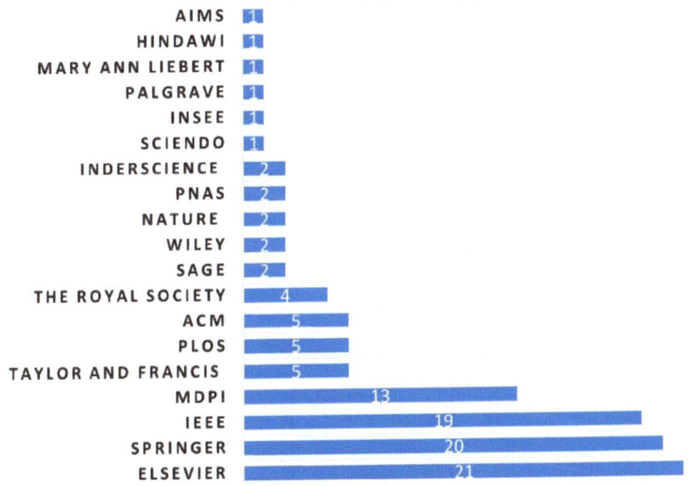

Figure 8. The distribution of publisher channels.

3.7. Mobile Phone Data Methods and Problems

Here, we detail mobile phone data problems and provide a comparison of different methods used to solve these problems (see Figure 9). Generally, mobile phone data problems can be divided into five main groups: classification, clustering, detection, estimation, and privacy problems. The relevant studies have mostly been related to clustering problems. Thus, many mobile phone data studies have been conducted to solve problems with clustering approaches in numerous applications. For example, in these references [20,37,38], inferring land use types was identified as a clustering problem. In References [39–42], the authors sought to identify users' habits. In Reference [19], the authors clustered users based on their weekly patterns.

Social network analysis techniques and metrics have been widely used to solve problems related to the community detection problem (CDP), community structure, and social network visualization. For example, detecting criminal networks has been seen as a CDP. In References [28,29,43,44], the authors applied different community detection algorithms to detect criminal networks. Novovic et al. [45] applied community detection techniques to infer a correlation between human dynamics and land use. Moreover, Shi et al. [46] applied a community detection algorithm to detect the spatial interactions of urban social communities.

Classification problems have been examined in studies aiming to identify suspects. In References [9,47], classification algorithms were used to differentiate suspects from non-suspects. Another example of a classification problem is predicting crime hotspots. Bogomolov et al. [13] applied classification algorithms to classify crime hotspots into two classes, high or low crime levels. Moreover, Ref. [48] used algorithms to classify land uses.

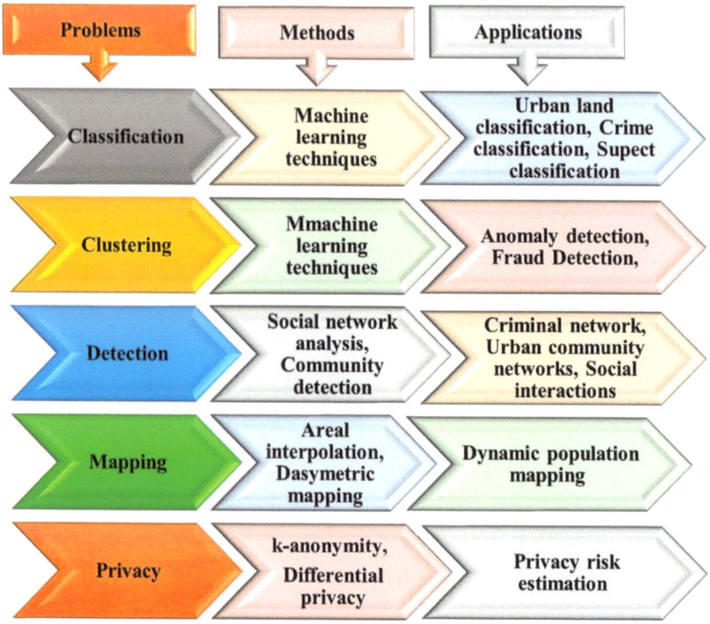

Figure 9. Comparison of different methods and problems used in mobile phone data studies.

Furthermore, k-anonymity techniques and approaches have been suggested as solutions to solve problems related to privacy risks and data protection, such as in [49,50]. For estimation problems and correlation analysis problems, statistical measurements, such as correlation coefficients and regression models, have been used; for example, Pearson's correlation coefficients were used in [6,51], whereas Spearman's correlation was used in [21,52]. Finally, areal weighting, dasymetric mapping, and Voronoi tessellation techniques have been used to solve problems related to spatial mapping and population mapping, such as in [53–55].

3.8. Analysis and Perspectives

The percentage of the selected articles studying human mobility patterns is higher than that of studies looking at communication behaviors. Figure 10 highlights this extraordinary growth in the number of studies that utilize mobile phone data to investigate human mobility and movement patterns, with these representing 66.4% (71 out of 107) of all studies, as compared to the mere 18.6% (20 out of 107) that used such data to study communication behaviors, and studies that investigate both human behaviors represented a further 15% (16 out of 107). This observation may be attributable to the fact that mobile phone data aggregated at the cell tower level were the most commonly investigated, as shown in Figure 6, which shows the distribution of mobile phone data types under investigation, and this type of data (mobile phone data at the cell tower level) reveals only information about human spatiotemporal patterns. As a result, several applications related to human mobility patterns can be derived from this data type, i.e., mobile phone data aggregated at cell tower level. This finding may also be explained by privacy concerns that restrict and increase the difficulty of accessing or acquiring mobile phone data at the individual level (CDRs data), which might contain sensitive details such as spatiotemporal trajectories and communication information about the receiving side of the communication, as opposed to mobile phone data at the cell tower level, which does not reveal communication details. Furthermore, due to difficulties seen in managing and processing CDRs data based on the nature of raw data, data cleansing and preprocessing, such as noise reduction and managing sparsity constraints, is required.

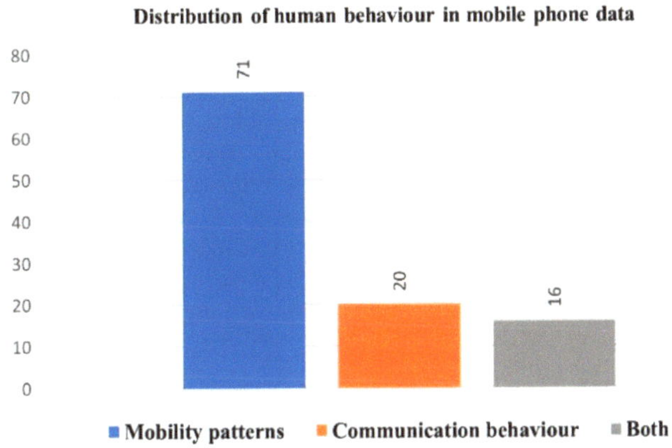

Figure 10. Distribution of analysis perspectives in mobile phone data.

3.9. Network Visualization of the Co-Authorship Analysis by Country in Mobile Phone Data

This network in Figure 11 visualizes the worldwide co-authorship of the countries that have published articles on mobile phone data by evaluating the performance of the participating countries and the degree of cooperation between countries to produce papers on mobile phone data. Each node inside the cluster represents a country, and the node size refers to the publication weight, while the total link strength reflects the degree of co-authorship links to other countries. For example, the United States, China, and the United Kingdom have the largest proportion of publications in mobile phone data with 203, 181, and 131 publications, respectively, and the total number of co-authorship ties to other countries is 229, 136, and 187.

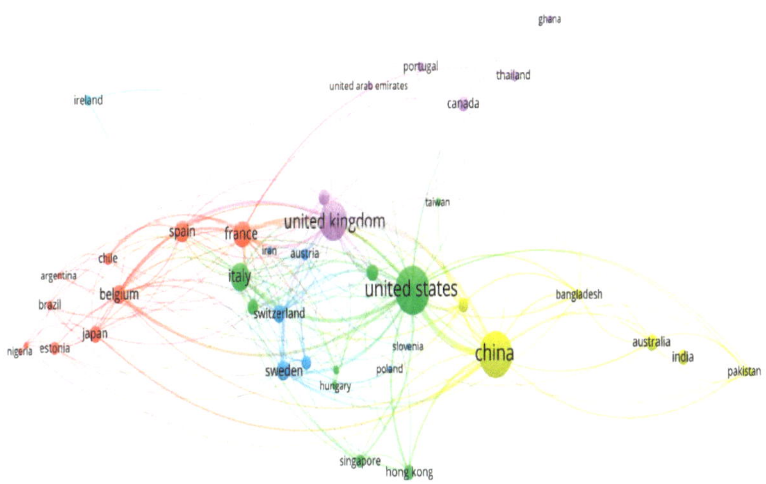

Figure 11. Network visualization of the co-authorship analysis by country.

3.10. Co-Occurrence Network Visualization of Keywords in Mobile Phone Data Studies

This section describes the construction of the keyword co-occurrence map, which is based on the co-occurrence data. The map in Figure 12 visualizes the co-occurrence network of the most frequently used keywords or search terms in mobile phone data studies. To further understand the relationships between different clusters within the network, each node has been made to represent its value or importance based on the occurrence weight of the node itself and the strength of its link with other nodes (each node represents a search term, links represent the occurrence of a pair of search terms, and the weight of the link is represented by the co-occurrence frequency of each pair of search terms). Node size refers to the frequency of the occurrence of a keyword (e.g., mobile phone data, detailed call records, etc.) in the selected publications, and it is measured by the number of articles that have used that keyword (or a corresponding term) in their list of keywords. The first cluster contains blue nodes and is the largest of all seven clusters. It depicts mobile phone data with a weight (occurrence) of 170 and 492 links, followed by human mobility with a weight (occurrence) of 84 and 219 links. The blue cluster includes the following human mobility pattern search terms: mobile phone data, human mobility, mobility patterns, mobile communication, urban area, and others. The second cluster contains red nodes representing terms related to mobile phone data types, such as CDRs data, and human communication behavior, such as social networking, social network analysis, criminal networks, big data, and economic and social effects. The third cluster contains green nodes that represent terms related to human mobility patterns and their applications, such as spatial–temporal analysis, population distribution and density, human activities, and geographic mapping.

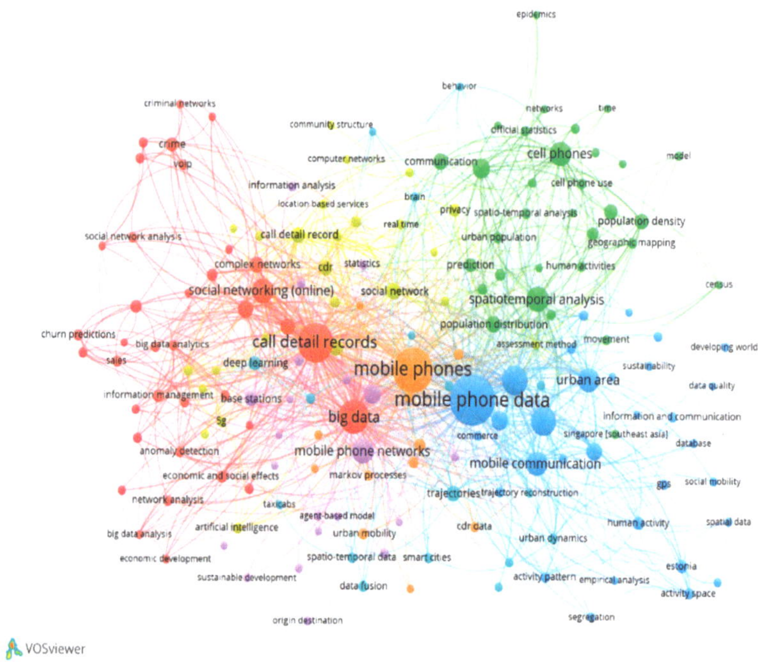

Figure 12. Network visualization of keywords in mobile phone data studies.

4. Study Taxonomy

This section describes the process by which the selected studies were organized and categorized into structured taxonomies in a way that helps address the research questions and sheds light on the current state of mobile phone data applications. A taxonomy is presented in Figure 13 that contains categories and subcategories of mobile phone data according to specific factors, including the analysis and processing techniques (analysis perspectives), the dataset level (i.e., individual, aggregated, cell tower), and types of applications. Generally, mobile phone data are utilized in the analysis of human communication behaviors and mobility patterns; thus, the taxonomy is categorized according to aspects related to the analysis perspectives of mobile phone data. Since the analysis of mobile phone data occurs at three different levels, which are the individual, aggregated, and cell tower levels, a new classification is made according to these levels. Similarly, the processing techniques used to analyze mobile phone data on each level embrace numerous applications that also require subcategorization as a new classification.

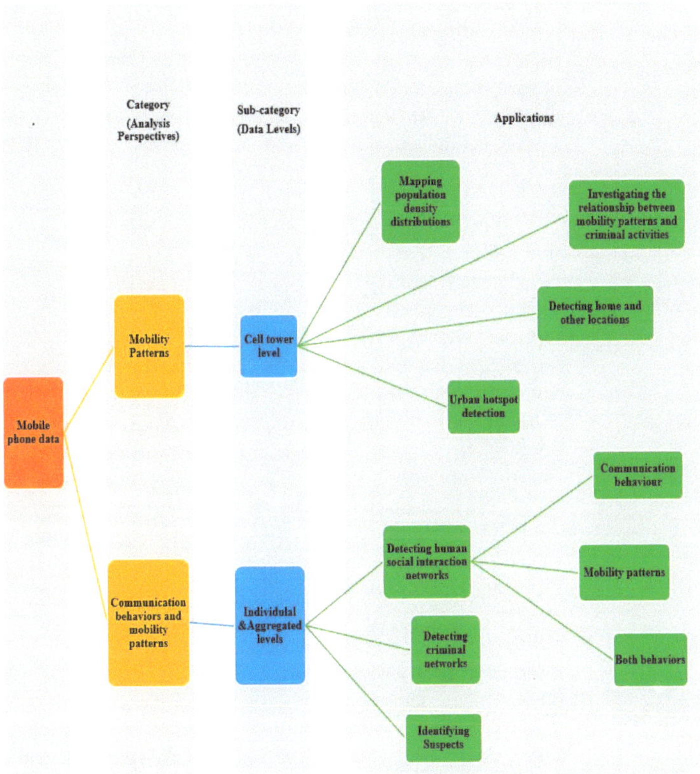

Figure 13. Taxonomy of research literature in mobile phone data.

4.1. *Mobility Patterns (Main Category): The First Leg*

The first leg of the study comprises mobility patterns, which are further classified into one of the mobile phone data types (levels), in this case, mobile phone data aggregated at the cell tower level. This leg discusses the studies that use mobile phone data aggregated at the cell tower level. Such data store records that detail the user ID (caller ID), the timestamp (e.g., call date, call time), and the location data (cell tower ID), where each record is geolocated based on the nearest BTS. These details allow us to capture users' spatiotemporal change patterns extracted from spatiotemporal information collected from this type of data. Thus, mobile phone data aggregated at the cell tower level have been used for several applications related to human mobility patterns, such as estimating populations, identifying home and work locations, and identifying land use types.

Mobile phone data at this level are usually used to capture human mobility patterns since only spatiotemporal information is recorded. Thus, the details at this level allow for the estimation of a population in a certain block or area based on the phones that are connected to the cell tower. Identification of visitors to and residents of an area is carried out by identifying nearby cell towers that the mobile devices are connected to most of the time. Furthermore, the activities of mobile phone users within a given geographical location can be captured by the cell towers connected to them, and changes in the location of mobile phone users from one place to another can allow them to be identified as residents or visitors based on human activity, represented by spatiotemporal characteristics obtained from mobile phone data. These spatiotemporal characteristics that explain human activities provide a wide array of applications, which will be discussed here.

Mobile phone data at the cell tower level cannot be used to investigate and study human communication behavior and social communication patterns because they do not contain information regarding calling patterns that illustrate details of the other side of the communication. A full description is added later for the second leg when discussing applications used for mobile phone data at the individual and aggregated levels, usually called CDRs and aggregated CDRs, respectively. Thus, this leg mostly discusses human mobility patterns based on mobile phone data aggregated at the cell tower level.

4.1.1. First Application: Estimating and Mapping Population Distributions

Many aspects of human activities are related to human mobility patterns, and investigating these patterns has become a common use of mobile phone data; this can be seen clearly in the number of applications derived from this analysis perspective based on the spatiotemporal characteristics that can be extracted from mobile phone data that cover multiple aspects of human life activities. Importantly, this includes estimating and mapping population distributions. Mobile phone data have been used to estimate population densities by mapping the hourly dynamics of population based on spatial-temporal trajectories extracted from mobile phone data. To map the population's presence at a cell tower, Deville et al. [6] applied an interpolation method of spatial mapping known as areal weighted interpolation (AWI), which allows the interpolation of coverage areas' spatial division and its attributes through areal intersection with spatial units, such as blocks or administrative areas. In this manner, the areal weight of a census block can be intersected with the cell tower coverage of a mobile network. This study led to further discoveries based on such applications. Sakarovitch et al. [56] aimed to estimate resident populations by using Voronoï tessellation, which partitions the geographical space of cell tower coverage into Voronoï polygons. By applying dasymetric mapping methods to enhance population mapping on a more fine-grained spatial scale, Ref. [53] applied a two-step floating catchment area method (2SFCAe) and land use regression (LUR).

However, as these methods only consider mapping populations based on spatial distribution, to improve the mapping of the spatial distribution of cell towers with respect to population and thus to map population dynamics, a more fine-grained spatial and temporal scale is required; various researchers [55,57,58] have thus applied dasymetric interpolation methods to map population distribution by integrating this with a temporal perspective. The aim of such work is to use multi-temporal function-based dasymetric (MFD) interpolation to enhance the accuracy of the spatiotemporal resolution of population dynamic distributions by capturing temporal patterns. However, Liu et al. [59] criticize previous work in mapping dynamic populations due to their failure to estimate a population distribution at a fine temporal scale due to capturing the temporal patterns of the population only over a given time period. Thus, they aimed to map population dynamics at hourly intervals by reconstructing time series trajectories of hourly population density. In their quest to enhance the accuracy of mapping population density distribution effectively, Ref. [60] determined that a lack of ground truth data for the dynamic population density distribution over various time scales might affect the estimation of a population at a finely grained temporal resolution; they thus used the Tencent positioning dataset with fine-grain temporal resolution as ground truth data for training in a deep learning model using a deep convolutional generative adversarial network (DCGAN).

However, Salat et al. [54] criticized previous methods because they required a large number of finely grained data sets in order to train a given model, such as census and satellite data. This is especially the case in some developing countries where census data are not always available to validate the models; thus, the authors sought to provide a model without requiring training datasets by applying a hierarchical clustering method (hierarchical cluster analysis). References [61,62] sought to solve these problems related to estimating the population density, such as data heterogeneity and multiple sources of mobile network operator (MNO) data. They performed this by proposing two novel methodological frameworks which they designed to correlate multiple mobile phone data sources (location area-level data, CDRs, aggregated CDRs, and mobile phone data) from multiple MNOs based on data fusion models and joint analysis techniques.

Taking this idea further, based on a similar aim to map and estimate population distribution, Shi et al. [63] not only attempted to estimate population density distribution but also strove to investigate the correlations between population density distribution and public service facilities such as retail stores, businesses, hotels, culture and art facilities, and parks. The findings of their study showed that the distribution of public service facilities is strongly linked to population density during the day (daytime population).

4.1.2. Second Application: Investigating the Relationship between Human Mobility Patterns and Criminal Activity Patterns

This application covers studies examining human behavioral activities as reflected in spatiotemporal mobility patterns extracted from mobile phone data and their associations with crime patterns. Empirically speaking, estimating how people move, measuring their presence at a given place, measuring population risks, and estimating the flow of the general population to provide information about criminals' movements—all these measures have been reconstructed or derived from spatial-temporal characteristics in mobile phone data as part of investigations into their relationship with crime patterns, as well as to develop a better understanding of spatiotemporal patterns of crime. References [13,14,64] were some of the first studies that investigated the correlation between human mobility patterns and criminal activity patterns in the mobile phone data domain. Traunmueller et al. [64] aimed to observe such a correlation based on testing Jacob's hypothesis, which suggests that high population density and population diversity (age diversity, ratio of visitors, and ratio of residents) reduce violent crime rates, and the results show that the relationship between crime activities and the diversity of age and ratio of visitors was negatively correlated. Bogomolov et al. [13,14] performed one of the first studies to investigate correlations between human behavioral activities as depicted in spatiotemporal patterns from mobile phone data and criminal activities. The authors used this data as a proxy to measure people's presence at a given place to predict the relevant crime levels (classifying crime levels) in terms of "low crime levels" or "high crime levels". These studies opened up the type of data used in such applications, and [21,51,65] then went a step further by using mobile phone data as a measure to estimate the ambient population (population at-risk). In the [21] study, the authors measured the ambient population to investigate its relationship with crime rates, with results that showed a strong correlation between the ambient population and theft crimes, based on identifying "people who might commit theft". Similarly, Ref. [51] estimated the ambient population as an alternative measurement of the population at risk to investigate the effects of the ambient population on the spatiotemporal patterns for migrants and natives in terms of violence committed by migrants and natives, and the results show that the ambient population has a positive relationship with migrant violent crimes. Finally, Ref. [65] aimed to examine the relationship between the ambient population and the spatial crime pattern of larceny-theft, with results showing that the ambient population has a positive link with larceny-theft crimes.

Instead of estimating the ambient population as in previous studies, References [52,66] investigated the correlations between exposed population-at-risk (population at risk of exposure to violence), which may be a mix of criminals and victims, and temporal–spatial patterns of violent crime in public to determine their impact on violent crime in public spaces. Haleem et al. [52] aimed to evaluate the influence of exposed and ambient populations-at-risk on violent crimes associated with the nighttime economy (NTE). Lee et al. [66] built on the same notion of "exposed population-at-risk", with the addition of the spatial and temporal characteristics of violent crime in public spaces. Finally, Song et al. [67] attempted to determine whether the daily mobility flows of the general population could provide a template for the daily mobility of criminals.

4.1.3. Third Application: The Detection of Homes and Other Meaningful Locations

Many studies have focused on identifying home and work locations, which can be part of the analysis of mobile phone data. Not all studies were primarily focused on the detection of home and work locations, but during the analysis phase, the detection of home and work locations may have been required for preprocessing prior to further analysis. This process has been widely followed with mobile phone data in many studies [5,12,13,18], while other studies have mainly focused on detecting home or work locations [35,68,69].

Human mobility patterns extracted from mobile phone data have been used to model daily human activities (for example, home, work, shopping, etc.) by parsing trajectory features from spatiotemporal information into fixed locations. This indicates where people conduct their activities or the locations where the most activity takes place. Therefore, daily activity patterns based on mobile phone data can identify and estimate home and work locations or other meaningful locations. Kung et al. [35] aimed to detect home and work locations based on human mobility patterns. Two criteria were chosen to define home and work locations, namely, the home location was identified as the most frequently visited location during nighttime, and the work location was the most frequently visited location during daytime hours. Empirically, identifying a person's home means that a single cell tower is allocated as their home location, so the most frequently used cell tower location during the night hours, for example, 7 p.m.–7 a.m., is the approximate location of residence. Tongsinoot and Muangsin [68] identified the home detection by correlating mobile phone data with Internet data usage containing attributes such as mobile numbers, timestamps, upload volume, download volume, cell tower ID, and network ID. This was conducted to improve the detection of home and work locations, claiming that previous identification methods are based on the time or duration criteria, where the proportion of staying time is calculated to estimate the location. Vanhoof et al. [69] aimed to improve home detection by defining five home criteria based on calling activities and mobility patterns.

4.1.4. Fourth Application: Urban Hotspot Detection

This part highlights scholarship that uses mobile phone data to detect urban hotspots (hotspots refer to regions with higher concentrations of people, the most congested places, or high-intensity crime areas) based on human mobility patterns. Louail et al. [34] sought to detect urban crowd hotspot areas (areas considered to be dense) in 31 Spanish cities. They performed this by extracting spatial–temporal characteristics, such as aggregating every hour (because hotspots fluctuate over time as a result of human mobility patterns) the total number of mobile users in each cell tower, which helps in turn to estimate population density. After depicting user density based on human mobility, the authors then established a threshold to identify hotspots by using a non-parametric method based on the logarithmic derivative of the Lorenz curve. The threshold density population describes each cell i (cell size is between 500 m and 2 km) with a density of users larger than the threshold δ for the density $\rho(i, t) > \delta$ as a hotspot cell at time t, while Yang et al. [70] set out to detect two types of urban human dynamics hotspots—convergent and dispersive hotspots—in the city of Shenzhen, China. In order to identify human mobility hotspots, they applied an unsupervised clustering algorithm, the X-means algorithm, statistical analysis, and kernel density estimation (KDE). Similarly, the KDE method was used by Ghahramani et al. [71], who aimed to detect hotspots in Macau, China. However, the authors extracted characteristics of calling behaviors to illustrate urban population density, such as the frequency of calls at different timestamps and the duration of calls, along with spatial and temporal characteristics such as spatial objects referring to cell towers.

4.2. Communication Behaviors and Mobility Patterns (Main Category): The Second Leg

The second leg comprises communication behaviors and mobility patterns at the individual and aggregate levels. This section thus discusses what applications remain to be derived from mobile phone data at the individual and aggregate levels, and what human behavioral patterns may be captured by these data types.

Mobile phone data at both the individual level and aggregate level can be used to investigate and study human communication behavior and social communication alongside mobility patterns due to the fact that mobile phone data at both the individual level and aggregate level contains both communication information and spatial–temporal information. Individual data do contain details that reflect attributes such as caller ID, callee ID, caller's connected cell tower ID, callee's connected cell tower ID, duration of call, and timestamp, which allows the development of applications related to communication behaviors such as the mobile social networks (detecting social networking), the detection of criminal relationships, inference of social ties, and various applications related to human mobility patterns, such as the identification of suspects based on spatiotemporal characteristics, and the detection of criminals based on their calling patterns and mobility behaviors, which all will be discussed in this section. A full description of the applications derived from mobile phone data at individual and aggregated levels is thus offered in the next section.

4.2.1. Social Network Applications

Mobile phone data's application to the investigation of mobile social networks is a solid and self-sufficient topic. The study of human sociality using mobile phone data has evolved into a distinct field of study that gives insight into the dynamics of human social networks [30], which explains the rapid expansion in the volume of such studies. Mobile phone data have thus been used to study a huge range of human sociality-related topics across various applications, including the investigation of social ties, the inference of relationships, the detection of social networking communities, and the detection of temporal or spatial social networks based on spatiotemporal characteristics.

4.2.2. First Application: Detecting Human Social Interaction Networks Based on Spatiotemporal Mobility Patterns

This application focuses on studies that use CDRs to identify social communities based on spatiotemporal mobility patterns, or, in other words, using mobility patterns as a means to detect communities. Shi et al. [46] constructed human social network interactions in a manner that aimed to discover spatiotemporal interaction communities arising from spatial human mobility patterns extracted from spatiotemporal information in CDRs data, such as the identification of the most frequented locations of users, identified as homes and workplaces, based on each user's most active cell tower. To achieve this aim, the authors applied two methods: the Newman method and the Moore community detection algorithm, which detects social communities and uses the kernel density estimation method to visualize the spatial distributions of different communities. Truică et al. [72] aimed to detect or cluster groups of nodes that reflected social interactions based on spatiotemporal information (mobility patterns) by applying the Louvain algorithm, a well-known community detection algorithm, while Xu et al. [73] aimed to detect communities across faculty members and students in a virtual campus mobile network based on spatiotemporal patterns of users' trajectories. Lind et al. [74] built social networks that aimed to detect spatial–temporal interaction communities based on human mobility patterns extracted from CDRs data; however, the authors also extracted one additional attribute from CDRs data, Internet usage, based on the fact that detecting communities based on just SMSs and phone calls limits the registration of spatiotemporal information to cell tower contacts; adding internet usage to represent interactions thus increases observed user events by allowing the visualization of additional areas (spatial information) based on user-triggered events, such as browsing or accessing the Internet. Sumathi et al. [75] aimed to build a social events network to detect and estimate the number of participants attending the Indian Institute of Science in Bengaluru based on their mobility patterns.

4.2.3. Second Application: Detecting Human Social Interaction Networks based on Human Communication Behaviors

This section discusses studies where human communication behavior is used as a means to construct social networks. Schläpfer et al. [36] constructed social networks of human interactions based on communication behavior extracted from CDRs data, such as the total number of contacts, call volume, and number of calls, where subscribers (contacts) are the nodes and the call volume and number of calls are measures of reciprocity to quantify social ties between nodes. Their aim was to investigate the relationship between the size of the city and human social interactions, which in turn scale superlinearly with the city population size. Filipowska et al. [76] aimed to build user social profiles based on call information, including the number and duration of phone calls, to visualize social network activities among groups of users to help differentiate or classify relationship levels based on defining weak or strong ties between individuals. Reference [77] constructed students' social networks based on communication behaviors represented by their calling characteristics, such as the total number of calls and SMSs as well as call duration. Their aim was to construct students' social networks based on identifying chronotypes, such as owls (evening-active) and larks (early risers and early sleepers). In this process, degrees were assigned to each node and weights were assigned to each link in order to assess network structures. Yu et al. [78] also constructed social networks of friend relationships based on communication behaviors extracted from calling information that included the total number of calls and SMSs, along with call duration, timestamps, and other measures, by applying a semi-supervised algorithm. Their aim was to classify user relationships based on the strength of social ties between two classes, "friends" and "non-friend". Similarly, Gaito et al. [79] built social networks to visualize human social interactions based on communication information that included the number of calls and SMSs, call durations, and call frequency. Their aim included investigating which communication channels, as represented by phone calls and text messages, users preferred for their interactions.

4.2.4. Third Application: Inferring Social Network Based on Mobility Patterns and Social Interactions

Other studies combined both types of human behaviors, such as [5,80], both of which sought to capture macroscale patterns of mobility and social interactions. They studied the interplay between human mobility patterns and human social interactions to investigate how human mobility patterns influence social interactions.

Deville et al. [80] aimed to capture any relationships between human mobility and social networks, based on combining three different mobile phone datasets simultaneously to capture two perspectives on human behavior, defined by human mobility and social networks. The results revealed that these two behaviors are not independent, as there is a strong relationship between human mobility and communication patterns within social networks: as distance increases, the average number of fluxes in social interactions increases (number of calls) given the same volume of mobility fluxes (number of jumps between two locations). Phithakkitnukoon and Smoreda [5] investigated the interplay between human mobility and sociality (in terms of social tie strength) by extracting human social behaviors from calling information such as the daily number of calls made and received, call duration, and human mobility patterns as extracted from spatiotemporal information such as the number of locations a person visited in a time period, the travel distance range, and the degree of variation. Finally, Morales et al. [81] constructed an ethnic interactions network based on mobility and communication patterns by correlating two datasets, in which the first contains calling information and the latter contains spatial–temporal information. The network aims to detect different ethnic and religious groups in Ivory Coast by mapping each community to its geographically closest ethnic group. To achieve this goal, the authors applied community detection techniques such as the Louvain community detection algorithm and a K-means clustering algorithm.

4.2.5. Fourth Application: Suspect Identification

This application arises from studies related to identifying suspects (people who are thought to be involved in certain criminal activities), based on detecting suspicious activities and movement patterns of all parties involved by examining the digital traces left by mobile phone devices that depict communication behaviors and mobility patterns. Digital traces left by people at locations where a crime has taken place, for example, can reveal a representative sample of the population present at a crime scene at a given time. Table 5 shows different features and analytical perspectives used to identify suspects.

Table 5. Prior research on suspect identification.

Reference	Features	Description
[82]	Spatiotemporal features	This study aimed to identify the most probable suspects in a given case by correlating CDRs with other data sources, such as digital video recorders (DVRs) and base transceiver station (BTS) log files to help investigators with otherwise insufficient evidence pinpoint hidden details about their suspects and gather further digital evidence to show how a crime is committed. The author extracted spatiotemporal information, such as the suspects' various trajectories, from CDR data and cell tower IDs that showed each suspect's home cell tower location along with other visited cell tower locations, to prove involvement in the crime.
[83]	Call features	This study aimed to identify suspects based on their calling characteristics, including any phone calls made or received by the suspects at the crime scene, in conjunction with archived CDRs data drawn from a central database that contained details on previously convicted criminals whose names had been recorded in older cases.
[84,85]	Call and spatiotemporal features	These studies aimed to improve the identification performance of suspects in terms of efficiency, effort, and scalability. To achieve this, they proposed a system-based big data analytic process to extract communication and mobility information from CDRs data, including aspects such as the most frequent caller, the number of times the suspect called other suspects, call frequency, suspect trajectories, and the most visited location based on the most frequently used cell tower.
[86]	Spatiotemporal features	This study proposed a terrorist detection system that aims to detect suspicious activities based on user trajectories.
[87]	Call and spatiotemporal features	This study aimed to investigate additional details by identifying suspects and their accomplices. To achieve this, the authors extracted calling and spatiotemporal information from the CDRs, such as calls made and received by suspects and suspects' trajectories near crime locations, then applied MariaDB, an open-source relational database management system (RDBMS), to analyze the CDRs data.
[9,47]	Call features	Rather than applying traditional methods, these studies proposed machine learning methods to tackle the identification process. They applied classification algorithms that aimed to separate suspects from non-suspects based on communication behaviors.
[88]	Call features	Going one step further, some studies discussed the challenges associated with analyzing CDRs to identify suspects. Marshall and Miller [88] aimed to present different techniques and scenarios suspects might use to avoid recording of their communication and mobility activities, such as stealth SIM, voice changing, roaming callback, and call obfuscation.

4.2.6. Fifth Application: Detecting Criminal Networks

This application is drawn from studies that utilize CDR data to detect criminal networks based on communication behaviors and mobility patterns.

With regard to the detection of criminal networks based on communication behaviors, Ferrara et al. [10] proposed a forensic analysis system named LogAnalysis, whose conceptual framework aimed to detect the most influential criminals in a criminal organization by applying social network analysis (SNA) tools and metrics such as degree centrality, closeness centrality, and "betweenness" centrality to identify both influential members and less-involved members of criminal networks and to quantify the degree of the relationships between vertices. Similarly, Refs. [11,28,29,43,44] proposed multiple forensic systems to detect criminal networks based on the calling characteristics of criminal communication behaviors, including outgoing and incoming calls between two identified vertices (criminals) and the maximum and minimum numbers of incoming or outgoing calls and messages. The SIIMCO system created by [11] aimed to detect lower-level criminals and their immediate leaders in a criminal network, as these are the most likely to be arrested, while the IICCC by [43] and CLDRI by [44] systems aimed to detect high-level criminals, identified as the most influential members in a criminal organization. ECLfinder [28] similarly aimed to detect and classify both high-level and lower-level criminals in a criminal network. Agreste et al. [29] aimed to uncover the underlying structure of Italian Mafia gangs and detect their key leaders.

Other work focused on the detection of criminals based on mobility patterns: Griffiths et al. [12] aimed to detect mobility patterns within specific terror networks (UK-based Islamist terrorists) by extracting various spatial and temporal features such as the locations most frequently visited by each criminal as reflected in their phones' connections with each cell tower, which would also allow the measurement of the relevant distances between criminals' home locations, crime locations, and other time-stamped locations.

5. Research Questions

The review introduces two research questions to cover the absence of data in the literature and what existing literature lacks in the fields of criminology and urban sensing.

5.1. RQ1: What Are the Current State-of-the-Art Methods and Techniques Regarding the Use of Mobile Phone Data to Identify Suspects and Predict Crimes?

Before reviewing the state-of-the-art methods and techniques that employ mobile phone data for the identification of suspects and prediction of crimes, we first give a brief discussion on why such data can be seen as a sensor for human activities and mobility in the context of criminology.

Mobile phone data contain different kinds of digital traces, such as mobility traces and communication traces, which can be used as evidence in criminal investigations [89]. Therefore, the digital traces left by a large number of mobile devices provide valuable information that facilitates the understanding of human behavior and mobility in the context of criminology, such as the prediction and identification of crimes and suspects. For example, Griffiths et al. [12] analyzed the mobility behaviors of criminals based on the digital traces they left at home and other meaningful locations, such as the crime scene. The mobility traces of criminals were identified by cell tower locations where they previously received a call. The traces were then analyzed to determine the regularities in the criminals' movements and to investigate whether the movements were not random. The authors subsequently concluded that there is a high degree of spatial regularity in the criminals' movements.

We report the state-of-the-art methods and techniques concerning the use of mobile phone data in identifying suspects and detecting criminals. A particular focus is given to existing scientific literature that has explored the use of mobile phone data in the context of criminal behavior from people- and place-centric perspectives. Our taxonomy of crime study concerning mobile phone data identified three applications: suspect identification, criminal network detection, and investigating the correlation between human mobility patterns and spatiotemporal crime patterns.

5.1.1. The First Group of Applications Deals with Using Mobile Phone Data to Identify Suspects

A suspect is defined as an individual who is suspected to be involved in a crime [47] based on the digital traces left at the crime scene. In criminal investigations, location-based mobile phone data can be used to indicate the presence of suspects in an area at a certain time when a crime has taken place, whereas communication traces can be used to identify accomplices in a criminal activity. As examples, References [82,84] collected mobility traces left at crime scenes to determine suspects' positions and presence at the crime scenes.

At the identification phase, the literature has shown that researchers used several parameters and attributes to determine suspects (e.g., outgoing calls, incoming calls, the start time of the call, location, duration, the number of calls made, and messages received). For example, in [9,83] methods, the researchers extracted communication information such as "outgoing calls", "incoming calls", frequent callers", and "maximum duration" to identify suspects. However, Reference [87] extracted spatiotemporal information along with communication details to identify suspects.

5.1.2. Suspect Identification Models Can Be Divided into Unsupervised and Supervised Models

Unsupervised models use unlabeled data and subjective definitions for the identification of suspects (e.g., "suspects are those who contacted previously contacted criminals and also made calls nearby the crime scene").

Supervised models use historical data where each user is labeled as suspect or non-suspect and try to find patterns in phone call data records that distinguish between those who were historically selected as suspects and non-suspects.

Khan et al. [85] used CDR data of various suspects and victims in order to extract associations between pairs of telephone numbers that can point out a few correct directions for identifying the most likely correspondence between suspects and victims. The methodology was based on the idea that frequent calls and the duration of calls may be indicative of a criminal–victim relationship.

The technical implementation was conducted using a combination of Hadoop (a framework for distributed processing of large data sets across clusters of computers) and Hive (a data warehouse architecture for querying data stored using Hadoop). The choice of tools was justified by the widely known efficiency of these tools for mining big data and by the security of Hadoop, which is important due to the use of highly confidential data. Even though it is a "simple implementation", some important weaknesses are worth mentioning. There is no evidence that frequency of calling or maximum call duration are helpful in identifying actual criminals, as authors do not validate their model against ground truth data. They used only a very limited set of call features, not including spatiotemporal characteristics, to identify suspects. The use of location data would have been useful in placing the suspects and their accomplices at the crime scene.

In Reference [83], the authors used CDRs data to identify links that exist among criminals and anti-social elements. Their analytic approach was based on the idea that anti-social elements have their own network of contacts, and the identification of those closely linked to previously convicted people is helpful in shortlisting suspects. That is why their network analysis methodology was based on graph theory as a tool for identifying otherwise hidden relationships.

While the authors demonstrated an actionable graph-based decision support tool for streamlining inference that would otherwise be difficult to achieve using slicing and dicing data in spreadsheets, the study has some weaknesses. The approach relies on looking for exact matches in long-term historical data and thus makes the unrealistic assumption that suspects do not change their mobile devices after communicating with convicted criminals, which looks like very incautious behavior for experienced criminals.

Authors in Reference [9] proposed a supervised machine learning method to identify suspects. The task was to classify users into suspects and non-suspects. The researchers turned the CDR-level dataset (13 million rows) into a user-level dataset (10,000 rows) through data aggregation and feature generation. As a result, each user was characterized by 30 discrete features derived by discretizing such features as the number of calls made, the average duration of calls, the proportion of calls and text messages, etc. A targeted Bayesian network learning (TBNL) model was applied that resulted in a descriptive network in which the selected features and their interactions were used to discriminate between positive (e.g., "suspects") and negative (e.g., "non-suspects").

The model was validated using 10-fold cross-validation, where the sample of 10,000 users was split into 10 random folds of 1000 users each, and every time, 9 folds were used for training and 1 fold was used as a holdout sample for testing purposes.

The main strengths of the study are the demonstration of the proposed model's strength relative to several competing algorithms and the fact that the model results in actionable empirical facts about factors that increase the probability of being a suspect. The main weaknesses of the study are as follows: The authors demonstrated that the text-to-call ratio is substantially lower for suspects than for non-suspects in the late morning hours but did not provide the same type of interpretation regarding other predictors. The cross-validation procedure was used for optimizing model parameters, feature selection, and measuring its performance, which could inflate performance metrics. This could have been avoided by keeping around 10% of the dataset for final model testing.

Hassan et al. [47] applied a Graph Convolutional Network model (GCN) in order to identify suspects from non-suspects. The authors built a straightforward undirected graph (G), represented by the input matrix A, which was to be applied to two-dimensional convolutional layers. Graph G contains six nodes, with the features of each node used to classify criminals from non-criminals. Hence, the authors performed a semi-supervised classification method on a small number of labeled data to train the classifier (seed nodes belong to convicted criminals to help train the classifier). The output then is a single binary for each node, indicating whether the corresponding node is predicted to be a suspect or not. Although the proposed method has yielded promising results, it is not without limitations. To begin with, CDRs were fully employed; therefore, the communication information only featured the node. Second, because the resulting network of CDRs data is rather sparse, modeling the network in the context of seed nodes may require domain-expert knowledge. Methods in deep learning are usually effective as long as they require large numbers of data. Thus, Hassan's model produced a sparse network due to the limited training sample.

5.1.3. The Second Application Deals with the Detection of Criminal Relationships Based on Communication Behaviors and Mobility Patterns

Once the suspects in a crime have been identified, it is important to investigate the roles that each criminal plays within a specific network. Connecting a suspect to other perpetrators and understanding their relationships with criminal networks is difficult, and thus the use of CDRs data has been increasingly exploited by social network analysis (SNA) tools and metrics, including degree centrality and betweenness centrality, all of which can be used to identify influential members and low-level members of criminal networks.

5.1.4. The Construction of a Social Network from Mobile Phone Data

A graph (or network) can be used to model mobile phone data (Graph $G = (V, E)$. A graph contains various nodes (or vertices) that represent different mobile phone users, and the edges E represent text messages and calls between two individual users.

Studies on social networking cover many topics, including community detection, social network structure, and measuring network modularity. Partitions or clusters of nodes in a graph are typically known as communities in network investigations. Communities can be structured in two ways: non-overlapping structuring, in which nodes belong to only one community, and overlapping structuring, in which nodes can be part of multiple communities. Several researchers have used social network analysis tools to solve the community detection problem [45,46,90]. SNA tools can also help investigators understand the hierarchical structure of criminal networks since it is difficult for forensic analysts to determine who belongs to a criminal organization and the relationships that exist within it. Thus, SNA can be harnessed to determine the relations and interactions between criminals by reconstructing the communication relationships that are obtained from mobile phone data as a network, where a node represents a criminal and an edge represents a communication (i.e., a phone call or a message). Using this approach in the analysis of criminal networks allows the investigators to understand the hierarchy and structural properties of the network.

5.1.5. Detecting Criminal Networks Based on Communication Information

In this section, research that has used social network analysis tools and measures to detect criminal networks based on mobile phone data will be presented.

In the relevant works of the literature, a collection of SNA techniques and algorithms have been used to detect and probe criminal networks. These approaches and algorithms have primarily been used to solve problems relating to community detection, while statistical metrics have been used to analyze relationships between vertices and assess structural centrality in networks. Several researchers [10,28,43] have developed many detection methods based on communication information extracted from mobile phone data to detect communities in criminal networks. Moreover, these researchers have employed the same analytical method used to investigate criminal networks (namely, social network analysis) but with different detection algorithms.

For instance, Ferrara et al. [10] proposed LogAnalysis, a criminal investigation expert system, to detect criminal networks. This system incorporates a well-known detection algorithm called the Girvan and Newman (GN) algorithm. They opted to use the GN algorithm due to its capacity to identify edges in networks lying between communities (when edges are less central, they are most likely to fall "between" communities). Subsequently, the system removes these edges, leaving the communities behind. The researchers have used this system to identify interconnected nodes that belong to different clusters and gradually remove them, which disconnects the clusters and ultimately reveals the community structure. Then, edge-betweenness and centrality metrics were calculated. The measurements focus on the less central edges, where the edges are most "between" communities. This is more effective than using a measure that focuses on the central edges. In the experimental setup, 381 nodes and 428 edges were identified. After the mobile phone network was configured and the detection algorithm was applied, a total of 16 communities were identified. The key objective was to identify edges from interconnected nodes that belong to different clusters (different communities) and progressively remove them. Therefore, the edge-betweenness centrality measure was incorporated in the algorithm, which facilitated the removal of 28 edges and the development of a community consisting of multiple groups (after each node is assigned to one cluster). The findings also reveal that all vertices are linked to a central vertex, which serves as a hub and generates a centralized network. This happens because GN is greedy in its approach to clustering and focuses on collecting vertices in the network [11]. To perform this, a number of rules are followed, after which vertices are merged to create a coherent division of the criminal community structure. Therefore, if the central vertex is removed, a hierarchical network can be created, which, in turn, enables subgroups to be identified through their interactions with other group members.

After the researchers had identified the social criminal network, they plotted it on a graph using a visualization tool. Visualization plays a major role in increasing investigators' comprehension of the complexity of the network; thus, it is a useful tool for visualizing and presenting complicated networks. They tried three different visual layouts, namely, the node–link diagram, the convex hull, and the force-directed layout.

Following that, Agreste et al. [29] worked with Italian law enforcement to collect mobile phone data that revealed the communication details of the Sicilian Mafia group. This work was similar to that of [5] in terms of detecting and structuring criminal networks but different when it came to describing how the criminal network functioned. The researchers created two networks, one of which was based on the mobile phone data and thus contained the identities of 1716 suspects (vertices) and 8481 contact logs in the form of phone calls, SMS, MMS, etc. (edges). On the other hand, the second network was based on the relationships between various individuals involved in criminal acts. The two networks were merged through an aggregated network that enabled all pairs of nodes to be connected by an edge in at least one network. Meanwhile, the results show that there are several criminals who can be identified by correlating mobile phone data with crime data.

Other criminal detection systems based on social network analysis are carried out by adopting Prim's Minimum Spanning Tree (MST) algorithm in [28], the Concept Space Approach (space algorithm) in [11], and Blondel's algorithm in [47]. The most significant variation between these systems are the metrics and measures used to identify key members of criminal networks (see Table 6 for more details).

Table 6. List of methods, analysis approaches, and metrics in crime applications.

Reference	Analysis Perspective	Analysis Approach	Algorithm/Measure	Network Metrics/ Parameter	Limitation
[28]	Communication behaviors	SNA	Detection algorithm (Prim's Minimum Spanning Tree Algorithm)	Edge-centric	Missing location data, greedy algorithm
[10]	Communication behaviors	SNA	Detection algorithm (Girvan–Newman and Fruchterman–Reingold)	Edge-betweenness centrality	Complex network, detection only based on communication information, greedy algorithm
[11]	Communication behaviors	SNA	The concept space approach (space algorithm)	Vertex-centric	Suitable for small networks, detection only based on communication information
[51]	Mobility patterns	Regression and Correlation Analysis	Akaike information criterion (AIC), spatial autocorrelation (SA) using Pearson's Correlation, and negative binomial regression model (NBM)	Offender anchor points.	Detection only based on spatiotemporal information
[12]	Mobility patterns	Statistical and Correlation Analysis	Spearman's rank coefficient (ρ) statistics, Pearson's correlations, and the cumulative distribution function	Offender anchor points.	Detection only based on spatiotemporal information

5.1.6. Detecting Criminals Based on Spatiotemporal Information

On the other hand, studies have investigated the use of spatiotemporal information to identify criminal relationships and activities.

For instance, Hassan et al. [47] identified suspects by monitoring the spatial–temporal movements of criminals, while the authors in Reference [12] carried out cumulative frequency analysis using various statistical functions, including cumulative distributions and cumulative probability distributions, to determine whether criminals have routine activity spaces. Thus, the authors extracted spatiotemporal characteristics of criminals, such as the distances between their homes and safe houses (i.e., bomb manufacturers or armories) and the most commonly visited locations. The findings indicated that the criminals frequented particular areas, with most of their activity clustered between their home and safe house (crime location). Ultimately, this implies that criminals do not select targets randomly and that their movements are routine and steady.

Furthermore, Feng et al. [51] studied spatial variations in crimes perpetrated by both native and migrant criminals by correlating multiple mobility datasets, including offender data, mobile phone data, and points-of-interest (POI) data. The authors selected anchor points to identify criminal spatial patterns as well as to understand what motivates criminals to carry out crimes in the proximity of their homes. The findings revealed that offender anchor points are more prominent in native violent crimes than those perpetrated by migrants. This is because criminals' homes and crime scenes share similar spatial patterns. This means that native offenders are much more likely to use their homes as anchor points when selecting targets for their crimes. On the other hand, migrant criminals are more likely to be impacted by crime attractors, crime generators (such as bars, clubs, etc.), and areas with vast populations.

The studies of both [12,51] explored criminal anchor points. Anchor points may be residences, workplaces, or any significant area that a criminal leaves to carry out a crime. Anchor points play a critical role in identifying places of importance to criminals and in detecting the spatial mobility of criminals. This is because criminals typically target areas near their residences to commit crimes. In other words, the probability of committing a crime decreases as one moves further away from their anchor points; thus, violent crimes tend to take place near the offender's anchor points.

To summarize, some studies detected criminal networks by analyzing communication behaviors based on extracting call information, whereas other studies analyzed criminals' activities based on spatial–temporal mobility patterns, as presented in Table 6; however, detecting criminal activities by taking into account both criminal communication behavior and mobility patterns may be extremely useful [4].

5.1.7. The Third Application Deals with Using Mobile Phone Data to Investigate Human Mobility Patterns and Spatial–Temporal Crime Patterns

The spatiotemporal patterns of crimes can be determined by extracting human routine activities and mobility patterns from mobile phone data [52] and then examining the correlations between the human dynamics and crime data [64]. Accordingly, mobile phone data have been widely used in crime analysis and predictions to identify crime hotspots [14], investigate the relationship between ambient population and crime hotspots [21], and measure population density at a certain place based on the number of mobile devices connected to a given cell tower located in the area where a crime has taken place.

Unlike previous applications, here, a large sample of the population is considered as a measure to investigate the relationship between human dynamics and crime patterns. Such a measure helps gain further insight into exploring whether mobility patterns of the population can help predict where criminals commit crimes [67] or serve as a measure of ambient population-at-risk [66]. Estimating the correlation between population dynamics and crime patterns was earlier investigated by Bogomolov et al. [13], who extracted users' locations to estimate population counts at a given location. However, these studies [51,52] have been interested recently in finding out the correlation between the spatiotemporal patterns of crimes and human routine activities, which may ultimately help to provide information about criminals' movements since the mobility patterns of the general population provide a template for the mobility of criminals [67]. The third application thus investigates the relationship between population dynamics and crime patterns.

Multiple types of data and different spatial units have been proposed to investigate the correlation between human dynamics and crime patterns. The spatial unit of analysis is different according to the format of the data and the official providers. Census units are used in Reference [67] because they are homogeneous in terms of population composition, and Lower Layer Super Output Area (LSOA) units are considered in these studies [13,21,52,66]. Additionally, spatiotemporal characteristics extracted from mobile phone data, such as "the number of times a mobile device communicates with the network", "timestamp", "cell ID", and "the most-contacted tower during daytime or nighttime", are used in different contexts. For example, the Mobile Phone Origin Destination (MPOD) dataset is used in References [52,66], and the lack of spatial granularity is marked as a weakness, as is the density of signal towers, which is higher in urban areas and lower in suburban ones. This may cause some errors in spatializing the data; thus, a geographical information system (GIS) is used in order to distribute MPOD data across LSOAs.

Statistical models were suggested in the literature to investigate such a correlation. As statistical methods are employed, multiple statistical scores and parameters are used to calculate correlations. For example, in Reference [21], the authors observed that there is no normal distribution; thus, Spearman's rank correlation coefficient (ρ) statistic was used over Pearson's product-moment coefficient (r) to calculate correlations between ambient population and crime rates. In Reference [67], the authors applied a discrete choice model to test this hypothesis and determine if the daily mobility flows of the general population can provide a template for the daily mobility of criminals.

The methods used for accomplishing each goal are different, but one aspect is common to all papers: the variables taken into account as 'crime generators' are: underground stations; schools (i.e., middle and primary schools); music venues; hotels; hospitals; restaurants; supermarkets; clubs; bars; subway stations; and banks. The mean, standard deviation, minimum, and maximum values were calculated and reported at the census level. For example, there are 11.15 restaurants on average per census unit. In Reference [67], primary schools, hospitals, basic stores, bus stops, supermarkets, banks, and restaurants are listed as 'crowded spaces'. Song et al. [67] then used the conditional logit model, which aims to analyze the effect of distance, crime generators, and the role population mobility patterns play in offenders' choice of locations for committing TFP (theft from person). The results showed that all facilities except schools, markets, and bars function as crime generators, and so their presence shows a high likelihood of offenders committing TFP. Furthermore, with larger facilities that have significant effects, such as subway stations, cinemas, or hospitals in the census unit, the odds of being chosen increase by 57.0, 15.4, and 13.5%, respectively. The results also showed that there is a strong correlation between a criminal's home and crime sites, where criminals often choose places nearby to commit crimes close to where they live.

In the study [21], the variables are residential population, workday population, geolocated Twitter messages, mobile phone activity counts, population 24/7 estimates, and theft from the person who committed the offense. Malleson and Andresen [21] used Getis–Ord Gi*statistics, which examine each location i (LSOAs in this case) together with its neighboring locations j, and then "it calculates whether or not the total number (or rate) of occurrences in i and j is greater or lesser than would be expected by chance when compared to surrounding locations up to a distance from i. If a difference is found, then the areas I and j are assumed to be associated with and different from their surroundings, i.e., a hotspot or coldspot." The results showed that there is a poor correlation between the residential and ambient populations. On the other hand, strong correlations are noticed between some of the measures of the ambient population (workday population, mobile phone data, and population 24/7 daytime estimates). Moreover, the correlation between thefts and ambient population is stronger than the one between thefts and residential population. Thus, for calculating the crime rate, the ambient population is more suitable than the residential population.

However, Haleem et al. [52] calculated both the ambient and exposed population-at-risk by correlating two datasets: census data to capture residential population counts and mobile phone data to capture transient population counts. This procedure allows for the estimation of the ambient and exposed populations for different time bins. The ambient population-at-risk, thus, was calculated by estimating the residential population at a given spatial unit, summing it with the population entering this unit at a certain period of the day, then subtracting the population existing in this area for the same time of the day. The Spearman's rank correlation coefficient (ρ as rho) statistic was used to evaluate the correlation between the ambient and exposed population-at-risk measures and violent crimes. The results showed that the exposed population is more significant than the ambient population, and the exposed population measure appears to be a more suitable denominator for exploring violent crimes in public space.

5.1.8. Recent Advances in Method

Recently, a variety of machine learning models and social network analysis techniques have employed mobile phone data [4] to improve criminal network detection, fraud activity detection, and crime prediction.

In recent years, the reconstruction of social networks from mobile phone data by means of graph theory and social network analytics (SNA) has become common in mobile phone data studies. Graph theory refers to the mathematical study of interactions between sets of nodes (otherwise known as vertices) linked by edges. Through the use of social network analysis tools and methods, computer and mobile social networks, including the internet and mobile communications, can be represented graphically in this manner, and graph theory techniques have thus been widely applied in the field of mobile phone data to identify various types of social networks, including the detection of criminal networks [91], the identification of ethnic communities [81], the development of specific socio-economic communities [92], and the determination of geographical networks [93]. This has become possible due to the fact that call data and spatial–temporal data acquired from mobile phones disclose multiple details about a variety of communication links and dynamic networks. The communications recorded on a mobile phone are assumed to constitute a representative part of a person's overall social networking, with mobile phone data creating a social network among those individuals making or receiving calls or messages, who are classified as actors (nodes) within the network; each link between the actors is then represented by the type of communication (call or message). Empirically, the resulting social networks are constructed based on both the communication behaviors (calling information) and the spatiotemporal information (mobility patterns) extracted from the mobile phone data, allowing observation of a range of social interactions. A network can thus be constructed based on the call patterns created by all the individuals making or receiving calls or messages in the network. A geographical network may, however, be based instead on spatiotemporal information, with the nodes set as geographic locations (e.g., cell towers) and the edges between nodes being represented by the interactions (mobile phone activity) between pairs of these cell towers.

Cavallaro et al. [91] reconstructed the criminal network of the Sicilian Mafia by applying SNA tools and matrices to identify key leaders and their reports, such as bosses and intermediaries. Ficara et al. [94] built a network of suspected criminals based on their calling information; here, the nodes were represented by suspected members and the edges were represented by phone call records. Dileep et al. [95] similarly proposed a forensic detection system to detect the development of suspicious communities based on extracted phone call records.

Statistical methods and machine learning techniques have also been employed to predict crime and detect fraud in other ways. For example, Bogomolov et al. [13] extracted human mobility patterns from mobile phone records to predict crime hotspots in London by using the Random Forest classifier to classify geographical areas into two classes based on whether they displayed high or low crime levels. However, Wu et al. [96] criticized previous data collection methods such as CDRs, Twitter, and Foursquare data in terms of errors in estimating mobility flows for crime prediction, choosing instead to estimate human origin–destination mobility flows using GPS data alongside applied deep learning models such as the gated recurrent units (GRU) model and the graph convolution network (GCN).

While many existing studies have used correlation and regression analysis to investigate the relationship between human dynamics and crime spatial–temporal patterns, Rummens et al. [22] used mobile phone data to investigate whether residential populations or ambient populations have the greatest positive impact on crime rates; the results showed a stronger correlation between ambient populations and crime rates, particularly those for bicycle theft and aggressive theft. Going further, while previous studies examined the impact of ambient and residential population on crime rates, Long et al. [97] aimed to investigate the impact of ambient populations on street robbery rates by applying correlation and regression analysis; they found that the ambient population has a significant effect in terms of reducing opportunistic street robbery and similar crimes. Long and Liu [98] also applied discrete choice models to investigate spatial differences in the patterns of two types of criminals committing street robberies, namely, migrant robbers and native robbers; those results suggested that migrant offenders tend to commit street robberies outside of the old town areas, in industrial areas, while native robbers prefer to commit crimes in villages and older urban areas due to their familiarity with the area, supported by the high mobility of the population and high socioeconomic heterogeneity.

Some studies have employed mobile phone data to detect suspicious and fraudulent behaviors for telecom companies, such as fraud call detection methods based on machine learning. Studies [99–101] proposed a range of deep learning models, such as deep neural networks (DNN), convolutional neural networks (CNN), and graph neural networks (GNN), to detect fraudulent phone calls, for example. Using unsupervised learning techniques, such as K-means, density-based spatial clustering of applications with noise (DBSCAN), and hierarchical clustering, References [102,103] also sought to detect fraudulent behaviors for telecom companies, such as fraudulent calls and suspicious call records. Finally, Reference [104] aimed to detect suspicious call behavior by using a range of supervised and unsupervised learning models, including K-means and Random Forest.

5.2. RQ2: *How Can Identifying Empirical Mobile Phone Data Studies to Predict Human Behavior and Mobility Patterns Contribute to a Clearer Understanding of the Dynamics of Criminal Behavior Contexts through a People- and Place-Centric Perspective?*

This question was designed to explore research that has utilized mobile phone data to gain a greater understanding of human behaviors and mobility patterns in urban environments. Experts need to explore people's actions and activities in the urban areas in which they live and socialize and classify individuals according to their mobility patterns so that the authorities can determine population flows in these zones before and during crimes and provide significant information about criminals' movements while they are engaging in criminal activities.

These approaches can generate significant information about the tools and methods previously used to analyze mobile phone data, as well as provide a broader understanding of human and/or individual actions and activities.

5.2.1. Human Mobility Patterns in Urban Environments

The spatiotemporal information provided by mobile phone data can help one understand population behavior and mobility patterns in several applications. Investigating population mobility patterns helps one to understand the way humans live, since such patterns reflect the places they visit and stay in the most, as well as their movements during working hours and weekends; thus, many studies use mobile phone data to understand human mobility patterns. To name a few, Thuillier et al. [19] classified individuals into 6 groups based on their daily mobility profiles to comprehend the mobility flows of individuals inside a territory in southwest Paris. These profiles were developed by leveraging the spatiotemporal characteristics extracted from mobile phone data. Ghahramani et al. [71] estimated the frequency of calls at each spatial object (cell towers) to construct a map of hotspots in China. These studies deal with human mobility in urban settings, where crimes are more likely to be committed. Therefore, it is highly important to analyze human mobility patterns inside the city since understanding where and how populations live and socialize and classifying individuals based on their mobility can help to understand population flow [31], which may ultimately help provide information about criminals' spatial–temporal patterns [67]. This section then reviews important contributions to the study of mobile phone data in urban settings.

5.2.2. Land Use Inference

Long-standing discussions in several disciplines have focused on the connection between land use and human mobility patterns [105] extracted from mobile phone data. This is because understanding the relationship between human activity and land use can help to provide valuable insights into human dynamics and interactions with their physical environment, such as depicting human lifestyles in urban areas and how humans interact and socialize, and investigating the impact of the land use characteristics (commercial, industrial, residential) on urban crime. Thus, many studies have acknowledged the importance of classifying land use to understand the relationship between land use patterns and human activities and interactions.

The classification of land use patterns for visitors or in residential or business areas can be conducted based on extracting human activity characteristics from mobile phone data. Specifically, spatiotemporal and call features extracted from mobile phone data can be used to depict human activity characteristics and infer land use types. For example, References [20,106] explored human activity patterns to infer land use based on spatiotemporal calling volume patterns. Novovic et al. [45] employed user activity variations in space and time to depict human activities; commuting flow patterns to infer land use types was investigated by [107]; and Lenormand et al. [90], along with Ríos and Muñoz [37], inferred land use based on the temporal changes in human activities.

5.2.3. Spatial Distribution of Mobile Phone Presence from Cell Towers to Census Spatial Units

Determining the spatial distribution of mobile phone presence in a cell tower's coverage area is an important step that needs to be resolved before conducting any further analysis, and it is a common standard procedure for mobile phone data processing. This requires that the spatial configuration of the base stations of a mobile network be matched with the census data. In order to match census data with mobile phone data, we must coincide the spatial scale because the use of different spatial units introduces difficulties when comparing the datasets [3]. Census data are collected according to geographical areas, such as blocks, tracts, or at the country level, whereas mobile phone data are collected at the base station. Thus, the spatial distribution of the base stations should be equal to the spatial units of the census data.

Identifying the position of a mobile device is based on the location of cell towers, which serve as a proxy for the mobile device. The cell towers are represented as Voronoi cells (polygons) using Voronoi tessellation [58]. Voronoi tessellation is used to visualize the position of mobile phones inside the cell towers' coverage area, which has been approximated as a Voronoi region of a cell tower. The Voronoi diagram contains a point for each cell tower, where the centroid of each point is based on the location of the corresponding cell tower. The resulting Voronoi cells can be viewed as a partition that corresponds to the optimum distribution of towers in a geographical area in a cellular network layout in the real world.

Hence, it is important to perform the spatial distribution of the base stations of a mobile network such that the bases correspond with predefined census units to obtain a fine-scale spatial resolution and to represent the spatial scale of the census data collected at a spatial unit with the mobile phone data collected at the cell towers. This entails the matching of the spatial configuration between the base stations and the census data, which represent the same geographical units.

6. Discussion

Previously, this work reviewed the mobile phone data domain and its applications in the areas of crime analysis and urban sensing, developing a consistent taxonomy based on a scientific approach in which studies can be classified at the first level based on human behavior analysis, then subcategorized based on the mobile phone data types used, before being finally classified based on applications derived from each mobile phone data type. This taxonomy helps to answer the research question, shed light on the current state of mobile phone data applications and the current investigation trends in mobile phone data, and highlight existing research gaps. This process was followed by the formulation of two research questions intended to investigate human behavior from both mobility and communication perspectives, the investigation of which helped to generate significant information about the tools and methods previously used to analyze mobile phone data as well as providing a broader understanding of human and individual actions and activities.

The purpose of this section is thus to discuss privacy concerns and investment behavior, and shed light on the emerging common challenges.

6.1. Privacy Concerns and Ethical Implications

Previously, this work discussed the benefits that such data can provide for the community and researchers in terms of fighting crime, detecting congestion zones, enhancing urban infrastructure design and urban planning, fighting epidemics, and preventing the spread of infectious diseases. However, mobile phone data are subject to various limitations, including the risk of privacy breaches due to their containing sensitive information about individuals' locations and their communication information. The potential for a breach of these sensitive details thus raises both privacy concerns and ethical questions about the use of such data.

Various privacy-preserving techniques have been suggested to address this issue. Arcolezi et al. [108] proposed the use of local differential privacy (LDP) techniques, in which each user's CDRs data are sanitized in the server held by the mobile network operator (MNO) before any data collection processes are performed, while Arfaoui et al. [50] proposed the application of specific anonymization techniques, such as suppression, k-anonymity, and L-diversity, to help guarantee anonymity and prevent personal identification of users. To protect mobile phone users' location privacy, Gramaglia et al. [109] also proposed a privacy model based on the application of generalization and suppression techniques to achieve k-anonymity in terms of mobile phone spatiotemporal trajectories.

Some authors have also provided recommendations with regard to the multiple ethical implications of such data use. Vespe et al. [110] suggested the development of an expert group of telecommunication engineers, data scientists, lawyers, and data protection and ethics experts, with the aim of addressing various scientific challenges to develop sound data security and protection protocols, alongside the establishment of an Ethical Committee to take on the mission of considering all ethical aspects of work in this field. Similarly, Cinnamon et al. [111] encouraged researchers to facilitate the development of global mobile data usage guidelines, regulations, and standards to provide rapid, secure data access for organizations and researchers that included rapid and efficient techniques for detecting gaps and biases in mobile phone data. Boenig-Liptsin et al. [112] developed a data science lifecycle framework that aims to educate data science students and researchers about the ethical elements of their work and teach or promote ethical principles for responsible data science.

However, privacy implications and ethical concerns still represent challenging obstacles in terms of the use of mobile phone data. In particular, most existing solutions and recommendations have been based on theoretical frameworks rather than empirical work, many of which are impractical and do not conform with national and international data protection regulations. Most existing privacy techniques thus rely heavily on anonymization solutions.

Recently, mobile phone data have been widely used to combat the spread of infectious diseases in emergency situations such as the COVID-19 pandemic and to prevent criminal activities such as terrorist attacks and street robberies. Ignoring the previous advantages of mobile phone data in enhancing quality of life and ensuring citizen safety, some fears and ethical implications have been raised about the violation of people's privacy and liberty and the imbalance of justice between the right to preserve personal data and law enforcement. However, during emergency situations and natural disasters, government surveillance operations serve to enforce laws against terrorism and serious crime and to place restrictions on people's basic liberties [113]. For example, in an emergency situation in which suspected terrorists are suspected of criminal activities, the acquisition of their mobile phone data is warranted for forensic analysis and real-time monitoring of their mobility dynamics. Thus, in some cases, it is difficult to strike a balance in data protection during dangerous situations such as terrorist attacks. In addition, creating a balance between controlling national security threats and preserving the personal privacy of suspected phone users is questionable when it comes to the public security and safety of citizens, which are more important than preserving users' data rights.

Mobile phone data are not the only critical data form that suffers from privacy complications; such concerns have been an ongoing topic with regard to other mobile sensing data. With the growing popularity of mobile wireless devices equipped with various kinds of sensing abilities and a plethora of on-board sensors, the emergence of a large variety of people-centric mobile crowd-sensing (MCS) systems has been rapid [114,115], raising additional concerns. As a result, MCS has become the main emerging sensing paradigm for large-scale sensing applications [116], and it is now used in a range of applications that includes urban dynamics mining, public safety, traffic planning, and environmental monitoring [117].

Mobile crowd-sensing systems are designed to collect city-wide spatiotemporal data [118] from a range of embedded and connected sensors such as GPS sensors, air quality sensors, cardio meters, and health care sensors [119]. However, although these recordings of valuable information offer various benefits for communities in terms of transportation planning and developing public health in communities, such data contain sensitive spatiotemporal information about individuals, such as home addresses, work locations, and health records, which may create possible threats to user privacy if such data are misused or re-identified [120] by attackers.

Privacy-preserving mobile crowd-sensing systems have thus been proposed to preserve and protect user privacy. Agir et al. [121] proposed a form of location privacy protection based on location obfuscation techniques while preserving worker location privacy. Jin et al. [122] designed an auction-based incentive mechanism for MCS systems that enabled data owners to sell location trace information and choose the level of location information to disclose to the MCS system. Chen et al. [123] also proposed a blockchain-based, decentralized framework for MCS systems that aimed to detect fake tasks input by malicious requesters as well as guarantee the task information was not tampered with.

This study contributes to addressing such aspects of privacy concerns in data formats such as MCSs by examining how the approaches proposed can preserve user privacy and protect their information. This is conducted to allow other privacy-preserving mechanisms to be adopted in a mobile phone data context, helping scholars discover new tools and mechanisms for protecting and preserving user privacy.

6.2. Investment Behavior

In recent years, advances in artificial intelligence and sensor technology as part of the technological revolution have influenced investment behavior and provided opportunities for corporate development [124,125]. Investments in the field of healthcare have produced highly advanced sensor technology, and many technology companies have invested in digital health products, such as new screening interventions and diagnostic testing. For example, the conversion from the Sanger sequencing method to parallel processing technologies in next-generation sequencing (NGS) has resulted in a significant decrease in the cost of whole-genome sequencing over the past 13 years [126]. There have also been advances in microfluidic technology and devices used to investigate cancer biology and cancer diagnostics. Microfluidic devices are favored for cancer cell detection because of their high sensitivity, control of fluids in the range of micro- to picoliters, and low cost [127]. Similarly, as the high-precision scientific research industry rapidly grows, so do the demands for extremely sophisticated sensor technologies [128]. For example, Reference [129] proposed a photonic spin Hall effect (PSHE) sensor with high sensitivity and the ability to detect both cancer cells and biomedical blood glucose. Thus, with the emergence of health technologies and technological advances in healthcare, individuals are able to make better decisions on how to invest in their health based on the available technology, and firms are able to make better decisions about investing in technology that is both profitable and effective at disease prevention, diagnosis, and cure.

Advances in agricultural technology have also played an important role in farmers' investment intentions and willingness to invest [130]. Despite the fact that investments in big data analytics solutions are still risky and the cost is substantial [131], firms that invest effectively can benefit from increased customer satisfaction and market performance [132].

6.3. *Challenges*

Although mobile phone data have proven useful in various domains and disciplines with respect to understanding human behavioral patterns, the literature shows a number of serious issues and challenges arising with regard to data access and analysis.

6.3.1. Data Acquisition Challenges

While accessing the required datasets in any study may be hard work, mobile phone data can be among the most difficult to access for several reasons. In particular, mobile phone data at the individual level (CDRs data) contain a wide range of sensitive details about personal characteristics that may expose a person's identity and personal characteristics. Calabrese et al. [3] thus recommended the use of mobile phone data at the aggregated level and at the cell tower level. However, mobile phone data at the cell tower level lacks calling information, and it is thus not suitable for applications related to human social interactions and communication behaviors. The literature also shows that the use of mobile phone data at the individual level in criminology studies always requires permission from police and law enforcement, which might be a long process.

Several studies have attempted to provide solutions and recommendations to protect and ensure data privacy so as to facilitate access to mobile phone data, such as that conducted by De Montjoye et al. [133], who proposed a remote access model wherein mobile phone data are held by mobile phone operators. However, accessibility remains a significant barrier to using mobile phone data, as governments and businesses are reluctant to make such data public due to privacy concerns [31].

6.3.2. Data Analysis Challenges

The first challenge arising during analysis is that mobile phone data are unlabeled, causing issues around later labeling, especially in supervised ML approaches that require a model to be trained with ground truth data. In the absence of ground truth data, some studies have turned to the use of semi-supervised models in which the model is primed with a small amount of labeled data, such as [20,78], with still others relying on a process of data annotation using domain experts and manual labeling, such as in [39,40,134]; however, the latter requires a lot of manual work.

6.3.3. Challenges Related to the Standardization of Mobile Phone Data Keywords (Terms)

In the literature, we found that there was misunderstanding and misuse of the correct terms for each mobile phone data type. For example, several studies have used various terms or keywords that refer to the mobile phone data, with "the mobile phone data" and "call detailed records data" being the most frequently used in the vast majority of articles. This makes it difficult to search for related papers on this topic. More specifically, when reporting relevant journals and conference proceedings that focus on the topic of mobile phone data in a systematic literature review (SLR), the inconsistent use of keywords by authors makes it infeasible to create search strings that cover all studies within the aforementioned domain. Furthermore, some papers completely omit the relevant "mobile phone data" terms from their keywords or abstracts, necessitating extra time to be dedicated to scanning the full text of such papers, which is a considerably tedious and inefficient job.

7. Problem Definition and System Model

This section defines the current problem in mobile phone data with regard to detecting criminal behaviors and proposes a system model to overcome the current challenge.

Problem definition:

The current state of mobile phone data in the context of detecting criminal behaviors and dynamics is still incomplete and inaccurate [4] due to challenges in mobile data preprocessing and analysis.

The problem with pre-processing mobile phone data comes from the fact that raw data can be rough, noisy, and sparse, making it hard to work with. Therefore, the data must be cleaned and preprocessed before being used [71,72,100]. Additionally, during analysis, the use of incomplete or partial mobile phone data, missing values, and partial information (incomplete mobility and calling information) can result in the misleading and inaccurate detection of criminal behaviors and a partial aspect of human behavior [4,79,135,136]. As a result, there is a need to address the issues of incomplete and inaccurate preprocessing and analysis of mobile phone data to improve the detection of criminal behaviors and dynamics.

Proposed model:

Based on a review of the available literature, a forensic analysis system for the detection of criminal behaviors and dynamic activity is proposed for future research. Figure 14 illustrates the different steps of the criminal detection model, which is composed of two stages.

The first stage incorporates mobile phone data at two levels (individual and cell tower levels) to capture different aspects of criminal behaviors (communication behaviors, social networks, and mobility patterns). This stage is then followed by data preprocessing and feature extraction, with several tools and techniques applied, including spatial mapping, feature dimensionality reduction, and uncertainty reduction methods.

The first step in the first stage is the data collection process, which includes the gathering of two types of mobile phone data: mobile phone data at the individual level (CDRs), which can represent the mobility and communication records of suspected criminals, and cell tower location data, which represents a larger sample of the general population, including victims, criminals, suspects, and visitors, with the latter indicating individuals' locations at the moment a crime takes place based on their use of cell towers in the crime area location. These can thus be used to investigate the relationships between human dynamics and interactions and spatiotemporal crime patterns, along with crime scene data that provides spatiotemporal information on crime incidents according to the official records. The second step is the preprocessing of mobile phone data, which includes the extraction of stay points to detect home location and other meaningful location and spatial mapping techniques to intersect or project mobile network cells into spatial units. The third step involves feature extraction, which helps describe criminal behaviors in terms of spatiotemporal features and call features.

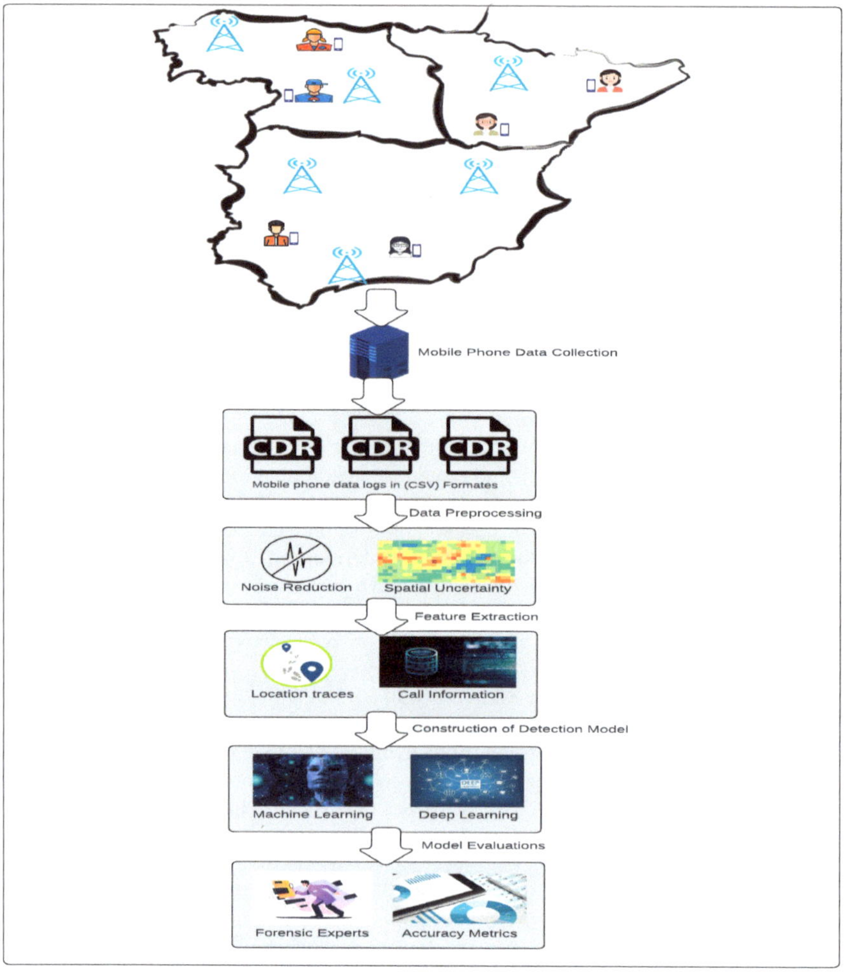

Figure 14. The proposed system model for detecting criminal behaviors.

The second stage itself is composed of two steps: analysis and validation. In the analysis step, the construction of a detection model is performed using multiple machine learning classifiers to classify individuals as criminals and non-criminals, thus constructing a classifier that enables the recognition of criminal activities based on spatiotemporal and phone use features. These algorithms are then evaluated to determine the most effective ones, which should yield better results than the other classification algorithms. The results are then used to build up a criminal network of suspected criminals based on applying social network analysis tools and metrics. The construction of a criminal network is conducted to assist law enforcement and crime agencies in identifying the most influential members (who issue commands) and low-level members in a criminal network, clarifying each member's role in the relevant criminal organizations. The final step is to evaluate the detection model in terms of its accuracy in detecting criminal activity, and this also involves evaluating the model results with the help of forensic experts.

8. Future Research Directions, Conclusions, and Limitations

The present review of the existing literature reveals possible directions for future research. The findings highlight aspects that should be considered with regard to data collection, data preprocessing, data analysis, and other considerations.

8.1. Data Collection

While collecting mobile phone information, researchers need to consider some points. First, these data should be obtained from leading mobile network operators with a minimum market share of 40–50% and a network providing spatial coverage for 95% of the target population, although these criteria can vary depending on the studies' goals. Second, researchers need to collect mobile phone data with a full range of users' attributes, including mobility and communication characteristics. Records that lack all or part of this information hamper analyses and make interpretations of human behavior difficult or even provide misleading evidence.

Last, anonymization is another important step that helps safeguard personal privacy. Before mobile phone data depart storage facilities, mobile telecom operators must anonymize subscribers' phone numbers and replace them with a unique security identity [5]. During analysis, k-anonymity techniques and approaches should be applied to avoid exposing personal characteristics or leaving enough patterns to reveal individual users' identities.

8.2. Pre-Processing Steps

8.2.1. Labeling Home and Other Meaningful Locations

Previous research has shown that identifying home and other meaningful locations is a crucial step in handling mobile phone data, which is part of pre-processing this information so that further analyses can be conducted using this information, as has been conducted in multiple studies [13,18,34,52,67,137]. Identifying these locations provides a better understanding of human mobility patterns and increases the comprehensiveness of the conclusions that can be drawn from the data. For example, Griffins et al. [12] first established the location of criminals' homes to clarify their involvement in terrorist attack plots since criminal activities often take place at or near frequently visited locations and criminals often commit crimes close to where they live.

8.2.2. Mapping Population Distribution

The geographical distribution of mobile phone users can be determined based on cell towers' coverage areas. This crucial step must be completed before proceeding with any further analysis. Defining geographical distribution is, more specifically, a standard technique used in mobile phone data, which has been conducted in multiple studies [57–63]. It requires a correspondence between census or land cover data and the spatial structure of a mobile network's base stations because unevenly distributed cell towers will hamper any attempt to map population distribution. Each mobile phone's geographical location is assigned to a specific cell tower that provides the network signal, so mobile phone location data's accuracy depends on the towers' coverage area. The literature shows that researchers often allocate their target population to 1 km- or 500-m-grid cells using methods such as Voronoi tessellation, areal weighting, and dasymetric interpolation.

For instance, Deville et al. [6] applied areal weighted interpolation to the spatial distribution of each cell tower's coverage area matching a specific spatial unit in order to map the relevant population's presence at that tower. The spatial unit used can vary between studies and can represent blocks, tracts, administrative units, or any other division that reflects how census data were collected. Deville et al. [6] calculated—for each cell tower j simulated and delineated as a Voronoi cell—the population density based on the number of calls or mobile phone presence per cell tower (σ_{ci}), in which c_i denotes the Voronoi cell associated with cell tower j. Equation (2) was used to estimate mobile phone presence σ_{c_j} for an area of unit c_i that intersects c_j:

$$\sigma_{ci} = \frac{1}{A_{c_i}} \sum_{c_j} \sigma c_j (c_i \cap c_j) \tag{1}$$

in which A_{ci} is spatial unit c_i's area and $A(c_i \cap c_j)$ is the intersection between unit c_i's area and Voronoi cell c_j.

8.3. General Recommendations

8.3.1. Recommendations for Improving Interpretation and Justification

Providing a theoretical explanation can play a key role in interpreting differences in results. Justification is absent from the existing literature due to the absence of validation data, so Vanhoof et al. [69] observed that researchers could have trouble discussing results on a theoretical level and determining which outcomes and approaches are better. Blondel et al. [1] also mentioned the need to provide theoretical explanations along with empirical evidence, which can facilitate the interpretation of variations in results and, subsequently, the determination of which findings are significant.

8.3.2. Recommendations for Considering Spatiotemporal Information

Extracting spatiotemporal characteristics to visualize the geographical location of nodes (to visualize the spatial distribution of nodes, or subscribers, in a social network) has been missing in many studies, and current studies rely either on communication information or spatial information to construct social networks. Thus, we recommend including spatiotemporal information with communication information to investigate the interplay between criminal mobility patterns and social interactions. For instance, previous studies [10,11,28] have not considered spatiotemporal information to detect criminals (i.e., the geographic position of nodes is unknown) and have overlooked the spatial position of a node that can connect it to a crime scene or area. Moreover, geographic proximity offers opportunities for face-to-face interactions between individuals. Thus, during graph partitioning into groups of nodes, their geographic locations should be considered to be where nodes have a geographic position.

In addition, the identification of important members was founded on features extracted from communication information and conducted by placing a weight on the edges between nodes (criminals), such as the maximum number of outgoing phone calls or messages and call duration. Therefore, location data will play an important role in the weighting in that some nodes may not reflect the importance value of a given node in criminal networks. Thus, weighting edges by considering criminals' mobility patterns could affect results, since the weights of edges reflect their relational strength between the network's vertices.

Furthermore, few studies have attempted to investigate the relationship and interplay between all aspects of human behavior (mobile communication behavior, social networks, and mobility patterns). This suggestion should arise more in the future for investigating the interplay between communications, social interactions, and mobility patterns through the lens of mobile phone data.

8.3.3. Recommendations to Build a Data-Driven Approach

The study results show that various spatiotemporal and call features have been extracted from mobile phone data to depict or capture criminal behaviors and activities [4]. However, there is no generalized approach in which mobility and social (communication) characteristics can be extracted to capture human behavior, as the literature shows that there are various and multiple spatiotemporal and temporal scales to characterize human and criminal behaviors. Thus, a data-driven framework is needed to determine which measurements and characteristics can be extracted from mobile phone data to visualize and depict different aspects of criminal behaviors, as well as to differentiate and generalize all features and their different functionalities.

8.3.4. Recommendations for Labeling Mobile Phone Data

As mobile phone data are unlabeled, semi-supervised approaches are needed to tackle this issue. A small number of labeled data can be obtained from surveys, censuses, or other geospatial data sources, such as training samples. Mobile phone data can also be labeled by domain experts.

8.4. Conclusions and Limitations

This study conducted an SLR to gain comprehensive, up-to-date insights into the current state-of-the-art methods and techniques utilizing mobile phone data in crime-control applications, as well as research that has used mobile phone data to investigate and predict human actions and mobility patterns with reference to urban sensing, which can significantly assist researchers in forming a complete picture of all related crime dimensions. By including studies that have utilized mobile phone data to understand and predict human behavior, the present review made an important contribution to what topics need to be included and discussed to provide a complete understanding of how such studies can help meet objectives in this area. Exploring the movements of human beings in urban areas enables researchers to gain more profound insights into how humans live and the places they most often frequent. The present investigation thus examined the current state of mobile phone data usage in criminology research and shed light on the methods employed to process these data in order to understand the dynamics of human behavior and mobility in the context of urban sensing applications. The latter include estimating and mapping population density, inferring the correlations between human dynamics and land use, and detecting home and work locations.

This study was the first to review the research focused on human mobility and communication behavioral patterns and to make both variables the SLR's main focus in crime applications, in combination with a lesser emphasis on urban zones. The review covered the most prominent results reported thus far, in particular, analyses of mobile phone data in criminology. The current research is concentrated on detailed data processing and analysis techniques used to understand mobile phone data. The results also include a list of recommendations regarding which techniques and features to use and a discussion of the extant lacunae and obstacles to help researchers and scholars better plan their studies. In addition, the SLR explored which applications have been derived based on human behavioral patterns extracted from mobile phone data.

Although the present research's approach was based on standard SLR methodology, this study was still subject to limitations. First, the review was intended to provide up-to-date comprehensive coverage of the chosen topic, but the results are neither complete nor should they be regarded as a definitive summary of all the related research. This limitation is primarily due to the exclusion of relevant academic material published in other languages. Nonetheless, the studies included in this review were carefully selected from eight databases and published in international journals. A number of other relevant journals may also have fallen outside the scope of the current review. Those excluded cover, among others, studies of churn prediction, the transportation sector (e.g., transportation planning, transportation mode detection, and commuter trips), anomaly detection, and the epidemiology of infectious diseases, such as COVID-19, in which human mobility patterns have been investigated. These publications were left out because of the large body of literature available and the chosen research objective.

Last, the findings provide unique and potentially useful contributions to the field of criminology, including supporting the conclusion that mobile phone data's applications in the crime domain still have great potential for further extending the existing knowledge. These approaches can be adopted to explore other domains. On a technical level, the existing analytical perspectives on mobile phone data are somewhat similar in all academic fields that mainly focus on human communication behaviors and mobility patterns. For example, human mobility analyses have been widely conducted in many domains as part of varied practical applications in which mobile phone data facilitated the capture of individuals' spatial–temporal mobility patterns in a range of human activities associated with urban zones, crime, transportation, and the COVID-19 pandemic. Researchers can adopt more of the techniques and approaches applied in other areas, as well as combine two or more methods, to develop the current understanding of human mobility and interactions further.

In addition, the experiments reported in the mobile phone data literature have often incorporated a variety of different setups and assumptions, each adjusted to complement the techniques applied. In other words, the empirical research conducted with these data has involved various contexts and applications designed to serve each study's purpose. The current SLR provided a broad overview that can help scholars decide which tools serve their purpose best and discover new uses, thus opening up the possibility of broader—and fewer limitations on—applications so that they can be tailored to serve each study's specific goals. This finding further justifies this SLR's consideration of other experiments conducted in different contexts.

Author Contributions: Conceptualization, M.O., L.Y.P. and T.F.A.; methodology, M.O., L.Y.P. and T.F.A.; formal analysis, M.O., L.Y.P., T.F.A., W.A.-H. and C.S.K.; writing—original draft preparation, M.O., L.Y.P. and T.F.A.; writing—review and editing, M.O., L.Y.P., T.F.A., W.A.-H. and C.S.K.; supervision, L.Y.P. and T.F.A.; project administration, L.Y.P. and T.F.A.; resources, T.F.A. and L.Y.P.; funding acquisition, C.S.K. All authors have read and agreed to the published version of the manuscript.

Funding: This research received no external funding.

Institutional Review Board Statement: Not applicable.

Informed Consent Statement: Not applicable.

Data Availability Statement: Not applicable.

Conflicts of Interest: The authors declare no conflict of interest.

References

1. Blondel, V.D.; Decuyper, A.; Krings, G. A survey of results on mobile phone datasets analysis. *EPJ Data Sci.* **2015**, *4*, 10. [CrossRef]
2. Dobra, A.; Williams, N.E.; Eagle, N. Spatiotemporal detection of unusual human population behavior using mobile phone data. *PLoS ONE* **2015**, *10*, e0120449. [CrossRef] [PubMed]
3. Calabrese, F.; Ferrari, L.; Blondel, V.D. Urban sensing using mobile phone network data: A survey of research. *ACM Comput. Surv. Csur.* **2014**, *47*, 1–20. [CrossRef]
4. Okmi, M.; Por, L.Y.; Ang, T.F.; Ku, C.S. Mobile Phone Data: A Survey of Techniques, Features, and Applications. *Sensors* **2023**, *23*, 908. [CrossRef] [PubMed]
5. Phithakkitnukoon, S.; Smoreda, Z. Influence of social relations on human mobility and sociality: A study of social ties in a cellular network. *Soc. Netw. Anal. Min.* **2016**, *6*, 42. [CrossRef]
6. Deville, P.; Linard, C.; Martin, S.; Gilbert, M.; Stevens, F.R.; Gaughan, A.E.; Blondel, V.D.; Tatem, A.J. Dynamic population mapping using mobile phone data. *Proc. Natl. Acad. Sci. USA* **2014**, *111*, 15888–15893. [CrossRef]
7. Park, S.; Xu, Y.; Jiang, L.; Chen, Z.; Huang, S. Spatial structures of tourism destinations: A trajectory data mining approach leveraging mobile big data. *Ann. Tour. Res.* **2020**, *84*, 102973. [CrossRef]
8. Xu, Y.; Li, J.; Xue, J.; Park, S.; Li, Q. Tourism geography through the lens of time use: A computational framework using fine-grained mobile phone data. *Ann. Am. Assoc. Geogr.* **2021**, *111*, 1420–1444. [CrossRef]
9. Gruber, A.; Ben-Gal, I. Using targeted Bayesian network learning for suspect identification in communication networks. *Int. J. Inf. Secur.* **2018**, *17*, 169–181. [CrossRef]
10. Ferrara, E.; De Meo, P.; Catanese, S.; Fiumara, G. Detecting criminal organizations in mobile phone networks. *Expert Syst. Appl.* **2014**, *41*, 5733–5750. [CrossRef]
11. Taha, K.; Yoo, P.D. SIIMCO: A forensic investigation tool for identifying the influential members of a criminal organization. *IEEE Trans. Inf. Secur.* **2015**, *11*, 811–822. [CrossRef]
12. Griffiths, G.; Johnson, S.D.; Chetty, K. UK-based terrorists' antecedent behavior: A spatial and temporal analysis. *Appl. Geogr.* **2017**, *86*, 274–282. [CrossRef]

13. Bogomolov, A.; Lepri, B.; Staiano, J.; Oliver, N.; Pianesi, F.; Pentland, A. November. Once upon a crime: Towards crime prediction from demographics and mobile data. In Proceedings of the 16th International Conference on Multimodal Interaction, Istanbul, Turkey, 12–16 December 2014; pp. 427–434.
14. Bogomolov, A.; Lepri, B.; Staiano, J.; Letouzé, E.; Oliver, N.; Pianesi, F.; Pentland, A. Moves on the street: Classifying crime hotspots using aggregated anonymized data on people dynamics. *Big Data* **2015**, *3*, 148–158. [CrossRef] [PubMed]
15. Zhang, F.; Wu, L.; Zhu, D.; Liu, Y. Social sensing from street-level imagery: A case study in learning spatio-temporal urban mobility patterns. *ISPRS J. Photogramm. Remote Sens.* **2019**, *153*, 48–58. [CrossRef]
16. Sekimoto, Y.; Sudo, A.; Kashiyama, T.; Seto, T.; Hayashi, H.; Asahara, A.; Ishizuka, H.; Nishiyama, S. Real-time people movement estimation in large disasters from several kinds of mobile phone data. In Proceedings of the 2016 ACM International Joint Conference on Pervasive and Ubiquitous Computing: Adjunct, Heidelberg, Germany, 12–16 September 2016; pp. 1426–1434.
17. Lai, S.; Erbach-Schoenberg, E.Z.; Pezzulo, C.; Ruktanonchai, N.W.; Sorichetta, A.; Steele, J.; Li, T.; Dooley, C.A.; Tatem, A.J. Exploring the use of mobile phone data for national migration statistics. *Palgrave Commun.* **2019**, *5*, 1–10. [CrossRef] [PubMed]
18. Hankaew, S.; Phithakkitnukoon, S.; Demissie, M.G.; Kattan, L.; Smoreda, Z.; Ratti, C. Inferring and modeling migration flows using mobile phone network data. *IEEE Access* **2019**, *7*, 164746–164758. [CrossRef]
19. Thuillier, E.; Moalic, L.; Lamrous, S.; Caminada, A. Clustering weekly patterns of human mobility through mobile phone data. *IEEE Trans. Mob. Comput.* **2017**, *17*, 817–830. [CrossRef]
20. Pei, T.; Sobolevsky, S.; Ratti, C.; Shaw, S.L.; Li, T.; Zhou, C. A new insight into land use classification based on aggregated mobile phone data. *Int. J. Geogr. Inf. Sci.* **2014**, *28*, 1988–2007. [CrossRef]
21. Malleson, N.; Andresen, M.A. Exploring the impact of ambient population measures on London crime hotspots. *J. Crim. Justice* **2016**, *46*, 52–63. [CrossRef]
22. Rummens, A.; Snaphaan, T.; Van de Weghe, N.; Van den Poel, D.; Pauwels, L.J.; Hardyns, W. Do mobile phone data provide a better denominator in crime rates and improve spatiotemporal predictions of crime? *ISPRS Int. J. Geo-Inf.* **2021**, *10*, 369. [CrossRef]
23. Hanaoka, K. New insights on relationships between street crimes and ambient population: Use of hourly population data estimated from mobile phone users' locations. *Environ. Plan. B Urban Anal. City Sci.* **2018**, *45*, 295–311. [CrossRef]
24. Szocska, M.; Pollner, P.; Schiszler, I.; Joo, T.; Palicz, T.; McKee, M.; Asztalos, A.; Bencze, L.; Kapronczay, M.; Petrecz, P.; et al. Countrywide population movement monitoring using mobile devices generated (big) data during the COVID-19 crisis. *Sci. Rep.* **2021**, *11*, 5943. [CrossRef]
25. Willberg, E.; Järv, O.; Väisänen, T.; Toivonen, T. Escaping from Cities during the COVID-19 Crisis: Using Mobile Phone Data to Trace Mobility in Finland. *ISPRS Int. J. Geo-Inf.* **2021**, *10*, 103. [CrossRef]
26. Lanza, G.; Pucci, P.; Carboni, L.; Vendemmia, B. Impacts of the Covid-19 pandemic in inner areas. Remote work and near-home tourism through mobile phone data in Piacenza Apennine. *TEMA* **2022**, *2*, 73–89.
27. Sakamanee, P.; Phithakkitnukoon, S.; Smoreda, Z.; Ratti, C. Methods for inferring route choice of commuting trip from mobile phone network data. *ISPRS Int. J. Geo-Inf.* **2020**, *9*, 306. [CrossRef]
28. Taha, K.; Yoo, P.D. Using the spanning tree of a criminal network for identifying its leaders. *IEEE Trans. Inf. Secur.* **2016**, *12*, 445–453. [CrossRef]
29. Agreste, S.; Catanese, S.; De Meo, P.; Ferrara, E.; Fiumara, G. Network structure and resilience of Mafia syndicates. *Inf. Sci.* **2016**, *351*, 30–47. [CrossRef]
30. Bhattacharya, K.; Kaski, K. Social physics: Uncovering human behaviour from communication. *Adv. Phys. X* **2019**, *4*, 1527723. [CrossRef]
31. Ghahramani, M.; Zhou, M.; Wang, G. Urban sensing based on mobile phone data: Approaches, applications, and challenges. *IEEE/CAA J. Autom. Sin.* **2020**, *7*, 627–637. [CrossRef]
32. Kitchenham, B.; Brereton, O.P.; Budgen, D.; Turner, M.; Bailey, J.; Linkman, S. Systematic literature reviews in software engineering–a systematic literature review. *Inf. Softw. Technol.* **2009**, *51*, 7–15. [CrossRef]
33. Liberati, A.; Altman, D.G.; Tetzlaff, J.; Mulrow, C.; Gøtzsche, P.C.; Ioannidis, J.P.; Clarke, M.; Devereaux, P.J.; Kleijnen, J.; Moher, D. The PRISMA statement for reporting systematic reviews and meta-analyses of studies that evaluate health care interventions: Explanation and elaboration. *J. Clin. Epidemiol.* **2009**, *62*, e1–e34. [CrossRef] [PubMed]
34. Louail, T.; Lenormand, M.; Cantu Ros, O.G.; Picornell, M.; Herranz, R.; Frias-Martinez, E.; Ramasco, J.J.; Barthelemy, M. From mobile phone data to the spatial structure of cities. *Sci. Rep.* **2014**, *4*, 5276. [CrossRef] [PubMed]
35. Kung, K.S.; Greco, K.; Sobolevsky, S.; Ratti, C. Exploring universal patterns in human home-work commuting from mobile phone data. *PLoS ONE* **2014**, *9*, e96180. [CrossRef] [PubMed]
36. Schläpfer, M.; Bettencourt, L.M.; Grauwin, S.; Raschke, M.; Claxton, R.; Smoreda, Z.; West, G.B.; Ratti, C. The scaling of human interactions with city size. *J. R. Soc. Interface* **2014**, *11*, 20130789. [CrossRef] [PubMed]
37. Ríos, S.A.; Muñoz, R. Land Use detection with cell phone data using topic models: Case Santiago, Chile. *Comput. Environ. Urban Syst.* **2017**, *61*, 39–48. [CrossRef]
38. Furno, A.; El Faouzi, N.E.; Fiore, M.; Stanica, R. Fusing GPS probe and mobile phone data for enhanced land-use detection. In Proceedings of the 2017 5th IEEE International Conference on Models and Technologies for Intelligent Transportation Systems (MT-ITS), Naples, Italy, 26–28 June 2017; IEEE: Piscataway, NJ, USA, 2017; pp. 693–698.

39. Gabrielli, L.; Furletti, B.; Giannotti, F.; Nanni, M.; Rinzivillo, S. Use of mobile phone data to estimate visitors mobility flows. In *Proceedings of the International Conference on Software Engineering and Formal Methods, York, UK, 7–11 September 2015*; Springer: Cham, Switzerland, 2015; pp. 214–226.
40. Gabrielli, L.; Furletti, B.; Trasarti, R.; Giannotti, F.; Pedreschi, D. City users' classification with mobile phone data. In *Proceedings of the 2015 IEEE International Conference on Big Data (Big Data), Santa Clara, CA, USA, 29 October–1 November 2015*; IEEE: Piscataway, NJ, USA, 2015; pp. 1007–1012.
41. Andrade, T.; Cancela, B.; Gama, J. Discovering locations and habits from human mobility data. *Ann. Telecommun.* **2020**, *75*, 505–521. [CrossRef]
42. Bianchi, F.M.; Rizzi, A.; Sadeghian, A.; Moiso, C. Identifying user habits through data mining on call data records. *Eng. Appl. Artif. Intell.* **2016**, *54*, 49–61. [CrossRef]
43. Taha, K.; Yoo, P.D. Shortlisting the influential members of criminal organizations and identifying their important communication channels. *IEEE Trans. Inf. Secur.* **2019**, *14*, 1988–1999. [CrossRef]
44. Taha, K.; Yoo, P.D. A system for analyzing criminal social networks. In Proceedings of the 2015 IEEE/ACM International Conference on Advances in Social Networks Analysis and Mining, Paris, France, 25–28 August 2015; pp. 1017–1023.
45. Novović, O.; Brdar, S.; Mesaroš, M.; Crnojević, V.; Papadopoulos, A.N. Uncovering the relationship between human connectivity dynamics and land use. *ISPRS Int. J. Geo-Inf.* **2020**, *9*, 140. [CrossRef]
46. Shi, L.; Chi, G.; Liu, X.; Liu, Y. Human mobility patterns in different communities: A mobile phone data-based social network approach. *Ann. GIS* **2015**, *21*, 15–26. [CrossRef]
47. Hassan, S.U.; Shabbir, M.; Iqbal, S.; Said, A.; Kamiran, F.; Nawaz, R.; Saif, U. Leveraging deep learning and SNA approaches for smart city policing in the developing world. *Int. J. Inf. Manag.* **2019**, *56*, 102045. [CrossRef]
48. Jia, Y.; Ge, Y.; Ling, F.; Guo, X.; Wang, J.; Wang, L.; Chen, Y.; Li, X. Urban land use mapping by combining remote sensing imagery and mobile phone positioning data. *Remote Sens.* **2018**, *10*, 446. [CrossRef]
49. Pratesi, F.; Gabrielli, L.; Cintia, P.; Monreale, A.; Giannotti, F. PRIMULE: Privacy risk mitigation for user profiles. *Data Knowl. Eng.* **2020**, *125*, 101786. [CrossRef]
50. Arfaoui, S.; Belmekki, A.; Mezrioui, A. Privacy increase on telecommunication processes. In Proceedings of the 2018 International Conference on Advanced Communication Technologies and Networking (CommNet), Marrakech, Morocco, 2–4 April 2018; IEEE: Piscataway, NJ, USA, 2018; pp. 1–10.
51. Feng, J.; Liu, L.; Long, D.; Liao, W. An examination of spatial differences between migrant and native offenders in committing violent crimes in a large Chinese city. *ISPRS Int. J. Geo-Inf.* **2019**, *8*, 119. [CrossRef]
52. Haleem, M.S.; Do Lee, W.; Ellison, M.; Bannister, J. The 'exposed'population, violent crime in public space and the night-time economy in Manchester, UK. *Eur. J. Crim. Policy Res.* **2020**, *27*, 335–352. [CrossRef]
53. Liu, L.; Peng, Z.; Wu, H.; Jiao, H.; Yu, Y. Exploring urban spatial feature with dasymetric mapping based on mobile phone data and LUR-2SFCAe method. *Sustainability* **2018**, *10*, 2432. [CrossRef]
54. Salat, H.; Smoreda, Z.; Schläpfer, M. A method to estimate population densities and electricity consumption from mobile phone data in developing countries. *PLoS ONE* **2020**, *15*, e0235224. [CrossRef]
55. Peng, Z.; Wang, R.; Liu, L.; Wu, H. Fine-Scale Dasymetric Population Mapping with Mobile Phone and Building Use Data Based on Grid Voronoi Method. *ISPRS Int. J. Geo-Inf.* **2020**, *9*, 344. [CrossRef]
56. Sakarovitch, B.; Bellefon, M.P.D.; Givord, P.; Vanhoof, M. Estimating the residential population from mobile phone data, an initial exploration. *Econ. Stat.* **2018**, *505*, 109–132. [CrossRef]
57. Zhang, G.; Rui, X.; Poslad, S.; Song, X.; Fan, Y.; Ma, Z. Large-scale, fine-grained, spatial, and temporal analysis, and prediction of mobile phone users' distributions based upon a convolution long short-term model. *Sensors* **2019**, *19*, 2156. [CrossRef]
58. Järv, O.; Tenkanen, H.; Toivonen, T. Enhancing spatial accuracy of mobile phone data using multi-temporal dasymetric interpolation. *Int. J. Geogr. Inf. Sci.* **2017**, *31*, 1630–1651. [CrossRef]
59. Liu, Z.; Ma, T.; Du, Y.; Pei, T.; Yi, J.; Peng, H. Mapping hourly dynamics of urban population using trajectories reconstructed from mobile phone records. *Trans. GIS* **2018**, *22*, 494–513. [CrossRef]
60. Zhang, G.; Rui, X.; Poslad, S.; Song, X.; Fan, Y.; Wu, B. A method for the estimation of finely-grained temporal spatial human population density distributions based on cell phone call detail records. *Remote Sens.* **2020**, *12*, 2572. [CrossRef]
61. Ricciato, F.; Lanzieri, G.; Wirthmann, A.; Seynaeve, G. Towards a methodological framework for estimating present population density from mobile network operator data. *Pervasive Mob. Comput.* **2020**, *68*, 101263. [CrossRef]
62. Ricciato, F.; Widhalm, P.; Pantisano, F.; Craglia, M. Beyond the "single-operator, CDR-only" paradigm: An interoperable framework for mobile phone network data analyses and population density estimation. *Pervasive Mob. Comput.* **2017**, *35*, 65–82. [CrossRef]
63. Shi, Y.; Yang, J.; Shen, P. Revealing the correlation between population density and the spatial distribution of urban public service facilities with mobile phone data. *ISPRS Int. J. Geo-Inf.* **2020**, *9*, 38. [CrossRef]
64. Traunmueller, M.; Quattrone, G.; Capra, L. Mining mobile phone data to investigate urban crime theories at scale. In Proceedings of the International Conference on Social Informatics, Barcelona, Spain, 11–13 November 2014; Springer: Cham, Switzerland; pp. 396–411.
65. He, L.; Páez, A.; Jiao, J.; An, P.; Lu, C.; Mao, W.; Long, D. Ambient population and larceny-theft: A spatial analysis using mobile phone data. *ISPRS Int. J. Geo-Inf.* **2020**, *9*, 342. [CrossRef]

66. Lee, W.D.; Haleem, M.S.; Ellison, M.; Bannister, J. The influence of intra-daily activities and settings upon weekday violent crime in public spaces in Manchester, UK. *Eur. J. Crim. Policy Res.* **2020**, *27*, 375–395. [CrossRef]
67. Song, G.; Bernasco, W.; Liu, L.; Xiao, L.; Zhou, S.; Liao, W. Crime feeds on legal activities: Daily mobility flows help to explain thieves' target location choices. *J. Quant. Criminol.* **2019**, *35*, 831–854. [CrossRef]
68. Tongsinoot, L.; Muangsin, V. Exploring home and work locations in a city from mobile phone data. In Proceedings of the 2017 IEEE 19th International Conference on High Performance Computing and Communications; IEEE 15th International Conference on Smart City; IEEE 3rd International Conference on Data Science and Systems (HPCC/SmartCity/DSS), Bangkok, Thailand, 18–20 December 2017; IEEE: Piscataway, NJ, USA, 2017; pp. 123–129.
69. Vanhoof, M.; Reis, F.; Ploetz, T.; Smoreda, Z. Assessing the Quality of Home Detection from Mobile Phone Data for Official Statistics. *J. Off. Stat.* **2018**, *34*, 935–960. [CrossRef]
70. Yang, X.; Zhao, Z.; Lu, S. Exploring Spatial-Temporal Patterns of Urban Human Mobility Hotspots. *Sustainability* **2016**, *8*, 674. [CrossRef]
71. Ghahramani, M.; Zhou, M.; Hon, C.T. Mobile phone data analysis: A spatial exploration toward hotspot detection. *IEEE Trans. Autom. Sci. Eng.* **2018**, *16*, 351–362. [CrossRef]
72. Truică, C.O.; Novović, O.; Brdar, S.; Papadopoulos, A.N. Community detection in who-calls-whom social networks. In Proceedings of the International Conference on Big Data Analytics and Knowledge Discovery, Regensburg, Germany, 3–6 September 2018; Springer: Cham, Switzerland, 2018; pp. 19–33.
73. Xu, K.; Zou, K.; Huang, Y.; Yu, X.; Zhang, X. Mining community and inferring friendship in mobile social networks. *Neurocomputing* **2016**, *174*, 605–616. [CrossRef]
74. Lind, A.; Hadachi, A.; Piksarv, P.; Batrashev, O. Spatio-temporal mobility analysis for community detection in the mobile networks using CDR data. In Proceedings of the 2017 9th International Congress on Ultra Modern Telecommunications and Control Systems and Workshops (ICUMT), Munich, Germany, 6–8 November 2017; IEEE: Piscataway, NJ, USA, 2017; pp. 250–255.
75. Sumathi, V.P.; Kousalya, K.; Vanitha, V.; Cynthia, J. Crowd estimation at a social event using call data records. *Int. J. Bus. Inf. Syst.* **2018**, *28*, 246–261.
76. Filipowska, A.; Mucha, M.; Perkowski, B.; Szczekocka, E.; Gromada, J. Towards social telco applications based on the user behaviour and relations between users. In Proceedings of the 2015 18th International Conference on Intelligence in Next Generation Networks, Paris, France, 17–19 February 2015; IEEE: Piscataway, NJ, USA, 2015; pp. 95–102.
77. Aledavood, T.; Lehmann, S.; Saramäki, J. Social network differences of chronotypes identified from mobile phone data. *EPJ Data Sci.* **2018**, *7*, 46. [CrossRef]
78. Yu, C.; Wang, N.; Yang, L.T.; Yao, D.; Hsu, C.H.; Jin, H. A semi-supervised social relationships inferred model based on mobile phone data. *Future Gener. Comput. Syst.* **2017**, *76*, 458–467. [CrossRef]
79. Gaito, S.; Quadri, C.; Rossi, G.P.; Zignani, M. Urban communications and social interactions through the lens of mobile phone data. *Online Soc. Netw. Media* **2017**, *1*, 70–81. [CrossRef]
80. Deville, P.; Song, C.; Eagle, N.; Blondel, V.D.; Barabási, A.L.; Wang, D. Scaling identity connects human mobility and social interactions. *Proc. Natl. Acad. Sci. USA* **2016**, *113*, 7047–7052. [CrossRef]
81. Morales, A.J.; Creixell, W.; Borondo, J.; Losada, J.C.; Benito, R.M. Characterizing ethnic interactions from human communication patterns in Ivory Coast. *Netw. Heterog. Media* **2015**, *10*, 87. [CrossRef]
82. Chemello, N. Correlating CDR with other data sources. In Proceedings of the 2016 IEEE International Conference on Cybercrime and Computer Forensic (ICCCF), Vancouver, BC, Canada, 12–14 June 2016; IEEE: Piscataway, NJ, USA, 2016; pp. 1–5.
83. Kumar, M.; Hanumanthappa, M.; Kumar, T.S. Crime investigation and criminal network analysis using archive call detail records. In *Proceedings of the 2016 Eighth International Conference on Advanced Computing (ICoAC), Chennai, India, 19–21 January 2017*; IEEE: Piscataway, NJ, USA, 2017; pp. 46–50.
84. Khan, E.S.; Azmi, H.; Ansari, F.; Dhalvelkar, S. Simple implementation of criminal investigation using call data records (CDRs) through big data technology. In Proceedings of the 2018 International Conference on Smart City and Emerging Technology (ICSCET), Mumbai, India, 5 January 2018; IEEE: Piscataway, NJ, USA, 2018; pp. 1–5.
85. Khan, S.; Ansari, F.; Dhalvelkar, H.A.; Computer, S. Criminal investigation using call data records (CDR) through big data technology. In Proceedings of the 2017 International Conference on Nascent Technologies in Engineering (ICNTE), Vashi, India, 27–28 January 2017; IEEE: Piscataway, NJ, USA, 2017; pp. 1–5.
86. Hoyos, I.; Esposito, B.; Nunez-del-Prado, M. DETECTOR: Automatic Detection System for Terrorist Attack Trajectories. In *Proceedings of the Annual International Symposium on Information Management and Big Data, Lima, Peru, 4–6 September 2018*; Springer: Cham, Switzerland, 2018; pp. 160–173.
87. Abba, E.; Aibinu, A.M.; Alhassan, J.K. Development of multiple mobile networks call detailed records and its forensic analysis. *Digit. Commun. Netw.* **2019**, *5*, 256–265. [CrossRef]
88. Marshall, A.M.; Miller, P. CaseNote: Mobile phone call data obfuscation & techniques for call correlation. *Digit. Investig.* **2019**, *29*, 82–90.
89. Zhang, A.; Bradford, B.; Morgan, R.M.; Nakhaeizadeh, S. Investigating the uses of mobile phone evidence in China criminal proceedings. *Sci. Justice* **2022**, *62*, 385–398. [CrossRef]
90. Lenormand, M.; Picornell, M.; Cantú-Ros, O.G.; Louail, T.; Herranz, R.; Barthelemy, M.; Frías-Martínez, E.; San Miguel, M.; Ramasco, J.J. Comparing and modelling land use organization in cities. *R. Soc. Open Sci.* **2015**, *2*, 150449. [CrossRef] [PubMed]

91. Cavallaro, L.; Ficara, A.; De Meo, P.; Fiumara, G.; Catanese, S.; Bagdasar, O.; Song, W.; Liotta, A. Disrupting resilient criminal networks through data analysis: The case of Sicilian Mafia. *PLoS ONE* **2020**, *15*, e0236476. [CrossRef] [PubMed]
92. Mao, H.; Shuai, X.; Ahn, Y.Y.; Bollen, J. Quantifying socio-economic indicators in developing countries from mobile phone communication data: Applications to Côte d'Ivoire. *EPJ Data Sci.* **2015**, *4*, 15. [CrossRef]
93. Andrea, C.; Lehmann, S.; Larsen, J.E. Inferring human mobility from sparse low accuracy mobile sensing data. In Proceedings of the 2014 ACM International Joint Conference on Pervasive and Ubiquitous Computing: Adjunct Publication, Seattle, DC, USA, 13–17 September 2014; pp. 995–1004.
94. Ficara, A.; Cavallaro, L.; Curreri, F.; Fiumara, G.; De Meo, P.; Bagdasar, O.; Song, W.; Liotta, A. Criminal networks analysis in missing data scenarios through graph distances. *PLoS ONE* **2021**, *16*, e0255067. [CrossRef]
95. Dileep, G.K.; Sajeev, G.P. A Graph Mining Approach to Detect Sandwich Calls. In Proceedings of the 2021 IEEE International Conference on Electronics, Computing and Communication Technologies (CONECCT), Bangalore, India, 8–10 July 2021; IEEE: Piscataway, NJ, USA, 2021; pp. 1–6.
96. Wu, J.; Abrar, S.M.; Awasthi, N.; Frias-Martinez, E.; Frias-Martinez, V. Enhancing short-term crime prediction with human mobility flows and deep learning architectures. *EPJ Data Sci.* **2022**, *11*, 53. [CrossRef]
97. Long, D.; Liu, L.; Xu, M.; Feng, J.; Chen, J.; He, L. Ambient population and surveillance cameras: The guardianship role in street robbers' crime location choice. *Cities* **2021**, *115*, 103223. [CrossRef]
98. Long, D.; Liu, L. Do Migrant and Native Robbers Target Different Places? *ISPRS Int. J. Geo-Inf.* **2021**, *10*, 771. [CrossRef]
99. Hu, X.; Chen, H.; Liu, S.; Jiang, H.; Chu, G.; Li, R. BTG: A Bridge to Graph machine learning in telecommunications fraud detection. *Future Gener. Comput. Syst.* **2022**, *137*, 274–287. [CrossRef]
100. Xing, J.; Yu, M.; Wang, S.; Zhang, Y.; Ding, Y. Automated fraudulent phone call recognition through deep learning. *Wirel. Commun. Mob. Comput.* **2020**, *2020*, 8853468. [CrossRef]
101. Chu, G.; Wang, J.; Qi, Q.; Sun, H.; Tao, S.; Yang, H.; Liao, J.; Han, Z. Exploiting Spatial-Temporal Behavior Patterns for Fraud Detection in Telecom Networks. *IEEE Trans. Dependable Secur. Comput.* **2022**, 1–13. [CrossRef]
102. Hilas, C.S.; Mastorocostas, A.; Rekanos, I.T. Clustering of telecommunications user profiles for fraud detection and security enhancement in large corporate networks: A case study. *Appl. Math. Inf. Sci.* **2015**, *9*, 1709–1718.
103. Jabbar, M.A. Fraud Detection Call Detail Record Using Machine Learning in Telecommunications Company. *Adv. Sci. Technol. Eng. Syst. J.* **2020**, *5*, 63–69. [CrossRef]
104. Kilinc, H.H. Anomaly Pattern Analysis Based on Machine Learning on Real Telecommunication Data. In Proceedings of the 2022 7th International Conference on Computer Science and Engineering (UBMK), Diyarbakir, Turkey, 14–16 September 2022; IEEE: Piscataway, NJ, USA, 2022; pp. 43–48.
105. Yang, X.; Fang, Z.; Yin, L.; Li, J.; Lu, S.; Zhao, Z. Revealing the relationship of human convergence–divergence patterns and land use: A case study on Shenzhen City, China. *Cities* **2019**, *95*, 102384. [CrossRef]
106. Mao, H.; Ahn, Y.Y.; Bhaduri, B.; Thakur, G. Improving land use inference by factorizing mobile phone call activity matrix. *J. Land Use Sci.* **2017**, *12*, 138–153. [CrossRef]
107. Liu, Y.; Fang, F.; Jing, Y. How urban land use influences commuting flows in Wuhan, Central China: A mobile phone signaling data perspective. *Sustain. Cities Soc.* **2020**, *53*, 101914. [CrossRef]
108. Arcolezi, H.H.; Couchot, J.-F.; Al Bouna, B.; Xiao, X. Improving the utility of locally differentially private protocols for longitudinal and multidimensional frequency estimates. *Digit. Commun. Netw.* 2022, *in press*. [CrossRef]
109. Gramaglia, M.; Fiore, M.; Furno, A.; Stanica, R. GLOVE: Towards privacy-preserving publishing of record-level-truthful mobile phone trajectories. *ACM/IMS Trans. Data Sci. (TDS)* **2021**, *2*, 1–36. [CrossRef]
110. Vespe, M.; Iacus, S.M.; Santamaria, C.; Sermi, F.; Spyratos, S. On the use of data from multiple mobile network operators in Europe to fight COVID-19. *Data Policy* **2021**, *3*, e8. [CrossRef]
111. Cinnamon, J.; Jones, S.K.; Adger, W.N. Evidence and future potential of mobile phone data for disease disaster management. *Geoforum* **2016**, *75*, 253–264. [CrossRef]
112. Boenig-Liptsin, M.; Tanweer, A.; Edmundson, A. Data Science Ethos Lifecycle: Interplay of ethical thinking and data science practice. *J. Stat. Data Sci. Educ.* **2022**, *30*, 228–240. [CrossRef]
113. Peter, K. Government surveillance, privacy, and legitimacy. *Philos. Technol.* **2022**, *35*, 8.
114. Qiu, F.; Wu, F.; Chen, G. Privacy and quality preserving multimedia data aggregation for participatory sensing systems. *IEEE Trans. Mob. Comput.* **2014**, *14*, 1287–1300. [CrossRef]
115. Jin, H.; Su, L.; Ding, B.; Nahrstedt, K.; Borisov, N. Enabling privacy-preserving incentives for mobile crowd sensing systems. In *2016 IEEE 36th International Conference on Distributed Computing Systems (ICDCS)*; IEEE: Piscataway, NJ, USA, 2016; pp. 344–353.
116. Li, H.; Li, T.; Wang, W.; Wang, Y. Dynamic participant selection for large-scale mobile crowd sensing. *IEEE Trans. Mob. Comput.* **2018**, *18*, 2842–2855. [CrossRef]
117. Guo, B.; Yu, Z.; Zhou, X.; Zhang, D. From participatory sensing to mobile crowd sensing. In Proceedings of the 2014 IEEE International Conference on Pervasive Computing and Communication Workshops (PERCOM WORKSHOPS), Budapest, Hungary, 24–28 March 2014; IEEE: Piscataway, NJ, USA, 2014; pp. 593–598.
118. Xu, S.; Chen, X.; Pi, X.; Joe-Wong, C.; Zhang, P.; Noh, H.Y. ilocus: Incentivizing vehicle mobility to optimize sensing distribution in crowd sensing. *IEEE Trans. Mob. Comput.* **2019**, *19*, 1831–1847. [CrossRef]

119. Capponi, A.; Fiandrino, C.; Kantarci, B.; Foschini, L.; Kliazovich, D.; Bouvry, P. A survey on mobile crowdsensing systems: Challenges, solutions, and opportunities. *IEEE Commun. Surv. Tutor.* **2019**, *21*, 2419–2465. [CrossRef]
120. Gisdakis, S.; Giannetsos, T.; Papadimitratos, P. Security, privacy, and incentive provision for mobile crowd sensing systems. *IEEE Internet Things J.* **2016**, *3*, 839–853. [CrossRef]
121. Agir, B.; Papaioannou, T.G.; Narendula, R.; Aberer, K.; Hubaux, J.-P. User-side adaptive protection of location privacy in participatory sensing. *GeoInformatica* **2014**, *18*, 165–191. [CrossRef]
122. Jin, W.; Xiao, M.; Li, M.; Guo, L. If you do not care about it, sell it: Trading location privacy in mobile crowd sensing. In Proceedings of the IEEE INFOCOM 2019-IEEE Conference on Computer Communications, Paris, France, 29 April–2 May 2019; IEEE: Piscataway, NJ, USA, 2019; pp. 1045–1053.
123. Chen, Z.; Gul, O.M.; Kantarci, B. Practical Byzantine Fault Tolerance-based Robustness for Mobile Crowdsensing. *Distrib. Ledger Technol. Res. Pract.* **2023**. [CrossRef]
124. Chen, S.; Li, Z. Research on Enterprise Innovation Behavior Based on the Regression Analysis Under Big Data Technology. In Proceedings of the 2022 3rd International Conference on Big Data and Social Sciences (ICBDSS 2022), Hulunbuir, China, 19–21 August 2022; Atlantis Press: Amsterdam, The Netherlands, 2022; pp. 665–673.
125. Jiang, H.; Wang, X.; Xiao, Q.; Li, S. Investment Behavior Related to Automated Machines and Biased Technical Change: Based on Evidence from Listed Manufacturing Companies in China. *Front. Psychol.* **2022**, *13*, 874820. [CrossRef] [PubMed]
126. Ungar, W.J. Next generation sequencing and health technology assessment in autism spectrum disorder. *J. Can. Acad. Child Adolesc. Psychiatry* **2015**, *24*, 123. [PubMed]
127. Zhang, Z.; Nagrath, S. Microfluidics and cancer: Are we there yet? *Biomed. Microdevices* **2013**, *15*, 595–609. [CrossRef] [PubMed]
128. Liu, S.; Yin, X.; Zhao, H. Dual-function photonic spin Hall effect sensor for high-precision refractive index sensing and graphene layer detection. *Opt. Express* **2020**, *30*, 31925–31936. [CrossRef]
129. Sui, J.-Y.; Liao, S.-Y.; Li, B.; Zhang, H.-F. High sensitivity multitasking non-reciprocity sensor using the photonic spin Hall effect. *Opt. Lett.* **2022**, *47*, 6065–6068. [CrossRef]
130. Wang, S.; Tian, Y.; Liu, X.; Foley, M. How Farmers Make Investment Decisions: Evidence from a Farmer Survey in China. *Sustainability* **2020**, *12*, 247. [CrossRef]
131. Cheng, Y.; Kuang, Y.; Shi, X.; Dong, C. Sustainable investment in a supply chain in the big data era: An information updating approach. *Sustainability* **2018**, *10*, 403. [CrossRef]
132. Raguseo, E.; Vitari, C. Investments in big data analytics and firm performance: An empirical investigation of direct and mediating effects. *Int. J. Prod. Res.* **2018**, *56*, 5206–5221. [CrossRef]
133. De Montjoye, Y.A.; Gambs, S.; Blondel, V.; Canright, G.; De Cordes, N.; Deletaille, S.; Engø-Monsen, K.; Garcia-Herranz, M.; Kendall, J.; Kerry, C.; et al. On the privacy-conscientious use of mobile phone data. *Sci. Data* **2018**, *5*, 180286. [CrossRef]
134. Zinman, O.; Lerner, B. Utilizing digital traces of mobile phones for understanding social dynamics in urban areas. *Pers. Ubiquitous Comput.* **2020**, *24*, 535–549. [CrossRef]
135. Sultan, K.; Ali, H.; Zhang, Z. Call detail records driven anomaly detection and traffic prediction in mobile cellular networks. *IEEE Access* **2018**, *6*, 41728–41737. [CrossRef]
136. Xu, Y.; Belyi, A.; Bojic, I.; Ratti, C. How friends share urban space: An exploratory spatiotemporal analysis using mobile phone data. *Trans. GIS* **2017**, *21*, 468–487. [CrossRef]
137. Hoteit, S.; Chen, G.; Viana, A.; Fiore, M. Filling the gaps: On the completion of sparse call detail records for mobility analysis. In Proceedings of the Eleventh ACM Workshop on Challenged Networks, New York, NY, USA, 3–7 October 2016; pp. 45–50.

MDPI AG
Grosspeteranlage 5
4052 Basel
Switzerland
Tel.: +41 61 683 77 34

Sensors Editorial Office
E-mail: sensors@mdpi.com
www.mdpi.com/journal/sensors

www.ingramcontent.com/pod-product-compliance
Lightning Source LLC
Chambersburg PA
CBHW041428210926
51850CB00013B/3380
9783725832255